Multiblock Data Fusion in Statistics and Machine Learning

Multiblock Data Fusion in Statistics and Machine Learning

Applications in the Natural and Life Sciences

Age K. Smilde
Swammerdam Institute for Life Sciences, University of Amsterdam, Amsterdam, NL and
Simula Metropolitan Center for Digital Engineering, Oslo, NO

Tormod Næs
Nofima
Ås, NO

Kristian Hovde Liland
Norwegian University of Life Sciences
Ås, NO

This edition first published 2022

Registered Offices
John Wiley & Sons, Inc., 111 River Street, Hoboken, NJ 07030, USA
John Wiley & Sons Ltd, The Atrium, Southern Gate, Chichester, West Sussex, PO19 8SQ, UK

Editorial Office
The Atrium, Southern Gate, Chichester, West Sussex, PO19 8SQ, UK

For details of our global editorial offices, customer services, and more information about Wiley products visit us at www.wiley.com.

Wiley also publishes its books in a variety of electronic formats and by print-on-demand. Some content that appears in standard print versions of this book may not be available in other formats.

A catalogue record for this book is available from the Library of Congress

Hardback ISBN: 9781119600961; ePDF ISBN: 9781119600985; epub ISBN: 9781119600992;
Obook ISBN: 9781119600978

Cover image: © Professor Age K. Smilde
Cover design by Wiley

Set in 10/12pt WarnockPro-Regular by Integra Software Services Pvt. Ltd, Pondicherry, India

Printed and bound by CPI Group (UK) Ltd, Croydon CR0 4YY

C9781119600961_230322

Contents

Foreword

It is a real honour to write a few introductory words about *Multiblock Data Fusion in Statistics and Machine Learning*. The book is maybe not timely! The subject has been around in chemometrics since the late 1980s; usually under the term multiblock analysis.

Let me take that back immediately–the book is definitely timely. Even though this subject has been discussed for decades, it has taken off dramatically lately. And not only in chemometrics, but in a variety of fields. There are many diverse and interesting developments and in fact, it is quite difficult to really understand what is going on and to filter or even just understand the literature from so many sources. Each field will have their own internal jargon and background. This may be the biggest obstacle right now. It is evident that there are many interesting developments but grasping them is next to impossible. This book fixes that. And not only that, this book provides a comprehensive overview across fields and it also adds perspective and new research where needed. I would argue that this is the place if you want to understand data fusion comprehensively.

That is, if you want to understand how to apply data fusion; or you want to develop new data fusion models; or learn how the algorithms and models work; or maybe you want to understand what the shortcomings of different approaches are. If you have questions like these or you simply want to know what is happening in this area of data science, then reading this book will be a nice and fulfilling experience.

To write a comprehensive book about such an enormous field requires special people. And indeed, there are three very competent persons behind this book. They have all worked within the area for many years and have each provided important research on both the theoretical and the application sides of things. And they represent both the experience of the old-timers and visions of the coming generations. I can say that without insulting (I hope) as I am in the same age group as the more ~~ex~~distinguished part of the authors.

I have the deepest respect and the highest admiration for the three authors. I have learned so many things from their individual contributions over the years. Reading this joint work is not a disappointment. Please do enjoy!

Rasmus Bro
Køge, Denmark, July 28, 2021

Preface

Combining information from two or possibly several blocks of data is gaining increased attention and importance in several areas of science and industry. Typical examples can be found in chemistry, spectroscopy, metabolomics, genomics, systems biology, and sensory science. Many methods and procedures have been proposed and used in practice. The area goes under different names: data integration, data fusion, multiblock analyses, multiset analyses, and others.

This book is an attempt to provide an up-to-date treatment of the most used and important methods within an important branch of the area; namely methods based on so-called components or latent variables. These methods have already obtained enormous attention in, for instance, chemometrics, bioinformatics, machine learning, and sensometrics and have proved to be important both for prediction and interpretation.

The book is primarily a description of methodologies, but most of the methods will be illustrated by examples from the above-mentioned areas. The book is written such that both users of the methods as well as method developers will hopefully find sections of interest. At the end of the book there is a description of a software package developed particularly for the book. This package is freely available in R and covers many of the methods discussed.

To distinguish the different types of methods from each other, the book is divided into five parts. Part I is an introduction and description of preliminary concepts. Part II is the core of the book containing the main unsupervised and supervised methods. Part III deals with more complex structures and, finally, Part IV presents alternative unsupervised and supervised methods. The book ends with Part V discussing the available software.

Our recommendations for reading the book are as follows. A minimum read of the book would involve chapters 1, 2, 3, 5, and 7. Chapters 4, 6 and 8 are more specialized and chapters 9 and 10 contain methods we think are more advanced or less obvious to use. We feel privileged to have so many friendly colleagues who were willing to spend their time on helping us to improve the book by reading separate chapters. We would like to express our thanks to: Rasmus Bro, Margriet Hendriks, Ulf Indahl, Henk Kiers, Ingrid Måge, Federico Marini, Åsmund Rinnan, Rosaria Romano, Lars Erik Solberg, Marieke Timmerman, Oliver Tomic, Johan Westerhuis, and Barry Wise. Of course, the correctness of the final text is fully our responsibility!

Age Smilde, Utrecht, The Netherlands
Tormod Næs, Ås, Norway
Kristian Hovde Liland, Ås, Norway
March 2022

List of Figures

1 We thank Frans van der Kloet for making these figures.

List of Tables

Part I

Introductory Concepts and Theory

1 Introduction

1.1 Scope of the Book

In many areas of the natural and life sciences, data sets are collected consisting of multiple blocks of data measured on the same or similar systems. Examples are abundant, e.g., in genomics it is becoming increasingly common to measure gene-expression, protein abundances and metabolite levels on the same biological system (Clish *et al.*, 2004; Heijne *et al.*, 2005; Kleemann *et al.*, 2007; Curtis *et al.*, 2012; Brink-Jensen *et al.*, 2013; Franzosa *et al.*, 2015). In sensory science, the interest is often in relations between the chemical and sensory properties of the samples involved as well as consumer liking of the same samples (Næs *et al.*, 2010). In chemistry, sometimes different types of instruments are utilised to characterise different properties the same set of samples (de Juan and Tauler, 2006). In cohort studies, it is increasingly popular to perform the same type of measurements in different cohorts to confirm results and perform meta-analyses. In (bio-)chemical process industry, plant-wide measurements are available collected by several sensors in the plant (Lopes *et al.*, 2002). Clinical trials are often supported by auxiliary measurements such as gene-expression and cytokines to characterise immune responses (Coccia *et al.*, 2018). Challenge tests to establish the health status of individuals usually contain multiple types of data collected for the same individuals as a function of time (Wopereis *et al.*, 2009; Pellis *et al.*, 2012; Kardinaal *et al.*, 2015). All these examples show that simple data sets are increasingly becoming less common.

Unfortunately, there is no consensus yet about terminology regarding the structure of such data sets and the related research questions. In bioinformatics, the terms data fusion or data integration are often used where the latter distinguishes also N- or P-integration (N means the same samples and P means the same variables), horizontal and vertical integration. In psychometrics, the terms multiset and multigroup data analysis are used; in chemometrics, multiblock data analysis is in use and in the computational sciences and machine learning the term multiview or multitable data analysis is used. We will encounter all these terms in this book but we will use the noun multiblock as much as possible.[1]

In Elaboration 1.1 we define the terms concerning data sets we will use throughout in this book. Sometimes, we will sidestep this to some extent to make connections between fields. At those places we will clarify exactly what we mean.

1 We will refrain from using hyphenation in words like multiblock and multiview to reduce confusion.

Multiblock Data Fusion in Statistics and Machine Learning: Applications in the Natural and Life Sciences,
First Edition. Age K. Smilde, Tormod Næs, and Kristian Hovde Liland.

ELABORATION 1.1

Glossary of terms

Data set: The total collection of all data that is under consideration for a particular problem.
Data block: One block of data organised in a matrix (array) with rows and columns as a part of a data set.
Multiblock data set: The organisation of the data set in blocks of data.
Multiblock data analysis: The process of analysing the whole multiblock data set simultaneously using multiblock methods.
Object, Subject, Sample: Entity for which measurements are obtained. They can be random drawings from a population and/or they can come from an experimental design. The general term is a sample but if these samples pertain to human beings they may be called subjects. They constitute the row entries of a matrix.
Variable: A measured property of an entity collected in the columns of a matrix; this is called a *feature* in machine learning.
Measurement scale: The scale on which a variable is measured (ratio, interval, ordinal, or nominal-scaled).
Homogeneous versus heterogeneous data: If a data set contains blocks of data all measured on the same scale then this is called homogeneous data; if not, then the data are called heterogeneous. In most cases, homogeneous data will refer to blocks containing quantitative data (at least interval-scaled).

Elaboration 1.1 suggests a consistent vocabulary to be used in the book. However, the difference between variables and objects is not always that clear (for examples, see Chapter 8 on complex relations). We will try, however, to remain as consistent as possible and give extra explanations of terms at the appropriate places. In the rest of this chapter we will delineate our potential audience. We will give some examples of why multiblock methods are necessary and give an overview of the types of problems encountered. Moreover, we will give some history and discuss briefly some fundamental concepts which we need in the rest of the book. We end by giving the notation which we will use in this book and a list of abbreviations.

1.2 Potential Audience

Our ambition is to serve different types of audiences. The first set of users consists of practitioners in the natural and life sciences, such as in bioinformatics, sensometrics, chemometrics, statistics, and machine learning. They will mainly be interested in the question how to perform multiblock data analysis and what to use in which data analysis situation. They may benefit from reading the main text and studying the examples. The second set of users are method developers. They want to know what is already available and spot niches for further development; apart from the main text and the examples they may also be interested in the elaborations. The final set of users are computer scientists and software developers. They want to know which methods are worthwhile to build software for and may also study the algorithms.

We will try to serve all groups. This means that we will explain most of the methods in a rather detailed manner (especially in Parts II and III) and will also pay attention to validation and visualisation to encourage proper interpretation. At the end of the book in Chapter 11, we describe multiblock toolboxes and packages in R, MATLAB and Python and showcase the accompanying R package `multiblock` which includes many of the methods described in this book.

1.3 Types of Data and Analyses

1.3.1 Supervised and Unsupervised Analyses

In any multiblock data analysis, we first have to choose between the main paradigms *unsupervised* and *supervised* analysis. *Unsupervised* analysis refers to explorative analysis looking for structure and connections in the data either in a single data block or across data blocks, typically using dimension reduction including maximisation/minimisation of some criterion combined with orthogonalisation, or by clustering techniques. It is crucial that the roles of the blocks are exchangeable: we can change the order of the blocks without changing the solution.

Supervised analysis refers to predictive data analysis, where emphasis is on a single block of data, $\mathbf{Y}$, (dependent block/response) which is connected to one or more blocks of data, $\mathbf{X}_m$, (independent block(s)/predictors) through regression or classification. The role of the blocks is now important: some blocks are regarded as dependent and some are regarded as independent.

There are also more *complex* structures where the multiblock problem is a mixture of these two (see, e.g., the L-shape problem in Figure 1.7).

1.3.2 High-, Mid- and Low-level Fusion

The data fusion literature (see, e.g., Mitchell (2012); Kedem *et al.* (2017); van Loon *et al.* (2020)) distinguishes between different ways of putting data blocks together. Here we will focus on one of these distinctions, namely between measurement level fusion, feature level fusion, and decision level fusion. Other names used for this are low-level, medium/intermediate/mid-level, and high-level fusion, respectively. For the case of supervised analyses, the three are illustrated in Figure 1.1. Low-level fusion means that the data blocks are simply concatenated and then analysed together as one single block. Mid-level fusion refers to first extracting features from each block before putting them together in a regression or classification model. High-level fusion means that predictions based on single input blocks are established before combining the results, see Elaboration 1.2. For unsupervised analyses, we can also distinguish between these different levels of fusion. In low-level fusion, the data blocks are used as such in an unsupervised multiblock data analysis method without any pre-selection of variables. In mid-level fusion, first the most important variables are selected per block and then an unsupervised fusion method is used. Such an approach is often taken in genomics where several thousands of variables are measured and filtered before entering any kind of modelling (see Section 1.4.2 for examples). High-level fusion would entail an unsupervised analysis per data block and then combining the results with visualisation tools. This approach is not taken very much. In this book we will focus on low-level and mid-level fusion.

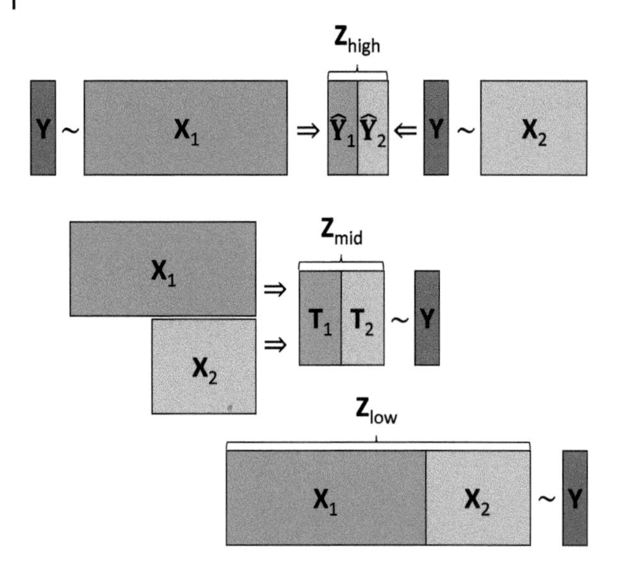

Figure 1.1 High-level, mid-level, and low-level fusion for two input blocks. The Z's represent the combined information from the two blocks which is used for making the predictions. The upper figure represents high-level fusion, where the results from two separate analyses are combined. The figure in the middle is an illustration of mid-level fusion, where components from the two data blocks are combined before further analysis. The lower figure illustrates low-level fusion where the data blocks are simply combined into one data block before further analysis takes place.

ELABORATION 1.2

High Level Supervised Fusion

High-level supervised fusion focuses on combining classification or prediction results for improved precision. Instead of using a method which takes the different data sets into account in building a predictor or classifier, high-level fusion combines results from individual predictions and combines them in the best possible way. In other words, high-level fusion refers to combining results from already established prediction or classification methods.

A possible drawback with this strategy as compared to low-level and feature-level fusion is that it does not provide further insight into how the different measurements relate to each other and how they can be combined in a good way in the prediction of the outcome. On the other hand, high-level fusion of prediction results for new samples does not generally require the individual predictors to be developed from the same samples. In other words, when two (or more) predictors are to be combined for a new sample, they do not need to come from the same data source. It is possible to simply plug in the new data and obtain predictions that can be combined as described below. In this sense it is more flexible (Ballabio *et al.* (2019)) than low- and feature-level fusion. It has been shown in Doeswijk *et al.* (2011) that fusing classifiers most often gives similar or improved prediction results as compared to using only one of them. An overview of the use of high-level fusion (and other methods) can be found in Borràs *et al.* (2015).

A simple way of combining classifiers is to use voting based on counting the number of times the classifiers agree. There are different types of voting schemes that are proposed in the literature. One of them is simple democratic majority voting which means that the group/class that gets the highest number of votes is chosen. In the case of ties, the result is inconclusive. An alternative strategy is 75% voting which means that 75% of the votes should be for the same class before a decision can be made.

Fusing quantitative predictors is most easily done using averages or weighted averages with weights depending on the prediction error of the different predictions, as determined by, for instance, cross-validation. This strategy has similarities with so-called bagging (see, e.g., Freund (1995)). In machine learning, high-level supervised fusion is found in the sub-domain 'ensemble learning'.

1.3.3 Dimension Reduction

There are many ways to perform multiblock data analysis. We restrict the focus of this book to approaches that use dimension reduction methods to tackle multiblock problems. The basic idea of dimension reduction methods is to extract components or latent variables from data blocks, see Figure 1.2.

In this figure, the matrix $\mathbf{X}$ $(I \times J)$ consists of J variables measured on I samples. The matrix $\mathbf{W}$ $(J \times R)$ of weights defines the scores $\mathbf{XW} = \mathbf{T}$ $(I \times R)$ where R is much smaller than J. This is the dimension reduction (or data compression) part and the idea is that $\mathbf{T}$ represents the samples in matrix $\mathbf{X}$ in a good way depending on the purpose. Likewise, the variables are represented in the loadings $\mathbf{P}$ $(J \times R)$ which can be connected to the scores in a least squares sense, e.g., in the model $\mathbf{X} = \mathbf{TP}^t + \mathbf{E}$. There are many alternatives to compute the weights, scores, and loadings depending on the specific situation; this will be explained in subsequent chapters.

The idea of dimension reduction by using components or latent variables is very old and has proven to be a very powerful paradigm, with many applications in the natural- life- and social sciences. When considering multiple blocks of data, each block is summarised by its components and the relationships between the blocks is then modelled by building relationships between those components. There are also many ways to build such relationships and we will discuss those in this book.

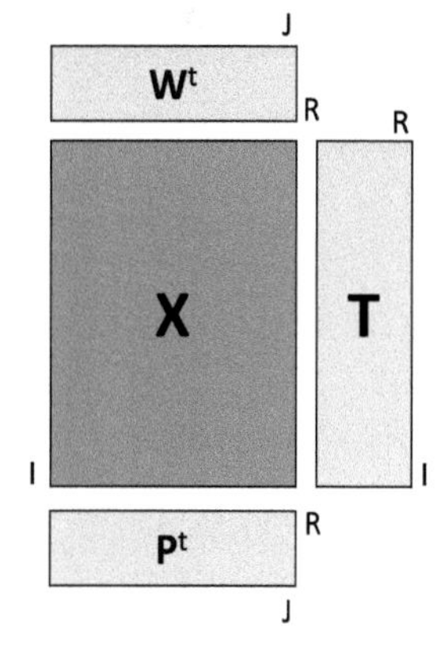

Figure 1.2 Idea of dimension reduction and components. The scores $\mathbf{T}$ summarise the relationships between samples; the loadings $\mathbf{P}$ summarise the relationships between variables. Sometimes weights $\mathbf{W}$ are used to define the scores.

There are many reasons for and advantages of using dimension reduction methods:

- The number of sources of variability in data blocks is usually (much) smaller than the number of measured variables.
- Component-based methods are suitable for interpretation through the scores and loadings associated with the extracted components.

- Underlying components and latent variables are appropriate for mental abstractions and interpretation.
- Multivariate data analysis becomes numerically stable and statistically robust if the components are chosen in a suitable way.
- Empirical validation of the models becomes manageable.
- The effect of measurement noise is reduced.
- Outliers can often be detected by visual inspection of the associated subspace projections provided by the extracted components.

1.3.4 Indirect Versus Direct Data

When discussing types of data, it is useful to distinguish between direct and indirect data. Direct data are always in the form of a matrix or table containing measurements of variables on a set of samples. Indirect data or derived data are always in the form of variables × variables or samples times samples matrices. Examples of such types of data are cross-products of matrices of direct data, covariances, distances and the like. The main focus in this book is on direct data, but we will discuss some indirect methods as well. First, to limit ourselves and, secondly, analyses on direct data are usually easier to understand and interpret. Thirdly, in many applications of multiblock data analysis in the natural and life sciences, direct data are available. For a more formal description of this distinction, see Section 2.2.1.

1.3.5 Heterogeneous Fusion

The final property of data we need to present is whether all blocks in the data set are measured on the same scale or not, i.e., if the data set is homogeneous or heterogeneous. These concepts are explained in more detail in Chapter 2 (Section 2.2.2). Briefly, if all blocks contain measurements on the same scale, e.g., they are all numerical or quantitative data, then the resulting problem will be called homogeneous fusion. If they are not of the same scale, e.g., a mixture of quantitative and binary measurements, then the problem is called heterogeneous fusion. We will discuss both of these in this book although most methods are made for homogeneous data.

1.4 Examples

This section contains some examples of multiblock data analysis problems in different fields of the natural and life sciences. It serves to give an idea about which types of questions are asked and which types of data sets are available. A full explanation of the methods used is given in the following chapters. These examples are only appetisers!

1.4.1 Metabolomics

Metabolomics is the part of life sciences concerned with measuring and studying the behaviour of metabolites (small biochemical compounds) in biological systems. The field has grown considerably in the last 20 years with conferences and dedicated journals. A large part of the applications concern finding biomarkers for diseases which translates into finding the metabolites that discriminate between groups of objects (e.g., control *versus* diseased subjects). Elaboration 1.3 shows some of the terms used in metabolomics research.

ELABORATION 1.3

Terms in metabolomics and proteomics

Biomarkers: Chemical compounds (e.g., metabolites) that mark a difference between conditions, e.g., between healthy and diseased persons.
GC-MS: Gas chromatography–mass spectrometry. A separation method coupled to a mass spectrometer used a lot in advanced chemical analyses of volatile compounds.
LC-MS: Liquid chromatography–mass spectrometry. A separation method coupled to a mass spectrometer used a lot in advanced chemical analyses for a large diversity of chemical compounds.
Metabolome: The set of all metabolites of a biological organism responsible for its metabolism.
NMR: Nuclear magnetic resonance. A fast chemical analysis method giving a fingerprint of a sample and concentrations of chemical compounds.
Proteomics: The study and measurements of proteins in biological organisms. Proteins are mostly enzymes catalysing metabolic reactions.

There are several multiblock data analysis challenges in metabolomics. It is increasingly popular to measure different sets of chemically related metabolites on the same samples using different instrumental protocols (Smilde *et al.*, 2005b; Pellis *et al.*, 2012; Kardinaal *et al.*, 2015). These blocks of data (each block pertaining to one instrumental protocol) then need to be combined to arrive at a global view on metabolism. Metabolites can also be measured in different compartments, such as in blood, urine, liver, muscle, kidney (Fazelzadeh *et al.*, 2016). This also generates multiblock data analysis problems. Metabolites are converted in biochemical reactions catalysed by enzymes (proteins). Hence, it is also worthwhile in some cases to measure proteins and combine those with metabolomics measurements (Wopereis *et al.*, 2009). Plants are complex organisms with a rich variety of metabolites. The metabolism of plants is influenced by environmental conditions, such as temperature and light. Example 1.1 illustrates this.

Example 1.1: Metabolomics example: plant science data

This metabolomics example comes from a larger study in plant sciences (Caldana *et al.*, 2011). The goal of the study was to investigate changes in metabolism and gene-expression of Arabidopsis related to growth under different light and temperature conditions. To this end, time-resolved experiments were performed. The design of the data set is shown in Figure 1.3. It is not a fully crossed design, but for each cell in the design gene-expression and metabolomics measurements were performed at 19 time points. We will only use the metabolomics measurements which comprised around 65 identified metabolites and use the part of 21^0C (the third line in the table below). This results in four blocks of data (21-D, 21-LL, 21-L and 21-HL) each consisting of 19 rows (time points) and 65 columns (measured

	0	75	85	150	300
4^0	4-D		4-L		
21^0	21-D	21-LL		21-L	21-HL
32^0	32-D			32-L	

Figure 1.3 Design of the plant experiment. Numbers in the top row refer to light levels (in μE m^{-2} sec^{-1}); numbers in the first column are degrees centigrade. Legend: D = dark, LL = low light, L = light and HL = high light.

metabolites). Hence, we only study the factors light and time (the factor temperature is kept constant).

A first impression of the variation in metabolite levels can be obtained by performing a principal component analysis (PCA) on the data, see Figure 1.4(a)), where we have concatenated all four blocks (21-D, 21-LL, 21-L and 21-HL) below each other. The colour coding is according to the light conditions and this figure shows that there is systematic variation associated with the factor light in the data. A more advanced analysis of this data is by using a multiblock data analysis method that takes into account the underlying experimental design, such as ANOVA-simultaneous component analysis (ASCA, see Chapter 6). Figure 1.4(b) shows the scores on the first ASCA interaction component and this clearly shows a time dependent contrast between dark and high light conditions. The original data set also comprises gene-expression measurements which makes the problem even more challenging.

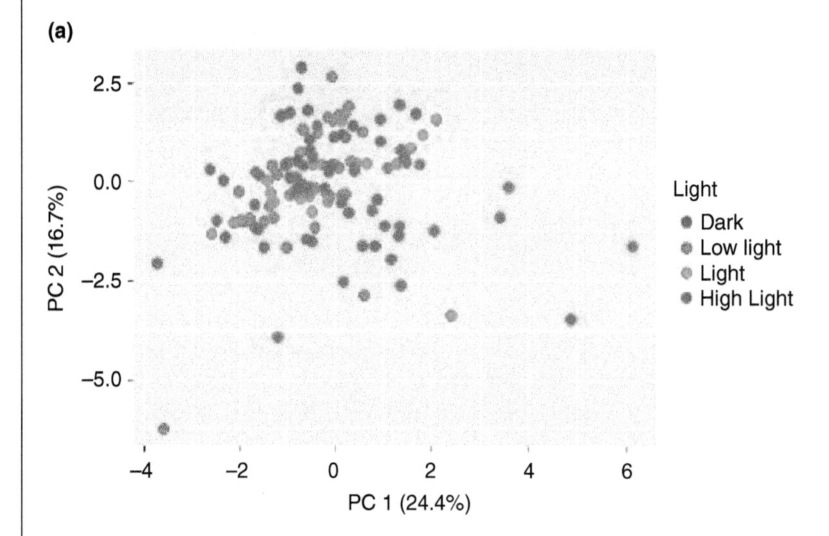

Figure 1.4 Scores on the first two principal components of a PCA on the plant data (a) and scores on the first ASCA interaction component (b). Legend: D = dark, LL = low light, L = light and HL = high light.

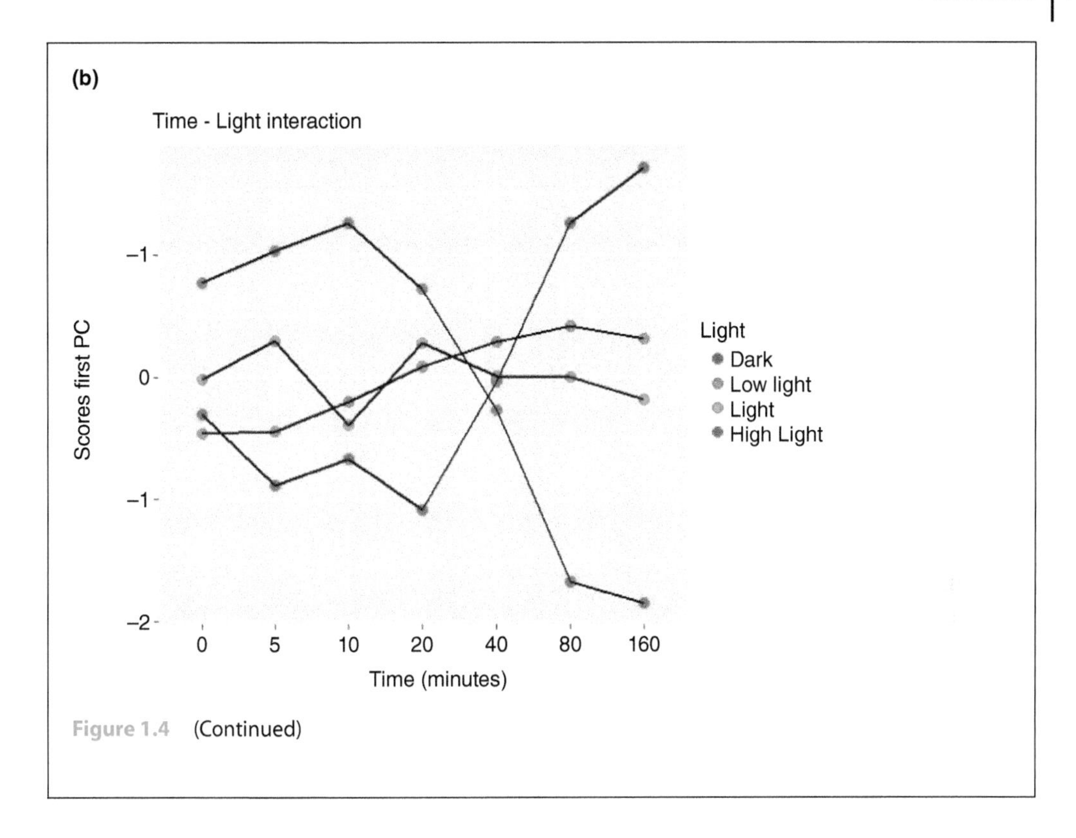

Figure 1.4 (Continued)

1.4.2 Genomics

Genomics covers the area of life science research related to the genomes of biological organisms. In a broad sense, it may cover very many aspects regarding the genome such as the transcriptome, genetics and epi-genetics. In a smaller context, it may encompass only the measurement of the genomic transcripts, e.g., using RNAseq (see Elaboration 1.4 for an explanation of terms).

ELABORATION 1.4

Terms in genomics

CNA: Many biological organisms have several copies of the same gene (see Figure 1.5). Copy number aberration (CNA) quantifies this (see Example 1.2).

Epi-genetics: DNA can be modified chemically thereby regulating expression of the corresponding genes. This chemical modification of the DNA is called epi-genetics (see Figure 1.5).

Genetics: Biological organisms have DNA encoding their genetic make-up. Genetics studies this DNA.

Methylation: A methyl group can be attached to the DNA, affecting transcription. This is a part of epi-genetics (see Figure 1.5).

Mutation: DNA consists of four types of nucleotides (A, T, G, and C) containing the genetic code. Some of these nucleotides may be mutated, e.g., change from A to T. If this happens for a single nucleotide then this is called a single nucleotide polymorphism (SNP, see Figure 1.5).

RNAseq: The modern way of measuring gene-expression or the RNA of a biological organism. There are many types of RNA of which messenger-RNA (mRNA) is the most studied one.

Transcriptomics: Genes are transcribed to RNA and transcriptomics concerns the analysis of these transcripts.

Genomics is a very active field with many multiblock data analysis challenges due to the rapid development of measuring techniques. Whereas in former days gene-expression was measured with micro-arrays, this technology has been overtaken by next generation sequencing (mRNAseq, miRNAseq, siRNAseq, scRNAseq to name a few). This has led to open-access repositories containing genomics data of very different types, e.g., in cancer research (Tomczak *et al.*, 2015) which is often the basis for generating new multiblock data analysis methods (Aben *et al.*, 2016, 2018; Song *et al.*, 2018). Other examples are combining genomics data with data from non-omics techniques like medical imaging, e.g., for treatment response predictions.

Example 1.2: Genetics example

In cancer research, often cell-lines are used derived from tumour tissue (Iorio *et al.*, 2016). Of these cell-lines many measurements are made available in public databases. Such measurements may consist of measured RNA-levels (ratio-scaled values), but also measurements related to mutations (so-called single nucleotide polymorphisms or SNPs) which are on/off measurements and intrinsic of a binary nature.

One of the possible genetic determinants is the copy number of a gene, see Figure 1.5(a); such a gene may be duplicated. An extra layer of gene-regulation is provided by methylation of certain nucleotides of the genome (see Figure 1.5(b)). If a nucleotide is methylated, then transcription of the corresponding gene cannot occur; this area of genetics is called epigenetics. There are different ways of expressing methylation, but the most simple one is a yes or no whether or not a specific site is methylated. At a certain position on the genome, one nucleotide may have been changed (see Figure 1.5(c)). This is obviously binary since there may be a SNP or no SNP at a certain position on the genome. Hence, treating such data in a multiblock fusion setting requires specialised methods, see Chapter 5.

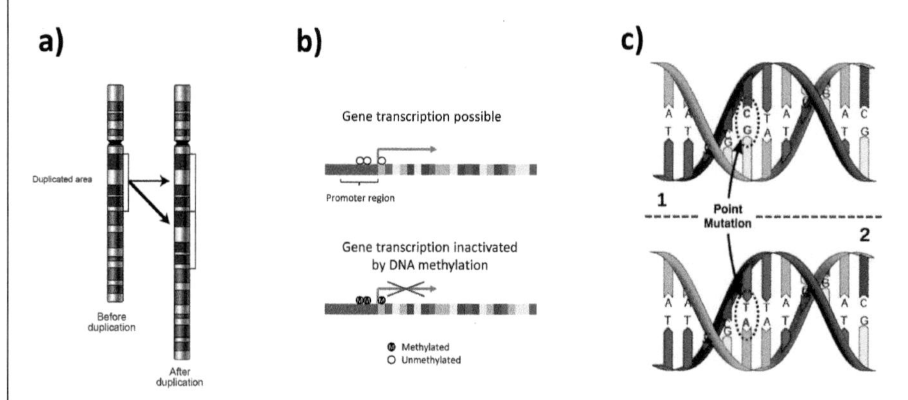

Figure 1.5 Idea of copy number variation (a), methylation (b), and mutation (c) of the DNA. For (a) and (c): Source: Adapted from Koch *et al.*, 2012.

1.4.3 Systems Biology

Taking it one step further in terms of omics measurements, we enter the area of systems biology. The general idea of systems biology is to describe biological systems as a network of interacting biochemical compounds. Often, the interactions in such networks show emerging behaviour which cannot be understood from studying single biochemical compounds (Bruggeman and Westerhoff, 2007).

There are basically two approaches to systems biology: top-down and bottom-up (Shahzad and Loor, 2012). In bottom-up approaches, fundamental models are made of parts of biochemical systems and, subsequently, parameters in those models are fitted to data. In top-down systems biology, many types of omics data are collected and these are combined into one holistic analysis. The latter goes under different names: intra- and inter-omics analysis, cross-omics analysis, statistical integration, statistical data fusion to name a few (Tayrac *et al.*, 2009; Richards *et al.*, 2010; Richards and Holmes, 2014). In all these top-down applications, multiblock data analysis is important. See also Elaboration 1.5 for more explanation.

ELABORATION 1.5

Terms in systems biology

Biological networks: In biological organisms, biochemical compounds act together in networks of activity. An example is a metabolic network describing all the conversions taking place in the metabolism of a cell.

Bottom-up: Approach in which detailed biochemical knowledge of a biological system is used to build mathematical models of that system (e.g., in terms of sets of differential equations). Such models are necessarily limited in size; they describe only a small part of the system.

Emerging property: Property of a system which cannot be understood from its single actors. Temperature is an example of an emerging property of a system containing a large number of molecules that interact.

Microbiome: The whole set of micro-organisms in and around a biological host. The gut-microbiome is the most famous example; essential for humans to metabolise food.

Top-down: Approach in which many measurements are performed on the same biological system and empirical modelling is subsequently used to model that system. These models usually contain many biochemical compounds but are much less detailed than the bottom-up models.

An intriguing new development in systems biology is to involve microbiome measurements of the biological system (Franzosa *et al.*, 2015). This has sparked many studies in different areas of medicine, such as inflammatory bowel disease (Huang *et al.*, 2014) and cancer (Weir *et al.*, 2013). It is also highly relevant for nutritional and food studies (Jacobs *et al.*, 2009; Van Duynhoven *et al.*, 2010; Moco *et al.*, 2012). In all these cases, the microbiome data are combined with other omics data generating multiblock data analysis problems.

1.4.4 Chemistry

Multiblock data analysis problems arise in different parts of chemistry. A very active area is analytical chemistry, with two very prominent topics. The first one is multivariate curve resolution where the general idea is to mathematically resolve chemical mixtures in underlying pure chemical components and their concentration profiles (Tauler *et al.*, 1995; de Juan and Tauler, 2006). Many different types of multiblock data analyses are performed in this area with a special emphasis on applying domain-specific constraints. The second application area is calibration where the purpose is to obtain concentrations from instrumental analysis methods. Also in this area multiblock data analysis methods are used (Næs *et al.*, 2013). A spectroscopy example is given in Example 1.3.

ELABORATION 1.6

Terms in chemistry

Multivariate curve resolution: Part of chemometrics that tries to mathematically resolve mixtures of chemicals into their individual compounds.

Multivariate calibration: Part of chemometrics that deals with predicting properties (e.g., concentrations) from spectroscopic measurement. The idea is to replace a slow, expensive measurement technique (the reference method) by a fast, cheaper, and often non-destructive one (a spectroscopic measurement).

Process chemometrics: Part of chemometrics devoted to processes; such as process analysis, multivariate process control and process monitoring.

Vibrational spectroscopy: Chemical measurement techniques that probe vibrational energies of molecules. There are different types of vibrational spectroscopy: infrared (IR), mid-infrared (MIR), near-infrared (NIR), ultraviolet (UV), visible (VIS) and Raman spectroscopy.

Another area in chemistry which is populated with multiblock data analysis problems is process chemometrics (MacGregor *et al.*, 1994; Wise and Gallagher, 1996; Kourti *et al.*, 1995; Lopes *et al.*, 2002). The general problem is how to combine multiple chemical process measurements for process understanding and statistical process monitoring.

Example 1.3: Chemistry example: Raman spectroscopy data

The data set was first published in a study containing both Raman and near infrared spectroscopy measurements of emulsions (Afseth *et al.*, 2005). For the Raman data, 1096 Raman shifts, from 1770 cm^{-1} to 675 cm^{-1}, were recorded for 69 emulsions containing a mixture of proteins, water, and fats (see Figure 1.6). Two reference values are used as responses: polyunsaturated fatty acids (PUFA) as percentage of total sample weight (0.3–11.5%) and as percentage of fats in sample (2.2–61.6%). The reference values have a correlation of $R = 0.73$, i.e. $R^2 = 0.54$, meaning that around half of the variation in PUFA content is due to the variation in total fat content. The aim of the original study was to be able to quantify the PUFA percentages using only spectroscopy to enable quick, cheap, and non-destructive measurements.

In this book, we will concentrate on the Raman block as this dominated completely in a previous multiblock data analysis study (Liland *et al.*, 2016), and rather split it into suitable wavelength regions, here splitting at 1350 cm^{-1} and 1100 cm^{-1}. This is done to explore the predictive power of the different wavelength regions. This data set will be analysed using several of the supervised methods in this book to see what is emphasised by each of them. In general, we see that the predictive models mostly leverage the variables corresponding to molecular vibrations associated with lipids and degrees of saturation, and that these models can reproduce the reference values with high precision.

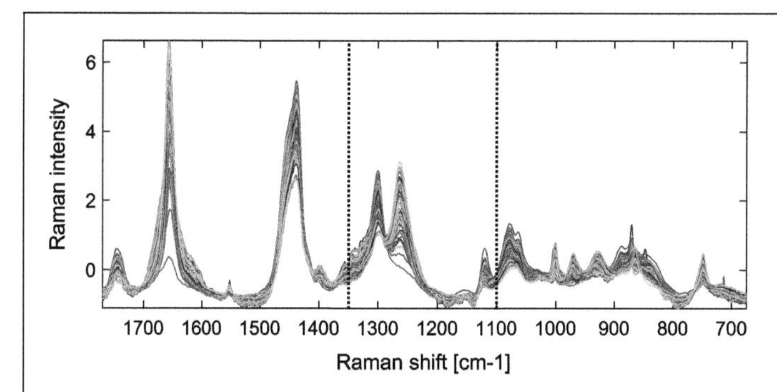

Figure 1.6 Plot of the Raman spectra used in predicting the fat content. The dashed lines show the split of the data set into multiple blocks.

1.4.5 Sensory Science

Sensory and consumer science is an important discipline in the assessment of food quality. It consists of a large number of measurement methods for determining the descriptive properties of products as well as the consumer liking of the same products (Lawless and Heymann, 2010). Often a product will be characterised by a number of different data types, ranging from classical descriptive sensory analysis using predefined attributes and a trained sensory panel to consumer based characterisation based on, for instance, the check-all-that-apply (CATA) method (Varela and Ares, 2012). The data sets will generally consist of a substantial number of attributes and a relatively moderate number of samples. Of special interest is estimating relations between data blocks related to liking and product characterisation. A large number of methods have been developed for this purpose as will be discussed in Chapters 7, 8 and 10 in this book (see Næs *et al.* (2010) for an overview). An example of a typical data structure and its related questions in sensory science is given in Example 1.4.

ELABORATION 1.7

Terms in sensory analysis

Consumer liking: For hedonic sensory methods, a consumer panel is used. The consumer can be asked about how much they like the different products and how willing they are to buy the products tested.

Sensory panel: For assessing product quality, it is common to use a sensory panel consisting of a number of trained assessors which assess the intensity on a predefined scale of a number of relevant sensory attributes.

Sensory attribute: The measurements, as performed by the sensory panel, such as sweetness, hardness, and acidity (depending on types of products).

Rapid sensory methods: There exist a number of so-called rapid sensory methods, for instance, projective mapping, sorting, and CATA. For the latter all participants are asked to tick, for each product, on the relevant attributes on a predefined list. This gives a table of 0s and 1s for each participant.

Example 1.4: Sensory example: consumer liking

A typical multiblock data structure that occurs in consumer science is depicted in Figure 1.7. The context is typically product development where interest is in understanding the relations between descriptive information of a number of prototype samples and the consumer liking of the same samples. In addition, interest is in interpreting the liking patterns in terms of consumer characteristics for better understanding of which consumer groups prefer which products (see e.g., Næs *et al.* (2018)). Based on this type of information, the product developer can more easily design products that better fit the consumer needs and liking patterns. As can be seen from the figure, both chemical attributes as well as sensory properties/attributes, obtained by a trained sensory panel, can be of interest for describing the products. A number of different liking scores can also be of interest, for instance related to taste and texture (Menichelli *et al.*, 2013), as depicted by the stack of data blocks for liking.

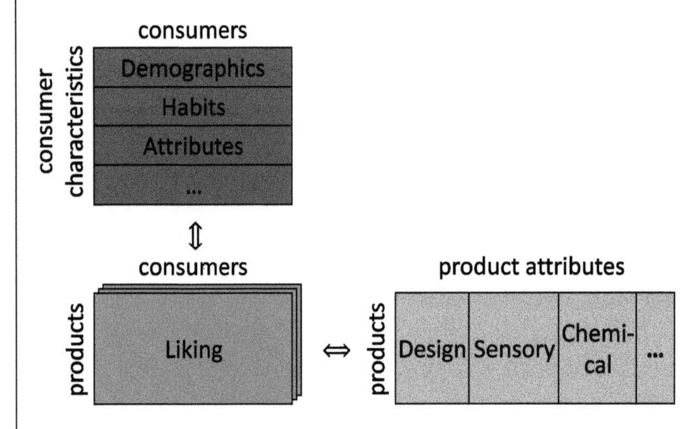

Figure 1.7 L-shape data of consumer liking studies.

Analysing this so-called L-shape data structure sheds light on, for instance, which are the sensory drivers of liking, which samples are the most liked, what characterises these samples, and what characterises the different consumer groups with different preference patterns.

1.5 Goals of Analyses

Many goals of multiblock data analysis can be envisaged. In current practice, these goals are usually implicit. By making these goals explicit it will become necessary to also make explicit the global optimisation criterion or, when such a criterion is difficult to formulate, to carefully think about the whole data analysis procedure and which method to choose. Several general goals will be discussed briefly.

Exploratory analysis: One of the most obvious goals of multiblock data analysis is exploration which is a part of unsupervised analysis. By plotting the weights, scores, and loadings, summaries of the data are obtained which can be interpreted and maybe further analysed using visualisation tools.

Predictive models: Another obvious goal is to try to predict the variation in one data block using several other data blocks; this is a part of supervised analysis. The idea is then that using multiple predictive blocks gives a better prediction for future samples.

Finding topologies: In the case of complex data, data blocks can be placed in different relationships. The arrangement of blocks as dependent or independent may be a purpose of the analysis. We call such an arrangement a *topology*. In that case, it would be useful to have a strategy for deciding on the topology that fits the data best.

Common versus distinct variation: There can be common and distinct variation in the multiple data blocks (see Section 1.8). This separation into types of variation greatly simplifies subsequent interpretation of the results.

Treatment effects: The effect of a treatment can be measured in different blocks of data. The interest is usually what the main effect of a treatment is on measurements in the different blocks of data.

Individual differences: Apart from group differences, also individual differences are useful. This can be for personalised medicine or nutritional interventions or consumer behaviour. Multiblock data analysis may help to find such differences and thereby facilitate population stratification and sub-typing.

Mixed goals: In real-life applications, a mixture of goals is usually present. It may be that a treatment has been given which expresses itself differently in the common and distinct variation. Moreover, interest may be in the main effects of treatments but also on individual treatment effect differences.

1.6 Some History

The history of multiblock data analysis methods goes back a long time. One of the starting points was principal component analysis (PCA, Pearson (1901)). Another early method in statistics was canonical correlation analysis (Hotelling, 1936a) which led to development of many related methods. In the social sciences, path-models were developed, such as LISREL and PLS (Jöreskog and Wold, 1982) which are also used in consumer science and marketing research. In psychometrics, inter-battery factor analysis (Tucker, 1958) was developed and later simultaneous component analysis methods (Ten Berge *et al.*, 1992) which have also sparked many alternatives. In parallel in chemometrics, methods such as consensus PCA and hierarchical PCA were developed (Westerhuis *et al.*, 1998). Ideas on the latter two methods already started in the 1970s at conferences. In the French data analysis school, multiple correspondence analysis was developed (Benzécri, 1980) and many other methods. In multiway data analysis (which can be regarded as a subset of multiblock data analysis methods) the earliest developments started with the work of Cattell (1944), Carroll and Chang (1970) and Harshman (1970). For a history of the latter, see Smilde *et al.* (2004). Figure 1.8 tries to systematically show many multiblock data analysis methods and how they relate to basic statistical methodology and subspace projection methods from various fields of applied data analysis.

1.7 Fundamental Choices

In any sort of multiblock data analysis, choices have to be made such as which method to use and what kind of pre-processing to apply. Two fundamental questions which always should be considered (and dealt with) are highlighted below.

Variation explained: Do we only want to explain variation between blocks or also within blocks?

Fairness: Should all blocks play a role in the final solution or can we allow some of the blocks to be dominant in this respect?

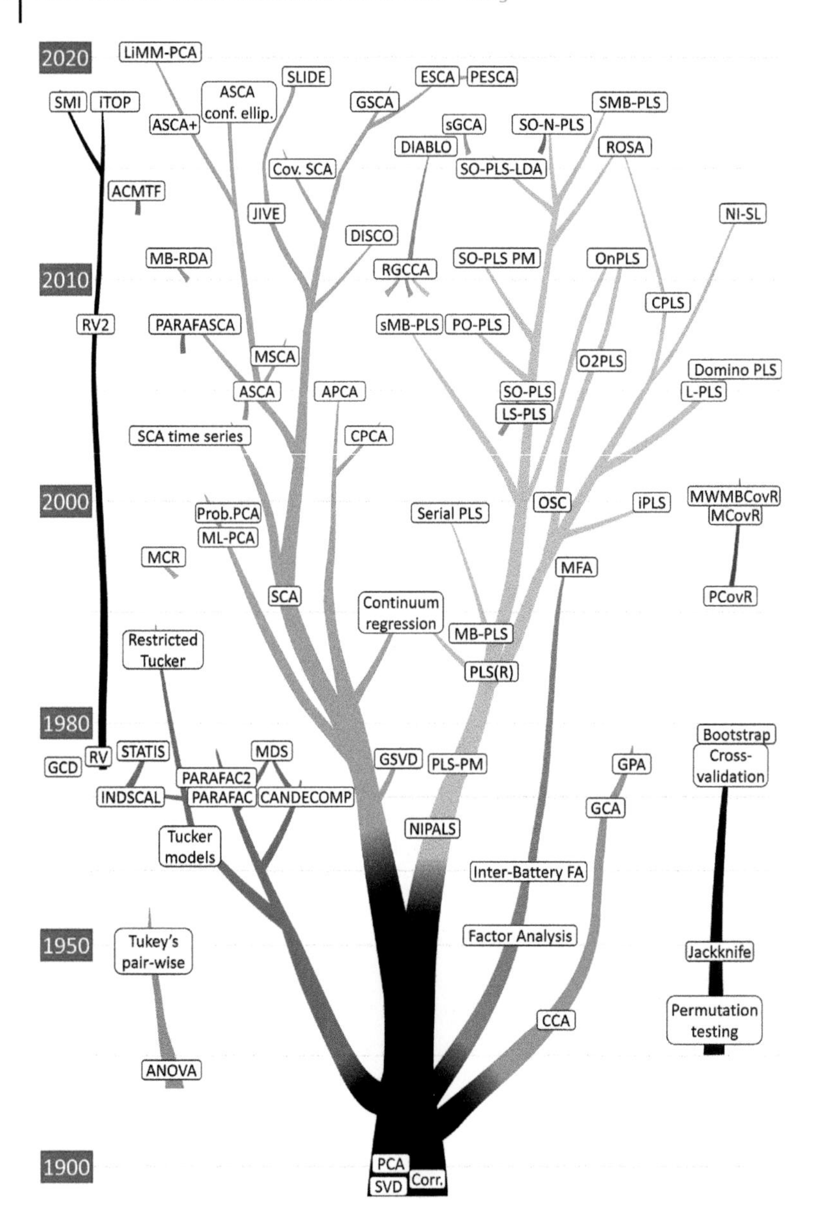

Figure 1.8 Phylogeny of some multiblock methods and relations to basic data analysis methods used in this book.

The first concept – variation explained – pertains to the choice that we can model the variation within the blocks and/or the variation between the blocks. Each multiblock data analysis method makes a different choice in this respect and thus it is up to the user to decide what aspect is the most important: between- or within-variation. A more detailed account is given in Section 2.4.

The concept of fairness relates to the notion that each block should participate in the final solution to a certain degree. Multiblock data analysis methods also differ in this respect: some methods are fair and some are 'block selectors' (Smilde *et al.*, 2003; Tenenhaus *et al.*, 2017). To some extent, fairness can be influenced by block-scaling (see Section 2.6), but some methods are invariant to this type of scaling. The method ROSA (see Section 7.5) builds on the

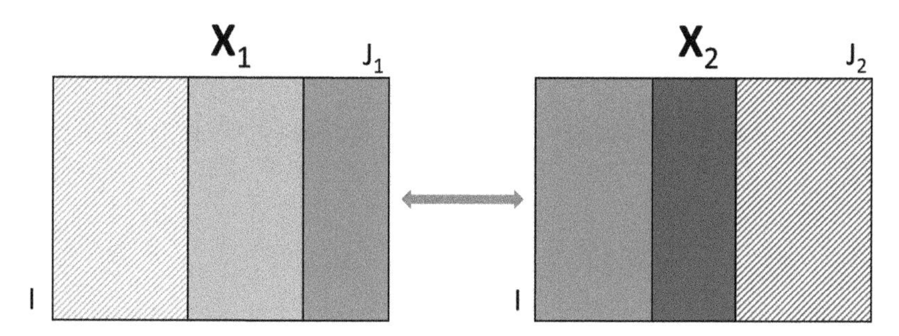

Figure 1.9 The idea of common and distinct components. Legend: blue is common variation; dark yellow and dark red are distinct variation and shaded areas are noise (unsystematic variation).

principle of fairness: each block is allowed to enter the solution in competition with the other blocks: if it is important then it will be included. The fairness concept has some relation to the concept of invariance discussed in Chapter 7.

1.8 Common and Distinct Components

A crucial concept that plays a role in almost all methods in this book is the idea of common and distinct subspaces and components (Smilde *et al.*, 2017). A schematic illustration of these concepts is shown in Figure 1.9 and a more detailed exposure of these concepts is given in Chapter 2.

Suppose there are two data blocks $\mathbf{X}_1$ and $\mathbf{X}_2$ sharing the same samples, i.e., different variables are measured on the same set of samples (see Chapter 3). Then these two blocks can have variation in common (the blue part). This common variation spans a subspace and the common components are then a basis for this subspace.

There is also a part in each block that contains still systematic variation (the dark yellow and dark red parts). These have nothing in common and are, therefore, called distinct parts. These also represent subspaces and the distinct components (two sets; one set for each block) are the bases for these subspaces. What is left in the matrices is unsystematic variation or noise (shaded parts).

The division of each data block in common, distinct, and unsystematic variation should not be read in terms of the individual variables being in common or being distinct but in terms of subspaces. Hence, a part of the variation of a variable in block 1 may be in common with variation of some variables in block 2 whereas the other part of that variable may be distinct, see Elaboration 1.8.

ELABORATION 1.8

Common and distinct in spectroscopy

Suppose that the same set of samples is measured in the UV-Vis regime (block $\mathbf{X}_1$) and with near-infrared (NIR, block $\mathbf{X}_2$). Also assume that this set of samples contains three chemical components (A,B,C): A absorbs both in UV-Vis and NIR; B only absorbs in the UV-Vis regime and C absorbs only in NIR. Then the common part is the absorption of A in both data blocks; the distinct parts are B in block 1 and C in block 2. However, at a particular wavelength in the NIR region there may be a contribution from both A and C. Hence, this wavelength, i.e., variable, has a common and a distinct part. The same can happen in block 1.

1.9 Overview and Links

In this book we will consider a multitude of methods. To streamline this a bit, we are going to give a summary, at the beginning of each chapter, of the methods and aspects which will be discussed. That will be done in the format of a table. We will specify the following aspects of the methods:

A A method for unsupervised (U), supervised (S) or complex (C) data structures.
B The method can deal with heterogeneous data (HET, i.e., different measurement scales) or can only deal with homogeneous data (HOM).
C A method that uses a sequential (SEQ) or simultaneous (SIM) approach.
D The method is defined in terms of a model (MOD) or in terms of an algorithm (ALG).
E A method for finding common (C); common and distinct (CD); or finding common, local and distinct components (CLD).
F Estimation of the model parameters is based on least squares (LS), maximum likelihood (ML), eigenvalue decompositions (ED) or maximising covariance or correlations (MC).

The first item (A) is used to organise the different chapters. Some methods can deal with data of different measurements scales (heterogeneous data) and some methods can only handle homogeneous data. The difference between the simultaneous and sequential method is explained in more detail in Chapter 2. Some methods are defined by a clear model and some methods are based on an algorithm. The already discussed topic of common and distinct variation is also a distinguishing and important feature of the methods and the sections in some of the chapters are organised according to this principle. Finally, there are different ways of estimating the parameters (weights, scores, loadings, etc.) of the multiblock models. This is also explained in more detail in Chapter 2.

Table 1.1 is an example of such a table for Chapter 6. This table presents a birds-eye view of the properties of the methods. Each chapter discussing methods will start with this table to set the scene. We will end most chapters with some recommendations for practitioners on what method to use in which situation.

Table 1.1 Overview of methods. Legend: U = unsupervised, S = supervised, C = complex, HOM = homogeneous data, HET = heterogeneous data, SEQ = sequential, SIM = simultaneous, MOD = model-based, ALG = algorithm-based, C = common, CD = common/distinct, CLD = common/local/distinct, LS = least squares, ML = maximum likelihood, ED =eigendecomposition, MC = maximising correlations/covariances. For abbreviations of the methods, see Section 1.11

		A			B		C		D		E			F			
	Section	U	S	C	HOM	HET	SEQ	SIM	MOD	ALG	C	CD	CLD	LS	ML	ED	MC
ASCA	6.1																
ASCA+	6.1.3																
LiMM-PCA	6.1.3																
MSCA	6.2																
PE-ASCA	6.3																

1.10 Notation and Terminology

Throughout this book, we will make use of the following generic notation. When needed, extra notation is explained in local paragraphs. For notational ease, we will not make a distinction between population and estimated weights, scores and loadings which is the tradition in chemometrics and data analysis. For regression equations, when natural we do make that distinction and there we will use the symbol $\widehat{b}$ or $\widehat{y}$ for the estimated parameters or fitted values.

x	a scalar
$\mathbf{x}$	column vector: bold lowercase
$\mathbf{X}$	matrix: bold uppercase
$\mathbf{X}^t$	transpose of $\mathbf{X}$
$\underline{\mathbf{X}}$	three-way array: bold uppercase underlined
$m = 1,\ldots,M$	index for block
$i_m = 1,\ldots,I_m$	index for first way (e.g., sample) in block m (not shared first way)
$i = 1,\ldots,I$	index for first shared way of blocks
$j_m = 1,\ldots,J_m$	index for second way (e.g., variable) in block m (not shared second way)
$j = 1,\ldots,J$	index for second shared way of blocks
$r = 1,\ldots,R$	index for latent variables/principal components
$\mathbf{R}$	matrix used to compute scores for PLS
$\mathbf{X}_m$	block m
$\mathbf{x}_{mi}$	i-th row of $\mathbf{X}_m$ (a column vector)
$\mathbf{x}_{mj}$	j-th column of $\mathbf{X}_m$ (a column vector)
$\mathbf{W}$	matrix of weights
$\mathbf{I}_L$	identity matrix of size $L \times L$
$\mathbf{T}$	score matrix
$\mathbf{P}$	loading matrix
$\mathbf{E}, \mathbf{F}$	matrices of residuals
$\mathbf{1}_L$	column vector of ones of length L
$diag(\mathbf{D})$	column vector containing the diagonal of $\mathbf{D}$
$\otimes$	Kronecker product
$\odot$	Khatri–Rao product (column-wise Kronecker product)
$*$	Hadamard or element-wise product
$\oplus$	Direct sum of spaces

When we discuss methods with only one X- and one Y-block we will use the indices J_X and J_Y for the number of variables in the X- and Y-block, respectively. When there are multiple X-blocks, we will differentiate between the number of variables in the X-blocks using the indices $J_m(m = 1, \ldots, M)$; for the Y-block we will then use simply the index J. We try to be as consistent as possible as far as terminology is concerned. Hence, we will use the terms scores, loadings, and weights throughout (see Figure 1.2 and the surrounding text). We will also use the term explained variance which is a slight abuse of the term variance, since it does not pertain to the statistical notion of variance. However, since it is used widely, we will use the term explained variance instead of explained variation as much as possible. Sometimes we need to use a predefined symbol (such as $\mathbf{P}$) in an alternative meaning in order to harmonise the text. We will make this explicit at those places.

1.11 Abbreviations

In this book we will use a lot of abbreviations. Below follows a table with abbreviations used including the chapter(s) in which they appear. A small character 's' in front of an abbreviation means 'sparse', e.g., sMB-PLS is the method sparse MB-PLS. For many methods mentioned below there are sparse versions; such as sPCA, sPLS, sSCA, sGCA, sMB-PLS and sMB-RDA. These are not mentioned explicitly in the table.

Table 1.2 Abbreviations of the different methods.

Abbreviation	Full Description	Chapter
ACMTF	Advanced coupled matrix tensor factorisation	5
ASCA	ANOVA-simultaneous component analysis	6
BIBFA	Bayesian inter-battery factor analysis	9
DIABLO	Data integration analysis biomarker latent component omics	9
DI-PLS	Domain-invariant PLS	10
DISCO	Distinct and common components	5
ED-CMTF	Exponential dispersion CMTF	9
ESCA	Exponential family Simultaneous Component Analysis	5
GAS	Generalised association study	4,9
GAC	Generalised association coefficient	4
GCA	Generalised canonical analysis	2,5,7
GCD	General coefficient of determination	4
GCTF	Generalised coupled tensor factorisation	9
GFA	Group factor analysis	9
GPA	Generalised Procrustes analysis	9
GSCA	Generalised simultaneous component analysis	5
GSVD	Generalised singular value decomposition	9
IBFA	Inter-battery factor analysis	9
IDIOMIX	INDORT for mixed variables	9
INDORT	Individual differences scaling with orthogonal constraints	9
JIVE	Joint and individual variation explained	5
LiMM-PCA	Linear mixed model PCA	6
L-PLS	PLS regression for L-shaped data sets	8
MB-PLS	Multiblock partial least squares	7
MB-RDA	Multiblock redundancy analysis	10
MBMWCovR	Multiblock multiway covariates regression	10
MCR	Multivariate curve resolution	5,8
MFA	Multiple factor analysis	5
MOFA	Multi-omics factor analysis	9
OS	Optimal-scaling	2,5
PCA	Principal component analysis	2,5,8
PCovR	Principal covariates regression	2
PCR	Principal component regression	2

Table 1.2 (Continued)

Abbreviation	Full Description	Chapter
PESCA	Penalised ESCA	9
PE-ASCA	Penalised ASCA	6
PLS	Partial least squares	2
PO-PLS	Parallel and orthogonalised PLS regression	7
RDA	Redundancy analysis	7
RGCCA	Regularized generalized canonical correlation analysis	5
RM	Representation matrix approach	9
ROSA	Response oriented sequential alternation	7
SCA	Simultaneous component analysis	2,5
SLIDE	Structural learning and integrative decomposition	9
SMI	Similarity of matrices index	4
SO-PLS	Sequential and orthogonalised PLS regression	7,10

2

Basic Theory and Concepts

2.i General Introduction

In the first chapter, we gave a background of types of data and models we are going to describe. To set the stage for the rest of this book, we will describe some basic models such as principal component analysis (PCA), principal component regression (PCR), partial least squares (PLS), simultaneous component analysis (SCA), and generalised canonical analysis (GCA) that can be considered building blocks for many of the multiblock data analysis methods to follow.

We will also discuss two general theories regarding data and measurements. These theories do not play a dominant role yet in multiblock data analysis but are nevertheless important for understanding the properties of the different methods. We will also expand the notions of common and distinct components (already briefly discussed in Chapter 1) and give a formal presentation of those concepts. Also the concepts of variance within and variance between blocks will be treated in more detail.

We will end with an extensive presentation of approaches of validation, which are crucial for data analysis. The Appendix of this chapter contains some background mathematics which is used in many multiblock data analysis methods. Since it is customary to (column-) centre the data, we will assume centring throughout in all chapters unless stated otherwise. As already stated in Chapter 1, we will only use hats to indicate estimates if necessary.

2.1 Component Models

2.1.1 General Idea of Component Models

In Chapter 1, Section 1.3, we advocated the use of component models since they give a very versatile framework for approaching all kinds of multiblock problems. Indeed, many methods in this area are based on components. In this chapter we first introduce briefly the mathematical concept of components and subsequently show how to use this concept in the different data structures of Chapter 3.

The basic idea of components or latent variables is visualised in Figure 2.1 (which is a repeat of Figure 1.2) and amounts to summarising or decomposing a matrix $\mathbf{X}$ $(I \times J)$ by scores $\mathbf{T}$ $(I \times R)$ and loadings $\mathbf{P}$ $(J \times R)$. Sometimes weights $\mathbf{W}$ $(J \times R)$ are used to define the scores.

Multiblock Data Fusion in Statistics and Machine Learning: Applications in the Natural and Life Sciences, First Edition. Age K. Smilde, Tormod Næs, and Kristian Hovde Liland.

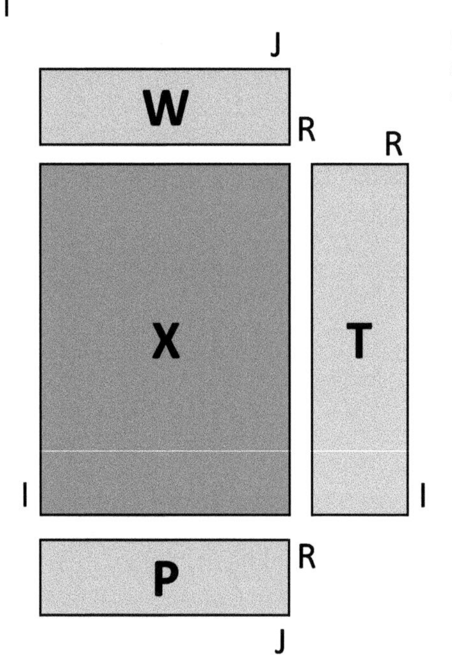

Figure 2.1 Idea of dimension reduction and components. Sometimes W is used to define the scores T which in turn define the loadings P.

The generic equation for such decompositions is

$$\mathbf{X} \approx \mathbf{TP}^t \tag{2.1}$$

where the symbol $\approx$ is used to indicate 'approximate', to be defined later. Alternatively, we may use

$$\mathbf{X} \approx \mathbf{XW}\widetilde{\mathbf{P}}^t = \widetilde{\mathbf{T}}\widetilde{\mathbf{P}}^t \tag{2.2}$$

which is not the same as Equation 2.1 since Equation 2.2 forces the components into the column-space of $\mathbf{X}$ whereas this is not necessarily the case in Equation 2.1. The scores collected in $\mathbf{T}$ are called components or latent variables (or unobserved variables), contrasting with the variables in the original matrix $\mathbf{X}$ which are sometimes called manifest (or observed) variables.

There are many choices in defining $\mathbf{T}$, $\mathbf{W}$ and $\mathbf{P}$. The matrices $\mathbf{T}$ and $\mathbf{P}$ can be seen as quantifications of the sampling and variable modes of $\mathbf{X}$, respectively, and there are several options (Van Mechelen and Schepers, 2007; Van Mechelen and Smilde, 2010). Also for defining the approximation rule ($\approx$) there are many options (Van Mechelen and Schepers, 2007). This will be discussed more extensively in Chapter 5.

2.1.2 Principal Component Analysis

The workhorse of multivariate data analysis is principal component analysis (PCA). There are different ways of introducing and defining PCA. The easiest one is to think of PCA as a dimension reduction method which tries to represent a matrix $\mathbf{X}$ $(I \times J)$ as a multiplication of two matrices of a much smaller size called scores $\mathbf{T}$ $(I \times R)$ and loadings $\mathbf{P}$ $(J \times R)$ where

R is much smaller than both I and J. The question is then how to choose $\mathbf{T}$ and $\mathbf{P}$. An obvious choice is to solve the following problem:

$$\min_{\mathbf{T},\mathbf{P}} \|\mathbf{X} - \mathbf{T}\mathbf{P}^t\|^2 \tag{2.3}$$

which minimises the residuals $\mathbf{E}$ of the model

$$\mathbf{X} = \mathbf{T}\mathbf{P}^t + \mathbf{E} \tag{2.4}$$

and this is the formulation as put forward by Pearson (Pearson, 1901) (note that $\|.\|$ is the default Frobenius norm of a matrix, see Section 2.8). To identify the solution, it is customary to normalise the scores and/or loadings. In chemometrics, usually $\mathbf{P}$ is chosen to be a matrix with orthonormal columns (orthogonal and of length one) and the columns of $\mathbf{T}$ to be orthogonal. This normalisation, however, does not change the solution essentially; the subspaces spanned by $\mathbf{T}$ and $\mathbf{P}$ remain the same. PCA has many nice optimality properties (Eckart and Young, 1936; Jolliffe, 2010) and it can also be shown that $\mathbf{T}$ is in the column-space of $\mathbf{X}$ (and, likewise, $\mathbf{P}$ is in the row-space of $\mathbf{X}$; see Section 2.8). Hence, solving the following problem

$$\min_{\mathbf{W},\mathbf{P}} \|\mathbf{X} - \mathbf{X}\mathbf{W}\mathbf{P}^t\|^2 \tag{2.5}$$

gives essentially the same solution as Equation 2.3 and $\mathbf{W} = \mathbf{P}$ (Ten Berge, 1993).

The fit of the PCA model can now be judged by calculating

$$100 \cdot [\frac{\|\mathbf{X}\|^2 - \|\mathbf{X} - \mathbf{T}\mathbf{P}^t\|^2}{\|\mathbf{X}\|^2}]\% = 100 \cdot [\frac{\|\mathbf{X}\|^2 - \|\mathbf{E}\|^2}{\|\mathbf{X}\|^2}]\% \tag{2.6}$$

which gives the percentage of explained variation. Apart from centring the matrix $\mathbf{X}$ prior to a PCA-analysis, in some situations it can also be useful to standardise the data by giving each column a standard deviation of one. If both preprocessing steps are combined, this is called autoscaling (Bro and Smilde, 2003). This will be discussed in more depth in Section 2.6. An example of PCA in the `multiblock` R-package is found in Section 11.5.2.

There is also a completely different route of explaining PCA (Hotelling, 1933). Apart from a constant $(1/I)$ which is immaterial in this context, the variance of a linear combination $\mathbf{t} = \mathbf{X}\mathbf{w}$ of the columns of $\mathbf{X}$ with unit-length $\mathbf{w}$ can be maximised by solving

$$\max_{\mathbf{w}} \mathbf{t}^t\mathbf{t} = \max_{\mathbf{w}} \mathbf{w}^t\mathbf{X}^t\mathbf{X}\mathbf{w}, \quad s.t. \ \ \mathbf{w}^t\mathbf{w} = 1 \tag{2.7}$$

where *s.t.* is the abbreviation of *subject to* which refers to a constraint on the solution. Note that centring $\mathbf{X}$ also centres $\mathbf{t}$ and thus (apart from a constant) $\mathbf{t}^t\mathbf{t}$ is the variance of $\mathbf{t}$. The solution of Equation 2.7 is the first eigenvector of $\mathbf{X}^t\mathbf{X}$ (Jolliffe, 2010). The resulting score $\mathbf{t} = \mathbf{X}\mathbf{w}$ is exactly the first column of the scores $\mathbf{T}$ from Equation 2.4 when these columns are ordered in decreasing length (see also Section 2.3.5). This shows that both approaches are essentially the same: maximising the *variance explained by* the principal component scores gives the same results as maximising the *variance of* the principal component scores. The next component can be obtained by deflating $\mathbf{X}$ for the first component and again performing a maximisation like Equation 2.7 but now for the deflated $\mathbf{X}$ (for an explanation of deflation, see Section 2.8). If the principal components are in the order of decreasing variance (i.e., in the order of decreasing eigenvalues of $\mathbf{X}^t\mathbf{X}$) then the solution is unique up to a sign permutation (between $\mathbf{t}$ and its corresponding $\mathbf{p}$); the components are then sometimes called *principal axes*. PCA can also be regarded as finding an optimal subspace of which the principal axes are just one choice of a basis.

The PCA solution can be rotated, which is essentially moving to another basis in that same subspace:

$$\mathbf{X} = \mathbf{T}\mathbf{P}^t + \mathbf{E} = \mathbf{T}\mathbf{Q}\mathbf{Q}^{-1}\mathbf{P}^t + \mathbf{E} = \widetilde{\mathbf{T}}\widetilde{\mathbf{P}}^t + \mathbf{E} \tag{2.8}$$

where $\mathbf{Q}(R \times R)$ is an arbitrary non-singular matrix; $\widetilde{\mathbf{T}} = \mathbf{T}\mathbf{Q}$ and $\widetilde{\mathbf{P}} = \mathbf{P}(\mathbf{Q}^{-1})^t$. This destroys the principal axis property, but the fitted subspace remains the same.

An easy way to calculate PCA scores and loadings is by using the singular value decomposition (SVD) of $\mathbf{X} = \mathbf{U}\mathbf{D}\mathbf{V}^t$ by taking $\mathbf{T} = \mathbf{U}\mathbf{D}$ and $\mathbf{P} = \mathbf{V}$, where $\mathbf{U}$ and $\mathbf{V}$ are orthogonal matrices (i.e., $\mathbf{U}^t\mathbf{U} = \mathbf{I}$ and $\mathbf{V}^t\mathbf{V} = \mathbf{I}$) and $\mathbf{D}$ is a diagonal matrix with the singular values of $\mathbf{X}$ in decreasing order on its diagonal. Sometimes it is useful to consider the correlation loadings which are defined as the correlations between the scores $\mathbf{T}$ and the original variables in $\mathbf{X}$; these are different from the loadings $\mathbf{P}$ (see, e.g., Martens and Martens (2001)). PCA also has a nice geometrical interpretation which is shown in Elaboration 2.1.

ELABORATION 2.1

Geometry of PCA

Each row in the matrix $\mathbf{X}(I \times J)$ represents a point (large blue dots in Figure 2.2 indicated by 1 and 2) in $\mathbb{R}^J$ and the loadings (the columns of $\mathbf{P}$) are also vectors in this space illustrated with the green lines. These loadings span a two-dimensional space (the grey plane). The points are projected orthogonally onto this plane (along the orange line) and these projections can be written as combinations of the two loadings indicated by the small blue dots on the green lines. These small blue dots are the scores as collected in $\mathbf{T}$. Finally, there are residuals (shown by the purple lines) which are collected in $\mathbf{E}$. The plane is now chosen such that all points are as close as possible to that plane.

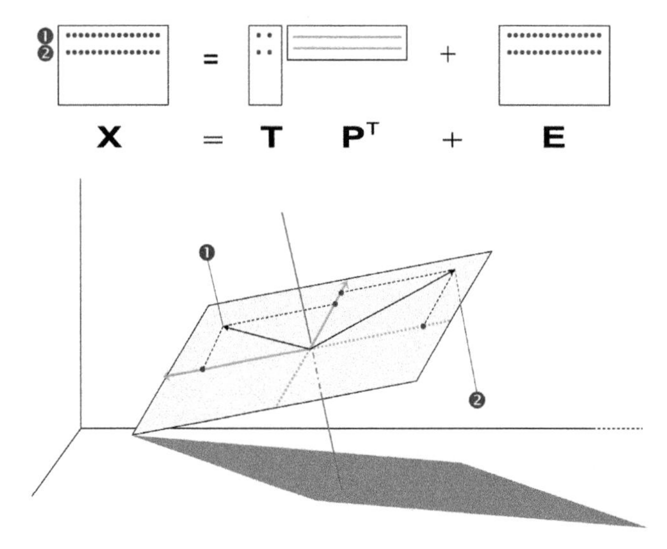

Figure 2.2 Geometry of PCA. For explanation, see text (with permission of H.J. Ramaker, TIPb, The Netherlands).

An illustration of PCA is given in Bro and Smilde (2014) of which a summary is shown in Example 2.1.

Example 2.1: PCA on wines

For 44 samples of Cabernet Sauvignon wines from four different regions in the world (Argentina, Australia, Chile, and South-Africa), 14 chemical constituents were measured (such as ethanol and methanol). Because these variables are measured in different units, the data were auto-scaled prior to the PCA analysis. Figure 2.3 shows the score and loading plots.

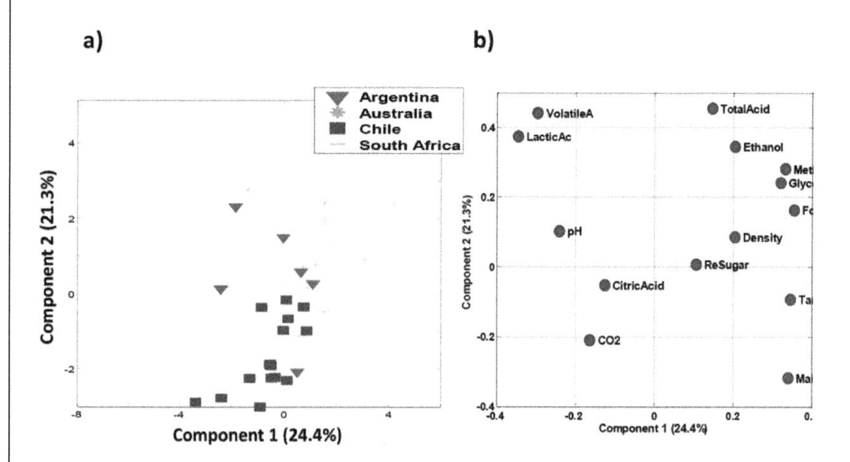

Figure 2.3 Score (a) and loading (b) plots of a PCA on Cabernet Sauvignon wines. Source: Bro and Smilde (2014). Reproduced with permission of Royal Society of Chemistry.

The first two principal components explain about 45% of the variation in data, which already shows a considerable dimension reduction. The score plot shows some grouping of the region which is mostly picked up by the second component. The loading plot shows the correlation, e.g., between methanol, ethanol, and glycerol. Several types of acids have relatively high loadings (in absolute values) on the second principal component, so probably the differences in regions are related to the aspect of acidity.

A crucial aspect of PCA is the selection of the number of principal components. There are very many methods proposed to perform such a selection and this also depends on the goal of the analysis. The simplest method is to study the explained variance as a function of the number of components and base the selection on the inspection of such a curve. The scree test is based on that (Raîche *et al.*, 2013). Cross-validation (see Section 2.7.5) is used a lot but needs to be done with care (Bro *et al.*, 2008). Also permutation testing can be used (Endrizzi *et al.*, 2013; Vitale *et al.*, 2017). In the end, a visual inspection of the scores and loadings always needs to be done to ensure appropriateness of the PCA model. PCA is often confused with factor analysis but the two are not the same, see Elaboration 2.2.

ELABORATION 2.2

The difference between PCA and factor analysis

PCA and factor analysis (FA) are often mixed up, yet, they are different methods. The model for factor analysis is

$$\mathbf{X} = \mathbf{F}\boldsymbol{\Lambda}^t + \mathbf{U} \tag{2.9}$$

$$\text{Cov}(\mathbf{X}) = \boldsymbol{\Lambda}\boldsymbol{\Lambda}^t + \boldsymbol{\Phi}$$

$$\boldsymbol{\Phi} = \text{Diag}(\phi_1, \ldots, \phi_J)$$

where the symbol $\text{Cov}(\mathbf{X})$ is used to indicate the variance–covariance matrix of the variables $\mathbf{x}$. The assumptions of the FA model are: (i) $\text{Cov}(\mathbf{F}) = \mathbf{I}$, (ii) the contributions $\mathbf{U}$ are stochastically independent of the factor scores $\mathbf{F}$, and (iii) $\text{Cov}(\mathbf{U}) = \boldsymbol{\Phi}$ which is a diagonal matrix of unique variances with not necessarily the same values on the diagonal. Hence, the differences with PCA are:

- FA concentrates on covariances and does not explain maximum variation as does PCA.
- FA provides explicit room for unique variances of the variables; PCA does not do this.
- The scores $\mathbf{F}$ are not linear combinations of the (manifest) variables.
- For maximum likelihood-based FA, the model is independent of within-block scaling (e.g., autoscaling the data, see Section 2.6) whereas this does not hold for PCA.
- The algorithms to fit the FA model differ from the ones of PCA. In FA, the emphasis is on fitting covariances (Mardia *et al.*, 1979).
- FA does not have the intrinsic axis property of PCA (i.e., there is no ordering of components like in PCA). Hence, it usually requires rotations to arrive at interpretable solutions.
- In practice, when the size of $\mathbf{U}$ is small relative to the model part $\mathbf{F}\boldsymbol{\Lambda}^t$ or if the unique variances are of similar size, the differences between FA and PCA are not very large and they describe similar subspaces (Jolliffe, 2010).

Thus the two should not be confused. FA is not used a lot in chemistry; there PCA is more dominant. In bioinformatics, FA is often used.

2.1.3 Sparse PCA

In machine learning and bioinformatics, the idea of sparsity is very popular. The basic concept is to make the loadings or weights sparse. In standard multivariate component based methods, weights are usually non-zero. Especially in high-dimensional data, such long weight vectors filled with non-zeros are difficult to interpret. Moreover, there may also be substance-related arguments for having weights that are more simple, i.e., have many zeros. This idea was already present in Varimax rotations of factor analysis solutions (Kaiser, 1958). This type of sparsity is usually imposed by special constraints (Murphy, 2012; Hastie *et al.*, 2015).

There exist several versions of sparse PCA with different flavours (Zou *et al.*, 2006; Witten *et al.*, 2009). The basic idea is explained in Elaboration 2.3. Such sparse versions of PCA are not without problems, especially in the deflation steps (Camacho *et al.*, 2020). There exists also a weighted sparse PCA version which can be used for heteroscedastic data (Van Deun *et al.*, 2019).

ELABORATION 2.3

Sparse PCA

Basically, there are two ways to impose sparsity in PCA: (i) on the loadings $\mathbf{P}$ (see Equation 2.4) or (ii) on the weights $\mathbf{W}$ (see Equation 2.5). These give different results and interpretations (Van Deun *et al.*, 2011). The sparse weights model was proposed by Zou *et al.* (2006) and explained more clearly in Van Deun *et al.* (2011); we will follow the latter description. This sparse PCA is found by solving

$$\min_{\mathbf{W}} \|\mathbf{X} - \mathbf{X}\mathbf{W}\mathbf{P}^t\|^2 + \lambda_L \|\mathbf{W}\|_1 + \lambda_R \|\mathbf{W}\|^2 \tag{2.10}$$

where the term $\lambda_L \|\mathbf{W}\|_1$ (with $\lambda_L \geqslant 0$) is the Lasso penalty (Tibshirani, 1996) that induces zeros in the weights and the term $\lambda_R \|\mathbf{W}\|^2$ (with $\lambda_R \geqslant 0$) is a Ridge-type penalty (using the Frobenius norm) which shrinks the weights (see Section 2.8 for definitions of norms of matrices). This combination of penalty terms is called 'elastic net' and has favourable properties for high-dimensional data (Zou and Hastie, 2005; Hastie *et al.*, 2015). The optimal values of λ_L and λ_R can, e.g., be found by cross-validation on $\mathbf{X}$ (i.e., by comparing the entries of $\mathbf{X}$ with their cross-validatory predictions).

Sparseness on the loadings $\mathbf{P}$ can be obtained by solving

$$\min_{\mathbf{P}} \|\mathbf{X} - \mathbf{T}\mathbf{P}^t\|^2 + \lambda_L \|\mathbf{P}\|_1 + \lambda_R \|\mathbf{P}\|^2 \tag{2.11}$$

where an elastic net is used and $\mathbf{T}$ is chosen to have orthonormal columns. Apart from differences in interpretation (see the discussion in Van Deun *et al.* (2011)) these two approaches also have different properties. The scores $\mathbf{T} = \mathbf{XW}$ in Equation 2.10 are by definition in the column-space of $\mathbf{X}$. The associated loadings $\mathbf{P}$ are in the row-space of $\mathbf{X}$ since they are obtained via a regression step after the optimal $\mathbf{W}$ is found. In Equation 2.11, however, the loadings $\mathbf{P}$ are not necessarily in the row-space of $\mathbf{X}$. Still, the associated scores $\mathbf{T}$ are in the column-space of $\mathbf{X}$ since these are obtained via a regression once the optimal $\mathbf{P}$ is found.

2.1.4 Principal Component Regression

In multiple regression problems we want to find a relationship between multiple x-variables and a response variable y. If the J x-variables are collected in $\mathbf{X}$ $(I \times J)$ and the I y-values are collected in $\mathbf{y}$ $(I \times 1)$ then a model for finding this relation can be stated as

$$\mathbf{y} = \mathbf{Xb} + \mathbf{f} \tag{2.12}$$

where the regression vector $\mathbf{b}$ contains the regression coefficients and $\mathbf{f}$ contains the residuals. The standard technique to find $\mathbf{b}$ is multiple linear regression (i.e., least squares) but that requires a full column-rank $\mathbf{X}$ with a good condition, i.e., without much multicollinearity (see Section 2.8). In many cases in the natural and life sciences we do not have a well conditioned $\mathbf{X}$ and, even more extreme, there are many cases in which $J > I$ and thus $\mathbf{X}$ does not have full column-rank; this is referred to as megavariate data or high-dimensional data (or fat and short data as opposed to thin and tall).

A simple method for these types of cases is to first perform a PCA on $\mathbf{X}$ and then perform the regression of $\mathbf{y}$ on the principal components of $\mathbf{X}$; this is called principal component regression (PCR) and can be formally stated as:

$$\begin{aligned}\mathbf{X} &= \mathbf{TP}^t + \mathbf{E}\\ \mathbf{T} &= \mathbf{XP}\\ \mathbf{y} &= \mathbf{Tq} + \mathbf{f} = \mathbf{XPq} + \mathbf{f} = \mathbf{Xb} + \mathbf{f}\end{aligned} \tag{2.13}$$

where in the first line of the equation a PCA model of $\mathbf{X}$ is formulated with a low number of principal components; in the second line the scores are defined (using the fact that we can choose $\mathbf{P}^t\mathbf{P} = \mathbf{I}$); in the third line the vector $\mathbf{y}$ is regressed (using LS) on the (orthogonal) scores collected in $\mathbf{T}$, $\mathbf{f}$ contains the residuals, and the regression coefficients $\mathbf{b}$ are defined implicitly. The stabilising property of PCR lies in the fact that the matrix $\mathbf{T}$ consists of the leading principal components which are associated with the largest singular values of $\mathbf{X}$ and thus avoids the unstable dimensions with small singular values. Thus, they provide a good and stable approximation of $\mathbf{X}$. They are also orthogonal thereby simplifying further calculations. An example of PCR in the `multiblock` R-package is found in Section 11.5.2.

Also for PCR it is important to select the number of components. In most cases this is performed with cross-validation whereby a growing number of principal components is used and tested for their predictive ability (Martens and Næs, 1989). Alternative strategies are possible for incorporating principal components: either using the principal components with decreasing variance or selecting among a set of principal components the ones that have the most predictive power. The latter strategy is sometimes chosen to circumvent a problem of PCR namely that the leading principal components of $\mathbf{X}$ are not necessarily informative for $\mathbf{y}$. Note that this strategy bears the risk of introducing unstable components of low variance. This is repaired to some extent with PLS which is treated in the next section.

2.1.5 Partial Least Squares

There are different ways to explain partial least squares (PLS; also sometimes called PLS regression (PLSR) (Wold *et al.*, 1984; Martens and Næs, 1989)), but perhaps the simplest way is to follow the approach of Höskuldsson (1988). Assuming a (preprocessed[1]) matrix $\mathbf{X}$ $(I \times J)$ and a (preprocessed) vector $\mathbf{y}$ $(I \times 1)$, PLS seeks a regression model to fit $\mathbf{y}$ to $\mathbf{X}$ by solving

$$\max_{\mathbf{w}} \text{cov}(\mathbf{Xw}, \mathbf{y}); \; s.t. \; \|\mathbf{w}\| = 1 \tag{2.14}$$

where cov(.,.) is the covariance between two vectors. Sometimes this problem is stated in terms of maximising cov^2 but note that the covariance can always be made positive by choosing the proper sign of $\mathbf{w}$. Having found the optimal $\mathbf{w}$, the scores are defined as $\mathbf{t} = \mathbf{Xw}$ and the matrix $\mathbf{X}$ is deflated with respect to this score $\mathbf{t}$ by solving the regression equation $\mathbf{X} = \mathbf{tp}^t + \mathbf{E}$ for $\mathbf{p}$ (i.e., orthogonalised, see Section 2.8) and the next component is calculated using $\mathbf{E}$. This can be repeated until a sufficient number of components (R) is obtained. Due to the restriction $\|\mathbf{w}\| = 1$, the scores $\mathbf{t} = \mathbf{Xw}$ are also a least squares solution for $\mathbf{X}$, see Elaboration 2.4.

1 Preprocessed includes at least a centring step.

ELABORATION 2.4

Constraint on weights w gives the scores t as a least squares (LS) solution

In the model

$$\mathbf{X} = \mathbf{t}\mathbf{w}^t + \mathbf{E};\ s.t.\ \|\mathbf{w}\| = 1 \tag{2.15}$$

for fixed **w**, the LS solution for **t** is

$$\mathbf{t} = \mathbf{X}\mathbf{w}(\mathbf{w}^t\mathbf{w})^{-1} = \mathbf{X}\mathbf{w} \tag{2.16}$$

due to the constraint on **w**. Hence, each expression $\mathbf{t} = \mathbf{X}\mathbf{w}$ is LS for **t** if **w** is fixed and of length one.

The final model can then be formulated as follows:

$$\begin{aligned} \mathbf{X} &= \mathbf{T}\mathbf{P}^t + \mathbf{E} \\ \mathbf{T} &= \mathbf{X}\mathbf{R} = \mathbf{X}\mathbf{W}(\mathbf{P}^t\mathbf{W})^{-1} \\ \mathbf{y} &= \mathbf{T}\mathbf{q} + \mathbf{f} = \mathbf{X}\mathbf{R}\mathbf{q} + \mathbf{f} = \mathbf{X}\mathbf{b} + \mathbf{f} \end{aligned} \tag{2.17}$$

where $\mathbf{T}(I \times R)$ contains the scores in its columns; **P** contains the loadings; **W** contains the weights **w** (constrained to length one) of the different components as its columns; **q** and **b** are regression coefficients (De Jong, 1993) (for simplicity we used the same symbols for PLS and PCR but the components are different!) and **R** is defined implicitly (see also Section 2.8). Note that the constraint on **t** in the form of $\mathbf{t} = \mathbf{X}\mathbf{w}$ is active now: it forces the scores to be in the column-space of **X**. Without this constraint the score **t** would not necessarily be in the column-space of **X**. The latter is, e.g., the case in latent root regression where no constraints on the vectors **t** are used (Webster *et al.*, 1974).

The PLS weight **w** can also be obtained as the first eigenvector of $\mathbf{S} = \mathbf{X}^t\mathbf{y}\mathbf{y}^t\mathbf{X}$ by solving

$$\max_{\mathbf{w}} \mathbf{w}^t\mathbf{S}\mathbf{w};\ s.t.\|\mathbf{w}\| = 1 \tag{2.18}$$

and the subsequent **w** are then obtained after proper deflation. Another way to look at PLS is that the **w** solves the problem

$$\min_{\mathbf{w}} \|\mathbf{X} - \mathbf{y}\mathbf{w}^t\|^2 \tag{2.19}$$

after which the weights are normalised to length one and (again) the subsequent **w** are then obtained after proper deflation.

The power of PLS is that it can handle collinear variables (in **X**), its ease of computation, that it also contains a model of **X** and, compared to PCR, also tries to fit **y** directly. Hence, it is widely used in data analysis and chemometrics. Note that taking the covariance as optimisation criterion is essentially different from taking the correlation. This can be understood by writing the covariance as

$$\text{cov}(\mathbf{X}\mathbf{w}, \mathbf{y}) = \sqrt{\text{var}(\mathbf{y})} \cdot \sqrt{\text{var}(\mathbf{X}\mathbf{w})} \cdot \text{corr}(\mathbf{X}\mathbf{w}, \mathbf{y}) \tag{2.20}$$

Given the fact that in Equation 2.20 the variance of **y** is constant, it is clear that PLS strikes a compromise by maximising $\text{var}(\mathbf{X}\mathbf{w})$ and $\text{corr}(\mathbf{X}\mathbf{w}, \mathbf{y})$. Thus also the variance of the scores play a role which stabilises the regression (Martens and Næs, 1989). Moreover, the scores **T** are different from the ones of a PCA on **X** and are also made relevant for fitting **y**. An example of PLS in the `multiblock` R-package is found in Section 11.5.2.

The most well-known algorithm for estimating parameters in a PLS model is non-linear iterative partial least squares (NIPALS, see Algorithm 2.1). There is also a version for a multivariate **Y**. Such a version would result in the following model:

$$\begin{aligned}\mathbf{X} &= \mathbf{T}\mathbf{P}^t + \mathbf{E}\\ \mathbf{T} &= \mathbf{X}\mathbf{R} = \mathbf{X}\mathbf{W}(\mathbf{P}^t\mathbf{W})^{-1}\\ \mathbf{Y} &= \mathbf{T}\mathbf{Q}^t + \mathbf{F} = \mathbf{X}\mathbf{R}\mathbf{Q}^t + \mathbf{F} = \mathbf{X}\mathbf{B} + \mathbf{F}\end{aligned} \tag{2.21}$$

where the matrix **R** is implicitly defined; this is called PLS2 in the literature (Martens and Næs, 1989). Also in this case, the NIPALS algorithm can be used to obtain the PLS weights. Alternatively, they can be obtained from the SVD of $\mathbf{X}^t\mathbf{Y}$ or ED of $\mathbf{X}^t\mathbf{Y}\mathbf{Y}^t\mathbf{X}$ (see also Equation 2.18) including deflation steps. An illustration of the use of PLS1 and PLS2 is given in Example 2.2. By far the most dominant way to select the PLS model complexity (i.e., the number of components) is by using cross-validation (Martens and Næs, 1989).

Algorithm 2.1

NIPALS

The NIPALS algorithm is a typical example of an iterative algorithm where components are calculated one at a time and a deflation step is used to go to the next round. There is a NIPALS version for multivariate **Y** and one for a univariate **y**. The latter goes as follows:

1:	Start with **X**, **y**	– start with original **X**, **y**
	Loop over components	– $r = 1, \ldots, R$
2:	$\mathbf{w} = \mathbf{X}^t\mathbf{y}/\|\mathbf{X}^t\mathbf{y}\|$	– calculate and normalise weights
3:	$\mathbf{t} = \mathbf{X}\mathbf{w}$	– solve $\mathbf{X} = \mathbf{t}\mathbf{w}^t + \mathbf{E}$ for **t**
4:	$\mathbf{p} = \mathbf{X}^t\mathbf{t}/(\mathbf{t}^t\mathbf{t})$; $q = \mathbf{y}^t\mathbf{t}/(\mathbf{t}^t\mathbf{t})$	– loadings of **X** and **y** found by regression
5:	$\mathbf{X}_{new} = \mathbf{X} - \mathbf{t}\mathbf{p}^t$; $\mathbf{y}_{new} = \mathbf{y} - \mathbf{t}q$	– deflate **X** and **y**
6:	$\mathbf{X} = \mathbf{X}_{new}$; $\mathbf{y} = \mathbf{y}_{new}$	– go to 2.
	End loop	

where step 5 contains the deflation steps (see Section 2.8). Steps 3 and 4 are both least squares steps (see Elaboration 2.4). There are different ways to perform the deflations which give essentially the same results, see also Section 2.8. The algorithm continues until the wanted number of components is reached. The NIPALS algorithm has inspired a lot of developments. One variation of the algorithm which spans the same subspaces and gives the same predictions, is to normalise the scores, **t**, and drop the scaling of the loadings **p**, e.g., used in the ROSA method (Section 7.5).

Example 2.2: Multivariate calibration using PLS

The Raman data set was introduced in Chapter 1, though here it is used in its original form without splitting into the three blocks. In this example the data are analysed using PLS regression with separate responses and PLS2 with both responses simultaneously. The responses are polyunsaturated fatty acids (PUFA) as (1) percentage of total sample weight ($PUFA_{sample}$) and as (2) percentage of fats in sample ($PUFA_{fat}$). The example shows how typical plots of scores and loadings look for spectroscopic data and serves as a yard stick for the explained variances obtained with supervised multiblock data analysis methods in Chapter 7.

In Figure 2.4 we see how the leave-one-out cross-validated explained variances develop through the first 10 components. It seems that the three models need roughly six components to reach the optimum. The models have maxima at 88.4%, 97.9% and 92.3% explained variance, respectively, for $PUFA_{sample}$, $PUFA_{fat}$ and the PLS2 model's average over the PUFA responses. In the latter case, both responses were standardised before analysis to give equal emphasis in the modelling. In this model system, Raman showed higher precision for lipid components than for other major constituents, leading to a higher explained variance for single fatty acids (in percentage of fat; $PUFA_{fat}$) and their degree of unsaturation than for the combined amount of fat (in percentage of sample; $PUFA_{sample}$).

We focus further on the single response PLS regression of $PUFA_{sample}$. In Figure 2.5 the two first score vectors and loading vectors are plotted. The validated explained variances for the two first components are 47.7% and 13.4%, respectively, and the large amount of explained variance is mirrored by the trend in colours representing the PUFA as percentage of sample ($PUFA_{sample}$). In the first component there is a positive peak corresponding to symmetric rock in *cis* (=C-H at $\sim$1260 cm^{-1}) and a negative peak in the in-phase methylene twist (C-H_2 at $\sim$1300 cm^{-1}). In the second component we observe peaks that can be associated with an increase in C-C stretch ($\sim$1075 cm^{-1}) and a decrease in C=C-related motions ($\sim$1000 cm^{-1}). These are all molecular vibrations that are connected with the amount of PUFA in the sample.

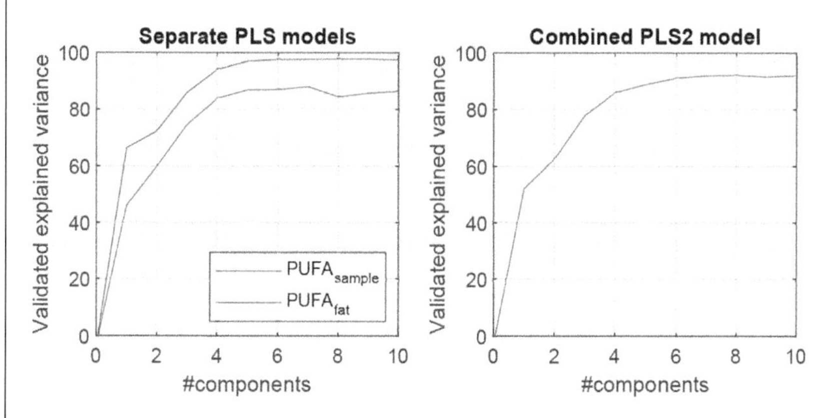

Figure 2.4 PLS validated explained variance when applied to Raman with PUFA responses. Left: PLSR on one response at a time. Right: PLS on both responses (standardised).

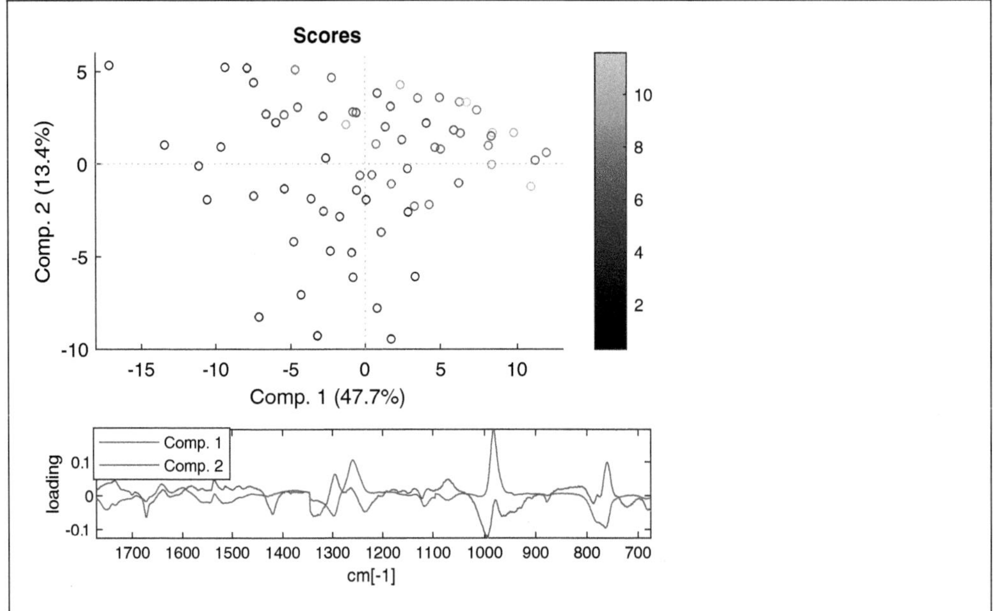

Figure 2.5 Score and loading plots for the single response PLS regression model predicting PUFA as percentage of total fat in the sample ($PUFA_{\text{sample}}$).

2.1.6 Sparse PLS

As stated already in Section 2.1.3, there are many multivariate analysis methods combined with sparsity, such as sparse canonical correlation (Waaijenborg and Zwinderman, 2009), sparse redundancy analysis (Csala *et al.*, 2017) and sparse PLS (Lê Cao *et al.*, 2008) to name a few. The basic idea of sparse PLS is to impose sparseness on the weights that define the components in the X-block:

$$\begin{aligned} \mathbf{X} &= \mathbf{T}\mathbf{P}^t + \mathbf{E} \\ \mathbf{T} &= \mathbf{X}\mathbf{W}_s(\mathbf{P}^t\mathbf{W}_s)^{-1} \\ \mathbf{y} &= \mathbf{T}\mathbf{q} + \mathbf{f} = \mathbf{X}\mathbf{W}_s(\mathbf{P}^t\mathbf{W}_s)^{-1}\mathbf{q} + \mathbf{f} = \mathbf{X}\mathbf{b} + \mathbf{f} \end{aligned} \tag{2.22}$$

where the weights $\mathbf{W}_s$ are sparse. This is essentially PLS1 (see Section 2.1.5) but now with sparse weights. The loadings in the first line of Equation 2.22 are simply obtained by regressing $\mathbf{X}$ on the scores $\mathbf{T}$. There are different ways to impose this sparseness, e.g. by truncating the weights (Liland *et al.*, 2013) or by combining PLS with soft-thresholding (Sæbø *et al.*, 2008a). The latter idea can be illustrated most insightfully with the adapted NIPALS algorithm, see Algorithm 2.2.

Algorithm 2.2

SPARSE NIPALS

1: Start with $\mathbf{X}, \mathbf{y}$ – start with original $\mathbf{X}, \mathbf{y}$

Loop over components – $r = 1, \ldots, R$

2: $\mathbf{w} = ST_\lambda(\mathbf{X}^t\mathbf{y})$ – calculate sparse weights for **X**

3: $\|\mathbf{w}\| = 1$ – normalise **w**

4: $\mathbf{t} = \mathbf{X}\mathbf{w}$ – solve $\mathbf{X} = \mathbf{t}\mathbf{w}^t + \mathbf{E}$ for **t**

5: $\mathbf{p} = \mathbf{X}^t\mathbf{t}/\mathbf{t}^t\mathbf{t};\ q = \mathbf{y}^t\mathbf{t}/\mathbf{t}^t\mathbf{t}$ – calculate X and y loadings

6: $\mathbf{X}_{new} = \mathbf{X} - \mathbf{t}\mathbf{p}^t;\ \mathbf{y}_{new} = \mathbf{y} - \mathbf{t}q$ – deflate X- and y-block

7: $\mathbf{X} = \mathbf{X}_{new};\ \mathbf{y} = \mathbf{y}_{new}$ – go to step 2

End loop

The crucial step in this algorithm is $\mathbf{w} = ST_\lambda(\mathbf{X}^t\mathbf{y})$ where ST is the abbreviation of soft-thresholding. This is defined for w, an element of the weight vector, as:

$$ST_\lambda(w) = sign(w)(|w| - \lambda)_+ \tag{2.23}$$

which means that if $|w| < \lambda$ it will become zero and otherwise it is shrunken toward zero with the value of λ and retaining the original sign. Note that this thresholding works on the individual elements w of the weight vector $\mathbf{w}$, thereby zeroing some of these weights. This is an instantiation of the Lasso penalty ($\lambda\|\mathbf{w}\|_1$; see Section 2.8) with tuning parameter $\lambda \geqslant 0$ (Murphy, 2012) which can be tuned with cross-validation. The 1-norm penalty induces zeros in w thereby making this sparse.

2.1.7 Principal Covariates Regression

Another regression method based on components is principal covariates regression (PCovR) (De Jong and Kiers, 1992). This method plays a role in other chapters of this book since it allows for several extensions relevant for multiblock and multiway data analysis. The PCovR model is defined as

$$\begin{aligned}\mathbf{X} &= \mathbf{T}\mathbf{P}^t + \mathbf{E}\\ \mathbf{T} &= \mathbf{X}\mathbf{W}\\ \mathbf{Y} &= \mathbf{T}\mathbf{Q}^t + \mathbf{F}\end{aligned} \tag{2.24}$$

where $\mathbf{Y}(I \times J_Y)$ is regressed on the components $\mathbf{T}(I \times R)$ of $\mathbf{X}(I \times J_X)$ and the $\mathbf{W}(J_X \times R)$ is estimated by

$$\min_{\mathbf{W}}[\alpha\|\mathbf{X} - \mathbf{X}\mathbf{W}\mathbf{P}^t\|^2 + (1 - \alpha)\|\mathbf{Y} - \mathbf{X}\mathbf{W}\mathbf{Q}^t\|^2] \tag{2.25}$$

where $\mathbf{P}(J_X \times R)$ and $\mathbf{Q}(J_Y \times R)$ are loading matrices for **X** and **Y**, respectively, which are calculated by solving the regression equations $\mathbf{X} = \mathbf{T}\mathbf{P}^t + \mathbf{E}$ and $\mathbf{Y} = \mathbf{T}\mathbf{Q}^t + \mathbf{F}$ for **P** and **Q**. The value of α is between 0 and 1. There are different ways of selecting the value of α, e.g., using cross-validation of predicting **Y** (Vervloet *et al.*, 2013). Also for PCovR, a sparse version is available (Van Deun *et al.*, 2018). PCovR is much in the same spirit as PLS; it can be seen as a 'simultaneous' version of PLS with the flexibility to focus more on **X** (i.e., stabilising components) or **Y** through changing the α-value.

2.1.8 Redundancy Analysis

Yet another basic multivariate regression model that has been developed in multiblock data analysis is redundancy analysis (RDA) or reduced rank regression (Van den Wollenberg, 1977; Ten Berge, 1993). This method and its multiblock extensions are used frequently in ecology and microbial research (Legendre and Legendre, 2012). The basic model is given by the solution of the problem:

$$\min_{\mathbf{W}} \|\mathbf{Y} - \mathbf{XWQ}^t\|^2 \tag{2.26}$$

with $\mathbf{W}(J_X \times R)$ a set of weights and $\mathbf{Q}(J_Y \times R)$ a set of Y-loadings. This is seen to be a restricted version of the multivariate regression model $\mathbf{Y} = \mathbf{XB} + \mathbf{F}$ with $\mathbf{B}$ having rank R. Sparse versions of RDA are also available (Csala *et al.*, 2017).

2.1.9 Comparing PLS, PCovR and RDA

From the previous it is clear that PLS, PCovR and RDA are very similar. However, there are also clear differences, which becomes clear when writing the fitted parts of $\mathbf{Y}$:

$$\begin{aligned}
\widehat{\mathbf{Y}}_{\text{PLS}} &= \mathbf{XW}(\mathbf{P}^t\mathbf{W})^{-1}\mathbf{Q}^t = \mathbf{XB}_{\text{PLS}} \\
\widehat{\mathbf{Y}}_{\text{PCovR}} &= \mathbf{XWQ}^t = \mathbf{XB}_{\text{PCovR}} \\
\widehat{\mathbf{Y}}_{\text{RDA}} &= \mathbf{XWQ}^t = \mathbf{XB}_{\text{RDA}}
\end{aligned} \tag{2.27}$$

Thus all methods can be seen as solving a multivariate multiple regression problem with a rank restriction on the regression coefficients. Note that all matrices $\mathbf{W}$ and $\mathbf{Q}$ are different but for simplicity the same symbols are used. Apart from the rank restriction, there is extra structure in PLS and PCovR. In PLS, the machinery imposes implicitly stability of the components in the column-space of $\mathbf{X}$. This stability is made explicit in PCovR by selecting the α value. Such a stabilising property is absent in RDA.

2.1.10 Generalised Canonical Correlation Analysis

Generalised canonical correlation analysis (GCA) is a generalisation of canonical correlation (Hotelling, 1936a) for two data sets to finding relationships between more than two data sets. More specific, the goal of GCA is to identify linear combinations of the blocks, $\mathbf{X}_m\mathbf{W}_m$, which fit as good as possible to a set of orthonormal components $\mathbf{T}$. This is done by minimising the criterion

$$\min_{(\mathbf{T},\mathbf{W}_m)} \sum_{m=1}^{M} \|\mathbf{X}_m\mathbf{W}_m - \mathbf{T}\|^2 \tag{2.28}$$

with respect to $\mathbf{T}$ ($\mathbf{T}^T\mathbf{T} = \mathbf{I}$) and $\mathbf{W}_m$ $(m = 1, \ldots, M)$ (Van der Burg and Dijksterhuis, 1996). The number of columns in $\mathbf{T}$, R, must be smaller than or equal to the number of columns in the $\mathbf{X}_m$ with the smallest number of columns. If the number of samples, I, is smaller than all J_m $(m = 1, \ldots, M)$, then GCA gives trivial solutions and cannot be used without further restrictions. There are several ways to generalise canonical correlation analysis (CCA) to three or more data blocks (Kettenring, 1971; Hanafi and Kiers, 2006) and GCA represents one of these. The method presented above is the most widely used and based on the work of Carroll (1968). An example of GCA in the `multiblock` R-package is found in Section 11.6.3.

2.1.11 Simultaneous Component Analysis

Simultaneous component analysis (SCA) was originally developed in psychometrics for combining data sets in which the same variables were measured (Ten Berge *et al.*, 1992; Timmerman and Kiers, 2003). This method will be explained for M blocks $\mathbf{X}_m \ (I_m \times J) \ ; m = 1, \ldots, M$. The basic model is

$$\mathbf{X}_m = \mathbf{T}_m \mathbf{P}^t + \mathbf{E}_m; \ m = 1, \ldots, M \tag{2.29}$$

with block-scores $\mathbf{T}_m \ (I_m \times R)$ and from which it is observed that the variable modes of the different blocks are modelled with the same loading matrix $\mathbf{P} \ (J \times R)$. The SCA model of Equation 2.29 can also be written as

$$\mathbf{X} = \begin{bmatrix} \mathbf{X}_1 \\ \cdot \\ \cdot \\ \cdot \\ \mathbf{X}_M \end{bmatrix} = \begin{bmatrix} \mathbf{T}_1 \\ \cdot \\ \cdot \\ \cdot \\ \mathbf{T}_M \end{bmatrix} \mathbf{P}^t + \begin{bmatrix} \mathbf{E}_1 \\ \cdot \\ \cdot \\ \cdot \\ \mathbf{E}_M \end{bmatrix} = \mathbf{TP}^t + \mathbf{E} \tag{2.30}$$

and can thus be estimated easily by a PCA on $\mathbf{X}$ (e.g., using an eigen-decomposition of $\mathbf{X}^t\mathbf{X}$ or an SVD of $\mathbf{X}$) and partitioning the scores $\mathbf{T}$.

Although not originally developed for this purpose, SCA can also be formulated for the case of shared samples (for an extensive description, see Sections 5.1.1.1 and 5.2.1.1). For the data blocks $\mathbf{X}_m \ (I \times J_m) \ ; m = 1, \ldots, M$ this becomes

$$\mathbf{X}_m = \mathbf{TP}_m^t + \mathbf{E}_m; \ m = 1, \ldots, M \tag{2.31}$$

where now the scores $\mathbf{T}$ are shared and the loadings $\mathbf{P}_m$ refer to the mth data block. An example of SCA in the `multiblock` R-package is found in Section 11.6.3.

2.2 Properties of Data

In this book we are dealing with data of very different types, and thus it is useful to discuss this briefly. There are two sets of theories relevant for understanding multiblock data: data theory and measurement theory. Data theory is concerned with the fundamental question of whether data are comparable. We can put many numbers in a data table or a matrix but to what extent can these numbers be analysed together? Do all these numbers have a similar meaning? If the answer is affirmative, then the next question is how such an analysis should look. Then we have to consider the nature of the data which is the realm of measurement theory. For heterogeneous data, that is, data of different scale-types to be explained below (and thus for heterogeneous multiblock data fusion) these questions are crucially relevant, but they may also be relevant for combining data of the same type (Smilde and Hankemeier, 2020). In the following, we will discuss both theories to some extent.

2.2.1 Data Theory

The first aspect in data analysis is the distinction between modes and ways of data (sets). This distinction is due to Tucker (Tucker, 1964) and is often used in describing data structures (Carroll and Arabie, 1980). A mode is defined as a particular class of entities (the sample mode, here indexed by $i = 1, \ldots, I$). An N-way array is defined as a function assigning (mostly real) values to elements of the Cartesian product of the number of ways. Formally, a way in an N-way array is an index set defining one of the directions of the N-way array. The number of ways is then the number of index sets defining all directions. The number of modes in an N-way array is the number of *different* index sets of the N-way array. Hence, the number of modes is always smaller than or equal to the number of ways. These definitions may seem a bit too formal, but they serve very well the purpose of structuring data and data analysis methods. In Chapter 1, Section 1.3 we discussed direct and indirect data. Using the above terminology, direct (two-way) data are the same as two-mode two-way data and indirect data are one-mode two-way data. A matrix containing distances between samples, has two ways but only one mode (samples). Almost all the multiblock data in this book will be of the two-mode two-way type, although we will also touch upon multiblock data containing three-mode three-way structures and two-mode three-way structures.

A second important aspect of data is called comparability and was pioneered by Coombs (1964) and explained for multiway analysis (Van Mechelen and Smilde, 2011). The first important notion in comparability is conditionality; where we can distinguish column-, row-, and matrix-conditionality. When considering numbers arranged in a matrix then different types of comparisons can be made: between numbers across rows in the same column and between numbers across columns in the same row. When such data can be compared meaningfully, the data are called column-conditional and row-conditional, respectively. Elaboration 2.5 gives an example of row-conditional data. When data can be meaningfully compared across rows and columns, then these data are called matrix-conditional.

ELABORATION 2.5

Normalisation of urine NMR metabolomics data

The prototypical example of row-conditional data are metabolomics measurements of urine, e.g., using NMR. Due to the different urine histories of the subjects, the urine can be more or less concentrated. This makes the values within one column of a data matrix incomparable since the (unknown) dilution factor of the subjects destroys the comparability. The typical solution of this problem is found in normalising the different samples thereby attempting to achieve matrix-conditionality, see Figure 2.6.

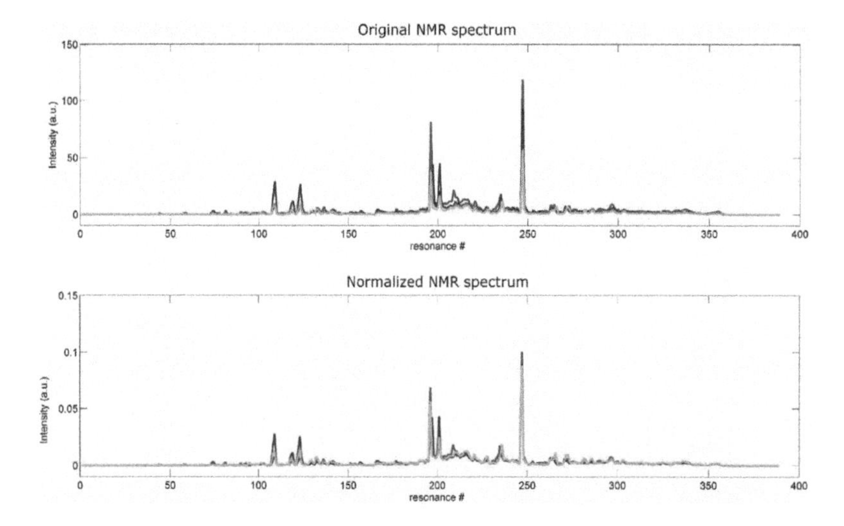

Figure 2.6 Raw and normalised urine NMR-spectra. Different colours are spectra of different subjects.

Another prime example of row-conditional data is microbiome data where RNAseq measurements are performed per sample. Such measurements may contain very different numbers of reads (sequence depth or total number of counts) per sample, making comparisons across samples difficult. There is no consensus yet in the literature as to how to deal with this problem (McMurdie and Holmes, 2014; McKnight *et al.*, 2019).

A more serious problem regarding comparability is lack-of-invariance[2]: the numbers in a single column do not have the same meaning. This problem is more fundamental than conditionality. Whereas in conditionality, numbers cannot be compared since there are (unknown) arbitrary differences, in lack-of-invariance the *meaning* of the variable changes within a column. The prototypical example is unsynchronised time series data where the time points are the variables in a matrix. The physical time points usually do not coincide with, e.g., the biological time and thus a specific physical time point changes meaning across the rows. Remedies of this problem are found in alignment procedures (e.g., using warping approaches (Christin *et al.*, 2010)) but this is not always feasible. If the data consist of multiple variables measured at multiple time points for multiple individuals, then the resulting three-way array may not be the best representation of the data. Instead, multiple blocks of data can be defined where each block describes the variation of one subject across time for all the variables. This leads naturally to a multiblock data analysis problem. Such a reformulation of the data structure has consequences for the types of analyses to consider (and the type of questions to be answered) (Smilde and Hankemeier, 2020).

The issues of comparability also carry over to the multiblock situation. This topic has received little attention. For multiway data, which can be considered a special case of multiblock data, some results have been reported (Van Mechelen and Smilde, 2011). However, this is also an important and fundamental issue for multiblock data analysis to consider. An example of lack-of-invariance in a multiblock problem is given in Elaboration 2.6. Before performing any multiblock data analysis, these types of issues need to be recognised and dealt with. There are different routes to take, e.g., using certain preprocessing methods or using specific methods (Van Mechelen and Smilde, 2011).

ELABORATION 2.6

Genomics and proteomics analysis of glucose starvation

E. coli was starved for glucose in a chemostat and then at time zero a glucose pulse was administered for 1 hour to initiate the glucose response (Borirak *et al.*, 2015). Gene-expression (mRNA) and protein concentrations (P) were measured at times 0, 5, 15, 30, and 60 minutes after the glucose pulse. Both biological and technical replicates were taken. This resulted in 557 measured protein concentrations and their related gene-expressions. The resulting matrices $\mathbf{X}_{\text{gen}}$ and $\mathbf{X}_{\text{prot}}$ can be partitioned in parts pertaining to the replicates. The sizes of the two matrices are $(I \times J_{\text{gen}})$ and $(I \times J_{\text{prot}})$, respectively, and a naive approach would be to run an SCA on the concatenated matrix $[\mathbf{X}_{\text{gen}}|\mathbf{X}_{\text{prot}}]$ since the first mode is shared. But this is problematic since time point 5 (for instance) does not have the same meaning in the two blocks. This can be understood by considering the underlying biology of translating mRNA into a protein:

$$\frac{d[P_{it}]}{dt} = k_i[mRNA_{it}] - k_{[\text{deg}_i}[P_{\text{it}}] \tag{2.32}$$

---

2 Unfortunately, the term 'invariant' has different meanings in different fields, e.g., in Chapter 7 solutions are said to be invariant to between-block scaling

where $[mRNA_{it}]$ and $[P_{it}]$ are the concentrations of the *i*-th mRNA and the corresponding *i*-th protein at time *t*; k_i is the formation constant and k_{deg_i} is the degradation constant. Hence, there is an integrative relationship between the concentrations of mRNA and the proteins at a certain time point. Note that this integrative relationship is different for the different mRNA, protein pairs. Thus, simply shifting one of the matrices relative to the other to match time points is not possible. The authors solved this problem by analysing the relationships between aggregated features of both data blocks (Borirak *et al.*, 2015). No easy solutions here!

2.2.2 Scale-types

The notion of scale-types was pioneered by Stevens (Stevens, 1946) and later taken up and further developed by several authors (Krantz *et al.*, 1971; Roberts, 1985; Narens and Luce, 1986; Luce and Narens, 1987) in theories of measurement. The dominant theory of measurement is representational theory (Hand, 1996) and this will be explained briefly in the following. There are two important notions in this theory: a *representation* of a system and *uniqueness* properties of the numerical representation of that system. A small example will be used to explain these ideas. Suppose we consider all sticks in the world; these are shown as an empirical relational system (ERS) in the upper left of Figure 2.7(a). Although the sticks may have different colours, we are only interested in their lengths. The relationships between the sticks can be represented numerically with the numbers in the upper right panel of Figure 2.7. An equally valid representation of the lengths of the sticks is given in the below right panel of Figure 2.7. Hence, we have two numerical representational systems (NRS1 and NRS2) that can both represent the same ERS. Although the two NRS are different, the *ratios* between the numbers within each of the NRS is the same: that property of the NRS is unique. Such NRS (and associated measurements) are therefore called ratio-scaled measurements. Note that the measurement unit is arbitrary.

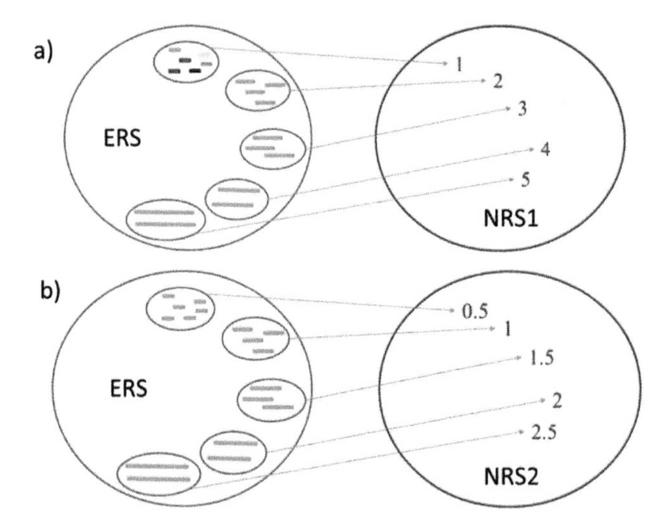

Figure 2.7 Numerical representations of the lengths of sticks: (a) left: the empirical relational system (ERS) of which only the length is studied, right: a numerical representation (NRS1), (b) an alternative numerical representation (NRS2) of the same ERS carrying essentially the same information.

Table 2.1 Formal treatment of types of data scales. The first column refers to the scale-type. The second column gives examples of such scale-types. The third column defines the scale-type in terms of permissible transformations (see text). Finally, the fourth column gives the permissible statistics for the types of scales.

Scale-type	Example	Permissible transformations	Permissible statistics
Nominal	Categories	One-to-one	Number of cases, frequencies
Ordinal	Survey data	Monotonic	Median, IQR
Interval	Degree Celsius	Positive linear transformation	Mean, Standard deviation
	Calendar time	$x' = \alpha x + \beta$ $(\alpha > 0)$	
Ratio	Length	Similarity transformation	Coefficient of variation
	Mass	$x' = \alpha x$ $(\alpha > 0)$	
Absolute	Counts	$x' = x$	All previous

A treatment of measurement scales following Stevens (1946) is given above (see Table 2.1) and a nice introduction is given by Hand (2004). The lowest measurement level is nominal or categorical data which are merely (exclusive) categories. Examples are different types of cars, different countries etc. The data are only used as class labels and these can be changed as long as each class receives a unique other label. Hence, the *permissible transformations* – the transformations between NRS that keep the relationships in the corresponding ERS intact (i.e., uniqueness) – are one-to-one transformations. The type of statistics to be used for this type of data are number of cases, frequencies, χ^2-tests, etc.

The next level of measurement scale are ordinal data. The prototypical example is survey data in which respondents can score on certain issues using the answers strongly disagree, disagree, neutral, agree, strongly agree. Obviously, there is an order in these answers; and these answers can be labelled 1–5. The difference between 2 and 1 on the one hand and between 3 and 2 on the other hand – although exactly equal – does not have a meaning. The ERS can also be represented using a different set of numbers, e.g., 2,4,7,8,9 but the transformation between the two NRSs needs to be strictly monotonic: the series 1,2,3,4,5 can be monotonically transformed in the series 2,4,7,8,9. The type of statistics to be employed are the ones for the lower-scaled measurement (i.e., nominal data) and in addition median, interquartile range (IQR) etc.

Interval-scale data is the next level. An example is degrees Celsius where the numbers 0 and 100 are arbitrarily chosen. Stated otherwise, this scale does not have a natural zero nor a natural unit. This means that another scale (x') can be used with $x' = \alpha x + \beta$ $(\alpha > 0)$ and this scale has the same meaning for the ERS; an example is Fahrenheit where $\alpha = 9/5$ and $\beta = 32$. Nevertheless, the ratio of differences between values of this scale has meaning in terms of the ERS, e.g., in using calendar times $\frac{1980-1960}{1945-1940} = 4$ can be interpreted in a meaningful way as the first period being four times as long as the second one. However, the ratio $\frac{1980}{990} = 2$ does not have a meaning: 1980 is not 'twice as old as' 990. Hence the name interval-scale. In addition to statistics at the lower measurement levels, also means and standard deviations can be used meaningfully for interval-scaled data.

The next level is ratio-scaled data with examples length and weight. As already explained in the main text, a ratio-scaled variable has no natural unit. Hence, the permissible transformation $x' = \alpha x$ $(\alpha > 0)$. It does have a natural zero, however, making ratios meaningful. In addition to the lower measurement levels, also coefficients of variation can be used meaningfully for ratio-scaled data.

The highest degree of measurability is absolute scale data, e.g., count data. Such data has a natural zero and a natural unit and the only permissible transformation is the identity. Apart from the measurement levels mentioned above, there are still other types of more esoteric scales (Krantz *et al.*, 1971). There has been an extensive discussion about permissible statistics and measurement scales in the statistics community with opposing opinions (Adams *et al.*, 1965; Michell, 1986; Velleman and Wilkinson, 1993; Hand, 1996).

The issue of different measurement scales in multiblock data are becoming more widespread in the natural and life sciences. In genomics, it is common to collect multiple blocks of data of the same objects of different scale-types, e.g., gene-expressions which are ratio-scaled and genetic data which may be binary (see Example 1.2). Even considering one type of omics measurement, such as metabolomics, may give rise to questions regarding scale-type (Smilde and Hankemeier, 2020). Other examples are from sensory science (Berget *et al.*, 2020).

In this book we will call ratio- and interval-scaled data *quantitative* data. Apart from that, we will use the terms categorical, ordinal, and binary data. As stated already in Chapter 1, multiblock data analysis methods for fusing data of different measurement scales will be called heterogeneous fusion, in contrast to methods that fuse data of the same measurement scale, which is called homogeneous fusion. Both types of fusion require different types of methods.

2.3 Estimation Methods

2.3.1 Least-squares Estimation

A choice often made in data analysis for fitting models is to use a squared error loss function (least squares or LS). This is almost the default in many data analysis methods. As an example, for PCA this would amount to solving for $\mathbf{T}$ and $\mathbf{P}$ in

$$\mathbf{X} = \mathbf{T}\mathbf{P}^t + \mathbf{E} \tag{2.33}$$

by minimising the sum of squared errors as collected in $\mathbf{E}$ (see also Section 2.1.2). This is easily solved using the SVD of $\mathbf{X}$ due to a famous theorem of Eckart and Young (1936).

In regression situations, least squares is usually performed by using the regression equation

$$\mathbf{y} = \mathbf{X}\mathbf{b} + \mathbf{f} \tag{2.34}$$

and the parameters $\mathbf{b}$ are then found by minimising the squared residuals contained in $\mathbf{f}$. One algorithm employing least squares is called alternating least squares (ALS) and is a very useful estimation method for more complex problems, see Algorithm 2.3.

Algorithm 2.3

ALTERNATING LEAST SQUARES

Alternating least squares is a generic estimation scheme if the set of parameters can be divided in subsets. This will be illustrated by a simple PCA model, solving

$$\min_{\mathbf{T},\mathbf{P}} \|\mathbf{X} - \mathbf{T}\mathbf{P}^t\|^2 \tag{2.35}$$

and estimation of **T** and **P** can now proceed by taking a starting value for **T** and minimising Equation 2.35 for **P**. Subsequently, this **P** is fixed and Equation 2.35 is solved for **T**. This alternating procedure is repeated until a preset convergence criterion for the loss function of Equation 2.35 is met. If the separate steps are all least squares then this ALS-scheme will monotonically decrease (at least not increase) the loss function of Equation 2.35. Each separate step does not have to be a linear regression step as it is in this example. Any (non-linear) least squares step will suffice. Although ALS ensures convergence of the LS loss function[3], there is no guarantee that it will converge to the global optimum. This depends on the complexity of the problem and the starting values. For complex optimisation problems, it is usually recommended to start the algorithm from different starting values and compare the resulting solutions. Of course, for certain problems the solutions can still be rotated and will not necessarily be unique.

In the case of PCA this ALS-scheme will give the same results as an SVD if also orthogonality is imposed in the separate regression steps. Hence, using the SVD for estimating PCA scores and loadings is more convenient.

2.3.2 Maximum-likelihood Estimation

Maximum likelihood estimation is a general principle used a lot in statistics. It starts with distributional assumptions for the stochastic part of the model; then derives the likelihood function for the parameters to estimate and estimates those parameters by maximising this likelihood. In this book, it will be used for multiblock data analysis methods for special data (i.e., categorical and count data) and for heterogeneous fusion.

We will take the case of PCA as an example. PCA has been introduced in Equation 2.33 without any stochastic assumptions. Although differently stated sometimes in the literature, traditional PCA does not need any distributional assumption. To justify PCA with the squared error loss function from a statistical point of view would require the residuals to be *iid* (independent identically distributed) around zero with the same variance. When this assumption is made stronger by assuming the errors to be *iid* and normally distributed around zero with the same variance, then the squared error loss functions amounts to a maximum likelihood method (Wentzell *et al.*, 1997). This idea can be extended to a full probabilistic PCA model (Tipping and Bishop, 1999) and gives room for defining other distributions than the Gaussian to accommodate different types of data, e.g., binary data (see Elaboration 2.7 (Song *et al.*, 2017)).

ELABORATION 2.7

PCA for binary data

By encoding the binary data with zeros and ones, it is possible to run a classical PCA on that data (after mean centring). This would be equivalent to using a least squares approach to PCA but this approach may be difficult to interpret, e.g., after mean-centring the data are no longer binary. Alternatively, consider the element x_{ij} of $\mathbf{X}$ $(I \times J)$ as being drawn from a Bernoulli distribution with parameter ϕ_{ij} which is an element of the probability matrix $\mathbf{\Phi}(I \times J)$. Specifically, the probability that $x_{ij} = 1$ equals ϕ_{ij}. The likelihood of observing a zero or a one is $\phi_{ij}^{x_{ij}}(1 - \phi_{ij})^{(1-x_{ij})}$ and given ϕ_{ij} all values x_{ij} are independent. Then the likelihoods for all elements can be multiplied, and taking

3 In some rare cases the LS loss function does not have a minimum but an infimum (Krijnen *et al.*, 2008)

logarithms then ends up in the log likelihood for observing $\mathbf{X}$ given the probability matrix $\boldsymbol{\Phi}$:

$$l(\boldsymbol{\Phi}) = \sum_{i}^{I} \sum_{j}^{J} [x_{ij}\log(\phi_{ij}) + (1 - x_{ij})\log(1 - \phi_{ij})] \tag{2.36}$$

The probability of observing a '1' relative to a '0' is called the odds and the log-odds are then defined as $\theta_{ij} = log(\frac{\phi_{ij}}{1-\phi_{ij}})$. These log-odds are quantitative (real) values, and to make the connection of the probabilities ϕ_{ij} with these quantitative values often the logistic function $\eta(.)$ defined as $\phi_{ij} = \eta(\theta_{ij}) = (1+e^{-\theta_{ij}})^{-1}$ is used, which is the inverse of $\theta_{ij} = log(\frac{\phi_{ij}}{1-\phi_{ij}})$. In terms of this logistic function, the likelihood becomes

$$l(\boldsymbol{\Theta}) = \sum_{i}^{I} \sum_{j}^{J} [x_{ij}\log(\eta(\theta_{ij})) + (1 - x_{ij})\log(1 - \eta(\theta_{ij}))] \tag{2.37}$$

which is now maximised under the constraint that a low-dimensional structure exists in the log-odds: $\boldsymbol{\Theta} = \mathbf{1}\boldsymbol{\mu}^t + \mathbf{A}\mathbf{B}^t$ with $\mathbf{A}$ and $\mathbf{B}$ of low-rank and $\mathbf{1}\boldsymbol{\mu}^t$ accounting for the offsets. This method is called logistic PCA, and using this machinery, a low-rank structure in the real domain can be imposed and 'transported' to the binary domain (Schein *et al.*, 2003; Song *et al.*, 2017)). A more detailed explanation is given in Chapter 5.

The following example is taken from Song *et al.* (2017). The data set is from Genomics Determinants of Sensitivity in Cancer (GDSC1000) (Iorio *et al.*, 2016) and contains measured genomics data from cancer cell lines. We focus on three cancer types: breast cancer (BRCA; 48 samples), lung cancer (LUAD, 62 samples) and skin cancer (SKCM, 50 samples). Hence, there are three data blocks containing different (numbers of) samples and the same variables are measured for those. This is an example of a shared variable mode (see Chapter 3).

The type of genomics data are mutation data with 198 binary variables and is very sparse, only 2% is labelled as '1'. We also show the results of the methylation data with 38 variables and relatively good balance (27% labelled as '1'). The score plots for the mutation data are shown in Figure 2.8. There is no strong grouping between the cancer types. The results of logistic PCA and classical PCA are different. However, the classical PCA shows some extreme points, and if these are not considered, the score plots show some resemblance.

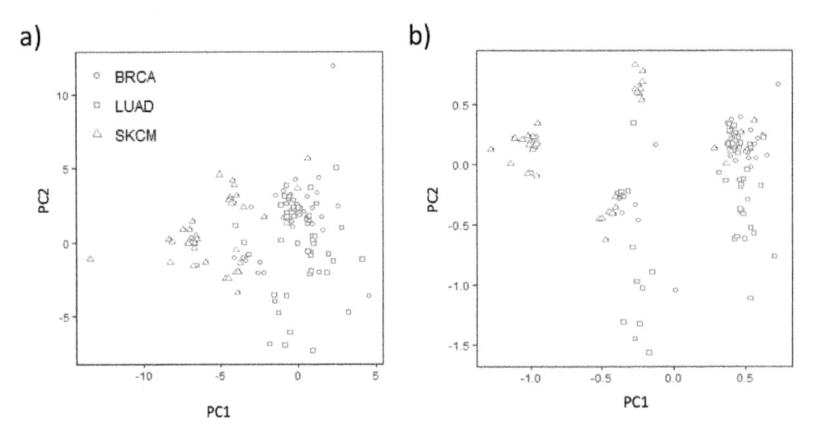

Figure 2.8 Classical (a) and logistic PCA (b) on the same mutation data of different cancers. Source Song *et al.* (2017). Reproduced with permission from Oxford Academic Press.

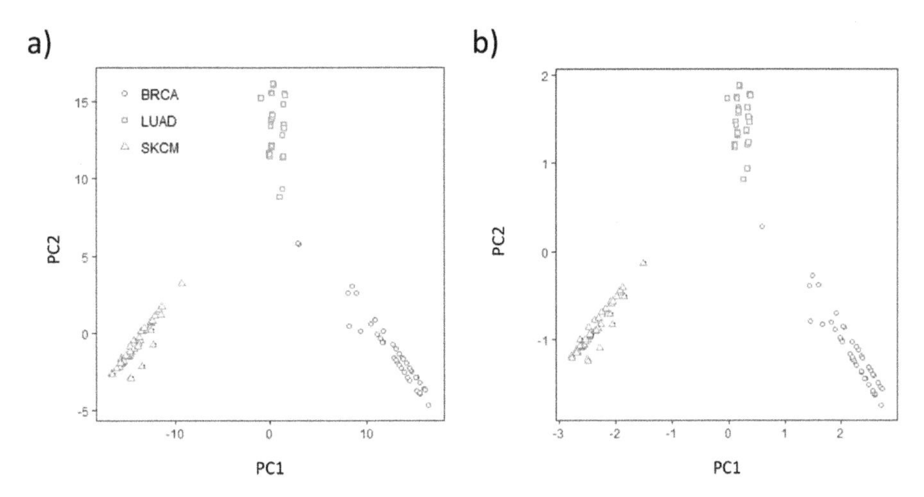

Figure 2.9 Classical (a) and logistic PCA (b) on the same methylation data of different cancers. Source Song *et al.* (2017). Reproduced with permission from Oxford Academic.

The score plots for the methylation data are shown in Figure 2.9. There is a strong grouping between the cancer types, and this grouping is shown both in the logistic as well as the classical PCA.

Clearly, the classical and logistic PCA can give different results but this is data set specific. Moreover, this depends also on the balancedness of the data set and whether or not a strong grouping structure is present. One important difference in the methods is that logistic PCA solutions are not nested for the different dimensionalities whereas this is the case for (ordinary) PCA (Landgraf and Lee, 2015) (see Section 2.3.5).

In the maximum likelihood context, for categorical data a multinomial distribution can be used and for ordinal data the beta-binomial distribution is an option (Stasinopoulos *et al.*, 2017) which has been used in psychometrics (Tellegen and Laros, 2017). Also Hinge-types of loss functions can be used for ordinal data (Rennie and Srebro, 2005). All these methods are maximum likelihood based.

2.3.3 Eigenvalue Decomposition-based Methods

Some of the multiblock data analysis methods use eigenvalue decompositions (ED) or singular value decompositions (SVD) to estimate parameters. This was already briefly discussed for PCA in Section 2.1.2. PCA scores and loadings of $\mathbf{X}$ can be computed from an SVD of $\mathbf{X}$ but also from an ED of $\mathbf{X}^t\mathbf{X}$ or from an ED of $\mathbf{X}\mathbf{X}^t$. Also PLS, RDA, GCA and SCA components can be calculated using an ED (see Sections 2.4.7 and 2.1.11). Hence, in some cases there are multiple ways of estimating the components.

2.3.4 Covariance or Correlation-based Estimation Methods

Many of the already discussed methods use completely different types of optimisation criteria for finding components. The seminal example of this is canonical correlation (Hotelling, 1936a) which maximises the correlation between two linear combinations of two sets of variables; one set for each data block. It then continues by extracting a second component orthogonal to the first one. An example of CCA in the `multiblock` R-package is found in Section 11.5.2. This principle of maximising correlations between sets of components has also been generalised to more than two data blocks (Van de Geer, 1984). In some cases, it is also possible to write such a correlation-based criterion in a least squares context (Van der Burg and Dijksterhuis, 1996) as in the GCA treatment of Section 2.1.10, but in general this is not possible.

As already shown in Section 2.1.5, PLS uses covariance as optimisation criterion for finding components. Partial results exists for the loss function of PLS in terms of least squares (Ter Braak and de Jong, 1998), and results are available in terms of eigenvalue decompositions (Höskuldsson, 1988). Also methods such as regularised generalised canonical correlation analysis (RGCCA, see Chapter 5) use covariance based optimisation criteria and these cannot easily be translated into either a least-squares or maximum likelihood methods (Tenenhaus *et al.*, 2017). Hence, the covariance or correlation based methods aim at finding components in the different blocks that correlate or covary maximally; they do not aim directly at fitting the blocks of data in some sense. The difference between covariance and correlation-based optimisation is very important and was already explained in Section 2.1.5.

2.3.5 Sequential Versus Simultaneous Methods

Apart from estimating all components of PCA simultaneously using an SVD or ALS scheme, it is also possible to estimate scores and loadings in PCA using a *sequential* approach (see also Section 2.1.2). Since such sequential approaches are very popular in many component-based multiblock data analysis methods, we will present this approach using the example of PCA. This approach starts by solving

$$\min_{\mathbf{t},\mathbf{p}} \|\mathbf{X} - \mathbf{t}\mathbf{p}^t\|^2 \tag{2.38}$$

which returns the score vector $\mathbf{t}_1$ and loading vector $\mathbf{p}_1$ and can be done by using, e.g., the NIPALS algorithm. Subsequently, the matrix $\mathbf{X}$ is orthogonalised or deflated using these vectors. There are different ways of deflating $\mathbf{X}$: either using $\mathbf{t}_1$ or using $\mathbf{p}_1$. When using $\mathbf{t}_1$ we are actually solving a regression problem:

$$\mathbf{X} = \mathbf{t}_1\mathbf{p}_1^t + \mathrm{E} \tag{2.39}$$

where the vector $\mathbf{t}_1$ is fixed and we have to solve for $\mathbf{p}_1$. Standard regression theory now gives

$$\mathbf{p}_1 = \mathbf{X}^t\mathbf{t}_1(\mathbf{t}_1^t\mathbf{t}_1)^{-1} \tag{2.40}$$

and because the residuals are orthogonal to the regressor, it holds that $\mathbf{E}^t\mathbf{t}_1 = 0$. The $\mathbf{E}$ of Equation 2.39 is now called $\mathbf{X}_1$ and the next principal component can be calculated. This is an instance of orthogonal deflation (see Section 2.8). Likewise, we can also deflate using the $\mathbf{p}_1$ vector by conceiving Equation 2.39 as a regression of $\mathbf{X}$ on the regressor $\mathbf{p}_1$ and solve for $\mathbf{t}_1$:

$$\mathbf{t}_1 = \mathbf{X}\mathbf{p}_1(\mathbf{p}_1^t\mathbf{p}_1)^{-1} \tag{2.41}$$

and because the residuals are orthogonal to the regressor, it holds that $\mathbf{E}\mathbf{p}_1 = 0$. For PCA, both kinds of deflation (using $\mathbf{t}$ or $\mathbf{p}$) give exactly the same result but this is a special case: for many cases in component models, this is not true (for PLS, see Elaboration 2.8). For PCA, both deflation solutions are also equivalent to the simultaneous solution.

A property which can be concluded from Equations 2.40 and 2.41 is that $\mathbf{p}_1 \in R(\mathbf{X}^t)$ and $\mathbf{t}_1 \in R(\mathbf{X})$ where the symbol $R(\mathbf{X})$ is used to indicate the column-space of $\mathbf{X}$. Hence, $R(\mathbf{E}^t) \subseteq R(\mathbf{X}^t)$ and $R(\mathbf{E}) \subseteq R(\mathbf{X})$ and thus deflation is a rank reducing step (see Section 2.8). Also these properties do not carry over easily to many multiblock methods.

Apart from the sequential nature of multiblock models related to estimating the components sequentially, there are also methods that use a sequential procedure at a higher aggregation level. Such methods usually consist of standard methods from multivariate data analysis used sequentially. A prototypical example is principal component regression. First, a PCA is performed on the data set and, subsequently, a regression is performed on the scores of this PCA. Hence, we have two types of sequentiality: one on the level of components and one on the level of (building blocks of) methods. What both types of sequentiality have in common is that they do not rely on a single objective function that is minimised or maximised. In contrast, simultaneous methods are exactly doing that. So:

A simultaneous method solves a global optimisation problem using a single objective function. All other methods are called sequential.

An example of a simultaneous method is generalised-SCA which calculates all components simultaneously by maximising a global objective function using maximum likelihood (see Chapter 5). An example of a sequential method is PLS which calculates successive components using deflation steps.

The classical tradition of statistics is based on setting up a full probabilistic model and estimating simultaneously all unknown parameters in the model using a criterion like for instance maximum likelihood. In connection with the component methods covered in this book, other strategies are usually chosen. First of all, focus is on geometrical structures and stable estimation of subspaces and not on probabilistic distributions. Distributional aspects are usually concentrated on standard deviations and prediction variances obtained by either cross-validation or bootstrapping. Some of the simultaneous methods do indeed set up a statistical model, especially in the context of fusing data sets containing measurements of different scale-types.

There are advantages and disadvantages regarding sequential strategies. Many methods to be discussed in this book work sequentially and include deflation steps. This is not always trivial (see Elaboration 2.8). A possible downside of sequential methods is that fair significance values are difficult to obtain since all steps after the first are conditional on choices made on that first step. On the other hand, model validation and interpretations may be simpler since each step is based on well understood methods and principles such as PCA and linear regression. Interpretations, model evaluations and validations can then be made sequentially and sometimes more safely than if all data are put into a more complex model based on sometimes questionable assumptions. For the latter, the output may look nice, but results may be difficult to evaluate. Another distinguishing property of methods that calculate components sequentially is nestedness of models: a model with $R-1$ components is nested within a model with R components. This is usually not the case for simultaneous models (unless specific restrictions are applied).

In this book, a number of methods have a sequential nature:

- An example is SO-PLS discussed in Chapter 7, which has a sequential and alternating use of PLS regression and orthogonalisation. In this case the contribution of each new block can be interpreted separately and judgements regarding the importance of new blocks is done sequentially.

- PLS-DA followed by LDA (see Section 7.2.2). In this case PLS is used for establishing a natural space for doing classification before LDA is used for doing the actual separation between classes.
- PO-PLS (Chapter 7) and PCA-GCA (Chapter 5) are examples where the focus is on finding common and distinct subspaces, either for prediction or for interpretation. In the former case, various PLS models are calculated and inspected for common and unique variability before regression. In the latter case, PCA is used for each block before a GCA is used for finding the common variability. Finally, distinct variation is found orthogonal to the common variability.
- Local regression analysis. Although this is not a major focus point in this book, it is touched upon in Chapter 10. In that case a splitting of the set of samples into groups with similar relations between X and Y is considered. Then a regression can be performed within each cluster.
- RGCCA is also typically a sequential method (see Chapter 5) as well as its derivative DIABLO (see Chapter 9).
- Yet another example is the L-structure analyses discussed in Chapter 8. In one of the strategies discussed, first the relation between two of the blocks is analysed before the third is related to the results from the first.

In all the situations mentioned above, a discussion will be given regarding interpretation and validation.

ELABORATION 2.8

Deflation in PLS

In PLS, the process of deflation is also very popular. Even stronger, there are no simultaneous methods known for calculating scores, loadings, and weights in PLS models. However, deflation can be done in different ways in PLS and using the scores, loadings or weights gives different residuals and properties of those residuals. This has led to some confusion (Pell *et al.*, 2007; Wold *et al.*, 2009; Manne *et al.*, 2009; Bro and Elden, 2009). In multiblock PLS methods, deflation is not trivial and can even lead to unexpected results (Westerhuis and Smilde, 2001) (see also Chapter 7).

2.3.6 Homogeneous Versus Heterogeneous Fusion

Having explained the different types of data we can have in multiblock data analysis models and the different ways of estimating the parameters, we can turn to different types of multiblock data fusion depending on the measurement scales of the data. We will explain this using multiblock data sharing the sample mode. In its most simple case, we would like to fuse two data blocks $\mathbf{X}_1(I \times J_1)$ and $\mathbf{X}_2(I \times J_2)$. One of the models doing so is SCA (see Section 2.1.11) with a shared sample mode which is repeated here for convenience:

$$[\mathbf{X}_1|\mathbf{X}_2] = \mathbf{T}[\mathbf{P}_1^t|\mathbf{P}_2^t] + [\mathbf{E}_1|\mathbf{E}_2] \tag{2.42}$$

and the parameters $\mathbf{T}$ and $\mathbf{P}_1$, $\mathbf{P}_2$ are found by minimising the sum of squared residuals in $\mathbf{E}_1$ and $\mathbf{E}_2$. When the data blocks are both quantitative (interval- or ratio-scaled) then there is no problem with this model.

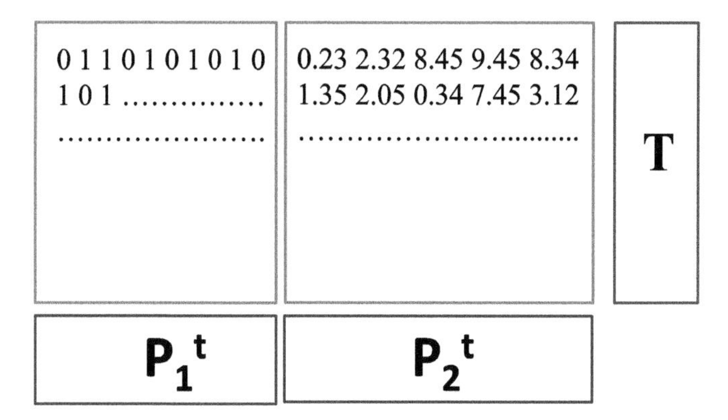

Figure 2.10 SCA for two data blocks; one containing binary data and one with ratio-scaled data.

In the case of heterogeneous data – data of different scale-types (see Section 2.2.2) – the use of the least squares methods such as SCA is problematic. This is illustrated with an example. Suppose that we want to simultaneously analyse two data sets; one with binary data ($\mathbf{X}_1$) and one with ratio-scaled data ($\mathbf{X}_2$), see Figure 2.10. When simply using the SCA approach by solving

$$\min_{\mathbf{T},\mathbf{P}_m} \sum_{m=1}^{2} \|\mathbf{X}_m - \mathbf{T}\mathbf{P}_m^t\|^2 \tag{2.43}$$

it is clear that the binary and quantitative values will be mixed in Equation 2.43. The squared errors of the binary block are summed in Equation 2.43 with the quantitative (real-valued) squared errors of fitting $\mathbf{X}_2$. This is obviously problematic: all these errors are not comparable. Hence, this is an example of block-conditional data (see Section 2.2.1). Stated differently, a value of 1 in the binary block has a different meaning than a value of 1 in the quantitative block: in the binary block it is an indicator and in the quantitative block it may be a concentration. This is the problem of comparability which was already described in Section 2.2.1 and holds in a broader sense for multiblock data fusion.

There is an abundance of multiblock data sets with blocks of data of different scale-types in the life sciences (see Example 1.2). In sensory science there are also examples, such as temporal check-all-that-apply (TCATA) data which contain yes or no scores on food and taste related questions (Castura *et al.*, 2016; Berget *et al.*, 2020) or ordinal-scaled data (Hirst and Næs, 1994). Also Likert data in surveys (strongly agree, agree, neutral, disagree, strongly disagree) are essentially ordinal-scaled. Seemingly quantitative data such as metabolomics data are not trivially ratio- or interval-scaled (Smilde and Hankemeier, 2020). Hence, it is worthwhile to discuss multiblock data analysis methods that can deal with such different data types.

There are several methods to deal with heterogeneous data; some of these methods are extensions of the homogeneous ones. Broadly speaking, the methods can be classified into four groups:

- using probability distributions (see Elaboration 2.7)
- using optimal-scaling (see Elaboration 2.9)
- using representation matrices (see Chapter 9)
- using copulas (see Chapter 9).

For heterogeneous fusion, there are only methods available for a shared sample mode and all methods work in a simultaneous fashion.

ELABORATION 2.9

Optimal-scaling

One set of methods for dealing with measurements of different scale-types is least squares based and is called optimal-scaling[4]. This method goes back some time (Takane *et al.*, 1977) and can also be used to merge data of different measurement levels (Young *et al.*, 1978). A nice summary is given by Michailidis and de Leeuw (1998) and the standard reference is Gifi (1990). The basic approach of optimal-scaling for PCA is

$$\min_{H(.),\mathbf{T},\mathbf{P}} \|H(\mathbf{X}) - \mathbf{T}\mathbf{P}^t\|^2 \tag{2.44}$$

where $\mathbf{X}$ is the matrix to be analysed; $\mathbf{T}$ and $\mathbf{P}$ are the scores and loadings, respectively. The function $H(.)$ is part of the optimisation problem and is such that it conforms with the nature of the measurements $\mathbf{X}$ (see Section 2.2). Estimation is usually done in an alternating fashion: update $H(.)$ given the current $\mathbf{T}$ and $\mathbf{P}$ and vice versa. This is a very versatile scheme allowing for many generalisations of multivariate analysis methods for different data types (Gifi, 1990) and will be discussed in more detail in Chapter 5.

2.4 Within- and Between-block Variation

In the section on fundamental choices (Section 1.7) we already mentioned the aspects of explaining variation within and between blocks. Many multiblock data analysis methods make implicitly or explicitly a choice on which type of variation is modelled. We will first give a definition of within- and between-block variation, illustrated with an example. Next, we will show how within- and between-block variation is handled by PLS, RDA and CCA. Along the way, we will also explain the rationale of using covariance or correlation as criteria for defining components in multiblock models. This will also entail connections between covariance and least squares based criteria. Hence, this section ties together many concepts already touched upon: covariance and least squares based estimation, fundamental choices (see Section 1.7) and relationships between several multiblock data analysis methods such as PLS, RDA, CCA and RGCCA.

2.4.1 Definition and Example

We are going to make the concepts of within- and between-block variation more precise now and we will use a simple example: two blocks containing measurements on the same set of five samples in an unsupervised analysis. The example is drawn from Van de Geer (1984) with some modifications. We will illustrate the concepts with equations and connected figures.

Assume two blocks of column-centred data $\mathbf{X}_1$ $(I \times J_1)$ and $\mathbf{X}_2$ $(I \times J_2)$ both of full column rank. We also assume these blocks to have the same Frobenius norms (i.e., by using proper block scaling) to ensure that the blocks have the same sums-of-squares before starting any

4 Not to be confused with scaling as preprocessing step.

modelling. Both assumptions are made for simplicity but without lack of generality. For illustrative purposes, we will consider only one component; subsequent components are usually obtained after deflation, which is discussed at the end of this section.

The singular value decompositions of $\mathbf{X}_1$ and $\mathbf{X}_2$ are, respectively, $\mathbf{X}_1 = \mathbf{U}_1\mathbf{D}_1\mathbf{V}_1^t$ and $\mathbf{X}_2 = \mathbf{U}_2\mathbf{D}_2\mathbf{V}_2^t$. Suppose that we choose one component from each block: $\mathbf{t}_1 = \mathbf{X}_1\mathbf{w}_1$ and $\mathbf{t}_2 = \mathbf{X}_2\mathbf{w}_2$ for normalised weights $\mathbf{w}_1$ and $\mathbf{w}_2$ (both of length one). These block scores, $\mathbf{t}_1$ and $\mathbf{t}_2$, can be presented on a line in which each sample i is represented by two numbers: the corresponding entries of $\mathbf{t}_1$ (with elements t_{1i}) and $\mathbf{t}_2$ (with elements t_{2i}). Figure 2.11 shows three alternative arrangements for five samples including the averages of the block scores per sample which will be collected in $\mathbf{t}_a$ (with typical elements t_{ai}). Both blocks' scores are centred due to the centring of $\mathbf{X}_1$ and $\mathbf{X}_2$. This is also visible in Figure 2.11 where for each block the corresponding scores (indicated with green squares and blue dots) have zero mean. We will use the following definitions: $\mathbf{X} = [\mathbf{X}_1|\mathbf{X}_2]$, $\mathbf{w}^t = [\mathbf{w}_1^t|\mathbf{w}_2^t]$ and $\mathbf{t} = \mathbf{X}\mathbf{w} = \mathbf{X}_1\mathbf{w}_1 + \mathbf{X}_2\mathbf{w}_2 = \mathbf{t}_1 + \mathbf{t}_2$. This also means that $\mathbf{t} = 2\mathbf{t}_a$.

Figure 2.11 can be used to read off many properties of multiblock data analysis methods and its interpretation is as follows. Due to the length one constraints on the weights $\mathbf{w}_m$, the block scores $\mathbf{t}_1$ and $\mathbf{t}_2$ are also the least squares solutions of the regressions $\mathbf{X}_m = \mathbf{t}_m\mathbf{w}_m^t + \mathbf{E}_m$ for fixed weights $\mathbf{w}_m$ (see Elaboration 2.4). Then $||\mathbf{t}_m\mathbf{w}_m^t||^2$ is the amount of fitted variation of $\mathbf{X}_m$ and this equals $\mathbf{t}_m^t\mathbf{t}_m$ which follows from working out $||\mathbf{t}_m\mathbf{w}_m||^2$ under the constraint $||\mathbf{w}_m|| = 1$. Hence, the variation of the scores for a block as shown in Figure 2.11 is indicative of its amount of explained variation within this block (note that we standardised the blocks to the same sum-of-squares to facilitate comparing explained variances across blocks). For illustration, in panel (a) of Figure 2.11 the solution is drawn towards explaining block $\mathbf{X}_1$ since its corresponding scores (green squares) are well-spread. Contrary, the blue dots (block $\mathbf{X}_2$) are less well-spread, thus, block $\mathbf{X}_2$ is not fitted very well. Hence, explained variation within blocks is readily visible in Figure 2.11.

It is illustrative to work out $\mathbf{t}^t\mathbf{t}$:

$$\mathbf{t}^t\mathbf{t} = \mathbf{w}^t\mathbf{X}^t\mathbf{X}\mathbf{w} = \mathbf{w}_1^t\mathbf{X}_1^t\mathbf{X}_1\mathbf{w}_1 + \mathbf{w}_2^t\mathbf{X}_2^t\mathbf{X}_2\mathbf{w}_2 + 2\mathbf{w}_1^t\mathbf{X}_1^t\mathbf{X}_2\mathbf{w}_2 \tag{2.45}$$

which has several interpretable terms. The terms $\mathbf{w}_m^t\mathbf{X}_m^t\mathbf{X}_m\mathbf{w}_m = \mathbf{t}_m^t\mathbf{t}_m$ are the variances of the block scores which represent the amounts of explained variation in their respective blocks (note that the weights $\mathbf{w}_m$ are constrained to length one). The term $\mathbf{w}_1^t\mathbf{X}_1^t\mathbf{X}_2\mathbf{w}_2 = \mathbf{t}_1^t\mathbf{t}_2$

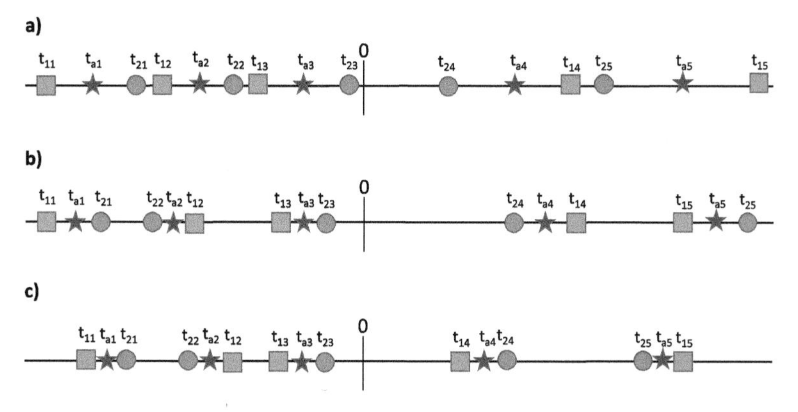

Figure 2.11 The block scores of the rows of the two blocks. Legend: green squares are block scores of the first block; blue circles are block scores of the second block and the red stars are their averages (indicated with t_a). Panel (a) favouring block $\mathbf{X}_1$, (b) the MAXBET solution, (c) the MAXNEAR solution.

is the covariance of the block scores. This covariance represents the amount of variation between the blocks but also describes to some extent the variation within the blocks, see Equations 2.50 and 2.52 and also the earlier discussion on using covariances instead of correlations (Equation 2.20).

2.4.2 MAXBET Solution

As a first criterion for fusing the two blocks it seems reasonable to maximise $\mathbf{t}^t\mathbf{t}$ under the constraints $\|\mathbf{w}_m\| = 1$. This is called the MAXBET solution (Van de Geer, 1984). In words, this means that we want to spread the points $\mathbf{t}$ along the line in Figure 2.11 as much as possible thereby also spreading out the average scores $\mathbf{t}_a$ and thus the variation explained by these average scores. Using $\mathbf{t} = \mathbf{t}_1 + \mathbf{t}_2$, maximising $\mathbf{t}^t\mathbf{t}$ can also be written as

$$\max_{\mathbf{w}_1,\mathbf{w}_2} \mathbf{t}^t(\mathbf{t}_1 + \mathbf{t}_2) = \max_{\mathbf{w}_1,\mathbf{w}_2}[\text{cov}(\mathbf{X}\mathbf{w}, \mathbf{X}_1\mathbf{w}_1) + \text{cov}(\mathbf{X}\mathbf{w}, \mathbf{X}_2\mathbf{w}_2)] \tag{2.46}$$

under the restriction that $\|\mathbf{w}_1\| = \|\mathbf{w}_2\| = 1$. This is a criterion we will encounter more in this book (e.g., the RGCCA method). Since $\mathbf{X}\mathbf{w} = \mathbf{t}_1 + \mathbf{t}_2 = 2\mathbf{t}_a$ this problem can be interpreted as trying to find components that co-vary maximally with the average scores. This criterion leads to a compromise of explaining between-block and within-block variation and is visualised in Figure 2.11 panel (b), see also Equation 2.45). Compared to panel (a) the variation explained in block $\mathbf{X}_1$ is less (lower variance of the green squares); the variation explained in block $\mathbf{X}_2$ is more (higher variance of the blue dots) and the variation of $\mathbf{t}_a$ is also higher (higher variance of the red crosses).

2.4.3 MAXNEAR Solution

Another reasonable criterion to measure similarity is by minimising the squared distance between $\mathbf{t}_1$ and $\mathbf{t}_2$, i.e., minimising the sum-of-squares of $(t_{1i} - t_{2i})$. This can be written as:

$$\min_{\mathbf{w}_1,\mathbf{w}_2} \|\mathbf{t}_1 - \mathbf{t}_2\|^2 = \min_{\mathbf{w}_1,\mathbf{w}_2} \|\mathbf{X}_1\mathbf{w}_1 - \mathbf{X}_2\mathbf{w}_2\|^2; \; s.t. \; \|\mathbf{w}_1\| = \|\mathbf{w}_2\| = 1 \tag{2.47}$$

which upon using the equality $\|\mathbf{X}\|^2 = tr(\mathbf{X}^t\mathbf{X})$ can also be written as:

$$\min_{\mathbf{w}_1,\mathbf{w}_2}(\mathbf{w}_1^t\mathbf{X}_1^t\mathbf{X}_1\mathbf{w}_1 + \mathbf{w}_2^t\mathbf{X}_2^t\mathbf{X}_2\mathbf{w}_2 - 2\mathbf{w}_1^t\mathbf{X}_1^t\mathbf{X}_2\mathbf{w}_2); \; s.t. \; \|\mathbf{w}_1\| = \|\mathbf{w}_2\| = 1 \tag{2.48}$$

or, equivalently,

$$\max_{\mathbf{w}_1,\mathbf{w}_2}(2\mathbf{w}_1^t\mathbf{X}_1^t\mathbf{X}_2\mathbf{w}_2 - \mathbf{w}_1^t\mathbf{X}_1^t\mathbf{X}_1\mathbf{w}_1 - \mathbf{w}_2^t\mathbf{X}_2^t\mathbf{X}_2\mathbf{w}_2); \; s.t. \; \|\mathbf{w}_1\| = \|\mathbf{w}_2\| = 1 \tag{2.49}$$

and which is known as the MAXNEAR solution (Van de Geer, 1984). This function is maximised by making the parts $\mathbf{w}_1^t\mathbf{X}_1^t\mathbf{X}_1\mathbf{w}_1$ and $\mathbf{w}_2^t\mathbf{X}_2^t\mathbf{X}_2\mathbf{w}_2$ small, and the part $\mathbf{w}_1^t\mathbf{X}_1^t\mathbf{X}_2\mathbf{w}_2$ large. Hence, this also optimises the between-block variation but *at the expense* of the within-block variation. This is very different from the MAXBET criterion and is visualised in Figure 2.11 panel (c). The variances of the block scores is clearly smaller than in the MAXBET solution (panel (b)) but these block scores are closer together than in the MAXBET solution. Both criteria (MAXBET and MAXNEAR) seem reasonable but they give very different solutions.

2.4.4 PLS2 Solution

Another obvious solution is found by solving

$$\max_{\mathbf{w}_1,\mathbf{w}_2} \mathbf{w}_1^t \mathbf{X}_1^t \mathbf{X}_2 \mathbf{w}_2; \; s.t. \; \|\mathbf{w}_1\| = \|\mathbf{w}_2\| = 1 \tag{2.50}$$

which gives the first component in a PLS2 model that maximises $cov(\mathbf{t}_1, \mathbf{t}_2)$ under the restriction $\|\mathbf{w}_1\| = \|\mathbf{w}_2\| = 1$. Note that we have switched from the usual PLS notation using $\mathbf{X}$ and $\mathbf{Y}$ to stay in line with the previous derivations. Using the SVDs of $\mathbf{X}_1$ and $\mathbf{X}_2$, this can also be written as

$$\max_{\mathbf{w}_1,\mathbf{w}_2} \mathbf{w}_1^t \mathbf{V}_1 \mathbf{D}_1 \mathbf{U}_1^t \mathbf{U}_2 \mathbf{D}_2 \mathbf{V}_2^t \mathbf{w}_2; \; s.t. \; \|\mathbf{w}_1\| = \|\mathbf{w}_2\| = 1 \tag{2.51}$$

and upon defining $\mathbf{z}_1 = \mathbf{V}_1^t \mathbf{w}_1$ and $\mathbf{z}_2 = \mathbf{V}_2^t \mathbf{w}_2$ this becomes:

$$\max_{\mathbf{z}_1,\mathbf{z}_2} \mathbf{z}_1^t \mathbf{D}_1 \mathbf{U}_1^t \mathbf{U}_2 \mathbf{D}_2 \mathbf{z}_2; \; s.t. \; \|\mathbf{z}_1\| = \|\mathbf{z}_2\| = 1 \tag{2.52}$$

since $\mathbf{z}_m^t \mathbf{z}_m = \mathbf{w}_m^t \mathbf{V}_m \mathbf{V}_m^t \mathbf{w}_m = \mathbf{w}_m^t \mathbf{w}_m = 1$. This shows that the PLS solution still considers the correlation structure of the blocks due to the involvement of the matrices $\mathbf{D}_1$ and $\mathbf{D}_2$. This is because $\mathbf{D}_m$ measures the strengths of the correlations in block $\mathbf{X}_m$ and the squares of the singular values are the amounts of explained variation in $\mathbf{X}_m$. For a geometric explanation of the role of $\mathbf{D}_m$, see Smilde *et al.* (2015) and Elaboration 4.1. This is a restatement of the difference between maximising correlations or covariances (see Equation 2.20). Hence, the PLS solution is stabilised using high correlation directions and thus also explains a certain amount of variation within the blocks which is a well-known property of PLS.

2.4.5 CCA Solution

Yet another solution is obtained by maximising $\mathbf{w}_1^t \mathbf{X}_1^t \mathbf{X}_2 \mathbf{w}_2$ under the constraint $\|\mathbf{t}_1\| = \|\mathbf{t}_2\| = 1$. Although this seems to be a small change relative to the PLS2 solution, it changes the solution completely. Writing out the constraint on $\mathbf{t}_1$ gives

$$\begin{aligned} \mathbf{t}_1^t \mathbf{t}_1 = \mathbf{w}_1^t \mathbf{X}_1^t \mathbf{X}_1 \mathbf{w}_1 &= \mathbf{w}_1^t \mathbf{V}_1 \mathbf{D}_1 \mathbf{U}_1^t \mathbf{U}_1 \mathbf{D}_1 \mathbf{V}_1^t \mathbf{w}_1 \\ &= \mathbf{w}_1^t \mathbf{V}_1 \mathbf{D}_1 \mathbf{D}_1 \mathbf{V}_1^t \mathbf{w}_1 = \mathbf{q}_1^t \mathbf{q}_1 = 1 \end{aligned} \tag{2.53}$$

where $\mathbf{q}_1$ is defined implicitly. A similar derivation can be made for $\mathbf{t}_2^t \mathbf{t}_2$ leading to an analogous $\mathbf{q}_2$. Hence, problem

$$\max_{\mathbf{w}_1,\mathbf{w}_2} \mathbf{w}_1^t \mathbf{X}_1^t \mathbf{X}_2 \mathbf{w}_2; \; s.t. \; \|\mathbf{t}_1\| = \|\mathbf{t}_2\| = 1 \tag{2.54}$$

can be written as

$$\max_{\mathbf{w}_1,\mathbf{w}_2} \mathbf{w}_1^t \mathbf{V}_1 \mathbf{D}_1 \mathbf{U}_1^t \mathbf{U}_2 \mathbf{D}_2 \mathbf{V}_2^t \mathbf{w}_2; \; s.t. \; \|\mathbf{t}_1\| = \|\mathbf{t}_2\| = 1 \tag{2.55}$$

or

$$\max_{\mathbf{q}_1,\mathbf{q}_2} \mathbf{q}_1 \mathbf{U}_1^t \mathbf{U}_2 \mathbf{q}_2; \; s.t. \; \|\mathbf{q}_1\| = \|\mathbf{q}_2\| = 1 \tag{2.56}$$

and the whole within-part disappears! Hence, this solution focuses completely on the between variation and is 'blind' for the within variation. An example of this is canonical correlation analysis (CCA).

Table 2.2 Different methods for fusing two data blocks, indicating the properties in terms of explained variation within and between the blocks. The last two columns refer to whether the methods favour explaining within- or between-block variation. For more explanation, see text.

Method	Criterion	Within variation	Between variation
MAXBET	$W_1 + W_2 + B(W)$	Strong	Some
MAXNEAR	$B(W) - W_1 - W_2$	Hardly	Strong
PLS	$B(W)$	Some	Some
CCA	B	None	Very strong

2.4.6 Comparing the Solutions

Table 2.2 summarises the four different options so far. In this table, W_m is shorthand notation for variation within block m; $B(W)$ is shorthand notation for the term $\mathbf{w}_1^t\mathbf{X}_1^t\mathbf{X}_2\mathbf{w}_2;\ \|\mathbf{w}_1\| = \|\mathbf{w}_2\| = 1$ which focuses on between-block variation with a flavour of within-block variation (see Equation 2.52) and, finally, B is shorthand notation for the term $\mathbf{w}_1^t\mathbf{X}_1^t\mathbf{X}_2\mathbf{w}_2;\ \|\mathbf{t}_1\| = \|\mathbf{t}_2\| = 1$ (see Equation 2.55). This table shows that the focus of the four methods is different. These are just four methods but there are many more!

2.4.7 PLS, RDA and CCA Revisited

Although the methods PLS, RDA and CCA seem very different they are in fact very related despite the fact that CCA is an unsupervised method and both PLS and RDA are supervised methods. However, they do handle the aspects of modelling variation within and between blocks differently. Nice overviews and frameworks are given in Burnham *et al.* (1996); Ter Braak and de Jong (1998); Tenenhaus *et al.* (2017). We will explain these relationships for two blocks $\mathbf{X}(I \times J_X)$ and $\mathbf{Y}(I \times J_Y)$ and only for the first component. Without loss of generality, we will assume for simplicity both $\mathbf{X}$ and $\mathbf{Y}$ to have full column rank. Note that we have switched to the more usual notation for PLS2 and RDA using $\mathbf{X}$ and $\mathbf{Y}$; for CCA this is not very usual but for the sake of simplicity we use this notation now also for CCA in this section.

The generic problem to solve for all three methods is

$$\max_{\mathbf{w},\mathbf{c}} \operatorname{cov}^2(\mathbf{t},\mathbf{u});\ \mathbf{t} = \mathbf{X}\mathbf{w};\ \mathbf{u} = \mathbf{Y}\mathbf{c} \tag{2.57}$$

and the crucial difference is now in the restrictions imposed on the solutions of this problem:

$$\begin{aligned} &\text{For PLS: } \|\mathbf{w}\| = \|\mathbf{c}\| = 1 \\ &\text{For RDA: } \|\mathbf{t}\| = \|\mathbf{c}\| = 1 \\ &\text{For CCA: } \|\mathbf{t}\| = \|\mathbf{u}\| = 1 \end{aligned} \tag{2.58}$$

which seems to represent small differences, but these restrictions are crucial and change the solution. This becomes clear when studying the iterative algorithms for these methods which are shown in Algorithm 2.4.

Algorithm 2.4

ITERATIVE ALGORITHMS FOR PLS, RDA AND CCA

For PLS, the version of the NIPALS algorithm for a multivariate **Y** can be used. The generic symbols **E** (or **e**) and **F** (or **f**) are used for residuals and are not indexed for simplicity but change every iteration.

1:	Start with $\mathbf{c}$ of length 1	– select starting weight for **Y**
2:	$\mathbf{u} = \mathbf{Y}\mathbf{c}$	– solve $\mathbf{Y} = \mathbf{u}\mathbf{c}^t + \mathbf{F}$ for $\mathbf{u}$
	Loop until convergence of u	
3:	$\mathbf{w} = \mathbf{X}^t\mathbf{u}/(\mathbf{u}^t\mathbf{u})$	– solve $\mathbf{X} = \mathbf{u}\mathbf{w}^t + \mathbf{E}$ for $\mathbf{w}$
4:	$\lVert\mathbf{w}\rVert = 1$	– normalise **w**
5:	$\mathbf{t} = \mathbf{X}\mathbf{w}$	– solve $\mathbf{X} = \mathbf{t}\mathbf{w}^t + \mathbf{E}$ for $\mathbf{t}$
6:	$\mathbf{c} = \mathbf{Y}^t\mathbf{t}/(\mathbf{t}^t\mathbf{t})$	– solve $\mathbf{Y} = \mathbf{t}\mathbf{c}^t + \mathbf{F}$ for $\mathbf{c}$
7:	$\lVert\mathbf{c}\rVert = 1$	– normalise **c**
8:	$\mathbf{u} = \mathbf{Y}\mathbf{c}$	– solve $\mathbf{Y} = \mathbf{u}\mathbf{c}^t + \mathbf{F}$ for $\mathbf{u}$
	End loop	

In this algorithm, it is seen that both **w** and **c** are scaled to length one (note that this makes the denominators of step 3 and step 6 redundant in practice). Step 3 and step 6 make sure that there will be a connection between **X** and **Y**. In PLS-path modelling terms these steps are called *reflective* or Mode A (Tenenhaus *et al.*, 2005).

The iterative algorithm of RDA (Fornell *et al.*, 1988) is very similar to the one for PLS and is as follows:

1:	Start with $\mathbf{c}$ of length 1	– select starting weight for **Y**
2:	$\mathbf{u} = \mathbf{Y}\mathbf{c}$	– solve $\mathbf{Y} = \mathbf{u}\mathbf{c}^t + \mathbf{F}$ for $\mathbf{u}$
	Loop until convergence of u	
3:	$\mathbf{w} = (\mathbf{X}^t\mathbf{X})^{-1}\mathbf{X}^t\mathbf{u}$	– solve $\mathbf{u} = \mathbf{X}\mathbf{w} + \mathbf{f}$ for $\mathbf{w}$
4:	$\mathbf{t} = \mathbf{X}\mathbf{w}$	– calculate component of **X**
5:	$\lVert\mathbf{t}\rVert = 1$	– normalise **t**
6:	$\mathbf{c} = \mathbf{Y}^t\mathbf{t}$	– solve $\mathbf{Y} = \mathbf{t}\mathbf{c}^t + \mathbf{F}$ for $\mathbf{c}$
7:	$\lVert\mathbf{c}\rVert = 1$	– normalise **c**
8:	$\mathbf{u} = \mathbf{Y}\mathbf{c}$	– solve $\mathbf{Y} = \mathbf{u}\mathbf{c}^t + \mathbf{F}$ for $\mathbf{u}$
	End loop	

In this algorithm, only steps 2 and 8 are least squares for **Y**; step 4 is not least squares for **X**. In fact, there is no least squares step for **X** and its component **t**. Again, steps 3 and 6 ensure a connection between **X** and **Y**. In terms of PLS-path modelling, step 3 is *formative* or Mode B for **X** and steps 2 and 8 are reflective or Mode A for **Y**.

Finally, the iterative algorithm for CCA comes down to a two-block mode-B PLS model (Wold, 1975)) and is as follows:

1: Start with $\mathbf{c}$ – select starting weight for $\mathbf{Y}$

2: $\mathbf{u} = \mathbf{Y}\mathbf{c}$ – calculate component of $\mathbf{Y}$

3: $\|\mathbf{u}\| = 1$ – normalise $\mathbf{u}$

Loop until convergence of u

4: $\mathbf{w} = (\mathbf{X}^t\mathbf{X})^{-1}\mathbf{X}^t\mathbf{u}$ – solve $\mathbf{u} = \mathbf{X}\mathbf{w} + \mathbf{f}$ for $\mathbf{w}$

5: $\mathbf{t} = \mathbf{X}\mathbf{w}$ – calculate component of $\mathbf{X}$

6: $\|\mathbf{t}\| = 1$ – normalise $\mathbf{t}$

7: $\mathbf{c} = (\mathbf{Y}^t\mathbf{Y})^{-1}\mathbf{Y}^t\mathbf{t}$ – solve $\mathbf{t} = \mathbf{Y}\mathbf{c} + \mathbf{e}$ for $\mathbf{c}$

8: $\mathbf{u} = \mathbf{Y}\mathbf{c}$ – calculate component of $\mathbf{Y}$

9: $\|\mathbf{u}\| = 1$ – normalise $\mathbf{u}$

End loop

and it is clear that in this case both **t** and **u** are normalised. There is no least squares step for either **X** or **Y**. Steps 4 and 7 ensure a connection between **X** and **Y** and these are both formative or Mode B.

Summarising, the normalisation makes a difference in the solution and also what is emphasised in the solution. Of particular importance is the notion of LS-steps in the algorithm. Such an LS-step ensures that within-block variation is considered. Clearly, the algorithms differ in this respect.

The algorithms in Algorithm 2.4 only give the solutions for the first component. Subsequent components can be obtained by deflating **X** and **Y** and starting the algorithms again. This deflation is done differently for each method: PLS deflates **X** and **Y** with the scores **t**; RDA deflates only **X** with the scores **t** and CCA deflates blocks **X** and **Y** with their corresponding scores **t** and **u**, respectively. It can be shown that the final solutions for **w** and **c** are eigenvectors of certain matrices (Höskuldsson, 1988; Burnham *et al.*, 1996) which is summarised in Table 2.3 for **w**. Note that this holds for the first component and subsequent components are obtained after deflation and thus the matrices **X** and **Y** will change.

We assume that the singular value decompositions of **X** and **Y** are, respectively, $\mathbf{P}_1\mathbf{S}_1\mathbf{Q}_1^t$ and $\mathbf{P}_2\mathbf{S}_2\mathbf{Q}_2^t$[5]. As explained in Elaboration 4.1, the matrices $\mathbf{P}_1$ and $\mathbf{P}_2$ define the basic structure

Table 2.3 The matrices of which the weights **w** are eigenvectors in its original form and using the SVDs of **X** and **Y**.

Method	Eigenproblem for w	Rewritten eigenproblem
PLS	$\mathbf{X}^t\mathbf{Y}\mathbf{Y}^t\mathbf{X}$	$\mathbf{Q}_1\mathbf{S}_1\mathbf{P}_1^t\mathbf{P}_2\mathbf{S}_2^2\mathbf{P}_2^t\mathbf{P}_1\mathbf{S}_1\mathbf{Q}_1^t$
RDA	$(\mathbf{X}^t\mathbf{X})^{-1}\mathbf{X}^t\mathbf{Y}\mathbf{Y}^t\mathbf{X}$	$\mathbf{Q}_1\mathbf{S}_1^{-1}\mathbf{P}_1^t\mathbf{P}_2\mathbf{S}_2^2\mathbf{P}_2^t\mathbf{P}_1\mathbf{S}_1\mathbf{Q}_1^t$
CCA	$(\mathbf{X}^t\mathbf{X})^{-1}\mathbf{X}^t\mathbf{Y}(\mathbf{Y}^t\mathbf{Y})^{-1}\mathbf{Y}^t\mathbf{X}$	$\mathbf{Q}_1\mathbf{S}_1^{-1}\mathbf{P}_1^t\mathbf{P}_2\mathbf{P}_2^t\mathbf{P}_1\mathbf{S}_1\mathbf{Q}_1^t$

5 We do not use the general form of the SVD ($\mathbf{U}\mathbf{D}\mathbf{V}^t$) here to avoid confusion with the latent variables **u**; we use now **P** and **Q** which are also used for loadings elsewhere but have a different meaning here.

of $\mathbf{X}$ and $\mathbf{Y}$, respectively; $\mathbf{S}_1$ and $\mathbf{S}_2$ summarise the correlations within the matrices $\mathbf{X}_1$ and $\mathbf{X}_2$ and, finally, $\mathbf{Q}_1$ and $\mathbf{Q}_2$ show how these correlations are distributed among the variables (Ramsay *et al.*, 1984) (for a graphical illustration, see Smilde *et al.* (2015)).

Table 2.3 contains in a very condensed form exactly the similarities and differences between PLS, RDA and CCA (for derivations of the entries, see Section 2.8). The most illuminating parts are shown in the last column of this table. Although the matrices shown in this last column look complicated they are in fact very structured. They all carry multiplications $\mathbf{P}_1^t\mathbf{P}_2$ or its transpose ensuring a connection between the blocks (i.e., explaining between-block variation). The matrix $\mathbf{Q}_1$ is simply a rotation and not important for the optimisation (see below). Of crucial importance is the presence of the matrices $\mathbf{S}_1$ and $\mathbf{S}_2$ since these contain the information about the correlations in $\mathbf{X}$ and $\mathbf{Y}$, respectively.

For PLS, the problem to be solved to obtain the weights $\mathbf{w}$ is

$$\max_{\|\mathbf{w}\|=1} \mathbf{w}^t\mathbf{Q}_1\mathbf{S}_1\mathbf{P}_1^t\mathbf{P}_2\mathbf{S}_2^2\mathbf{P}_2^t\mathbf{P}_1\mathbf{S}_1\mathbf{Q}_1^t\mathbf{w} \tag{2.59}$$

which upon writing $\mathbf{v} = \mathbf{Q}_1^t\mathbf{w}$ becomes

$$\max_{\|\mathbf{v}\|=1} \mathbf{v}^t\mathbf{S}_1\mathbf{P}_1^t\mathbf{P}_2\mathbf{S}_2^2\mathbf{P}_2^t\mathbf{P}_1\mathbf{S}_1\mathbf{v} \tag{2.60}$$

since $\mathbf{v}^t\mathbf{v} = \mathbf{w}^t\mathbf{Q}_1\mathbf{Q}_1^t\mathbf{w} = \mathbf{w}^t\mathbf{w} = 1$ and the rotation matrix $\mathbf{Q}_1$ is immaterial for the solution. To maximise the function in Equation 2.60 it is important that the vector $\mathbf{v}$ is 'aligned' with $\mathbf{S}_1$. Hence, the vector $\mathbf{v}$ is affected by large singular values of $\mathbf{X}$, thus finding a direction with high variance in $\mathbf{X}$. This is exactly the stabilising property of PLS. Note that also the explained variance in $\mathbf{Y}$ plays a role due to the presence of $\mathbf{S}_2^2$ in the criterion.

The case for RDA is different since the problem to solve is:

$$\max_{\|\mathbf{v}\|=1} \mathbf{v}^t\mathbf{S}_1^{-1}\mathbf{P}_1^t\mathbf{P}_2\mathbf{S}_2^2\mathbf{P}_2^t\mathbf{P}_1\mathbf{S}_1\mathbf{v} \tag{2.61}$$

where $\mathbf{v} = \mathbf{Q}_1^t\mathbf{w}$. To maximise this function the sizes of the singular values of $\mathbf{X}$ do not matter due to the presence of both $\mathbf{S}_1^{-1}$ as well as $\mathbf{S}_1$ in the criterion. Hence, RDA does not consider the explained variation in $\mathbf{X}$. Like PLS, RDA does consider the explained variation in $\mathbf{Y}$ due to the presence of $\mathbf{S}_2^2$ in the criterion.

The weights $\mathbf{c}$ in RDA are obtained from the eigenvectors of the matrix $\mathbf{Y}^t\mathbf{X}(\mathbf{X}^t\mathbf{X})^{-1}\mathbf{X}^t\mathbf{Y}$ (see Section 2.8). This matrix can also be written as $\mathbf{Y}^t\mathbf{X}\mathbf{X}^+\mathbf{Y}$ where $\mathbf{X}\mathbf{X}^+$ projects orthogonally on the column-space of $\mathbf{X}$ (Schott, 1997). Thus, RDA finds a direction in the range of $\mathbf{X}$ such that the projections of $\mathbf{Y}$ on this subspace explain the maximum variation in $\mathbf{Y}$. This is why RDA is also sometimes called a restricted PCA: it is a PCA of $\mathbf{Y}$ where the principal components are restricted to be in the column-space of $\mathbf{X}$ (Legendre and Legendre, 2012).

The CCA method solves the following problem:

$$\max_{\|\mathbf{v}\|=1} \mathbf{v}^t\mathbf{S}_1^{-1}\mathbf{P}_1^t\mathbf{P}_2\mathbf{P}_2^t\mathbf{P}_1\mathbf{S}_1\mathbf{v} \tag{2.62}$$

where $\mathbf{v} = \mathbf{Q}_1^t\mathbf{w}$. Hence, like RDA, CCA does not consider explained variance in $\mathbf{X}$. Moreover, since the term $\mathbf{S}_2^2$ is lacking CCA is also not considering the explained variance in $\mathbf{Y}$: CCA focuses completely on the between-group variation as exemplified in the term $\mathbf{P}_1^t\mathbf{P}_2\mathbf{P}_2^t\mathbf{P}_1$ in the criterion.

The second column of Table 2.3 shows that for finding $\mathbf{w}$ in PLS there is no matrix inversion; for RDA there is one matrix inversion (for $\mathbf{X}$) and for CCA there are two matrix inversions. Usually, matrix inversion leads to instabilities and this relates to the discussion above regarding the $\mathbf{S}$ matrices in the third column of Table 2.3. A completely analogous treatment can be given for finding the linear combinations for $\mathbf{Y}$ (the weights $\mathbf{c}$), we show that for RDA in Section 2.8.

2.5 Framework for Common and Distinct Components

The concepts of common and distinct components was already briefly touched upon in Chapter 1 but since it plays a central role in many multiblock data analysis methods it deserves a more elaborate treatment, which follows below in a unified framework.

The framework for common and distinct components has a geometrical origin (Smilde *et al.*, 2017) and can be explained using subspaces and direct sums of subspaces (see Section 2.8). The framework has been developed for the sample-sharing multiblock situation but can also – with some modifications – be used for the variable sharing situation. For the case of shared samples, the basic idea is to look at column-spaces of the matrices involved. In the case of two matrices $\mathbf{X}_1$ and $\mathbf{X}_2$ where each matrix has three rows (which are shared, i.e., the sample mode) the situation is visualised in Figure 2.12. In this special case for geometric reasons (the total space has dimension three and both column-spaces have dimension two) there must be an intersecting space of at least rank one. This is the red line indicated with $\mathbf{X}_{12C}$ in Figure 2.12 which, mathematically, can be described as $R(\mathbf{X}_{12C}) = R(\mathbf{X}_1) \cap R(\mathbf{X}_2)$. Obviously, for spaces with large dimensions in which low-dimensional column-spaces live, such an intersection may only consist of zero.

Given this one-dimensional common component, there are still components 'to give away' in the separate blocks of data. When both ranges of $\mathbf{X}_1$ and $\mathbf{X}_2$ are two-dimensional, then in each range there is still one free direction since the common components already take up one direction. Such a free component in the two ranges are the distinct ones and they can be chosen in different ways, see Figure 2.13. The subspaces spanned by these distinct components are denoted as $R(\mathbf{X}_{1D})$ and $R(\mathbf{X}_{2D})$. We can choose the respective distinct components to be orthogonal to the common component (see Figure 2.13(a)) or choose the distinct components orthogonal to each other (see Figure 2.13(b)). Whether or not this is an exclusive choice depends on the dimensions of the different spaces involved (Smilde *et al.*, 2017)

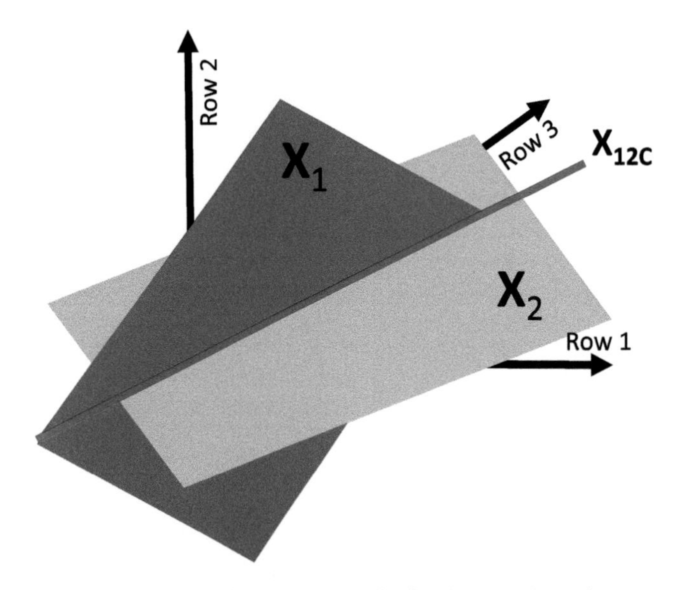

Figure 2.12 Two column-spaces each of rank two in three-dimensional space. The blue and green surfaces represent the column-spaces and the red line indicated with $\mathbf{X}_{12C}$ represents the common component. Source: Smilde *et al.* (2017). Reproduced with permission of John Wiley and Sons.

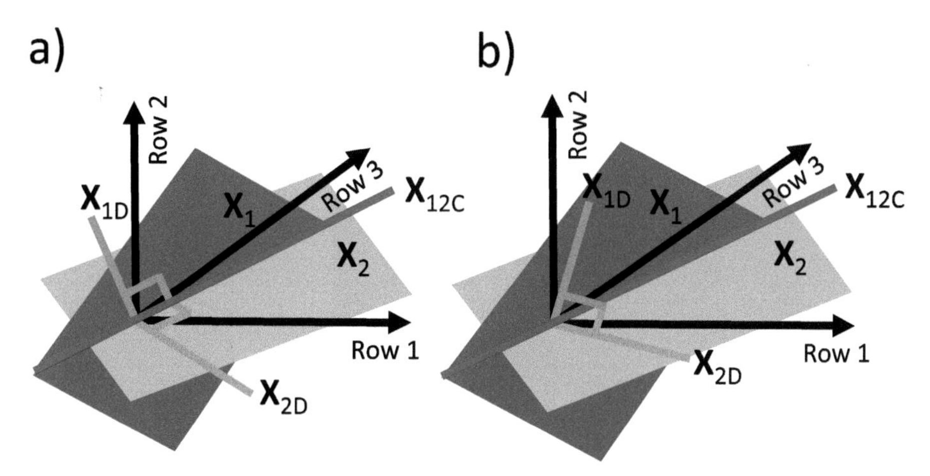

Figure 2.13 Common and distinct components. The common component is the same in both panels. For the distinct components there are now two choices regarding orthogonality: (a) both distinct components orthogonal to the common component, (b) distinct components mutually orthogonal. Smilde *et al.* (2017). Reproduced with permission of John Wiley and Sons.

but all methods separating common from distinct variation have to make a choice regarding orthogonality. In the current example, the common and distinct components describe the complete column-spaces of the respective matrices, but that does not need to be the case. In general, there may also be room for residuals. Hence, in the current case we can write the decomposition in common and distinct components as direct sums of spaces, i.e., $R(\mathbf{X}_1) = R(\mathbf{X}_{12C}) \oplus R(\mathbf{X}_{1D})$ and $R(\mathbf{X}_2) = R(\mathbf{X}_{12C}) \oplus R(\mathbf{X}_{2D})$ (see Section 2.8 for a description of direct sums of spaces). Note that this decomposition holds both for Figure 2.13(a) and Figure 2.13(b) since a direct sum does not require orthogonality.

The concepts of common and distinct as explained so far is for an idealised situation. In practice, the situation is more complicated because of noise, and the concepts apply only approximately. In case of a true common component in the noiseless case, adding noise to this case will result in column-spaces that do not necessarily have an intersection other than zero. See Figure 2.14 where we now assume that the column-spaces are embedded in the I dimensional space (with I much larger than three) and the blue and green column-spaces do not intersect (apart from the origin). There is a fundamental choice to make: do we want the common components in the respective column-spaces (Figure 2.14(a)) or not (Figure 2.14(b))? Two examples are canonical correlation analysis (CCA) (a) or simultaneous component analysis (SCA) (b) which both have advantages and disadvantages.

With CCA the common components are in the respective column-spaces and this facilitates finding the distinct components. Simply remove (i.e., regress out) the common direction from each column-space and perform a PCA on the remainders. For the SCA solution this is more complicated since the common components are in the range of $[\mathbf{X}_1|\mathbf{X}_2]$ and not necessarily in the separate column-spaces. A better name for such components may be *consensus* components. Hence, first these consensus components need to be projected onto these separate column-spaces; then these projected common components need to be removed from the separate column-spaces and then a PCA of the remainders gives the distinct components (these last steps are the same as in CCA).

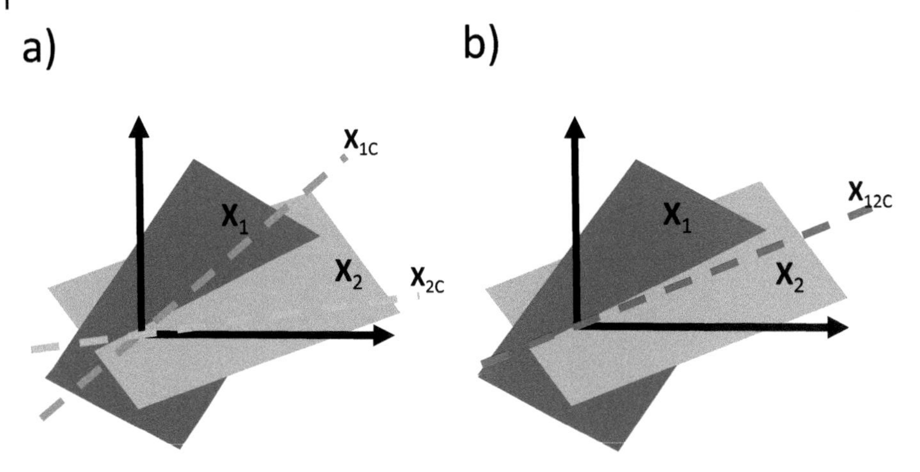

Figure 2.14 Common components in case of noise: (a) maximally correlated common components within column-spaces; (b) consensus component in neither of the columns-spaces. Smilde *et al.* (2017). Reproduced with permission of John Wiley and Sons.

The CCA common components are completely determined by correlations between the blocks. There is no guarantee that these components explain variation within the blocks (see also the discussion on fundamental choices in Section 1.7 and Section 2.4). Since the SCA (consensus) components focus completely on explaining variation in $[\mathbf{X}_1|\mathbf{X}_2]$, they also tend to explain variation in the separate blocks (if the solution is reasonably fair, see again Section 1.7). But the projected SCA common components (see above, these projected common components are in the separate blocks) are not necessarily highly correlated.

When considering more than two blocks of data, there are also local components. These are components that describe commonality between some of the blocks but not all. This framework can be extended to also describe common, local and distinct components. The crucial idea is the decomposition of column-spaces in direct sums of spaces. Of course, for real data (with noise) many choices have to be made on where to place the common and local components (within column-spaces, within sums of column-spaces etc.). Also, choices regarding orthogonality of the components are more challenging.

The nomenclature of common and distinct components is messy. Some authors call these common and distinctive components (Van Deun *et al.*, 2012) or shared and unshared components (Acar *et al.*, 2014) or joint and individual components (Lock *et al.*, 2013) or global, local and unique local components (Löfstedt *et al.*, 2013). This is not helping the field, and it would be nice to have a clear vocabulary. We propose the following:

Consensus component: a component that describes common variation between data blocks but is not necessarily in the column-space of the separate blocks.

Common component: a component that is located in the column-spaces of the separate blocks and describes the common variation between the blocks.

Local component: component that describes the common variation between some but not all of the separate blocks.

Distinct component: component that describes the variation within a block that is unique for that block.

Such a vocabulary is not unequivocal. It is clear that there are also two kinds of local components: those that are within the column-spaces of the participating blocks and those for which this is not the case. We will not use different names for these but from the description and context it will be clear what kind of local components are meant. Since the term 'common component' is abundantly used in the literature, we will not be very strict in the terminology *common* versus *consensus*. At places where this is needed we will make a remark to point out exactly what the status of such a component is.

There are interesting relationships between the notions of common, local, and distinct components on the one hand, and (partial) matrix correlations on the other hand. This will be explained to some detail in Chapter 4, but much more work is needed in this area to understand these relationships.

We would like to end this section by broadening the concepts of common, local, and distinct components. Whereas earlier, we gave a more formal exposition in terms of direct sums of subspaces, the concepts can be defined in a broader context. In general, data blocks can have variation in common, but they often have also distinct variation. This is a very general idea which will play a role in many methods in this book. In the examples above, we restricted ourselves to unsupervised methods, but these concepts certainly also play a role in supervised methods, even explicitly such as in PO-PLS (see Chapter 7). We will also touch upon these concepts in the case of complex topologies (see Chapter 8) and in those cases, although the concepts can still be used, clear definitions are lacking at this point.

2.6 Preprocessing

All multiblock data analysis methods as discussed in this book depend on preprocessing of the data. This traditionally encompasses operations such as centring and scaling but in the case of multiblock data we also have to consider block-scaling. All these aspects have repercussions for the end-result of the data analysis. We will not discuss all kinds of data pretreatments such as used in spectroscopy (e.g., standard normal variate, multiple scatter correction) nor what is used in omics studies (e.g., normalisation, rarification).

We will restrict ourselves to three aspects of preprocessing: centring, within-block scaling and between-block scaling. Centring (or sometimes called column-centring) is the process in which each column of a matrix is centred by subtracting the column-means from the corresponding entries. The idea of centring is that usually the absolute sizes of the measured variables are not relevant, but their differences across the samples. Almost all methods in this book centre the data across the samples. Note that for interval-scaled data it is also very reasonable to centre the data since a natural zero is lacking for such data.

Scaling can be divided into *within-block* and *between-block* scaling. Within-block scaling pertains to the usual scaling as performed for a single data block, e.g., by scaling each variable to standard deviation of one which – together with centring – is called auto-scaling. This is usually done to remove differences of measurement units or to make the sizes of the measurements more comparable. Block-scaling means that a whole block of data is simultaneously scaled, hence, block $\mathbf{X}_1$ is multiplied by a block-scaling weight v_1 to arrive at $v_1\mathbf{X}_1$ as the new data block. There are many ways to select the block-weights v_m and they all have special properties (Van Deun *et al.*, 2009). The basic idea is to somehow correct for different sizes of the blocks; otherwise the block with the largest size may affect the solution too much, but sometimes this is not clear cut (Wilderjans *et al.*, 2009). Some methods are invariant to certain types of scaling. Canonical correlation, is invariant to within- and between-block scaling and multiple linear regression is invariant to within-block scaling. Note that there is an interplay between block scaling and fairness (see Chapter 1).

2.7 Validation

When working with data analysis, it is important to have an overview of the concept of validation as it serves several purposes and comes in many flavours. In general, validation means confirmation, justification or proof, and in data analysis these terms typically refer to the underlying assumptions and performance of models. In this book, the most important use of validation of supervised analysis will be for assessing generalisability, predictive model performance, and for assessing significance of contributions to a model by factors, blocks or variables. For unsupervised analysis the focus is on stability of found patterns. A nice overview of different levels of validation is given in Harshman (1984):

Theoretical appropriateness: Is the correct model used – one that fits the goal and the data structure?

Computational correctness: Has the algorithm converged to the global minimum? Are there local minima?

Statistical reliability: Are the assumptions appropriate? Are the solutions stable under resampling? What are the confidence bounds for the found parameters or modelled values?

Explanatory validity: Is the model appropriate? Do the residuals indicate lack-of-fit? Do we gain new knowledge from the model?

There are several tools available for validation which will briefly be discussed in the sections following our definitions of the concepts of *residuals* and *outliers*. We will use linear regression analysis as an example of supervised analysis and PCA as an example of unsupervised analysis. Most of the tools can be directly transferred to more complicated models.

2.7.1 Outliers

2.7.1.1 Residuals

The simplest of all metrics is the residual, i.e., the difference between the observation and its prediction. For the regression of $\mathbf{y}$ on $\mathbf{X}$ this is:

$$f_i = y_i - \hat{y}_i \tag{2.63}$$

where $\hat{y}_i$ are the fitted values from the regression model, as seen in Equation 2.12. Throughout the rest of this section, hats will signify fitted/predicted values. In Figure 2.15, we illustrate

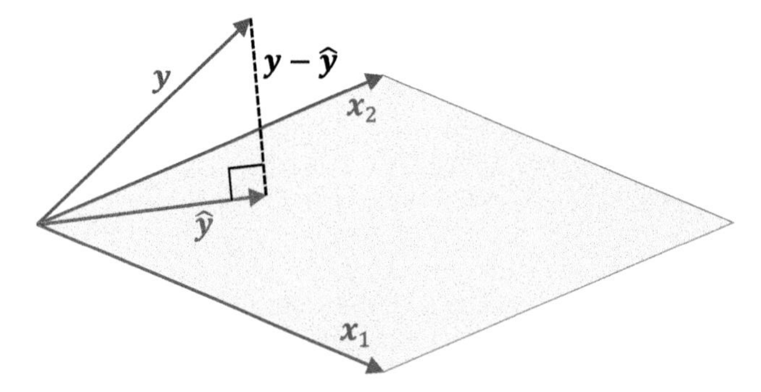

Figure 2.15 Visualisation of a response vector, $\mathbf{y}$, projected onto a two-dimensional data space spanned by $\mathbf{x}_1$ and $\mathbf{x}_2$.

how the fitted value is the projection of the observed response onto the space spanned by the input variables. The residuals can be assessed directly, i.e., looking at their relative sizes, or assessed in standardised form, where they have been divided by the estimated standard deviations ($\hat{\sigma}$, see below). The assessment is usually done with the goal of finding observations that are outliers. Typical rules of thumb indicate objects as outliers if their standardised residuals are larger than 2 or 3 in absolute values. Several variants of the standardised residuals exist, e.g., studentised residuals: $t_i = \frac{f_i}{\hat{\sigma}\sqrt{1-h_{ii}}}$, where h_{ii} is the leverage (influence, see below) of the sample.

Residuals are also commonly used in plots. One such plot is the so-called fitted-versus-residuals plot, where the residuals are plotted as a function of the fitted/predicted values (Figure 2.16). The plot can reveal systematic residual patterns that indicate the need for transformations of the response, **y**, or input data, **X**, or the need for inclusion of variables connected to the patterns of variation. For models that assume normally distributed errors, the residuals can be plotted against a normal distribution for assessment. However, most linear models are quite robust to deviations from normality, so the use of this type of plot is usually limited to applications of analysis of variance, with focus on hypothesis tests, and not included here.

Residuals can also be defined for unsupervised analysis. In PCA, these can be defined as:

$$e_{ij} = x_{ij} - \sum_{r=1}^{R} t_{ir} p_{jr}, \tag{2.64}$$

where t_{ir} are elements of **T** and p_{jr} are elements of **P** (see Equation 2.4). The values e_{ij} can be squared and summed across the index j for each sample i which gives an impression about the fit of that sample (sometimes termed a Q residual or Q statistic). Analogously, the sum-of-squares can be obtained per variable. The sum of squared residuals per sample is sometimes called the SPE (squared prediction error) in statistical process control applications for which there are also approximate distributions available allowing for statistical tests (Kourti and MacGregor, 1995).

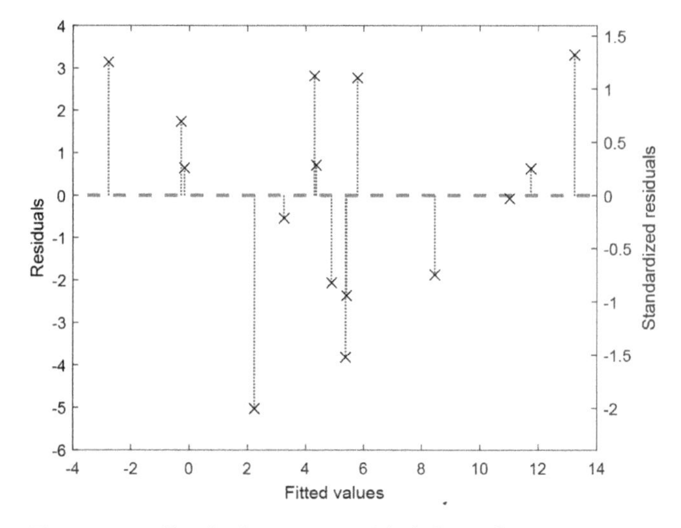

Figure 2.16 Fitted values versus residuals from a linear regression model.

2.7.1.2 Leverage

Outliers in regression models can be y values that deviate strongly from the expected mean of the model, as shown in Figure 2.16. Such outliers can be spotted by plots and also by deletion diagnostics, that is, redoing the analysis without that data point. There is ample literature of this in the regression context (Belsley *et al.*, 2005). The severity of an outlier is related to the leverage of the observation, i.e., how much the observation contributes to the model. In linear regression, given a data matrix $\mathbf{X}$ of dimension $I \times J$, leverage can be found on the diagonal of the projection matrix:

$$\mathbf{H} = \mathbf{X}(\mathbf{X}^t\mathbf{X})^{-1}\mathbf{X}^t, \tag{2.65}$$

where the diagonal elements, h_{ii}, sum to J. A rule of thumb is that observations having leverage larger than two times the average leverage, are considered high leverage points, i.e., $h_{ii} > 2J/I$. In more complicated models, leverage is not necessarily easy to compute, but the concept remains. In Figures 2.17 and 2.18, we see how leverage increases with distance from the mean value of the input variables in simple and multiple linear regression.

Outliers in the predictor space are observations that are far from the centre of mass of the data set, i.e., unusually far from the origin for centred variables. For component based methods like PCA and PLS, we can look for outliers in the score plots (see Figure 2.19) which may again be combined with deletion diagnostics by redoing the analysis without the suspicious point(s). Also in this case we use the definition from Equation 2.65, but exchange $\mathbf{X}$ with scores, $\mathbf{T}$, to define a measure for the importance of samples in a PCA model:

$$\mathbf{h}_T = \text{diag}(\mathbf{T}(\mathbf{T}^t\mathbf{T})^{-1}\mathbf{T}^t) \tag{2.66}$$

where $\mathbf{T}$ are the scores of the PCA model and $\mathbf{h}_T$ is a $(I \times 1)$ vector containing the squared Mahalanobis distances of the samples. Unlike Euclidean distances, Mahalanobis distances compensate for the covariance of the data space. The D-statistic used in multivariate process control is based on these distances and gives a test for how far from the origin a sample is in the model space (Kourti and MacGregor, 1995). Similar equations can be derived for the scores and loadings in PLS models (Martens and Næs, 1989). This strategy can also be used

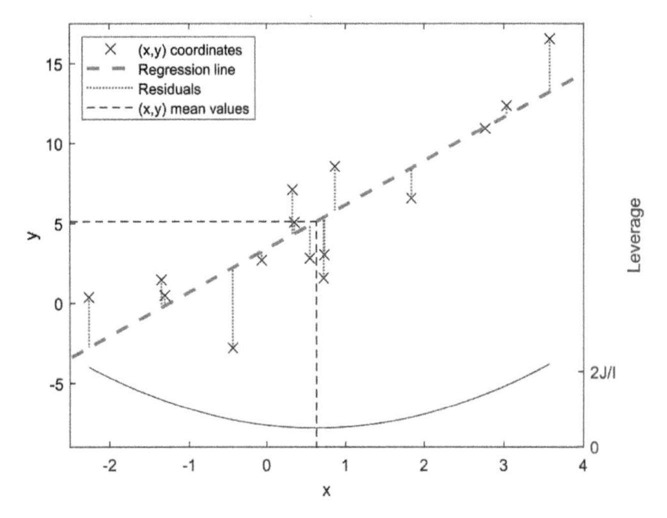

Figure 2.17 Simple linear regression: $\hat{y} = ax + b$ (see legend for description of elements). In addition, leverage is indicated below the regression plot, where leverage is at a minimum at $\bar{x}$ and increases for lower and higher x-values.

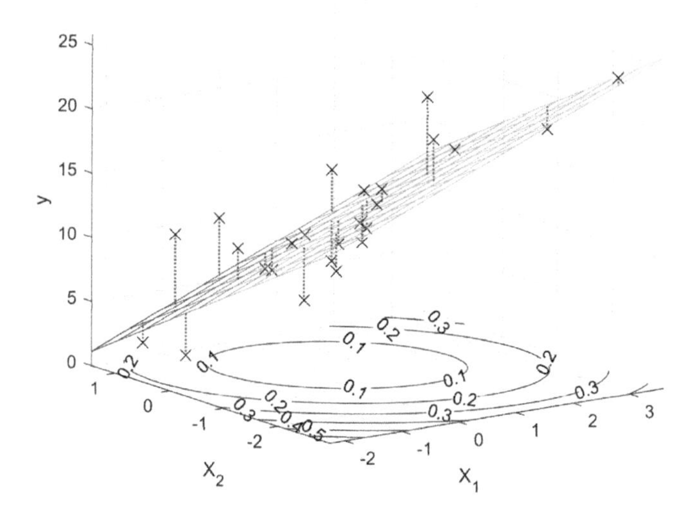

Figure 2.18 Two-variable multiple linear regression with indicated residuals and leverage (contours below regression plane).

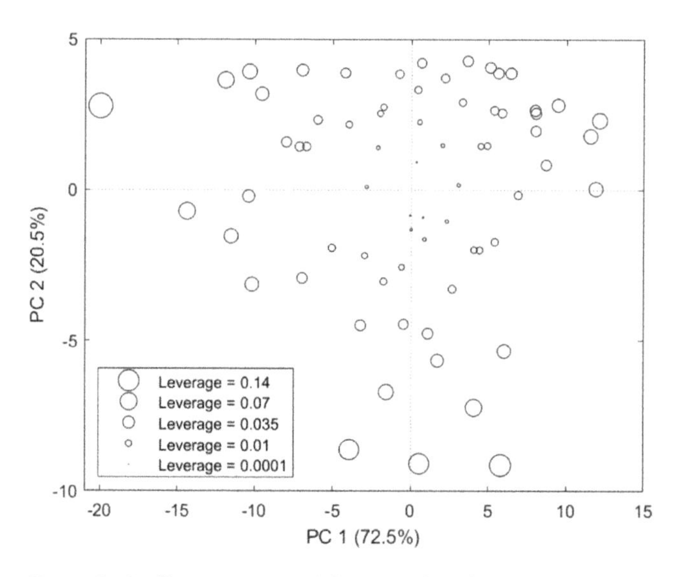

Figure 2.19 Two component PCA score plot of concatenated Raman data. Leverage for two components is indicated by the marker size.

for detecting validation samples that are outliers in the **X** space. This is done by projecting them into the score space before using Equation 2.66 on each projected sample, $\mathbf{t}_p^t$, to calculate $\mathbf{t}_p^t(\mathbf{T}^t\mathbf{T})^{-1}\mathbf{t}_p$.

2.7.2 Model Fit

Supervised analysis

Model fit is often visualised simply by plotting the true response values versus the predictions, as shown in Figure 2.20. Using the residuals defined above it is possible to define fit measures

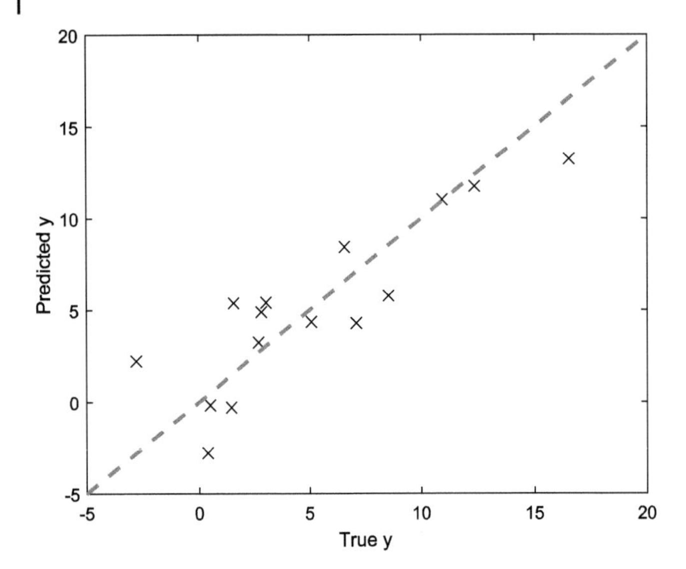

Figure 2.20 Illustration of true versus predicted values from a regression model. The ideal line is indicated in dashed green.

for models. The first one is the mean squared error of calibration (MSEC) for a regression model:

$$\text{MSEC} = \frac{1}{I}\sum_{i=1}^{I}(y_i - \hat{y}_i)^2 \tag{2.67}$$

and its close relative the root MSEC (RMSEC) which is the square root of the MSEC. These are both unvalidated measures using only the training data and its fitted values. The RMSEC has the advantage of being expressed in the original units of **y**. For multiple linear regression, the denominator in MSEC can be changed from I to $I - J$ (J being the number of estimated parameters) to obtain an unbiased estimate for the model error $Var(f) = \sigma^2$.

A set of metrics closely related to the (R)MSEC metrics above is the R^2 family. The one based on fitted values is called the coefficient of determination in statistics and is defined as:

$$\begin{aligned} R^2 &= 1 - \frac{SS_{\text{res}}}{SS_{\text{tot}}} \\ &= 1 - \frac{\sum_i (y_i - \hat{y}_i)^2}{\sum_i (y_i - \bar{y})^2} \\ &= 1 - \frac{\text{MSEC}}{\text{var}_I(y)}, \end{aligned} \tag{2.68}$$

where $\text{var}_I(y) = 1/I\sum_i (y_i - \bar{y})^2$ is variance weighted by I instead of $I - 1$. Note that $\bar{y}$ is redundant in the calculations above when y is centred, as is done in this book. In regression, R^2 is used as a measure of fit for the regression model. It can be noted that this is equal to the squared correlation coefficient between the observed **y** and the predicted **y**. The major disadvantage of the R^2 value is its lack of penalisation of model complexity: the R^2 value will always increase if more variables are added to the regression model. This can be circumvented to some extent by using the adjusted R^2-value (Draper and Smith, 1998).

If the variance, $\text{var}_I(y)$, of the response is known, one can easily calculate back and forth between the RMSEC-based measures and the R^2 based measures. The benefits of R^2 are that explained variances are scale free and more intuitive as proportions of response variation explained by the model. As with other fit measures, comparisons between R^2 values for different models, with the aim of selecting a model, should only be done with care. Although the R^2 values are on the same scale, and thus handy for comparing across data sets, they are affected by non-linearities, outliers, heterogeneous sampling, etc. the same way as (R)MSEC.

Unsupervised analysis

For unsupervised methods, fit measures are less developed. The MSEC and RMSEC can be defined as follows in case of the PCA model of $\mathbf{X}$ ($\mathbf{X} = \mathbf{TP}^t + \mathbf{E}$):

$$\text{MSEC} = \frac{1}{IJ}||\mathbf{E}||^2 = \frac{1}{IJ}tr(\mathbf{E}^t\mathbf{E}) = \frac{1}{IJ}\sum_{i=1}^{I}\sum_{j=1}^{J} e_{ij}^2 \tag{2.69}$$

and the RMSEC is the square-root of this. Correspondingly, the variance explained can be calculated as:

$$\text{VarExpl} = 1 - \frac{||\mathbf{E}||^2}{||\mathbf{X}||^2} \tag{2.70}$$

which can also be expressed in percentages by multiplying by 100 (see also Section 2.1.2). The variance explained is often plotted as a function of the number of principal components, as shown in Figure 2.24. All measures suffer from the same shortcoming as the MSEC, RMSEC and R^2: they do not penalise complexity. Taking more components in a PCA model will always result in more favourable metrics. One way to circumvent this would be to divide the error sum-of-squares by the corresponding degrees of freedom which are not well defined for PCA models – let alone for the more complicated unsupervised models in this book.

2.7.3 Bias-variance Trade-off

As the MSEC measure is a fit measure based only on the training data, it does not give insight into the bias-variance trade-off which is central to modelling. Starting from the regression equation $y = g(x) + f$, where $g(x)$ is most often a linear function in this book and f is the error term, the bias-variance trade-off can be explained as follows. Assume that we have collected data and based on that have estimated the function g (denoted with $\widehat{g}$). If we want to predict at a new point x_0 then the following expectation is relevant:

$$\begin{aligned} &E((y-\widehat{g}(x_0))^2|x=x_0) = \\ &= E((y-g(x_0)) + (g(x_0) - E\widehat{g}(x_0)) + (E\widehat{g}(x_0) - \widehat{g}(x_0)))^2 \end{aligned} \tag{2.71}$$

where the expectation is over data sets used for obtaining $\widehat{g}(x_0)$. We have here added and subtracted both the true function value $g(x_0)$ and the expectation of the estimated function value $E\widehat{g}(x_0)$ without changing the value of the expression. The terms on the right-hand side of this equation are independent, and this thus results in:

$$\begin{aligned} &E((y-\widehat{g}(x_0))^2|x=x_0) = \\ &= E((y-g(x_0))^2 + (g(x_0) - E\widehat{g}(x_0))^2 + (E\widehat{g}(x_0) - \widehat{g}(x_0)))^2 \\ &= \sigma_f^2 + \text{Bias}^2(\widehat{g}(x_0)) + Var(\widehat{g}(x_0)). \end{aligned} \tag{2.72}$$

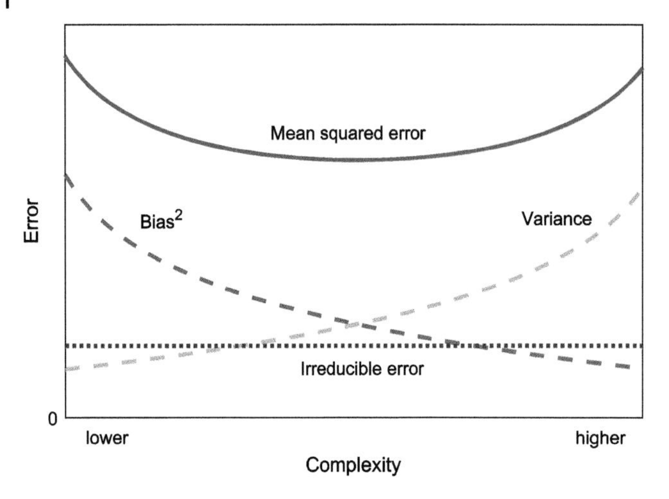

Figure 2.21 Visualisation of bias variance trade-off as a function of model complexity. The observed MSE (in blue) is the sum of the bias^2 (red dashed), the variance (yellow dashed) and the irreducible error (purple dotted).

The first term on the right-hand side is the irreducible error, i.e., the measurement error which cannot be modelled. The second term is the squared bias, i.e., the fixed offset between the true function and the estimated function. And the third term is the variance, i.e., the variation/stability of the estimated function. Typically, when increasing the complexity of a model the bias decreases and the variance increases as the offset between the true function and the estimated function decreases at the expense of a more unstable estimate. This is called the bias variance trade-off and is illustrated in Figure 2.21.

In machine learning a popular metric that balances bias and variance is the AIC (Akaike information criterion) (Murphy, 2012):

$$\text{AIC} = \text{MSEC} + 2\left(\frac{d}{I}\right)\widehat{\sigma_f^2} \tag{2.73}$$

where d is the number of parameters in the model and $\widehat{\sigma_f^2}$ is an estimate of the error in the (linear) model. Obviously, for (too) complex models where d becomes large, the AIC is penalising this complexity by giving high values.

For some methods in this book, distributions are assumed for the underlying parameters and residuals. In those cases, inference is possible based on maximum likelihood principles. Bayesian methods are also sometimes used, and then inference can be based on posterior distributions. All these methods rely on the validity of the assumptions of the underlying distributions. A good introduction to fit measures is given by Hastie *et al.* (2009).

2.7.4 Test Set Validation

The golden standard in validation is test set validation (also called external validation), where a separate data set is kept aside for final validation of the modelling task. In unsupervised analyses, a test set can be used to confirm that structures that were identified on the main data set are generalisable to new data. In supervised analyses, a test set is most commonly used to assess the final predictive performance of a model that has been tuned using internal validation (see cross-validation below). Tuning is here the process of selecting model complexity in component-based models, selecting blocks or variables, selecting penalty parameters or adjusting models in other ways to improve performance, i.e., tuning the hyper-parameters of the model.

An ideal test set has been generated completely independent of the main data set, i.e., on a different day, with different people involved, with fresh materials or samples, etc. For practical and economical reasons, test sets are more often the result of splitting a data set in two (randomly, stratified or sequentially) and saving one part for final validation.

The counterpart of the RMSEC for a test set is the root mean squared error of prediction (RMSEP) :

$$\text{RMSEP} = \sqrt{\frac{1}{I_p}\sum_{i=1}^{I_p}(y_i - \hat{y}_{(i)})^2}, \tag{2.74}$$

where $\hat{y}_{(i)}$ denotes the prediction of y_i when observation i was not included in the modelling. This looks almost identical to the RMSEC. The difference is that the I_p observations in RMSEP are new observations that were not used in the model fitting. The difference between fitted and predicted values can be used to construct learning curves, see Elaboration 2.10.

The test set equivalent to R^2 is called either Q^2, R^2_{pred} or validated explained variance and is defined as:

$$\begin{aligned} Q^2 &= 1 - \frac{\sum_i (y_i - \hat{y}_{(i)})^2}{\sum_i (y_i - \bar{y})^2} \\ &= 1 - \frac{\text{MSEP}}{var_{I_p}(y)}. \end{aligned} \tag{2.75}$$

As opposed to its calibration version, Q^2 can become negative, specifically when the model performance is worse than predicting the response simply as $\bar{y}$. The measure can be seen as a proportion of variance explained when positive, thus making it ideal for interpretation, given the same limitation of dependence on response variance as its calibration counterpart.

In the spectroscopic literature (see, e.g., Bro *et al.* (2005)), the above definition of RMSEP is sometimes termed apparent RMSEP, or $\text{RMSEP}_{\text{app}}$, and a corrected RMSEP, or $\text{RMSEP}_{\text{cor}}$ is defined as $\text{RMSEP}_{\text{cor}} = \sqrt{MSEP - V_{\Delta y}}$. Here $V_{\Delta y}$ is the measurement error variance associated with the reference method used for obtaining $\mathbf{y}$. The $\text{RMSEP}_{\text{cor}}$ is thus a measure of the error caused by the lack of fit by the model that is not due to the error in measurement (see also Equation 2.72).

Here, the various measures are based on the response from a supervised model. However, they can just as well be based on independent input data, e.g., using $\mathbf{X}_{\text{new}}$ and its estimate $\widehat{\mathbf{X}}_{\text{new}}$ from a PCA model. In this, unsupervised, setting RMSEP can be a variable-wise measure or it can be summarised over all variables in a data block. In the case of PCA, if performed naively, the RMSEP will monotonously decrease with increasing number of components (see Elaboration 2.11 for the cross-validation case), so validation must be done with care.

In classical statistics, the assessment of model assumptions plays an equally, or possibly more important role than validation, because asymptotic theory 'guarantees' that data conforming to model assumptions will lead to valid statistical tests and reproducibility under the same assumptions. With more complicated modern methods, highly multivariate data and changing sampling conditions, emphasis has shifted to validation based on reproducibility with new data. In laymen's terms: One may be as creative in the modelling process as one wishes, as long as the results are readily reproducible with new data.

ELABORATION 2.10

Sample size

The number of samples available for analysis is an important aspect to consider when collecting vast amounts of variables and creating complex models. A useful way of assessing the relationship between sample size and model performance is to repeatedly subsample data sets and summarise model performance from the subsamples in a plot. In Figure 2.22, this has been done for a simulated data set with a linear regression, subsampling training data from 50 up to 1000 samples 1000 times, showing medians as lines and shades between 2.5th and 97.5th percentiles. Blue and red curves show explained variance for training and test data, respectively, in 10-fold cross-validation. In machine learning the result is called a 'learning curve'.

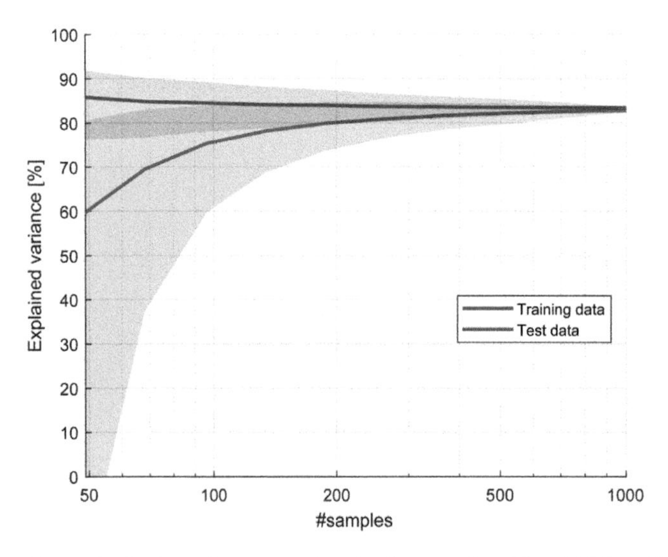

Figure 2.22 Learning curves showing how median R^2 and Q^2 from linear regression develops with the number of training samples for a simulated data set.

This type of analysis can reveal several characteristics of the data and models used. A large gap in explained variance between the training and test data indicates that the model does not generalise well, thus suffers from overfitting. This may be remedied by reducing model complexity, increasing regularisation or collecting more data. If both curves flatten out at a level below the performance level one expects, it could be that there are phenomena that affect the response that we either have not modelled well enough or that we do not have information about. It could also be because of more noise than expected in the data. Finally, if the training and test curves do not flatten out, but still point toward each other at maximum number of samples, it indicates that more samples could lead to better modelling.

2.7.5 Cross-validation

In fields like chemometrics and machine learning, the most common way of internally validating the performance of models based on only the main data set, is to use cross-validation

(Stone, 1974). Cross-validation is the process of repeatedly splitting a data set into training set and test set, building a model on the training data and assessing performance using the test set, and finally summarising over all the splits that were performed. The simplest forms are the two extremes leave-one-out cross-validation, where every sample is held aside once each to serve the role of test sample, and split-half cross-validation, where data are split into two sets, each serving as training and test set once. The splitting process can either be deterministic, e.g., with consecutive or interleaved (Venetian blinds) cross-validation segments, or random splitting. In the latter case, we can even reuse the data set and split in several random ways to increase the variation in the segments. A strategy which is especially popular in machine learning, is to stratify the cross-validation segments, i.e., taking into account groups or classes in the data to achieve balanced representation of these in the segments. If there are sets of replicates, or other groups that should be kept together, this can also be included in the segment selection strategy. In machine learning, the performance is usually assessed per segment in addition to summarising all segments. An illustration of a five-fold cross-validation with consecutive segments is shown in Figure 2.23.

While the amount of folds (number of cross-validation segments) is a matter of discussion in the literature (Hastie *et al.*, 2009), there is no doubt that cross-validation results should be interpreted with care (Rubingh *et al.*, 2006). For instance, if cross-validation is used for determining the number of components in a model, the magnitude of the assessed statistic should be interpreted with care and ideally exchanged with external validation to avoid over-optimism. In other words, the final assessment of model fit should be done on separate data after assessing model complexity using cross-validation. If resources and time allows for it, a double cross-validation (Stone, 1974; Westerhuis *et al.*, 2008) can be applied as an alternative to estimate the future performance. Here an inner cross-validation loop is used for model-tuning/complexity estimation, while an outer loop is used for validation, i.e., separating the tuning and validation processes.

Similar to the RMSEC value, its cross-validation counterparts can be calculated:

$$\mathrm{RMSECV} = \sqrt{\frac{1}{I}\sum_{i=1}^{I}(y_i - \hat{y}_{(i)})^2} \tag{2.76}$$

where the notation $\hat{y}_{(i)}$ means the measurement y_i has not been used in building the model. Contrary to its fit-counterpart RMSEC, the RMSECV does not necessarily reduce when model complexity increases. In classical statistics the predicted sum-of-squares (PRESS)

Splitting of training data

Segment 1

X_1 + X_2 ~ Y

Segment 5

Repeated modelling and prediction

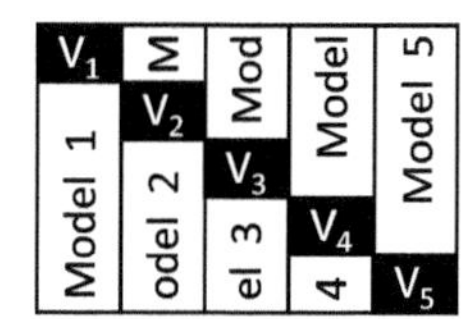

Figure 2.23 Visualisation of the process of splitting a data set into a set of segments (here chosen to be consecutive) and the sequential hold-out of one segment ($\mathbf{V}_k$) for validation of models. All data blocks $\mathbf{X}_m$ and the response $\mathbf{Y}$ are split along the sample direction and corresponding segments removed simultaneously.

measure is often used. PRESS can be derived from RMSECV as PRESS $= \sum_{i=1}^{I}(y_i - \hat{y}_{(i)})^2 = I \cdot \text{RMESCV}^2$ (usually based on leave-one-out cross-validation).

The Q^2-statistic for cross-validation is defined similarly to the test set version, except for MSEP being replaced with MSECV:

$$\begin{aligned} Q^2 &= 1 - \frac{\sum_i (y_i - \hat{y}_{(i)})^2}{\sum_i (y_i - \bar{y})^2} \\ &= 1 - \frac{\text{MSECV}}{\text{var}_I(y)}. \end{aligned} \tag{2.77}$$

Combining cross-validated residuals with ANOVA (Fisher, 1921), we can make a statistical test for the differences in model performance called CVANOVA, either across different methods or for various complexities of a single model, e.g., for varying numbers of components (Indahl and Næs, 1998). The idea is to use cross-validated residuals (absolute values or squared values, $D_{ik} = |y_{ik} - \hat{y}_{ik}|$) as the response in a two-way ANOVA model (Equation 2.78) with a fixed model/complexity effect (γ_k for model k), and a random effect (ξ_i) accounting for the possibly systematic variation in the samples:

$$D_{ik} = \mu + \xi_i + \gamma_k + f_{ik}. \tag{2.78}$$

Here, μ is the model mean and f_{ik} is the ANOVA model's error term. We have used the block counter k as a general index for model comparison, but in the context of this book this could also be exchanged with m for block comparisons, e.g, in the case of SO-PLS where blocks are added to the model thereby also increasing model complexity. The ANOVA model can either be interpreted on its own, testing for effect of model/complexity, or connected to Tukey's test for honestly significant differences (Tukey, 1949) to find pair-wise differences between factor levels and possibly a 'compact letter display' to summarise the results.

Cross-validation can also be used for unsupervised methods and similar metrics as for supervised analysis can be calculated. Taking for instance PCA, we can calculate the RMSECV for a matrix $\mathbf{X}(I \times J)$ using:

$$\text{RMSECV} = \sqrt{\frac{1}{IJ}\sum_{i=1}^{I}\sum_{j=1}^{J}(x_{ij} - \widehat{x}_{(ij)})^2} \tag{2.79}$$

where it is essential that the values x_{ij} and $\widehat{x}_{(ij)}$ are independent of each other. Stated differently, the value of $\widehat{x}_{(ij)}$ should be obtained without using x_{ij}. The same holds true for Equation 2.76 but by predicting $\hat{y}_{(i)}$ a new $x_{(i)}$ is used which was not involved in the model building. For unsupervised analysis, this is more involved (see Elaboration 2.11).

ELABORATION 2.11

Cross-validation of unsupervised methods

Cross-validation of unsupervised methods will be explained using PCA as an example. A naive way of performing cross-validation is to leave out samples, build a model and then use those samples to find predicted x-values and calculate the RMSECV according to Equation 2.79.

Suppose that the matrix $\mathbf{X}(I \times J)$ is split into a part $\mathbf{X}_{\text{in}}(I_{\text{in}} \times J)$ for model building and a part $\mathbf{X}_{\text{out}}(I_{\text{out}} \times J)$ for obtaining predictions. A PCA model for $\mathbf{X}_{\text{in}}(I_{\text{in}} \times J)$ is then $\mathbf{X}_{\text{in}} = \mathbf{TP}^t$ where we have used the maximum number of components. The first step to obtain predictions for $\mathbf{X}_{\text{out}}$ is to project $\mathbf{X}_{\text{out}}$ onto the loadings $\mathbf{P}_R$ to obtain the scores $\mathbf{T}_{R,out} = \mathbf{X}_{\text{out}}\mathbf{P}_R$ where we have now used R components. Next, the predictions can be obtained using $\widehat{\mathbf{X}}_{\text{out}} = \mathbf{T}_{R,\text{out}}\mathbf{P}_R^t$, but note that these predictions are not independent of $\mathbf{X}_{\text{out}}$ which becomes clear by writing $\widehat{\mathbf{X}}_{\text{out}} = \mathbf{T}_{R,out}\mathbf{P}_R^t$ as $\widehat{\mathbf{X}}_{\text{out}} = \mathbf{X}_{\text{out}}\mathbf{P}_R\mathbf{P}_R^t$. Hence, we use the values of $\mathbf{X}_{\text{out}}$ to obtain predictions and thus the assumption of x_{ij} being independent of $\widehat{x}_{(ij)}$ in Equation 2.79 is violated. Viewing the RMSECV in matrix notation:

$$\text{RMSECV}_R = \sqrt{\frac{1}{IJ}\|\mathbf{X}_{\text{out}} - \mathbf{X}_{\text{out}}\mathbf{P}_R\mathbf{P}_R^t\|_F^2} \tag{2.80}$$

we see that it will indeed decrease monotonically upon using more components in $\mathbf{P}_R$. If we take the maximum number of components then RMSECV is zero for the case where $R = J$, $\mathbf{P}_R\mathbf{P}_R^t = \mathbf{I}$.

An example of this naive cross-validation (only leaving out samples) of PCA, repeatedly estimating the 10% samples left out is shown in Figure 2.24, where calibrated (R^2) and cross-validated (Q^2) explained variances are displayed. Here, PCA is applied to the Raman data introduced in Example 1.3, where the Raman blocks have been concatenated before analysis.

Cross-validation in PCA thus is not trivial and should be done with care to avoid $\text{RMSECV}_{R=J} = 0$ described above. Strategies like the ones suggested by (Wold, 1978; Bro *et al.*, 2008; Camacho and Ferrer, 2012, 2014) do avoid these problems and will ensure the familiar behaviour where RMSECV does not go to zero and, depending on the data set, starts increasing when too many components are included (overfitting).

An aspect which should also be considered is how to deal with preprocessing of the data (e.g., centring and scaling). If the centring and scaling constants are based on the whole data set and subsequently a cross-validation is performed, then also the assumption of x_{ij} being independent of $\widehat{x}_{(ij)}$ is violated. Hence, such preprocessing should be a part of the cross-validation procedure. If there are many samples then this is of less importance.

In this elaboration we took PCA as an example of an unsupervised method but these results carry over to all other unsupervised methods. The key point is that the assumption of x_{ij} being independent of $\widehat{x}_{(ij)}$ should not be violated. This entails sometimes carefully thinking about cross-validation in more complex unsupervised models.

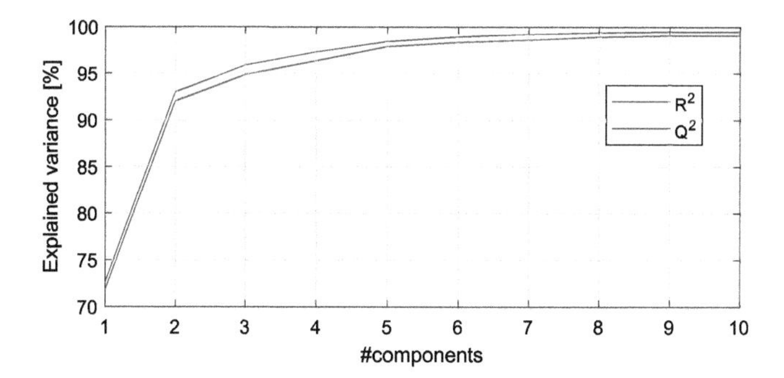

Figure 2.24 Cumulative explained variance for PCA of the concatenated Raman data using naive cross-validation (only leaving out samples). R^2 is calibrated and Q^2 is cross-validated.

2.7.6 Permutation Testing

When an analytical solution to a hypothesis test is not easily available or test statistic distributions are unknown, permutation testing (Pitman, 1937; Welch, 1937) can be used to

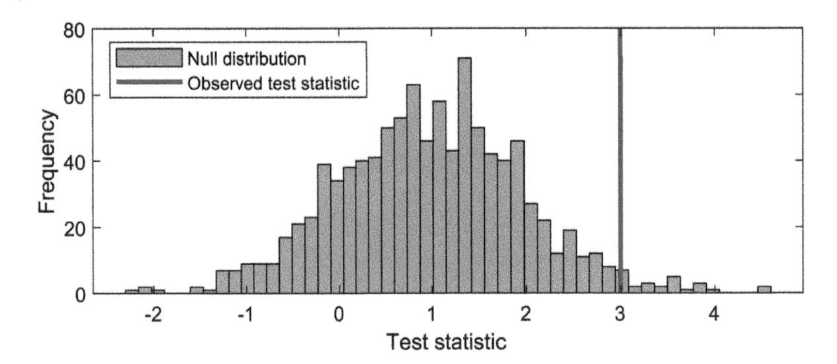

Figure 2.25 Null distribution and observed test statistic used for significance estimation with permutation testing.

simulate a null distribution. The idea is that permuting labels (response/data block) of a set of data points will result in a non-informative test statistic, mimicking the distribution of test statistics associated with the null hypothesis (see Figure 2.25). The test statistic computed from the full data set is deemed significant if it is outside a chosen set of percentiles of the null distribution, typically the 5th/95th for a one-sided test or 2.5th/97.5th percentile for at two-sided test, resulting in a 5% level of significance. Permutation testing can also be used to judge whether a fit measure is significant, e.g., when a Q^2 value of a model is significant, which has proven to be very useful in systems biology (Szymańska *et al.*, 2012). Two notable treatments of component selection in PCA using permutation testing are found in Endrizzi *et al.* (2013); Vitale *et al.* (2017).

2.7.7 Jackknife and Bootstrap

Jackknifing is a technique originally used to get a rough estimate of variance and bias (Quenouille, 1949, 1956; Tukey, 1958), which predates cross-validation by about two decades. Jackknifing works by holding out one sample at a time and redoing the estimation. Through this, many estimates of the parameters are obtained, all based on a slightly changed training set. From this, variances and higher-order moments of the parameters can be estimated and compared to the moments estimated from the full data set, e.g., for bias correction (Efron and Tibshirani, 1993).

Later the jackknife has been extended to estimating distributions of model parameters like regression coefficients. Assuming distributions on the coefficients, the jackknife estimates can be used for variable assessment or selection by constructing hypotheses and testing these per variable. This can be used, e.g., in PLS modelling (Westad and Martens, 2000).

Another very useful resampling technique is the bootstrap (Efron and Tibshirani, 1993). The idea is to mimic a population distribution by repeated sampling with replacement from an empirical distribution (i.e., the data). For each resampled data set (number of samples, I, equal to the number of samples in the full data set), the model is built and parameters estimated. Then the distribution of these parameters can be studied and inferences can be made for those. These methods can also be used in PCA (Timmerman *et al.*, 2007) or more complicated models such as multilevel simultaneous component analysis (Timmerman *et al.*, 2009), but this requires careful tuning and is not trivial.

2.7.8 Hyper-parameters and Penalties

In machine learning and multivariate statistics any parameter that cannot be estimated from data as part of the modelling process is called a hyper-parameter, or sometimes a metaparameter. Typically, hyper-parameters control the complexity of the model, for instance defining the number of components in methods like PCA (Section 2.1.2) or PLS (Section 2.1.5) or the degree of regularisation in methods using Lasso or Ridge penalties like PESCA (Section 5.2.2.5) or MOFA (Section 9.2.2.7). In certain cases the number of blocks and even the choice of method can be considered hyper-parameters. Especially in the emerging field of automatic machine learning, including 'method' as a hyper-parameter is becoming common. In the context of this book number of components, degree of regularisation, order of blocks, and various model constraints are what we mean by hyper-parameters.

Some hyper-parameters are in a grey area where automatic parameter selection is possible by minimising or maximising an objective criterion or metric, though often domain knowledge and human interpretation is unavoidable to obtain good results. An illustration of this can be found in Example 7.1, where choice of number of components in MB-PLS (Section 7.1) can be assessed from a graph of validated explained variance. As mentioned in different wording, concerning double cross-validation in Section 2.7.5, tuning of hyper-parameters is a process that must be handled with care to avoid data torture and information bleeding. It is therefore advisable to keep test data completely separate from the hyper-parameter tuning. Internal validation in the tuning process can be based on any of the sampling techniques mentioned above, e.g., setting aside a separate validation set from the training data, using cross-validation, bootstrapping, etc.

The simplest criterion to optimise is the average performance of the model, for instance using metrics like classification accuracy or prediction R^2. Especially for classification tasks with highly imbalanced class representation, metrics need to be selected that do not convey over-optimism. An example could be a two-class problem where the negative class has 99% of the samples, leading to almost perfect accuracy if all samples are classified as the majority class. In machine learning one typically exchanges accuracy with better suited metrics based on TP = true positive, TN = true negative, FP = false positive and FN = false negative. Two common examples are F_1 (Van Rijsbergen, 1979), the harmonic mean between precision ($\frac{\text{TP}}{\text{TP+FP}}$) and sensitivity ($\frac{\text{TP}}{\text{TP+FN}}$), or Matthew's correlation coefficient (MCC) (Yule, 1912)

$$\begin{aligned} F_1 &= \frac{2TP}{2TP + FP + FN} \\ \text{MCC} &= \frac{\text{TP} \times \text{TN} - \text{FP} \times FN}{\sqrt{(\text{TP+FP})(\text{TP+FN})(\text{TN+FP})(\text{TN+FN})}}, \end{aligned} \tag{2.81}$$

a correlation measure scaling from -1 to 1. In the 99% negative case, if all samples are classified as negative, TP = FP = 0, and thus F_1 = MCC = 0[6]. The upshot is that both classes need to achieve good classifications to obtain a high F_1 or MCC, even in cases of high class imbalance (for F_1 this is only true if the negative class dominates).

In addition to selecting based on average performance, tuning for model stability can be considered. In machine learning one typically calculates model metrics per cross-validation

6 In this case the numerical value of MCC= $\frac{0}{0}$, but it is defined to be zero in all cases that have a zero numerator.

segment and then inspects these metrics or their standard deviation to get an impression of stability.

2.8 Appendix

Column- and row-spaces

Some basic notions from linear algebra will play a role in multiblock data analysis. A vector $\mathbf{a}$ $(I \times 1)$ is said to be in the column-space of $\mathbf{X}$ $(I \times J)$ if there exists a vector $\mathbf{w}$ $(J \times 1)$ such that $\mathbf{Xw} = \mathbf{a}$. The space generated by all such vectors $\mathbf{a}$ is called the column-space or range of $\mathbf{X}$ (notation: $R(\mathbf{X}) \subseteq \mathbb{R}^I$).[7] Similarly, the row-space of $\mathbf{X}$ is the column-space of $\mathbf{X}^t$. The rank of $\mathbf{X}$ (notation: $r(\mathbf{X})$) is the dimension of the column-space. For two-way matrices the ranks of the column-space and row-space of a matrix are the same and the maximum of this rank is the minimum of (I, J). These properties do not hold for three-way arrays.

Direct sum of spaces

We will illustrate the concept of direct sum of spaces using the column-space of $\mathbf{X}$ $(I \times J)$. The column-space is a direct sum of two subspaces $R(\mathbf{X}_1)$ and $R(\mathbf{X}_2)$ if $R(\mathbf{X}) = R(\mathbf{X}_1) + R(\mathbf{X}_2)$ and $R(\mathbf{X}_1) \cap R(\mathbf{X}_2) = \{0\}$. The requirement $R(\mathbf{X}) = R(\mathbf{X}_1) + R(\mathbf{X}_2)$ means that any vector in $R(\mathbf{X})$ can be written as $\mathbf{x}_1 + \mathbf{x}_2$ with $\mathbf{x}_1 \in R(\mathbf{X}_1)$ and $\mathbf{x}_2 \in R(\mathbf{X}_2)$. If the requirement $R(\mathbf{X}_1) \cap R(\mathbf{X}_2) = \{0\}$ is added to this then this decomposition is unique: any vector $\mathbf{x}$ in $R(\mathbf{X})$ can be written uniquely as $\mathbf{x} = \mathbf{x}_1 + \mathbf{x}_2$ where $\mathbf{x}_1 \in R(\mathbf{X}_1)$ and $\mathbf{x}_2 \in R(\mathbf{X}_2)$. Note that the two subspaces $R(\mathbf{X}_1)$ and $R(\mathbf{X}_2)$ do not need to be orthogonal. The shorthand notation for direct sums is $R(\mathbf{X}) = R(\mathbf{X}_1) \oplus R(\mathbf{X}_2)$.

Positive definite matrices

A matrix $\mathbf{X}(I \times I)$ is symmetric if $\mathbf{X} = \mathbf{X}^t$. In addition, a matrix $\mathbf{X}$ is positive definite iff $\mathbf{z}^t\mathbf{Xz} > 0$ for all $\mathbf{z} \neq 0$. A matrix $\mathbf{X}$ is positive semi-definite iff $\mathbf{z}^t\mathbf{Xz} \geqslant 0$ for all $\mathbf{z}$. Note that a cross-product or covariance matrix is always positive (semi-)definite since $\mathbf{z}^t\mathbf{X}^t\mathbf{Xz} = (\mathbf{Xz})^t(\mathbf{Xz}) \geqslant 0$.

Singular value decomposition and Eigen decomposition

Any matrix $\mathbf{X}$ $(I \times J)$ can be decomposed according to the singular value decomposition (SVD):

$$\mathbf{X} = \mathbf{UDV}^t \tag{2.82}$$

where $\mathbf{U}$ $(I \times J)$ is the orthogonal matrix of left-singular vectors ($\mathbf{U}^t\mathbf{U} = \mathbf{I}$), $\mathbf{V}(J \times J)$ is the orthogonal matrix of right-singular vectors ($\mathbf{V}^t\mathbf{V} = \mathbf{VV}^t = \mathbf{I}$) and $\mathbf{D}$ is a diagonal matrix containing the non-negative singular values. It is customary to arrange the latter in decreasing order. This is the SVD for the case of $J \leqslant I$ but similar SVDs are available for the opposite case. Sometimes we will use other symbols for the matrices involved in the SVD (e.g., $\mathbf{PSQ}^t$) if we are already using the symbols $\mathbf{U}$, $\mathbf{D}$ and/or $\mathbf{V}$ for other purposes. This is unfortunately unavoidable. Many useful properties of $\mathbf{X}$ can be read off from its SVD, such as its rank (the number of non-zero singular values) and $R(\mathbf{X}) = R(\mathbf{U})$.

---

7 Note that we use the symbol R in different ways. The typeface will make clear what is meant.

Closely related to the SVD is the eigenvalue decomposition (ED) which is defined for square matrices $\mathbf{Z}$ $(J \times J)$. If this matrix has distinct eigenvalues $\lambda_j; j = 1, \ldots, J$ then:

$$\mathbf{Z} = \mathbf{K\Lambda K}^{-1} \tag{2.83}$$

where $\mathbf{K}$ $(J \times J)$ is a matrix containing the eigenvectors and $\mathbf{\Lambda}$ contains the eigenvalues. If the matrix $\mathbf{Z}$ is symmetric then all the eigenvalues are real and $\mathbf{K}$ $(J \times J)$ is an orthogonal matrix (i.e., $\mathbf{K}^t\mathbf{K} = \mathbf{I}$); thus the eigenvalue decomposition becomes then $\mathbf{Z} = \mathbf{K\Lambda K}^t$. If additionally $\mathbf{Z}$ is a (semi-) positive definite matrix (a Gramian matrix) then all eigenvalues are nonnegative (Schott, 1997).

The connection between the SVD and the ED becomes clear when considering the cross-product matrix $\mathbf{X}^t\mathbf{X}$ which is Gramian and (semi-) positive definite. Then it holds that:

$$\mathbf{X}^t\mathbf{X} = \mathbf{VDU}^t\mathbf{UDV}^t = \mathbf{VD}^2\mathbf{V}^t \tag{2.84}$$

by using the SVD of $\mathbf{X}$; this is exactly the ED of $\mathbf{X}^t\mathbf{X}$. Hence, the eigenvalues of $\mathbf{X}^t\mathbf{X}$ are the squares of the singular values of $\mathbf{X}$.

Trace and vec

The trace of a square matrix is the sum of its diagonal elements. A very useful property of the trace is cyclic permutation:

$$\mathrm{tr}(\mathbf{ABC}) = \mathrm{tr}(\mathbf{CAB}) = \mathrm{tr}(\mathbf{BCA}) \tag{2.85}$$

if the sizes of the involved matrices permit the multiplications.

The Vec operator puts all the columns of a matrix on top of each other:

$$Vec(\mathbf{X}) = \begin{bmatrix} \mathbf{x}_1 \\ \cdot \\ \cdot \\ \cdot \\ \mathbf{x}_J \end{bmatrix} \tag{2.86}$$

where $\mathbf{X}$ $(I \times J)$ has columns $\mathbf{x}_1, \ldots, \mathbf{x}_J$.

Norms of vectors and matrices

At several places in this book we will use norms of vectors and matrices. Norms are measures of the 'size' of vectors and matrices. Vector norms should fulfill the following criteria;

$$\begin{aligned} &\|\mathbf{x}\| \geq 0 \\ &\|\mathbf{x}\| = 0 \ \mathit{iff}\ \mathbf{x} = 0 \\ &\|c\mathbf{x}\| = |c|\|\mathbf{x}\| \text{ for any scalar } c \\ &\|\mathbf{x} + \mathbf{y}\| \leq \|\mathbf{x}\| + \|\mathbf{y}\| \end{aligned} \tag{2.87}$$

where the fourth requirement is often called the triangle inequality.

The norms we will use for vectors are defined as follows:

$$\|\mathbf{x}\|_F = \sqrt{\sum_{i=1}^{I} x_i^2} \tag{2.88}$$

$$\|\mathbf{x}\|_1 = \sum_{i=1}^{I} |x_i|$$

$$\|\mathbf{x}\|_q = \left(\sum_{i=1}^{I} |x_i|^q\right)^{\frac{1}{q}};\ 0 < q \leq 1$$

$$\|\mathbf{x}\|_0 = \text{number of non-zero values in } \mathbf{x}$$

where the subscript 'F' is used to indicate the Frobenius norm (also called Euclidean norm and sometimes denoted as $\|\mathbf{x}\|_2$; the 2-norm); the subscripts '0', 'q' and '1' are used to indicate the L_0, L_q and L_1 norms, respectively. Strictly speaking, $\|\mathbf{x}\|_q$ is not a norm since it does not satisfy the triangle inequality. In most cases, we use the Frobenius norm and then we skip the subscript 'F'.

For matrices we will also use different norms. The requirements for matrix norms are similar to those for vectors norms. The most used matrix norm is again the Frobenius norm:

$$\|\mathbf{X}\|_F = \sqrt{\sum_{i=1}^{I}\sum_{j=1}^{J} x_{ij}^2} = \sqrt{\text{tr}(\mathbf{X}^t\mathbf{X})} \tag{2.89}$$

where the symbol tr($\mathbf{Z}$) is defined above. The Frobenius norm also equals the sum of the squared singular values of $\mathbf{X}$. The square of the Frobenius norm also equals the sum of the squared singular values of $\mathbf{X}$. The other norm which is sometimes used is the nuclear norm which is the sum of the singular values of $\mathbf{X}$ (denoted as $\|\mathbf{X}\|^*$).

Deflation and orthogonalisation

In many cases in multiblock data analysis, components are calculated sequentially and a deflation step is included. The basic equation for deflating $\mathbf{X}$ $(I \times J)$ is:

$$\mathbf{X}_{new} = \mathbf{X} - \mathbf{t}\mathbf{p}^t \tag{2.90}$$

where $\mathbf{t}$ and $\mathbf{p}$ are vectors of conformable size to $\mathbf{X}$. Of course, this process can be continued for the following components.

The properties of the deflation step depends on the specific situation. If either $\mathbf{t}$ or $\mathbf{p}$ or both is/are the solution of a regression step, e.g., $\mathbf{p}^t = (\mathbf{t}^t\mathbf{t})^{-1}\mathbf{t}^t\mathbf{X}$, then this is called orthogonalisation because then it can be shown that $\mathbf{t}$ is orthogonal to $\mathbf{X}_{\text{new}}$ ($\mathbf{X}_{\text{new}}^t\mathbf{t} = 0$ for the case of regressing $\mathbf{X}$ on $\mathbf{t}$) which is a general property of regression. Likewise, when regressing $\mathbf{X}$ on $\mathbf{p}$ it holds that $\mathbf{X}_{\text{new}}\mathbf{p} = 0$.

The rank reducing property of the deflation depends on the size of $\mathbf{X}$ and the position of the vectors $\mathbf{t}$ and $\mathbf{p}$. As an example, if $\mathbf{t}$ is in the column-space of $\mathbf{X}$ and $\mathbf{X}$ is regressed on this $\mathbf{t}$ and if $J \leq I$, then the rank of $\mathbf{X}$ is reduced by one. Regressing $\mathbf{X}$ on a vector outside its range, does not give a rank reduction and may even lead to problems (Westerhuis and Smilde, 2001; Camacho *et al.*, 2020). Hence, orthogonalisation and deflation in general should be done with care.

In the NIPALS-PLS2 algorithm, both **X** and **Y** are deflated using score vectors **t** and loading vectors **p** and **q** after each component has been calculated. It is trivial to show that deflation of **Y** can be skipped while still obtaining the same decomposition. It is also possible to only deflate **Y**, however this requires also orthogonalising scores $\mathbf{t}_{r+1}$ on previous scores, **T**. For **X** data with many variables, avoiding deflation of **X** is beneficial with regard to computational efficiency as the score orthogonalisation works on smaller vector/matrix dimensions than the **X** deflations. This is also true for methods in Chapter 7 working on several blocks one block at the time, e.g., SO-PLS and ROSA, or concatenated blocks, e.g., MB-PLS. The intuition that deflation of only **Y** is equivalent to full deflation, when also orthogonalising scores, is not as apparent, but the linear algebra showing that this is correct is straightforward.

As usual **X** and **Y** are assumed centred. Matrices of the first r vectors of loadings, **P**, Y-loadings, **Q**, and orthogonal scores, **T**, are connected to the original, centred, data through:

$$\begin{aligned} \mathbf{X} &= \mathbf{T}\mathbf{P}^t + \mathbf{E} \\ \mathbf{Y} &= \mathbf{T}\mathbf{Q}^t + \mathbf{F} \end{aligned} \tag{2.91}$$

which is also found in Equation 2.21. Here, the repeated deflations in NIPALS ensure that **T** is orthogonal to both **E** and **F** as $\mathbf{TP}^t$ contains the components that have been removed from **X** to obtain **E** (and likewise for **Y** and **F**). We show the equivalence between deflation strategies sequentially for weights, scores and loadings.

Weights

$$\begin{aligned} \mathbf{w}_{r+1} &<= SVD(\mathbf{E}^t\mathbf{F}) && -\ r{+}1\text{-th weight vector as first left singular vector of } \mathbf{E}^t\mathbf{F} \\ \mathbf{X}^t\mathbf{F} &= (\mathbf{T}\mathbf{P}^t + \mathbf{E})^t\mathbf{F} && -\ \text{substitute deflated with original } \mathbf{X} \\ &= \mathbf{P}\underbrace{\mathbf{T}^t\mathbf{F}}_{=0} + \mathbf{E}^t\mathbf{F} && -\ \mathbf{T} \text{ orthogonal to } \mathbf{F} \\ &= \mathbf{E}^t\mathbf{F} && -\ \blacksquare \end{aligned}$$

The equations above show that the weights, $\mathbf{w}_{r+1}$, will be exactly the same if they are calculated based on $\mathbf{E}^t\mathbf{F}$ (with deflation) or $\mathbf{X}^t\mathbf{F}$ (without **X** deflation).

Scores

$$\begin{aligned} \mathbf{t}_{r+1} &= \mathbf{E}\mathbf{w}_{r+1} && -\ \text{NIPALS PLS scores} \\ \mathbf{t}^*_{r+1} &= \mathbf{X}\mathbf{w}_{r+1} && -\ \text{substitute deflated with original } \mathbf{X} \\ &= (\mathbf{T}\mathbf{P}^t + \mathbf{E})\mathbf{w}_{r+1} && -\ \text{use Equation 2.91} \\ \mathbf{t}^*_{r+1} &- \mathbf{T}(\mathbf{T}^t\mathbf{T})^{-1}\mathbf{T}^t\mathbf{t}^*_{r+1} && -\ \text{orthogonalise on previous scores} \\ &= \overbrace{\mathbf{T}\mathbf{P}^t\mathbf{w}_{r+1}}^{A} + \mathbf{E}\mathbf{w}_{r+1} \\ &\quad - \overbrace{\mathbf{T}\underbrace{(\mathbf{T}^t\mathbf{T})^{-1}\mathbf{T}^t\mathbf{T}}_{=I}\mathbf{P}^t\mathbf{w}_{r+1}}^{B} && -\ A = B,\ \text{thus cancelling out} \\ &\quad - \mathbf{T}(\mathbf{T}^t\mathbf{T})^{-1}\underbrace{\mathbf{T}^t\mathbf{E}}_{=0}\mathbf{w}_{r+1} && -\ \mathbf{T} \text{ orthogonal to } \mathbf{E} \\ &= \mathbf{E}\mathbf{w}_{r+1} && -\ \blacksquare \end{aligned}$$

The equations above show that the scores, $\mathbf{t}_{r+1}$, will be exactly the same if they are calculated as $\mathbf{E}\mathbf{w}_{r+1}$ (with deflation) or $\mathbf{X}\mathbf{w}_{r+1}$ (without $\mathbf{X}$ deflation).

Loadings

$$
\begin{aligned}
&\mathbf{p}_{r+1} = \mathbf{E}^t\mathbf{t}_{r+1}/(\mathbf{t}_{r+1}^t\mathbf{t}_{r+1}) && \text{– NIPALS PLS loadings}\\
&\mathbf{X}^t\mathbf{t}_{r+1}/(\mathbf{t}_{r+1}^t\mathbf{t}_{r+1}) && \text{– substitute deflated with original } \mathbf{X}\\
&\quad = (\mathbf{T}\mathbf{P}^t + \mathbf{E})^t\mathbf{t}_{r+1}/(\mathbf{t}_{r+1}^t\mathbf{t}_{r+1}) && \text{– use Equation 2.91}\\
&\quad = \mathbf{P}\underbrace{\mathbf{T}^t\mathbf{t}_{r+1}}_{=0}/(\mathbf{t}_{r+1}^t\mathbf{t}_{r+1}) && \text{– } \mathbf{T} \text{ orthogonal to } \mathbf{t}_{r+1}\\
&\qquad + \mathbf{E}^t\mathbf{t}_{r+1}/(\mathbf{t}_{r+1}^t\mathbf{t}_{r+1}) &&\\
&\quad = \mathbf{E}^t\mathbf{t}_{r+1}/(\mathbf{t}_{r+1}^t\mathbf{t}_{r+1}) && \text{– } \blacksquare
\end{aligned}
$$

The equations above show that the loadings, $\mathbf{p}_{r+1}$, will be exactly the same if they are calculated as $\mathbf{E}^t\mathbf{t}_{r+1}/(\mathbf{t}_{r+1}^t\mathbf{t}_{r+1})$ (with deflation) or $\mathbf{X}^t\mathbf{t}_{r+1}/(\mathbf{t}_{r+1}^t\mathbf{t}_{r+1})$ (without $\mathbf{X}$ deflation). The argument for the equivalence of loadings also holds for Y-loadings if one exchanges $\mathbf{P}$ with $\mathbf{Q}$ and $\mathbf{E}$ with $\mathbf{F}$. A side-effect of the deflation strategy is that loadings, $\mathbf{P}$, do not play an active role in the NIPALS loop and can just as well be computed after the loop as $\mathbf{P} = \mathbf{X}^t\mathbf{T}(\mathbf{T}^t\mathbf{T})^{-1}$.

After obtaining a score vector, $\mathbf{t}_{r+1}$, through the NIPALS algorithm as shown above, one can add a normalisation step such that $\|\mathbf{t}_{r+1}\| = 1$. This would make the orthogonalisation and calculation of loadings simpler and more efficient as all uses of $(\mathbf{T}^t\mathbf{T})^{-1}$ and $(\mathbf{t}_{r+1}^t\mathbf{t}_{r+1})^{-1}$ become redundant. Using normalised scores means that the loadings will absorb the norms that the scores usually have, but regression coefficients are unchanged and plots will look identical except for different scales. The numerical stability of this strategy has been thoroughly tested and confirmed in Liland *et al.* (2020).

Explained sum-of-squares

A very useful statistic in data analysis is the explained sum-of-squares (see also Section 2.7.2). The generic equation for this is:

$$
\begin{aligned}
\mathbf{X} &= \widehat{\mathbf{X}} + \mathbf{E}\\
\|\mathbf{X}\|^2 &= \|\widehat{\mathbf{X}}\|^2 + \|\mathbf{E}\|^2
\end{aligned}
\tag{2.92}
$$

where $\widehat{\mathbf{X}}$ is the fitted model. A sufficient condition for the second line in Equation 2.92 to hold is orthogonality of $\widehat{\mathbf{X}}$ and $\mathbf{E}$:

$$
\begin{aligned}
&\|\mathbf{X}\|^2 = \mathrm{tr}(\mathbf{X}^t\mathbf{X}) = \mathrm{tr}(\widehat{\mathbf{X}} + \mathbf{E})^t(\widehat{\mathbf{X}} + \mathbf{E}) =\\
&\mathrm{tr}(\widehat{\mathbf{X}}^t\widehat{\mathbf{X}}) + \mathrm{tr}(\mathbf{E}^t\mathbf{E}) + 2\mathrm{tr}(\widehat{\mathbf{X}}^t\mathbf{E})
\end{aligned}
\tag{2.93}
$$

and since $\mathrm{tr}(\mathbf{X}^t\mathbf{X}) = \mathrm{tr}(\mathbf{X}\mathbf{X}^t)$, a similar condition can be derived for $\mathbf{E}\widehat{\mathbf{X}}^t$. Hence, either $\widehat{\mathbf{X}}^t\mathbf{E} = 0$ and/or $\mathbf{E}\widehat{\mathbf{X}}^t = 0$ are sufficient for Equation 2.92 to hold. In regression situations (e.g., the orthogonal deflation step above) this is fulfilled, but in general this does not necessarily hold. For sparse methods, or otherwise methods that penalise parameters, this may result in non-trivial solutions (e.g., for sparse PCA, (Camacho *et al.*, 2020)). There is no general requirement that either the columns of $\mathbf{E}$ or of $\widehat{\mathbf{X}}$ are in the range of $\mathbf{X}$ (or a similar statement for the rows of $\mathbf{E}$ or of $\widehat{\mathbf{X}}$ and the row-space of $\mathbf{X}$).

When the conditions for Equation 2.92 are fulfilled, then the fraction of explained sum-of-squares can be calculated as

$$\left[1 - \frac{||\mathbf{E}||^2}{||\mathbf{X}||^2}\right] \tag{2.94}$$

which is a restatement of Equation 2.70. For the situations that the second line in Equation 2.92 does not hold the interpretation of the fraction explained sum-of-squares becomes problematic. It is still meaningful to use the fit error as $||\mathbf{E}||^2$ or its normalised form $\frac{||\mathbf{E}||^2}{||\mathbf{X}||^2}$ but it is unclear what Equation 2.94 means in that case. If it holds that $||\mathbf{X}||^2 < ||\widehat{\mathbf{X}}||^2 + ||\mathbf{E}||^2$ (which often happens empirically) then Equation 2.94 can be considered a lower bound[8] since then

$$\left[1 - \frac{||\mathbf{E}||^2}{||\mathbf{X}||^2}\right] < \frac{||\widehat{\mathbf{X}}||^2}{||\mathbf{X}||^2} \tag{2.95}$$

but this measure should be handled with care.

Multicollinearity

(Multi)collinearity is a serious problem in regression analysis and refers to large correlation between variables (or more precisely close to linear dependence among variables) in the matrix $\mathbf{X}$ $(I \times J)$ which is used in the regression equation:

$$\begin{aligned} \mathbf{y} &= \mathbf{Xb} + \mathbf{f} \\ \widehat{\mathbf{b}} &= (\mathbf{X}^t\mathbf{X})^{-1}\mathbf{X}^t\mathbf{y} \end{aligned} \tag{2.96}$$

where the solution of the regression problem is given on the second line. Clearly, when the columns of $\mathbf{X}$ have high correlation, the inverse of $(\mathbf{X}^t\mathbf{X})$ becomes unstable. This can be understood by considering the SVD of $\mathbf{X} = \mathbf{UDV}^t$. If some columns of $\mathbf{X}$ are highly correlated (i.e., are collinear) then there are very small singular values in $\mathbf{D}$. The inverse of $\mathbf{X}^t\mathbf{X}$ can be written as $\mathbf{VD}^{-2}\mathbf{V}^t$ by using the orthogonality properties of the SVD. Hence, these small singular values are squared (that is, even made smaller) and inverted which makes them very large. Thus, they start to dominate the solution (Næs *et al.*, 2001). In practice, the directions with the smallest variability in $\mathbf{X}$ will have a tendency of being less stable than the rest (due to noise) and therefore collinearity will give unstable regression coefficients and predictions. For cases where $I < J$ the inverse of $(\mathbf{X}^t\mathbf{X})$ does not even exist. This is one of the reasons why component-based methods are very popular in data analysis.

Moore–Penrose inverse

The Moore–Penrose (MP) or pseudo-inverse of $\mathbf{X}$ $(I \times J)$, denoted as $\mathbf{X}^+(I \times J)$, is a generalisation of the 'regular' inverse with the following properties:

1. $\mathbf{XX}^+\mathbf{X} = \mathbf{X}$
2. $\mathbf{X}^+\mathbf{XX}^+ = \mathbf{X}^+$
3. $(\mathbf{XX}^+)^t = \mathbf{XX}^+$
4. $(\mathbf{X}^+\mathbf{X})^t = \mathbf{X}^+\mathbf{X}$

which are called the MP conditions (Schott, 1997). The MP inverse is unique. A specific expression for the MP inverse is $\mathbf{X}^+ = \mathbf{VD}^{-1}\mathbf{U}^t$ where $\mathbf{UDV}^t$ is the SVD of $\mathbf{X}$. The matrix

8 We thank Marieke Timmerman for pointing this out.

$\mathbf{D}^{-1}$ contains the reciprocal values of the non-zero singular values and the (original) zero singular values remain zero. The MP inverse allows for many compact expressions (e.g., for the OLS regression coefficients $\widehat{\mathbf{b}} = \mathbf{X}^{+}\mathbf{y}$). The generalised inverse (notation $\mathbf{X}^{-}$) only satisfies the first MP condition (Schott, 1997).

Regression coefficients in NIPALS based algorithms

There are simple expressions for the weight matrix, **R**, transforming **X**-data into scores, **T**, and the regression coefficients, **B**, in PLS:

$$\begin{aligned} \mathbf{R} &= \mathbf{W}(\mathbf{P}^t\mathbf{W})^{-1} \\ \mathbf{B} &= \mathbf{W}(\mathbf{P}^t\mathbf{W})^{-1}\mathbf{Q}^t = \mathbf{R}\mathbf{Q}^t. \end{aligned} \tag{2.97}$$

These expressions, also found in Equation 2.21, were originally developed for weights, **W**, based on **X** and **Y**, obtained through the NIPALS algorithm. In Chapter 7 both SO-PLS and ROSA are examples of methods that make use of constrained **W**s in the NIPALS framework. It is straightforward to show that the expressions in Equation 2.97 hold for constrained **W**s and in fact for any arbitrarily chosen non-trivial weights. We will show this in two steps starting with the weights, **R**, and expanding further to regression coefficients. We assume centred **X** and **Y**, scores and loadings obtained using the NIPALS algorithm and an arbitrary weight matrix, **W**. Here, **W** is kept fixed and scores and loadings follow from NIPALS as $\mathbf{t}_{r+1} = \mathbf{X}_r\mathbf{w}_{r+1}$ and $\mathbf{p}_{r+1} = \mathbf{X}_r^t\mathbf{t}_{r+1}/(\mathbf{t}_{r+1}^t\mathbf{t}_{r+1})$.

Scores from weights, R

$$\begin{aligned} \mathbf{XR} &= \mathbf{XW}(\mathbf{P}^t\mathbf{W})^{-1} && \text{– transform } \mathbf{X} \text{ using } \mathbf{R} \\ &= (\mathbf{TP}^t + \mathbf{E})\mathbf{W}(\mathbf{P}^t\mathbf{W})^{-1} && \\ &= \mathbf{T}\underbrace{\mathbf{P}^t\mathbf{W}(\mathbf{P}^t\mathbf{W})^{-1}}_{=\mathbf{I}} && \text{– collapse to identity} \\ &\quad + \underbrace{\mathbf{EW}}_{=0}(\mathbf{P}^t\mathbf{W})^{-1} && \text{– } \mathbf{E} \text{ orthogonal to } \mathbf{W} \text{ due to deflation}^{\star} \\ &= \mathbf{T} && \text{- } \blacksquare \end{aligned}$$

The equations above show that the properties $\mathbf{X} = \mathbf{TP}^t + \mathbf{E}$ and **E** orthogonal to **W** (see below*), both ensured by using the NIPALS algorithm, guarantee that any choice of non-trivial **W** results in weight matrices **R** and scores **T** related to **X** through $\mathbf{XR} = \mathbf{T}$. The deflation process also leads to orthogonal columns in **T**. In other words, restricted weights, **W**, obtained by multiblock methods like SO-PLS or ROSA (Chapter 7), still lead to valid **R** and **T** with desirable properties.

Regression coefficients

$$\begin{aligned} \mathbf{TB_T} &= \widehat{\mathbf{Y}} && \text{– regression } \mathbf{Y} \sim \mathbf{T} \\ \mathbf{B_T} &= (\mathbf{T}^t\mathbf{T})^{-1}\mathbf{T}^t\mathbf{Y} = \mathbf{Q}^t && \text{– regression coefficients for } \mathbf{T} \\ \mathbf{XB} &= \widehat{\mathbf{Y}} && \text{– regression } \mathbf{Y} \sim \mathbf{X} \\ \mathbf{B} &= \mathbf{RB_T} && \text{- transform } \mathbf{X} \text{ using } \mathbf{R}\text{, then apply } \mathbf{B_T} \\ &= \mathbf{W}(\mathbf{P}^t\mathbf{W})^{-1}\mathbf{Q}^t && \text{- } \blacksquare \end{aligned}$$

The properties ensured by the NIPALS algorithm are exploited again in the equations above to obtain regression coefficients, **B**, defined the same way as for PLS. This means that the same formula for regression coefficient used in PLS can be applied in situations with arbitrary weight matrices, **W**, or simply by restricted versions of **W** like in SO-PLS or ROSA (Chapter 7).

$$\begin{aligned}
\star : \mathbf{E}\mathbf{w}_1 &= (\mathbf{X} - \mathbf{t}_1\mathbf{p}_1^t)\mathbf{w}_1 = \mathbf{X}\mathbf{w}_1 - \mathbf{t}_1\mathbf{p}_1^t\mathbf{w}_1 \\
&= \mathbf{X}\mathbf{w}_1 - \underbrace{\overbrace{\mathbf{X}\mathbf{w}_1}^{\mathbf{t}_1}\overbrace{(\mathbf{w}_1^t\mathbf{X}^t\mathbf{X}\mathbf{w}_1)^{-1}\mathbf{w}_1^t\mathbf{X}^t\mathbf{X}}^{\mathbf{p}_1^t}\mathbf{w}_1}_{=\mathbf{I}} \\
&= 0
\end{aligned} \tag{2.98}$$

This holds for any $\mathbf{w}_r$, meaning that **W** is orthogonal to **E** in NIPALS based algorithms.

Eigenvalue equations for PLS, RDA and CCA

The proofs of the eigenvalue equations in PLS, RDA and CCA (see Section 2.4.7) follow by carefully going through the iterative algorithm for those methods omitting the normalisation steps. For PLS, this becomes:

$$\mathbf{w} \propto \mathbf{X}^t\mathbf{u} = \mathbf{X}^t\mathbf{Y}\mathbf{c} \propto \mathbf{X}^t\mathbf{Y}\mathbf{Y}^t\mathbf{t} = \mathbf{X}^t\mathbf{Y}\mathbf{Y}^t\mathbf{X}\mathbf{w} \tag{2.99}$$

where the symbol $\propto$ means 'proportional to'. This gives the eigenequation

$$\mathbf{X}^t\mathbf{Y}\mathbf{Y}^t\mathbf{X}\mathbf{w} = \lambda\mathbf{w} \tag{2.100}$$

and using the SVDs of **X** and **Y** the matrix in Equation 2.100 becomes $\mathbf{Q}_1\mathbf{S}_1\mathbf{P}_1^t\mathbf{P}_2\mathbf{S}_2^2\mathbf{P}_2^t\mathbf{P}_1\mathbf{Q}_1^t$. Similar derivations can be made for RDA and CCA leading to the matrices reported in Table 2.3. The derivation for the eigenequation for **c** in RDA follows from

$$\mathbf{c} \propto \mathbf{Y}^t\mathbf{t} = \mathbf{Y}^t\mathbf{X}\mathbf{w} = \mathbf{Y}^t\mathbf{X}(\mathbf{X}^t\mathbf{X})^{-1}\mathbf{X}^t\mathbf{Y}\mathbf{c} \tag{2.101}$$

which gives the eigenequation

$$\mathbf{Y}^t\mathbf{X}(\mathbf{X}^t\mathbf{X})^{-1}\mathbf{X}^t\mathbf{Y}\mathbf{c} = \lambda\mathbf{c} \tag{2.102}$$

as shown in the main text. Note that a quadratic form can be maximised using the eigenequations. The problem

$$\max_{\|\mathbf{v}\|=1} \mathbf{v}^t\mathbf{A}\mathbf{v} \tag{2.103}$$

for a symmetric matrix **A** can be solved by taking the eigenvector associated with the largest eigenvalue of **A**.

3

Structure of Multiblock Data

3.i General Introduction

In the first chapter, we gave the background of the types of data and models we are going to describe. In the second chapter, we presented some basic theory to be used in the rest of this book. In the current chapter, we will describe the general structures of multiblock data sets. It is difficult to give a true taxonomy because a systematic classification of all the different multiblock methods is impossible to give. Indeed, not even the vocabulary is settled completely as explained in Chapter 1. Nevertheless, we will try to structure the methods according to some characteristics. In this chapter we will explain this taxonomy, and we will use this throughout the book.

3.1 Taxonomy

We will use the terminology from graph theory. In that context, a multiblock data set can be regarded as a graph where the data block is a node and a relation between two blocks is an edge. Subsequently, we can distinguish three levels of detail in such a graphical representation:

Skeleton of the graph: The entities and type of information associated with the modes of the blocks in the data set. This defines the possible connections between blocks, and this set of possible connections is the skeleton of the data set. This skeleton only describes possible connections and does not consider any model. It is visualised by an undirected graph.

Topology of the graph: Given the structure of the skeleton, which of the possible connections should be included in a model and is there an unsupervised or supervised relationship in the connections?

Linking structure of the graph: Given the topology of the graph, what is the proposed shape (i.e., the function) of the sought relationships between the blocks?

We will discuss and illustrate these characteristics in the next sections.

3.2 Skeleton of a Multiblock Data Set

As explained in Section 2.2.1, data blocks have ways and modes. In almost all methods dealt with in this book, the number of modes equals the number of ways. We will thus explain the skeleton in terms of modes.

Multiblock Data Fusion in Statistics and Machine Learning: Applications in the Natural and Life Sciences,
First Edition. Age K. Smilde, Tormod Næs, and Kristian Hovde Liland.

3.2.1 Shared Sample Mode

Shared sample mode means that a set of samples occupy the same rows in two or more blocks. Stated differently, the same samples are measured for a different set of variables. An example of three data blocks sharing the sample mode is given in Figure 3.1. In this figure, the data blocks $\mathbf{X}_1(I \times J_1)$, $\mathbf{X}_2(I \times J_2)$, and $\mathbf{X}_3(I \times J_3)$ share the sample mode for the three sets of variables. Hence, it is worthwhile to consider analysing this shared mode using all variables simultaneously thereby taking into account also the correlation between the sets of variables. There can also be more complicated structures, such as in Figure 3.2 which shows the skeleton for four blocks. Some of the connections may be removed in the data analysis (see next section).

3.2.2 Shared Variable Mode

Shared variable mode means that a set of variables occupy the same columns across two or more blocks. Stated differently, the same set of variables is measured in different groups of samples. An example of three data blocks sharing the variable mode is given in Figure 3.3. In this figure, the data blocks $\mathbf{X}_1(I_1 \times J)$, $\mathbf{X}_2(I_2 \times J)$, and $\mathbf{X}_3(I_3 \times J)$ share the variable mode for the three sets of samples.

3.2.3 Shared Variable or Sample Mode

Structures of more complex arrangements are also starting to emerge, e.g., with blocks sharing either samples, variables or both. Examples of such structures are the L-shape of Figure 3.4 and the U-shape of Figure 3.5(a).

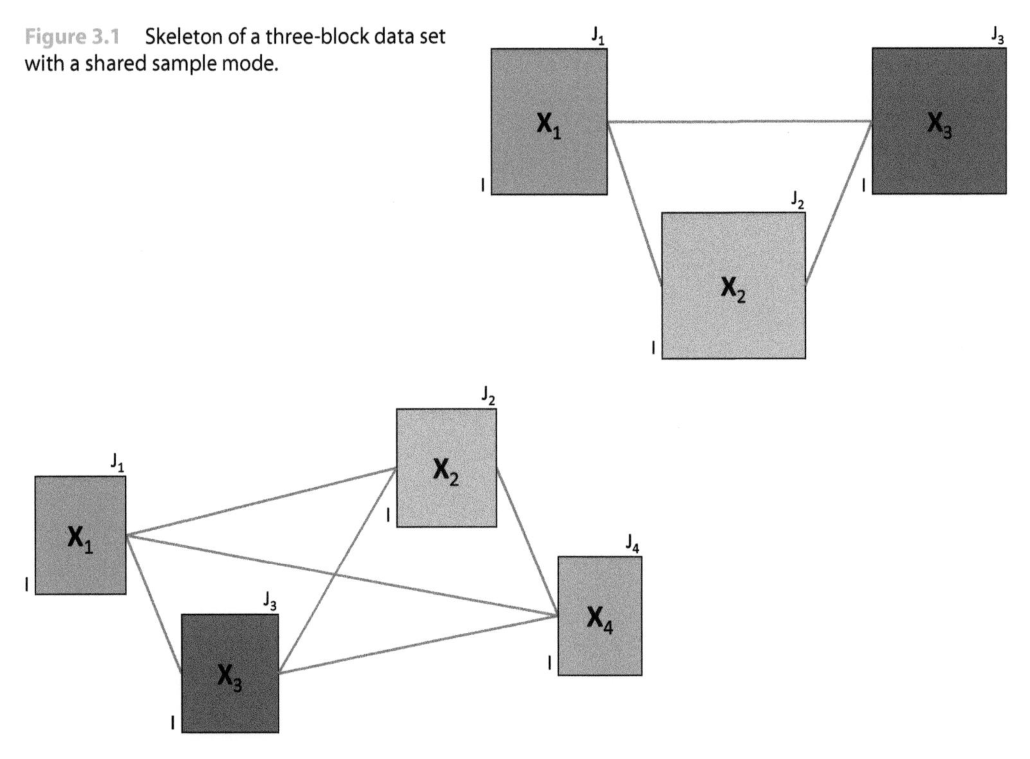

Figure 3.1 Skeleton of a three-block data set with a shared sample mode.

Figure 3.2 Skeleton of a four-block data set with a shared sample mode.

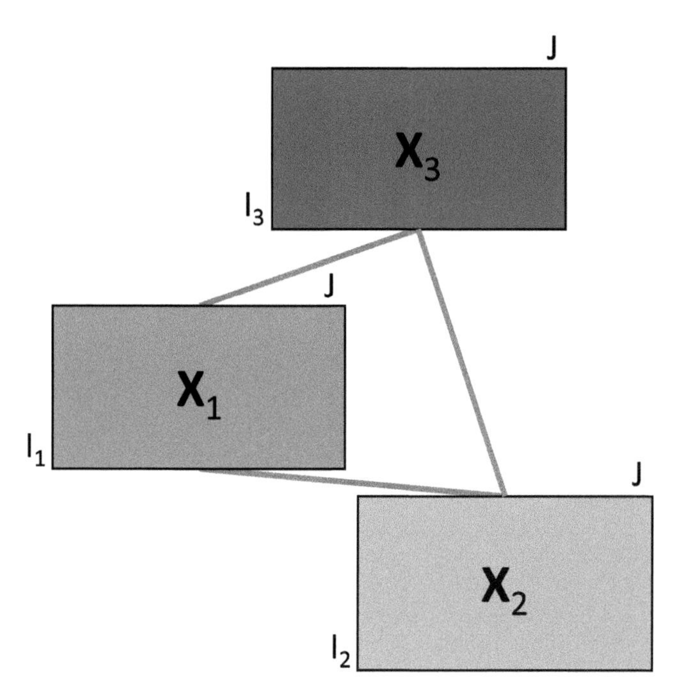

Figure 3.3 Skeleton of a three-block data set with a shared variable mode.

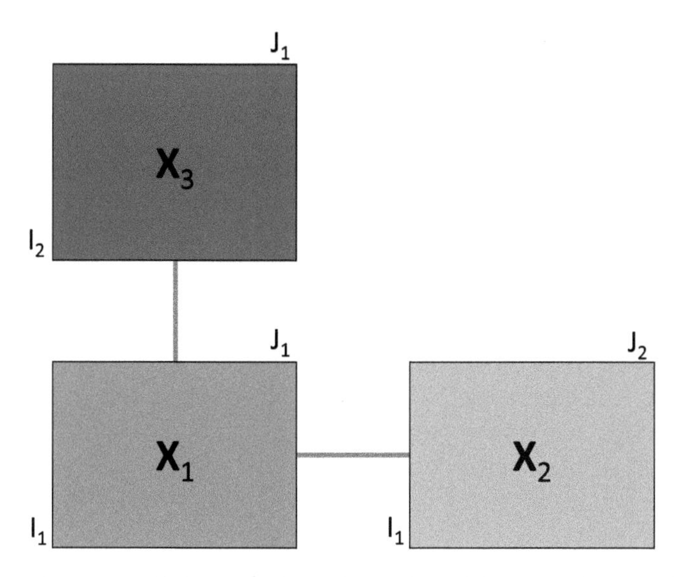

Figure 3.4 Skeleton of a three-block L-shaped data set with a shared variable or a shared sample mode.

3.2.4 Shared Variable and Sample Mode

There are also structures with both shared variable and sample modes. The skeleton of such a structure is shown in Figure 3.5(b) where for simplicity some obvious connections have been deleted. There can also be complicated structures with some blocks sharing samples and other blocks sharing variables, and some blocks sharing both.

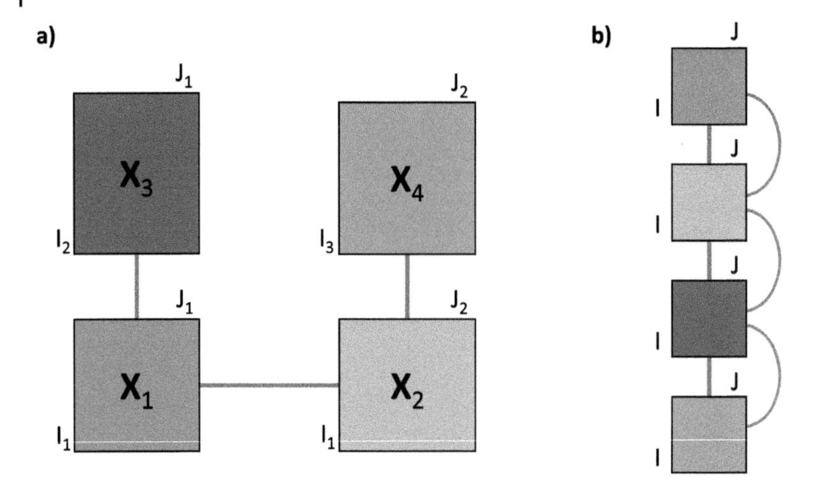

Figure 3.5 Skeleton of a four-block U-shaped data set with a shared variable or a shared sample mode (a) and a four-block skeleton with a shared variable and a shared sample mode (b). This is a simplified version; it should be understood that all sample modes are shared as well as all variable modes.

3.3 Topology of a Multiblock Data Set

Once the skeleton has been established, a choice has to be made regarding which blocks should be analysed simultaneously. This means that some links may be removed based on domain knowledge and the goal of the data analysis. After that, there is a basic choice to make for each link: is it unsupervised or supervised? Also this choice depends on domain knowledge and the goal of the analysis.

3.3.1 Unsupervised Analysis

In unsupervised analysis all blocks are treated on the same footing or stated differently, the blocks are exchangeable. There are different possibilities depending on the skeleton.

SHARED SAMPLE MODE

For the three-block situation this looks like Figure 3.6. The two-sided arrows in Figure 3.6 indicate an unsupervised analysis. For data sets with many blocks, the full topology

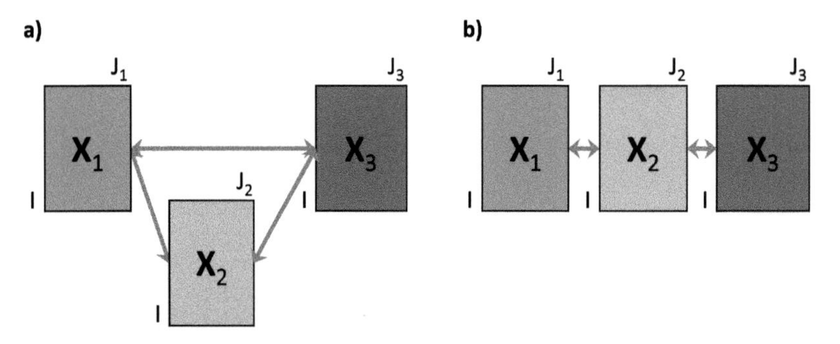

Figure 3.6 Topology of a three-block data set with a shared sample mode and unsupervised analysis: (a) full topology and (b) simplified representation.

(Figure 3.6(a)) can become a bit crowded and thus a more simplified representation can be given (Figure 3.6(b)) but notice that this simplified representation obscures some connections.

There are many examples of this type of multiblock topology. In genomics, it is becoming increasingly common to measure multiple omics data on the same set of samples (Iorio *et al.*, 2016; Balcke *et al.*, 2017; Gomez-Cabrero *et al.*, 2019). In chemistry, measurements may be available from different instruments on the same set of samples, e.g., see Afseth *et al.* (2005).

This type of multiblock data topology and the corresponding data analysis has different names in different fields. These can be called N-type integration (Rohart *et al.*, 2017b), horizontal integration (Ulfenborg, 2019), multiblock analysis (Westerhuis *et al.*, 1998) or multi-view data analysis (Gaynanova and Li, 2019).

Shared Variable Mode

There are also unsupervised topologies with a shared variable mode, see Figure 3.7 where we used the simplified representation (but all variable modes are connected). In this case, for three groups of samples the same set of variables are measured. Examples of these types of data structures are also abundant. In genomics, the same set of genes may be measured on different but similar biological systems, or of the same system under different conditions. In sensory science, this may encompass the same sensory attributes scored by the same panel on different sets of products. In chemistry, an example may be of the same spectroscopic measurements collected for different sets of samples.

Also in this case, there are several names for these types of structures and the corresponding data analysis, such as vertical integration (Ulfenborg, 2019), P-type integration (Rohart *et al.*, 2017b) or multi-group analysis (Tenenhaus and Tenenhaus, 2014).

Shared sample or variable mode

There are many other possible mode-sharing unsupervised topologies. One of these is the unsupervised L-shape structure (see Figure 3.4 (Martens *et al.*, 2005)). A primary set of data ($\mathbf{X}_1$) is connected with another set of variables measured on the same samples ($\mathbf{X}_2$) but there is also a data set available measured on the primary set of variables but in different samples ($\mathbf{X}_3$). In genomics, $\mathbf{X}_1$ may be a set of genes measured on a biological systems, for which also the proteome is measured on the same samples ($\mathbf{X}_2$). The relationships between the genes is also present in a data set ($\mathbf{X}_3$) of a very similar origin. The latter may help to focus the analysis on functionally related sets of genes in $\mathbf{X}_1$. In consumer science, it is important to find relationships between descriptive sensory data of a number of samples and the consumer liking of the same samples (see Example 1.4). In addition, it is important to understand how this pattern relates to consumer characteristics such as for instance gender and age. In such cases it is natural to denote the descriptive data by $\mathbf{X}_2$, the consumer liking data by $\mathbf{X}_1$ and consumer characteristics by $\mathbf{X}_3$.

Also the U-shape skeleton can have an unsupervised topology (see Figure 3.5 (Martens *et al.*, 2005)). Examples can be found in genomics and sensory data. Obviously, there are many more possible arrangements of data sets.

Shared variable and sample mode

A special type of arrangement follows when two of the modes are shared for all data blocks, as shown in Figure 3.8(a) where the same set of variables J and K are shared among all samples in the data sets. Because of this special structure, the data sets can also be arranged in a

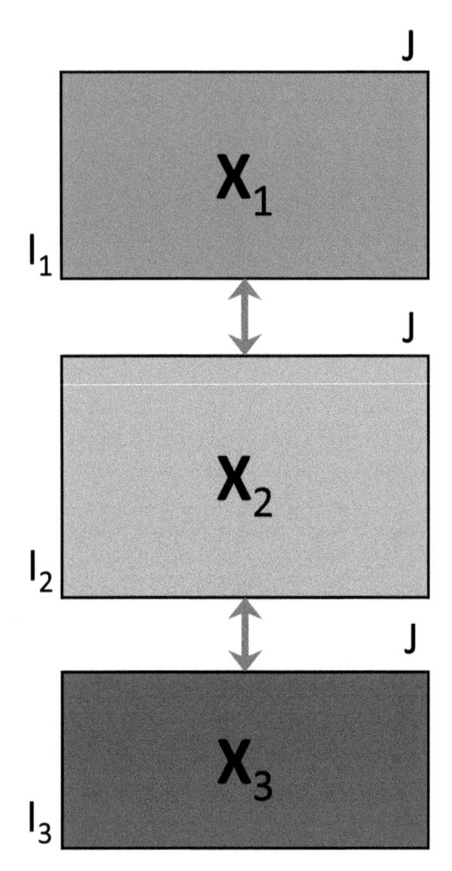

Figure 3.7 Topology of a three-block data set with a shared variable mode and unsupervised analysis.

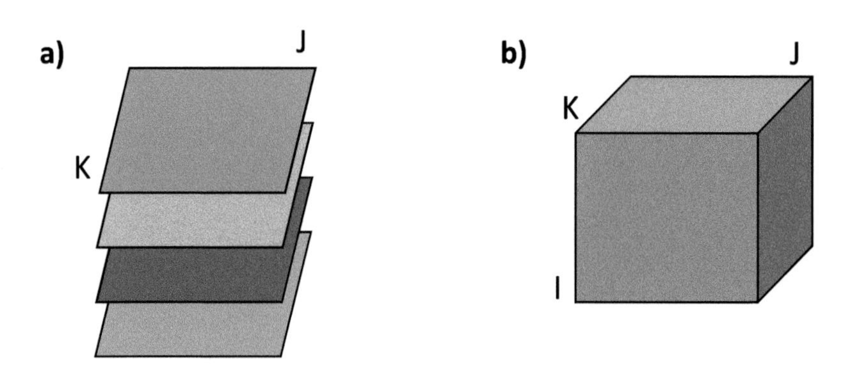

Figure 3.8 Different arrangements of data sharing two modes. Topology (a) and multiway array (b).

three-way array, as shown in Figure 3.8(b). Such data structures can be analysed with multiway methods. Monographs exist for those methods (Smilde *et al.*, 2004; Kroonenberg, 2008), and we will thus only briefly discuss those methods (see Chapter 9). Another set of methods for such structures is generalised Procrustes analysis (GPA) which we will also briefly discuss (see Section 9.3.1).

Obviously, it is also possible to combine multiway arrays and two-way arrays, as shown in Figure 3.9. Applications of multiblock data analysis with this structure start to emerge (Acar *et al.*, 2014, 2015; Biancolillo *et al.*, 2017). Unsupervised methods for all these different cases will be discussed in Chapters 5 and 9.

3.3.2 Supervised Analysis

In supervised analysis, the skeleton of the multiblock data set contains one-headed arrows indicating directionality. Hence, the blocks are no longer exchangeable. Such an imposed structure depends on the domain and the goal of the analysis. An example of a three-block set sharing the sample mode is shown in Figure 3.10. Supervised analysis is done exclusively for data sharing the sample mode.

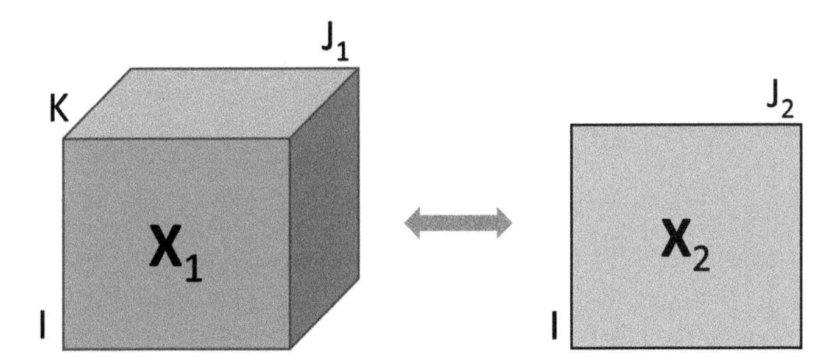

Figure 3.9 Unsupervised combination of a three-way and two-way array.

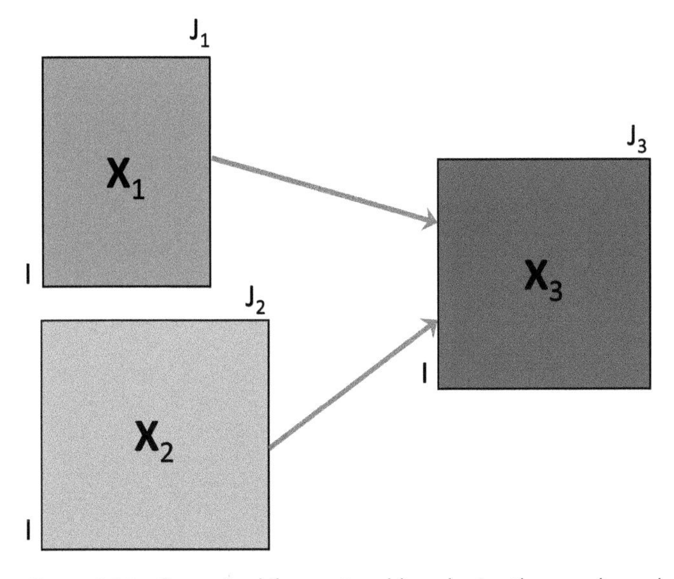

Figure 3.10 Supervised three-set problem sharing the sample mode.

The arrows can have different meanings. They can refer to *predictive* relationships or to *causal* relationships. In the case of predictive relationships there is a wish to predict blocks of variables using the other blocks. This is mostly done using regression types of approaches. In genomics, this may be proteomics data ($\mathbf{X}_1$) and gene-expression ($\mathbf{X}_2$) predicting drug response of cancer cell-lines ($\mathbf{X}_3$) (Aben *et al.*, 2016). In sensory science, an important issue is to understand how sensory attributes collected by a trained panel relate to various types of chemical measurements, typically measured by different instruments. In such cases, the sensory attributes will be the dependent variables in $\mathbf{X}_3$ and the different chemical instruments will be represented by the input blocks $\mathbf{X}_1$ and $\mathbf{X}_2$. It is also possible to have mixtures of unsupervised and supervised, see Figure 3.11. We will treat such structures as supervised ones.

The arrows can also indicate a proposed causal relationship. Such a representation goes back a long time (Wright, 1918, 1934) and is used a lot in path modelling or structural equation modelling (SEM). An example is given in Figure 3.12. The focus of these models is finding parameters describing the causal relationships. Of course, it is also possible to predict

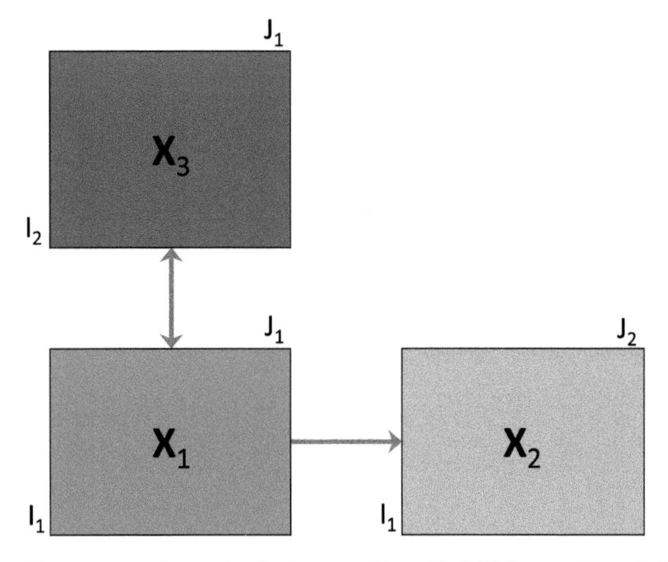

Figure 3.11 Supervised L-shape problem. Block $\mathbf{X}_1$ is a predictor for block $\mathbf{X}_2$ and extra information regarding the variables in block $\mathbf{X}_1$ is available in block $\mathbf{X}_3$.

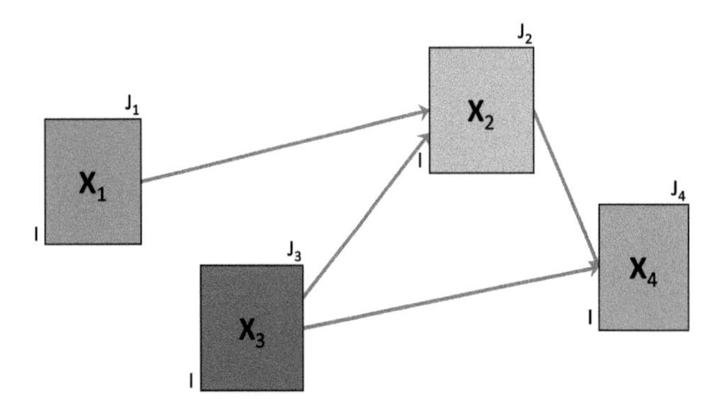

Figure 3.12 Path model structure. Blocks are connected through shared samples and a causal structure is assumed.

with such models and even predict the effect of interventions (Pearl, 2009). In this book, we will briefly touch upon this type of modelling using SO-PLS-PM in Section 10.5.4.

3.4 Linking Structures

Once the topology of the multiblock data set has been established, then the final ingredient is the choice of the linking structure. We have to make a distinction between unsupervised and supervised analysis and – for unsupervised analysis – between shared samples or variables. The linking structure for unsupervised analysis is based on components for all blocks; for supervised analysis the linking structure can be more involved. We will discuss these in the following sections.

3.4.1 Linking Structure for Unsupervised Analysis

As stated above, the linking structure of unsupervised analysis is based on components for each block. We will explain the general idea for two blocks sharing the sample mode. The starting point is as shown in Figure 3.13. The weights $\mathbf{W}_1$ $(J_1 \times R_1)$ and $\mathbf{W}_2$ $(J_2 \times R_2)$ are used to define scores $\mathbf{T}_1$ $(I \times R_1)$ and $\mathbf{T}_2$ $(I \times R_2)$ and these are sometimes used to obtain loadings $\mathbf{P}_1$ $(J_1 \times R_1)$ and $\mathbf{P}_2$ $(J_2 \times R_2)$. Note that R_1 is not necessarily the same as R_2. A new ingredient in the figure is the character L which is the abbreviation of linking structure[1]. The basic idea is that the scores $\mathbf{T}_1$, $\mathbf{T}_2$ and the loadings $\mathbf{P}_1$, $\mathbf{P}_2$ give a good approximation of $\mathbf{X}_1$ and $\mathbf{X}_2$ (see Chapter 5 for a formal treatment of such an approximation) and the linking function models the relationship between the blocks $\mathbf{X}_1$ and $\mathbf{X}_2$ through the scores $\mathbf{T}_1$ and $\mathbf{T}_2$. There are choices to make regarding the linking structures (Van Mechelen and Smilde, 2010).

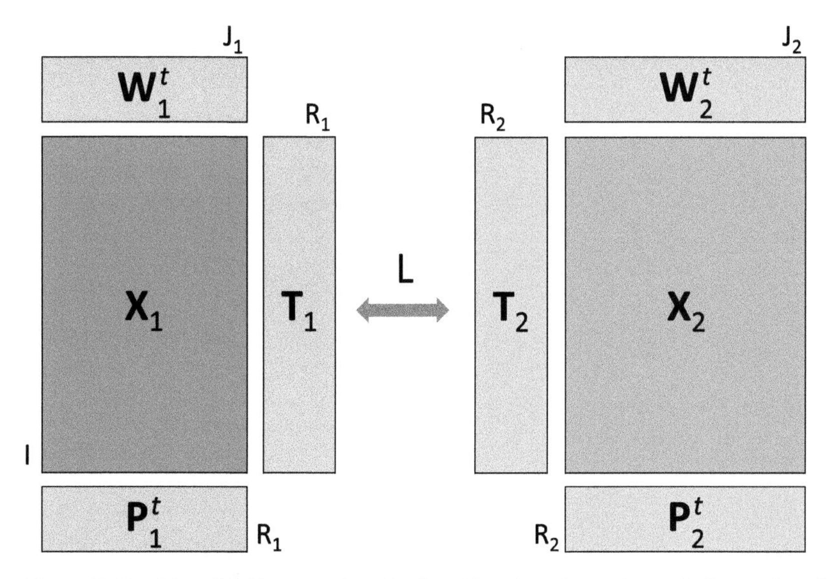

Figure 3.13 Idea of linking two data blocks with a shared sample mode. For explanation, see text.

1 Not to be confused with a link function in generalised linear models.

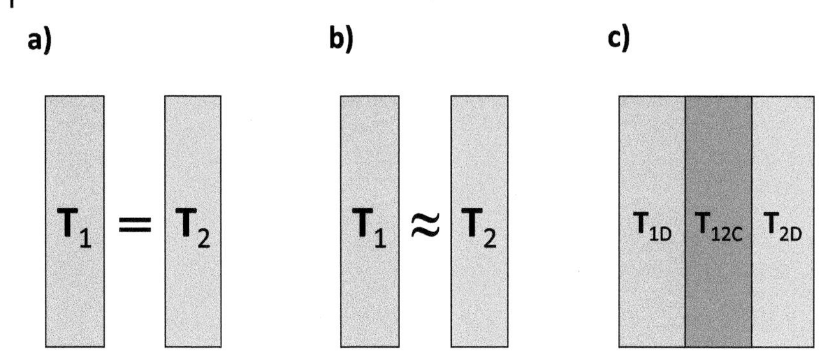

Figure 3.14 Different linking structures: (a) identity link, (b) flexible link, (c) partial identity link: common ($\mathbf{T}_{12C}$) and distinct ($\mathbf{T}_{1D}$, $\mathbf{T}_{2D}$) components.

The most used linking structures for unsupervised analysis are shown in Figure 3.14. The first one (Figure 3.14(a)) is the identity link in which $\mathbf{T}_1 = \mathbf{T}_2$ and thus, necessarily, $R_1 = R_2$. Simultaneous component analysis (SCA) (Timmerman and Kiers, 2003) is the prototypical example as well as multiple factor analysis and STATIS (see Section 5.2.1.2, (Van Deun *et al.*, 2009)). A second type of link (Figure 3.14(b)) which is often used is the one maximising the correlation between corresponding columns in $\mathbf{T}_1$ and $\mathbf{T}_2$ (and thereby also assuming $R_1 = R_2$). This is the link used in canonical correlation which is an example of a flexible link (Farias *et al.*, 2015). The third type of link is used for distinguishing common and distinct components which has been explained in more detail in Chapter 2. However, many other types of links are possible (e.g., non-linear relationships, shifts in a time/sample mode to include dynamics) and can be chosen domain specific.

The two-block case can easily be generalised to the multiblock case. We simply have to choose components and the linking structures (which can be different between the blocks). The SCA-method already allows for multiple data blocks and uses an identity link for all of them. Generalised canonical correlation maximises correlation between all blocks and several versions of those exist depending on how this maximisation is performed (Tenenhaus *et al.*, 2017).

When the variable mode is shared, then a similar setup can be chosen for unsupervised multiblock data analysis; for the two-block case this is shown in Figure 3.15. The idea of this type of analysis is very much the same as in the case of a shared sample mode. Again, the ingredients are components and linking structures. In this case, we present only loadings since that is mostly used in practice. However, it is also possible to use weights per block and describe the linking in terms of those weights. Actually, the original development of SCA was for the variable-mode sharing case including a version linking the weights (SCA-W, (Ten Berge *et al.*, 1992; Timmerman and Kiers, 2003)).

3.4.2 Linking Structures for Supervised Analysis

For supervised analysis there are also different linking structures, and we will explain this using two predictor blocks $\mathbf{X}_1$ and $\mathbf{X}_2$ and one predicted block $\mathbf{Y}$, see Figure 3.16. Note that we have here changed notation from the generic $\mathbf{X}_3$ to $\mathbf{Y}$, which is most common in supervised regression analysis (see Chapter 7 and Chapter 10). These structures are always from the regression type, but different possibilities exist. The first distinguishing feature is whether

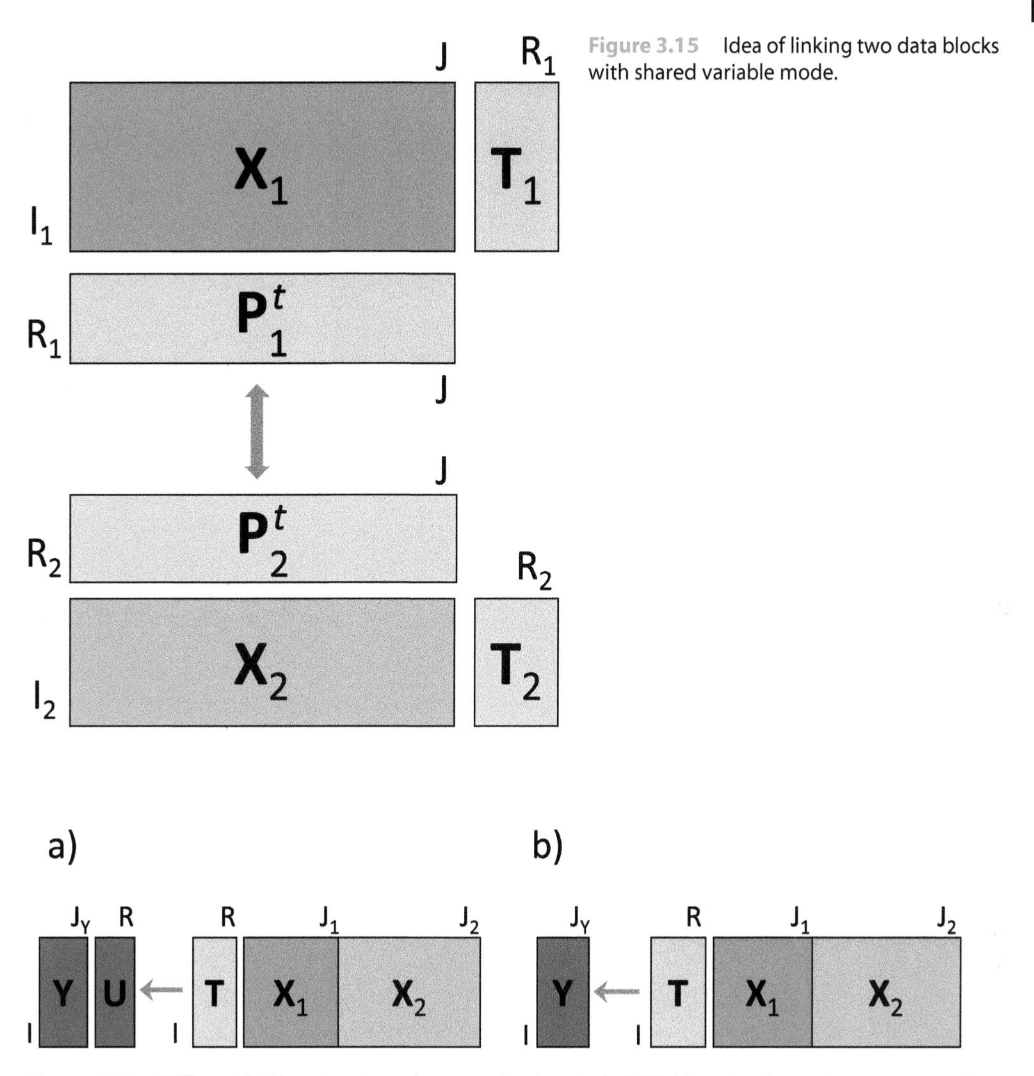

Figure 3.15 Idea of linking two data blocks with shared variable mode.

Figure 3.16 Different linking structures for supervised analysis: (a) linking structure where components are used both for the X-blocks and the Y-block; (b) linking structure that only uses components for the X-blocks.

components are used both in the X-blocks and the Y-block (3.16(a)) or only in the X-blocks (3.16(b)). The first type of linking is the one used for some PLS-type models and the latter type of linking is the one of all RDA-type models (see also Chapter 2).

The second distinguishing feature is whether in the X-block separations are made between common and distinct components as predictors for the Y-block, see Figure 3.17. For simplicity the components for the Y-block are not shown in Figure 3.17 but obviously those can be present depending on the linking structure as shown in Figure 3.16. There is a possibility to predict $\mathbf{Y}$ from $\mathbf{X}_1$ and $\mathbf{X}_2$ by using components from both X-blocks and thus use all types of information of $\mathbf{X}_1$ and $\mathbf{X}_2$ simultaneously to predict $\mathbf{Y}$ (see Figure 3.17(a)). This is the most simple and straightforward method (e.g., MB-PLS, see Section 7.2). However, this does not give much insight into exactly which type of information is used from both blocks, e.g., are only common parts used or also distinct parts?

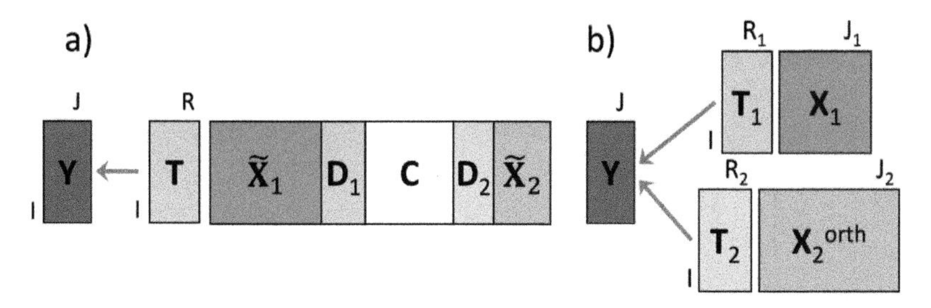

Figure 3.17 Treating common and distinct linking structures for supervised analysis: (a) Linking structure with no differentiation between common and distinct in the X-blocks (**C** is common, $\mathbf{D_1}$, $\mathbf{D_2}$ are distinct for $\mathbf{X}_1$ and $\mathbf{X}_2$, respectively; $\tilde{\mathbf{X}}_1$ and $\tilde{\mathbf{X}}_2$ represent the unsystematic parts of $\mathbf{X}_1$ and $\mathbf{X}_2$); (b) first $\mathbf{X}_1$ is used and then the remainder of $\mathbf{X}_2$ after removing common (predictive) part $\mathbf{T}_1$ of $\mathbf{X}_1$.

An alternative is to start with block $\mathbf{X}_1$ as predictor block; subsequently remove the common predictive part of $\mathbf{X}_1$ and $\mathbf{X}_2$ ($\mathbf{T}_1$) from $\mathbf{X}_2$ and continue using this reduced $\mathbf{X}_2$ as predictor block (see Figure 3.17(b)). This gives insight in the additional variation explained by $\mathbf{X}_2$ in $\mathbf{Y}$ on top of what $\mathbf{X}_1$ has already explained (e.g., SO-PLS, see Section 7.3). Of course, this method assumes an *a priori* hierarchy between the predictor blocks.

Yet another alternative is to use the concepts of common and distinct components (as explained in Chapter 2) and first separate common from distinct parts in the two predictor blocks $\mathbf{X}_1$ and $\mathbf{X}_2$ and subsequently predict $\mathbf{Y}$ from those common and distinct parts (e.g., PO-PLS, see Section 7.4). Obviously, there are many alternatives and this becomes even more rich for three or more blocks of data. Which linking structure to choose is an *a priori* decision of the data analyst preferably with domain-specific input and depends on the type of information requested from the data analysis.

3.5 Summary

In this chapter, we argue for a structured way to approach multiblock data analysis. For convenience, we repeat the ingredients.

Skeleton of the graph: This shows which modes the data blocks have in common. It visualises which blocks can be connected in which ways and gives the background for formulating research questions.

Topology of the graph: This shows which connections of the skeleton will be explored and how. Goals of the data analysis will guide the topology together with available domain knowledge.

Linking structure of the graph: This shows which linking structures are used for the selected topology. It gives a direction for which method to use to achieve the goals of the data analysis.

Apart from this, many choices have to be made regarding the specific data analysis method to use. These will be explained in the forthcoming chapters.

In this book we will mainly focus on simple topologies (link between sample modes or link between variable modes). We will also focus on the three linking structures as shown in Figures 3.14, 3.16 and 3.17 since these are used mostly in practice.

4

Matrix Correlations

4.i General Introduction

A much used metric in data analysis to express the relationship between variables is the correlation coefficient. Such a coefficient measures the linear relation between two variables (Pearson correlation) or its non-metric version which measures the relationship between the ranks of the variables (Spearman correlation). Such coefficients give a quick overview of the relationships between a set of variables, and it would be nice to also have such measures for relationships between data blocks. Such measures exist indeed and are called matrix correlations. The most common ones will be discussed in this chapter.

4.1 Definition

A correlation coefficient measures the linear relation between two sets of measurements collected in the vectors $\mathbf{x}_1$ and $\mathbf{x}_2$. This correlation coefficient $r(\mathbf{x}_1, \mathbf{x}_2)$ has certain properties such as $r(a\mathbf{x}_1, \mathbf{x}_2) = r(\mathbf{x}_1, \mathbf{x}_2) = r(\mathbf{x}_1, b\mathbf{x}_2)$, $r(\mathbf{x}_1, \mathbf{x}_2) = r(\mathbf{x}_2, \mathbf{x}_1)$, and $r(\mathbf{x}_1, \mathbf{x}_2) = 1$ if $\mathbf{x}_1 = b\mathbf{x}_2$ (for $a, b \neq 0$). Similar types of properties can also be formulated as requirements for correlations between matrices.

There is a variety of matrix correlation measures depending on which aspects of the relationship between the two data blocks to emphasise (Ramsay *et al.*, 1984). The most used matrix correlations are inner product correlation (Ramsay *et al.*, 1984), the GCD coefficient (Yanai, 1974), the RV-coefficient (Robert and Escoufier, 1976) and the SMI (Indahl *et al.*, 2018). These will be briefly discussed below and are available in the R-package `MatrixCorrelation` (Table 11.1). This exposition is inspired by the seminal paper of Ramsay *et al.* (1984). We will assume centred matrices throughout.

A matrix correlation is a function $r(\mathbf{X}_1, \mathbf{X}_2)$ that assigns a value between zero and one to a set of matrices $\mathbf{X}_1(I \times J)$ and $\mathbf{X}_2(I \times J)$ such that

$$\begin{aligned} & r(a\mathbf{X}_1, \mathbf{X}_2) = r(\mathbf{X}_1, b\mathbf{X}_2) = r(\mathbf{X}_1, \mathbf{X}_2) \\ & r(\mathbf{X}_1, \mathbf{X}_2) = r(\mathbf{X}_2, \mathbf{X}_1) \\ & r(\mathbf{X}_1, \mathbf{X}_2) = 1 \; \textit{if} \; \mathbf{X}_1 = b\mathbf{X}_2 \\ & r(\mathbf{X}_1, \mathbf{X}_2) = 0 \; \textit{iff} \; \mathbf{X}_1^t \mathbf{W} \mathbf{X}_2 = 0 \end{aligned} \tag{4.1}$$

where a, b are nonzero scalars, *iff* is the abbreviation of *if and only if* and $\mathbf{W}$ is a metric for the column-space (which can be taken as the identity $\mathbf{I}$ leading to $\mathbf{X}_1^t \mathbf{X}_2 = 0$). For the moment,

Multiblock Data Fusion in Statistics and Machine Learning: Applications in the Natural and Life Sciences,
First Edition. Age K. Smilde, Tormod Næs, and Kristian Hovde Liland.

we assume that both blocks have the same number of variables but some matrix correlations allow for different numbers of variables; this will be discussed in the following sections. The stated requirements (4.1) show that similarity can be defined in very many different ways. Rule three, e.g., can be changed by using an orthogonal matrix instead of a single scalar b thereby stating that two matrices that only differ in orientation are considered perfectly correlated.

The singular value decomposition is a crucial tool in data analysis and was explained in Section 2.8. A useful interpretation of the SVD for matrix correlation purposes is given in Elaboration 4.1.

ELABORATION 4.1

An interpretation of the SVD

The SVD can be understood in different ways; one way which is useful for understanding matrix correlations is as follows (and based on Ramsay *et al.* (1984); Smilde *et al.* (2015)). Assume that $\mathbf{X}$ is centred and scaled in such a way that $\mathbf{X}^t\mathbf{X} = \boldsymbol{\Sigma}$ is the correlation matrix apart from a constant. The SVD of $\mathbf{X} = \mathbf{UDV}^t$. Ramsay *et al.* (1984) provides the following interpretation:

- the matrix $\mathbf{U}$ is the basic component
- the matrix $\mathbf{D}$ is the scale component
- the matrix $\mathbf{V}$ is the orientation component

and this can be understood as follows. The matrix $\mathbf{U}$ contains the positioning of the samples irrespective of the correlation structure. This can be made more precise by considering two rows $\mathbf{u}_i^t$ and $\mathbf{u}_k^t$ of $\mathbf{U}$. Then it can be shown that the inner product of these rows $\mathbf{u}_i^t\mathbf{u}_k$ is $\mathbf{x}_i^t\boldsymbol{\Sigma}^{-1}\mathbf{x}_k$ (up to a constant) where $\mathbf{x}_i^t$ and $\mathbf{x}_k^t$ are rows of $\mathbf{X}$. Hence, $\mathbf{U}$ contains information about $\mathbf{X}$ in terms of Mahalanobis distances and thus is corrected for the variances and covariances of $\mathbf{X}$.

Figure 4.1(a) visualises what is happening in the case of two variables. This figure shows that if the singular values start to differ in size (matrix $\mathbf{D}$, orange part in panel (a)), the correlation between

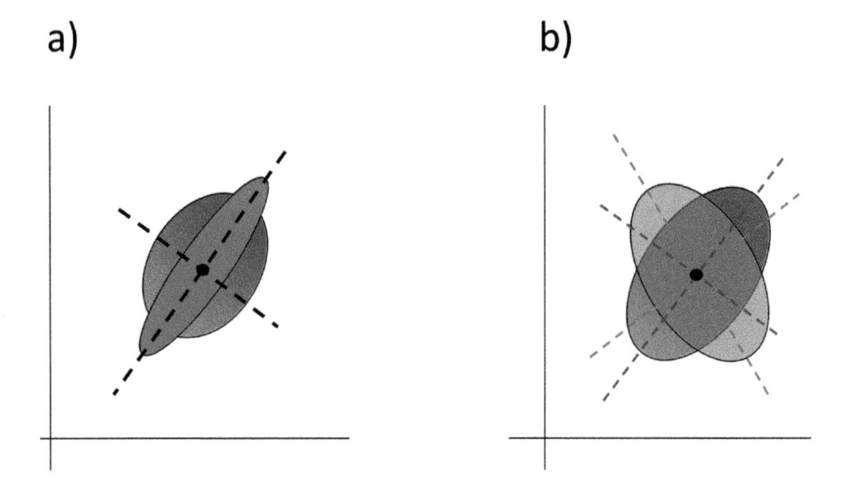

Figure 4.1 Explanation of the scale (a) and orientation (b) component of the SVD. The axes are two variables and the spread of the samples are visualised including their contours as ellipsoids. Hence, this is a representation of the row-spaces of the matrices. For more explanation, see text. Source: Smilde *et al.* (2015). Reproduced with permission of John Wiley and Sons.

the two variables become more pronounced but the orientation remains the same. The orientation is changed by using another matrix, **V**, as shown in panel (b) with different colours, but the sizes and thus the strength of the correlations remain the same (the ellipsoids do not change shape, only orientation). Summarising, the matrices **D** (size) and **V** (orientation) define the correlations between the variables and **U** defines the positioning of the samples corrected for these correlations.

4.2 Most Used Matrix Correlations

4.2.1 Inner Product Correlation

The most basic and simple matrix correlation is based on the inner product $\langle \mathbf{A}, \mathbf{B} \rangle = tr(\mathbf{A}^t\mathbf{B})$ on the space of matrices of equal size. Thus, the inner product matrix correlation R_{in} is:

$$R_{in}(\mathbf{X}_1, \mathbf{X}_2) = tr(\mathbf{X}_1^t\mathbf{X}_2)/\sqrt{tr(\mathbf{X}_1^t\mathbf{X}_1)tr(\mathbf{X}_2^t\mathbf{X}_2)} \tag{4.2}$$

where $\mathbf{X}_1\,(I \times J)$ and $\mathbf{X}_2\,(I \times J)$ are matrices that share the sample mode. Because of the Cauchy–Schwartz inequality the absolute value of R_{in} satisfies the rules of Requirements 4.1. As will be shown later, several other matrix correlations can be considered as special cases of R_{in}.

4.2.2 GCD coefficient

Yanai's general coefficient of determination (GCD, Yanai (1974)) is defined as follows:

$$\text{GCD} = \frac{1}{J} tr(\mathbf{X}_1(\mathbf{X}_1^t\mathbf{X}_1)^{-1}\mathbf{X}_1^t\mathbf{X}_2(\mathbf{X}_2^t\mathbf{X}_2)^{-1}\mathbf{X}_2^t) \tag{4.3}$$

where it is assumed that $\mathbf{X}_1$ and $\mathbf{X}_2$ are both of full column-rank and both have J columns and I rows. Under these conditions, Equation 4.3 can also be written as

$$\text{GCD} = \frac{\langle \mathbf{P}_{X_1}, \mathbf{P}_{X_2} \rangle}{||\mathbf{P}_{X_1}||\,||\mathbf{P}_{X_2}||} = \frac{1}{J}\langle \mathbf{P}_{X_1}, \mathbf{P}_{X_2} \rangle = \frac{1}{J} tr(\mathbf{P}_{X_1}, \mathbf{P}_{X_2}) \tag{4.4}$$

where $\mathbf{P}_{X_1}$ and $\mathbf{P}_{X_2}$ are the projection operators on the column-space of $\mathbf{X}_1$ and $\mathbf{X}_2$; $\langle \mathbf{A}, \mathbf{B} \rangle = tr(\mathbf{A}^t\mathbf{B})$ is again the inner product ($||.||$ is the default Frobenius norm). The denominator is J since the norm of a projection operator is the square root of the number of columns of the original matrices. Hence, the GCD is a special case of Equation 4.2 using the inner product of the projection operators $\mathbf{P}_{X_1}$ and $\mathbf{P}_{X_2}$.

It can be shown that the GCD is also the average of the squared canonical correlations between $\mathbf{X}_1$ and $\mathbf{X}_2$ which gives it a nice geometrical interpretation and provides a link to the GCA-method (see Chapter 2, Section 2.1.10). Hence, the GCD coefficient compares subspaces: the column-spaces of $\mathbf{X}_1$ and $\mathbf{X}_2$. For cases where $J > I$ and/or that one of the blocks is rank deficient, an approximation can be envisioned that first calculates PCA models of each block and then calculates the GCD among the principal components of those blocks. This is in the same spirit as the PCA-GCA method (to be discussed in Chapter 5.2.2.3). Note that the idea of first calculating PCA of each block and subsequently calculating the GCD of the resulting scores can also be used if the blocks have a different number of variables.

4.2.3 RV-coefficient

The RV-coefficient (Robert and Escoufier, 1976) is defined as follows:

$$\mathrm{RV} = \frac{tr(\mathbf{X}_2^t\mathbf{X}_1\mathbf{X}_1^t\mathbf{X}_2)}{\sqrt{tr[(\mathbf{X}_1^t\mathbf{X}_1)^2]tr[(\mathbf{X}_2^t\mathbf{X}_2)^2]}} \tag{4.5}$$

which can also be written as (Smilde *et al.*, 2009)

$$\mathrm{RV} = \frac{tr(\mathbf{X}_1\mathbf{X}_1^t\mathbf{X}_2\mathbf{X}_2^t)}{\sqrt{tr[(\mathbf{X}_1\mathbf{X}_1^t)^2]tr[(\mathbf{X}_2\mathbf{X}_2^t)^2]}} \tag{4.6}$$

by using cyclic permutations within the $tr(.)$ (see Section 2.8). This shows that the RV-coefficient can handle blocks with different numbers of variables since it only requires the cross-products $\mathbf{X}_m\mathbf{X}_m^t$ (also called the configuration matrices) which always have size $I \times I$. Hence, the RV coefficient compares configurations as induced by the matrices $\mathbf{X}_1$ and $\mathbf{X}_2$.

Another useful way of expressing the RV is

$$\mathrm{RV} = \frac{Vec(\mathbf{X}_1\mathbf{X}_1^t)^t Vec(\mathbf{X}_2\mathbf{X}_2^t)}{\sqrt{[Vec(\mathbf{X}_1\mathbf{X}_1^t)^t Vec(\mathbf{X}_1\mathbf{X}_1^t)][Vec(\mathbf{X}_2\mathbf{X}_2^t)^t Vec(\mathbf{X}_2\mathbf{X}_2^t)]}} \tag{4.7}$$

where $Vec(.)$ is the Vec-operator (see Section 2.8). Comparing this to the Pearson correlation or cosine between two (centred) vectors $\mathbf{x}_1$ and $\mathbf{x}_2$:

$$r(\mathbf{x}_1, \mathbf{x}_2) = \frac{\mathbf{x}_1^t\mathbf{x}_2}{\sqrt{(\mathbf{x}_1^t\mathbf{x}_1)(\mathbf{x}_2^t\mathbf{x}_2)}} \tag{4.8}$$

shows the interpretation of the RV as a correlation coefficient between $Vec(\mathbf{X}_1\mathbf{X}_1^t)$ and $Vec(\mathbf{X}_2\mathbf{X}_2^t)$.

For high-dimensional data (where the number of variables is much higher than the number of objects), the RV-coefficient breaks down and gives systematically too large values (Smilde *et al.*, 2009). The problem can be circumvented – based on theoretical arguments and supported by simulations – by ignoring the diagonal values of $\mathbf{X}_1\mathbf{X}_1^t$ and $\mathbf{X}_2\mathbf{X}_2^t$ thereby arriving at the modified RV-coefficient :

$$\mathrm{RV}_{\mathrm{mod}} = \frac{Vec(\widetilde{\mathbf{X}_1\mathbf{X}_1^t})^t\, Vec(\widetilde{\mathbf{X}_2\mathbf{X}_2^t})}{\sqrt{[Vec(\widetilde{\mathbf{X}_1\mathbf{X}_1^t})^t\, Vec(\widetilde{\mathbf{X}_1\mathbf{X}_1^t})][Vec(\widetilde{\mathbf{X}_2\mathbf{X}_2^t})^t\, Vec(\widetilde{\mathbf{X}_2\mathbf{X}_2^t})]}} \tag{4.9}$$

where the symbol '$(\widetilde{\mathbf{X}_1\mathbf{X}_1^t})$' indicates that the diagonals are set to zero for that matrix. This modified RV seems to work well in practice (Smilde *et al.*, 2009). There are more variants reported of the original RV coefficient (El Ghaziri and El Qannari, 2015; Mayer *et al.*, 2011), also trying to improve its properties in high dimensions.

4.2.4 SMI-coefficient

The similarity of matrices index (SMI) starts by computing PCA models of both $\mathbf{X}_1$ and $\mathbf{X}_2$:

$$\begin{aligned} \mathbf{X}_1 &= \mathbf{U}_1\mathbf{P}_1^t + \mathbf{E}_1 \\ \mathbf{X}_2 &= \mathbf{U}_2\mathbf{P}_2^t + \mathbf{E}_2 \end{aligned} \tag{4.10}$$

where $\mathbf{U}_1$ and $\mathbf{U}_2$ are both centred ($\mathbf{X}_1$ and $\mathbf{X}_2$ are assumed to be centred), $\mathbf{U}_1^t\mathbf{U}_1 = \mathbf{I}_{(R_1 \times R_1)}$ and $\mathbf{U}_2^t\mathbf{U}_2 = \mathbf{I}_{(R_2 \times R_2)}$ (Indahl *et al.*, 2018). These $\mathbf{U}_1$ and $\mathbf{U}_2$ are the basic components of $\mathbf{X}_1$

and $\mathbf{X}_2$, respectively (see Elaboration 4.1). Subsequently, a linear relationship between $\mathbf{U}_1$ and $\mathbf{U}_2$ is assumed:

$$\mathbf{U}_1 = \mathbf{U}_2\mathbf{B}_{U_1} + \mathbf{E}_{U_1} = \widehat{\mathbf{U}}_1 + \mathbf{E}_{U_1} \tag{4.11}$$
$$\mathbf{U}_2 = \mathbf{U}_2\mathbf{B}_{U_2} + \mathbf{E}_{U_2} = \widehat{\mathbf{U}}_2 + \mathbf{E}_{U_2}$$

where $\mathbf{B}_{U_1}$ and $\mathbf{B}_{U_2}$ are regression coefficients obtained according to the chosen regression method (OP or PR, see below). Now, the explained variances in both regressions of Equation 4.11 are, respectively, $||\widehat{\mathbf{U}}_1||^2/R_1$ and $||\widehat{\mathbf{U}}_2||^2/R_2$ since $||\mathbf{U}_1||^2 = R_1$ and $||\mathbf{U}_2||^2 = R_2$. Now the $SMI_{OP,PR}(\mathbf{U}_1, \mathbf{U}_2)$ is defined as:

$$\text{SMI}_{\text{OP,PR}}(\mathbf{U}_1, \mathbf{U}_2) = max(\frac{||\widehat{\mathbf{U}}_1||^2}{R_1}, \frac{||\widehat{\mathbf{U}}_2||^2}{R_2}) \tag{4.12}$$

which is a value between zero and one. In other words, $\text{SMI}_{\text{OP,PR}}(\mathbf{U}_1, \mathbf{U}_2)$ is the maximum of the explained variances of Equation 4.11.

There are different choices to make for the regression method OP, PR in Equation 4.11. The first one is simply an orthogonal projection (OP, which is also the least squares solution) leading to the simplification:

$$\text{SMI}_{\text{OP}}(\mathbf{U}_1, \mathbf{U}_2) = \frac{||\mathbf{S}||^2}{R} = \frac{1}{R}\sum_{r=1}^{R} s_r^2 \tag{4.13}$$

where $\mathbf{U}_1^t\mathbf{U}_2 = \mathbf{VSW}^t$ is the SVD of $\mathbf{U}_1^t\mathbf{U}_2$; $R = \min(R_1, R_2)$ and $s_1, \ldots, s_R$ are the singular values as collected in $\mathbf{S}$. These singular values are also identical to the canonical correlations obtained from the cross-correlation analysis of $\mathbf{U}_1$ and $\mathbf{U}_2$ showing the similarity between SMI_{OP} and the GCD coefficient (Indahl *et al.*, 2018).

An alternative choice to OP for the regression method is using Procrustes rotation (PR) and thus choosing the regression coefficient matrices to be of the form $\mathbf{B}_{U_1} = \mathbf{B}_{U_2}^t = g\mathbf{Q}$, where $\mathbf{Q} = \mathbf{VW}^t(R \times R)$ is the orthogonal matrix found by the SVD of $\mathbf{U}_1^t\mathbf{U}_2 = \mathbf{VSW}^t$ and $g = \bar{s}$ (mean of the singular values in $\mathbf{S}$) is the required scaling constant. For simplicity, we assume the same number of components in Equation 4.10 but this does not need to be the case. Now, the regression coefficient matrix $\mathbf{B}_{U_1} = g\mathbf{Q}$ represents a Procrustes transformation performing a scaled rotation. For this choice the corresponding SMI-measure simplifies to (Indahl *et al.*, 2018)

$$\text{SMI}_{\text{PR}}(\mathbf{U}_1, \mathbf{U}_2) = \max(\frac{||\mathbf{B}_{U_1}||^2}{R}, \frac{||\mathbf{B}_{U_2}||^2}{R}) = \frac{||\bar{s}\mathbf{Q}||^2}{R} = \bar{s}^2\frac{||\mathbf{Q}||^2}{R} = \bar{s}^2. \tag{4.14}$$

Hence, the SMI_{OP} gives the similarity between matrices in the least squares sense (i.e., compares column-spaces) whereas the SMI_{PR} restricts its focus to scaled rotations and preserves the configuration of the points. Note that these are different instances of formulating requirement (iii) in Requirements 4.1.

Which version of the SMI to use is a matter of choice based on prior knowledge regarding the context of the problem at hand. Note that the SMI operates on the orthogonal $\mathbf{U}_1$ and $\mathbf{U}_2$, hence, the singular values of $\mathbf{X}_1$ and $\mathbf{X}_2$ do not play a role which is similar to the GCD but different from the RV-coefficient (see also Ramsay *et al.* (1984)). This is important for sensory applications where the first principal component usually explains much more than the second, thus leading to a domination of RV by the (squared) singular value of the first dimension. However, we may be equally interested in the second and the first dimension for comparing panels or individuals. This is also a matter of choice governed by prior insights of the problem at hand. The original publication (Indahl *et al.*, 2018) gives a procedure for testing the significance of the SMI based on permutations. A similar procedure can be used for the other matrix correlations.

4.3 Generic Framework of Matrix Correlations

An inner product of the space of matrices **A** and **B** of equal size was defined as $\text{tr}(\mathbf{A}^t\mathbf{B})$ in Section 4.2.1 but it holds that $\text{tr}(\mathbf{A}^t\mathbf{B}) = \text{vec}(\mathbf{A})^t vec(\mathbf{B})$ and thus the latter also represents this inner product. Using these definitions, we can formulate one generic framework for the previously described matrix correlations.

This framework is based on the work of Ramsay *et al.* (1984) and Elaboration 4.1 and uses the singular value decompositions of $\mathbf{X}_1$ and $\mathbf{X}_2$:

$$\begin{aligned}\mathbf{X}_1 &= \mathbf{U}_1\mathbf{D}_1\mathbf{V}_1^t \\ \mathbf{X}_2 &= \mathbf{U}_2\mathbf{D}_2\mathbf{V}_2^t\end{aligned} \tag{4.15}$$

where $\mathbf{U}_1$, $\mathbf{U}_2$, $\mathbf{V}_1$ and $\mathbf{V}_2$ are orthogonal matrices; $\mathbf{D}_1$ and $\mathbf{D}_1$ are non-negative diagonal matrices containing the singular values of $\mathbf{X}_1$ and $\mathbf{X}_2$, respectively.

Using Equation 4.3, the SVDs of $\mathbf{X}_1$ and $\mathbf{X}_2$, and Equation 4.2, the GCD coefficient can also be written as:

$$\text{GCD} = R_{in}(\mathbf{U}_1\mathbf{U}_1^t, \mathbf{U}_2\mathbf{U}_2^t) \tag{4.16}$$

and thus only concerns the basic component; correlations within the matrices do not play a role. This is consistent with the idea of average canonical correlations (see Section 2.1.10). Non-singular transformations of the column-spaces of $\mathbf{X}_1$ and $\mathbf{X}_2$ do not affect the GCD.

Using Equation 4.6 and the SVDs of $\mathbf{X}_1$ and $\mathbf{X}_2$, the (uncorrected) RV can be written as:

$$\text{RV} = R_{in}(\mathbf{U}_1\mathbf{D}_1^2\mathbf{U}_1^t, \mathbf{U}_2\mathbf{D}_2^2\mathbf{U}_2^t) \tag{4.17}$$

and this concerns the basic component and the scale, but not the orientation of the correlations. Hence, the correlations only play a role regarding their sizes, but not their orientation. Orthogonal transformations of the matrices do not change their RV-coefficient. Note the similarity of the difference between the GCD and RV on the one hand and the GCA and SCA method on the other hand (see Section 5.2.1.3).

The inner product formulation of SMI_{OP} is similar to GCD but the matrices $\mathbf{U}_1$ and $\mathbf{U}_2$ are truncated to chosen numbers of components. When applying GCD, defined using R_{in}, on truncated **U** matrices with the same number of components in $\mathbf{U}_1$ and $\mathbf{U}_2$ then the GCD equals the SMI_{OP}. The case of unequal numbers of components is more involved, see Elaboration 4.2.

ELABORATION 4.2

SMI_{OP}* for the case of *$R_1 \neq R_2$

When using R_1 and R_2 components, respectively, in $\mathbf{U}_1$ and $\mathbf{U}_2$ in the GCD defined using R_{in} (see Equation 4.2), the denominator becomes $\sqrt{R_1 \cdot R_2}$, while the denominator in SMI_{OP} is $min(R_1, R_2)$. The latter choice is due to the formulation of SMI as the proportion of the smallest subspace (fewest components) explained by the larger subspace. Thus, if the number of components chosen for the two blocks is not equal, a correction is done as follows:

$$\text{SMI}_{\text{OP}} = R_{in}(\mathbf{U}_1\mathbf{U}_1^t, \mathbf{U}_2\mathbf{U}_2^t) \cdot \sqrt{R_1 \cdot R_2}/\min(R_1, R_2). \tag{4.18}$$

The effect of this correction is that the denominator of GCD, $\sqrt{R_1 \cdot R_2}$, is exchanged with the denominator of SMI_{OP}, $\min(R_1, R_2)$. For instance, if $R_1 = 4$ and $R_2 = 5$, GCD's denominator becomes $\sqrt{4 \cdot 5} \approx 4.47$, while the denominator of SMI_{OP} becomes $\min(4, 5) = 4$, i.e., $\sqrt{4 \cdot 5}/4 \approx 1.12$ times higher coefficient value for SMI_{OP} than GCD in this case. This shows how the choice of formulating SMI_{OP} as 'the proportion of the smallest subspace (fewest components) explained by the larger subspace' affects its numeric value.

All the mentioned coefficients (GCD, RV, RV_{mod} and SMI_{OP}) can also be expressed in terms of the vectorised form of the inner product. The two RV coefficients have already been stated this way, while GCD can be defined similarly as the RV following Equation 4.19:

$$\text{GCD} = \frac{\text{Vec}(\mathbf{U}_1\mathbf{U}_1^t)^t\text{Vec}(\mathbf{U}_2\mathbf{U}_2^t)}{\sqrt{[\text{Vec}(\mathbf{U}_1\mathbf{U}_1^t)^t\text{Vec}(\mathbf{U}_1\mathbf{U}_1^t)][\text{Vec}(\mathbf{U}_2\mathbf{U}_2^t)^t\text{Vec}(\mathbf{U}_2\mathbf{U}_2^t)]}}. \tag{4.19}$$

As explained in the previous section, SMI_{OP} with $R_1 = R_2 = R$ can be seen as GCD on a subset of the same left singular vectors. This means that the formulation similar as the RV coefficient of vectorised matrices also holds for SMI_{OP}.

4.4 Generalised Matrix Correlations

So far, matrix correlations have been defined for quantitative variables. There are also matrix correlations for variables of different scale-types which are useful for analysing heterogeneous multiblock data. In this section, we will describe two such generalisations.

4.4.1 Generalised RV-coefficient

The RV coefficient defined in Section 4.2.3 and its equivalent formulation in Equation 4.7 can alternatively be expressed as:

$$\text{RV} = \frac{\text{Vec}(\mathbf{S}_1)^t\text{Vec}(\mathbf{S}_2)}{\sqrt{[\text{Vec}(\mathbf{S}_1)^t\text{Vec}(\mathbf{S}_1)][\text{Vec}(\mathbf{S}_2)^t\text{Vec}(\mathbf{S}_2)]}} \tag{4.20}$$

where we defined $\mathbf{S}_1 = \mathbf{X}_1\mathbf{X}_1^t$ and $\mathbf{S}_2 = \mathbf{X}_2\mathbf{X}_2^t$. These are configuration matrices containing similarities between the rows (i.e., samples) of both blocks. This way of writing gives room for generalisations to heterogeneous data.

For binary data there are several similarity measure alternatives and one of those is the Jaccard similarity which is defined as follows. Consider two binary vectors $\mathbf{x}$ and $\mathbf{y}$. If a is the number of elements where both $\mathbf{x}$ and $\mathbf{y}$ are zero (for the same index); b is the number of elements where $\mathbf{x}$ is one and $\mathbf{y}$ is zero; c is the number of elements where $\mathbf{x}$ is zero and $\mathbf{y}$ is one and d is the number of elements where both $\mathbf{x}$ and $\mathbf{y}$ are one then

$$J(\mathbf{x}, \mathbf{y}) = \frac{d}{b + c + d} \tag{4.21}$$

and if $b+c+d$ is zero then $J(.,.)$ is defined as zero. Thus, this measure finds the relative amount of matches of ones in both vectors. Note that this a matter of choice; it is also possible to stress negative matches in an alternative formulation. The Jaccard similarities can be collected in a matrix $\mathbf{S}$ describing the similarities between two samples based on this Jaccard similarity.

Suppose now that we have two data sets, one containing quantitative data ($\mathbf{X}_1$) and one containing binary data ($\mathbf{X}_2$). The two sets of similarity measures $\mathbf{X}_1\mathbf{X}_1^t = \mathbf{S}_1$ and $\mathbf{S}_2$ based on the Jaccard similarity can be used in Equation 4.20 to give the matrix correlation of the quantitative block $\mathbf{X}_1$ and the binary block $\mathbf{X}_2$. This can also be used in the context of the modified RV. An application of these types of matrix correlations is shown later in Example 4.2. This generalised RV can also be used for other types of heterogeneous data (beyond quantitative and binary, e.g., for ordinal data) as long as proper similarity measures are used.

4.4.2 Generalised Association Coefficient

Another generalisation of the matrix correlation concept is based on a specific multiblock data analysis model for both data blocks which will be treated more extensively in Chapter 9. This model is called the generalised association study (GAS) (Li and Gaynanova, 2018) and for two blocks $\mathbf{X}_1(I \times J_1)$ and $\mathbf{X}_2(I \times J_2)$ this model is defined by:

$$
\begin{aligned}
\mathbf{X}_1 &= \mathbf{1}\boldsymbol{\mu}_1^t + \mathbf{T}_C\mathbf{P}_{1C}^t + \mathbf{T}_{1D}\mathbf{P}_{1D}^t + \mathbf{E}_1 = \boldsymbol{\Theta}_1 + \mathbf{E}_1 \\
\mathbf{X}_2 &= \mathbf{1}\boldsymbol{\mu}_2^t + \mathbf{T}_C\mathbf{P}_{2C}^t + \mathbf{T}_{2D}\mathbf{P}_{2D}^t + \mathbf{E}_2 = \boldsymbol{\Theta}_2 + \mathbf{E}_2
\end{aligned}
\tag{4.22}
$$

where the matrices $\mathbf{T}_C$ and $\mathbf{P}_{1C}, \mathbf{P}_{2C}$ are the scores and loadings of the consensus parts of the data blocks; $\mathbf{T}_{1D}, \mathbf{T}_{2D}$ and $\mathbf{P}_{1D}, \mathbf{P}_{2D}$ represent the distinct parts (scores and loadings; see also Chapter 2). The dimensionality of these matrices have to be assumed *a priori* and some identifiability requirements are needed to identify the model (see Section 9.2.2.6). The data sets $\mathbf{X}_1 = \boldsymbol{\Theta}_1 + \mathbf{E}_1$ and $\mathbf{X}_2 = \boldsymbol{\Theta}_2 + \mathbf{E}_2$ now have their own error distributions which may differ according to the type of data.

The generalised association coefficient (GAC) is now defined as

$$
r(\mathbf{X}_1, \mathbf{X}_2) = \frac{\|\overline{\boldsymbol{\Theta}}_1^t\overline{\boldsymbol{\Theta}}_2\|_*}{\|\overline{\boldsymbol{\Theta}}_1\|_F\|\overline{\boldsymbol{\Theta}}_2\|_F}
\tag{4.23}
$$

where the symbol $\|.\|_*$ indicates the nuclear norm (the sum of the singular values, see Section 2.8) and the overline indicates centred matrices (hence, subtracting the column averages $\boldsymbol{\mu}_1$ and $\boldsymbol{\mu}_2$ in Equation 4.22). This coefficient has certain properties:

1. $0 \leq r(\mathbf{X}_1, \mathbf{X}_2) \leq 1$
2. $r(\mathbf{X}_1, \mathbf{X}_2) = 0$ if and only if the column-spaces of $\overline{\boldsymbol{\Theta}}_1$ and $\overline{\boldsymbol{\Theta}}_2$ are mutually orthogonal
3. $r(\mathbf{X}_1, \mathbf{X}_2) = 1$ if $\overline{\boldsymbol{\Theta}}_1$ and $\overline{\boldsymbol{\Theta}}_2$ have the same left singular vectors and proportional singular values.

The first property specifies the range of the coefficient which is conveniently between zero and one. The second property describes when the coefficient is exactly zero and is intuitive since when the column-spaces of two matrices are orthogonal they have nothing in common. The third property (note that it provides only sufficiency) is less intuitive but will be explained in the following.

This GAC is related to the GAS model of Equation 4.22 in the following way. When the GAS model has the correctly specified ranks for the common and distinct parts and all identifiability requirements are fulfilled then:

1. $r(\mathbf{X}_1, \mathbf{X}_2) = 0$ if and only if $\mathbf{T}_C = 0$ and $\mathbf{T}_{1D}^t\mathbf{T}_{2D} = 0$
2. $r(\mathbf{X}_1, \mathbf{X}_2) = 1$ if $\mathbf{T}_{1D} = 0$, $\mathbf{T}_{2D} = 0$, $\mathbf{P}_{1C}^t\mathbf{P}_{1C} = c\mathbf{I}$ and $\mathbf{P}_{2C}^t\mathbf{P}_{2C} = (1-c)\mathbf{I}$ for some constant $0 < c < 1$

which provides a nice relationship between the concepts of common and distinct parts and matrix correlations. Conceptually, when two matrices have nothing in common and their distinct parts are orthogonal they have correlation zero. The second requirement is important since this makes a clear-cut difference between distinct and common parts. This can be explained as follows. An important identifiability constraint for the GAS model is that the loading matrices are orthogonal (Li and Gaynanova, 2018). This means that

$$\begin{aligned} \begin{bmatrix} \mathbf{P}_{1C} \\ \mathbf{P}_{2C} \end{bmatrix}^t \begin{bmatrix} \mathbf{P}_{1C} \\ \mathbf{P}_{2C} \end{bmatrix} &= \begin{bmatrix} \mathbf{P}_{1C}^t & \mathbf{P}_{2C}^t \end{bmatrix} \begin{bmatrix} \mathbf{P}_{1C} \\ \mathbf{P}_{2C} \end{bmatrix} \\ &= \mathbf{P}_{1C}^t\mathbf{P}_{1C} + \mathbf{P}_{2C}^t\mathbf{P}_{2C} = \mathbf{I}. \end{aligned} \tag{4.24}$$

However, this does not translate into orthogonality of the individual matrices $\mathbf{P}_{1C}$ and $\mathbf{P}_{2C}$, so it can happen that the first and second column of $\mathbf{P}_{1C}$ have very different norms and those of $\mathbf{P}_{2C}$ vice versa. Hence, in effect these two columns describe *distinct* components and not common ones which should result in a low matrix correlation. The second requirement prevents this from happening and has the same background as the third property of the GAC coefficient above. A small numerical example (Example 4.1) shows how this works and also sheds light on the complexity of a decomposition in common and distinct components.

Example 4.1: GAC and common/distinct components

This small numerical example is taken from the supplementary material of Li and Gaynanova (2018). In this simple example there are three samples and both blocks have two variables. The number of common components is two and there are no distinct components assumed. Suppose that

$$\mathbf{T}_C = \begin{bmatrix} 2 & 1 \\ -2 & 1 \\ 0 & -2 \end{bmatrix} \tag{4.25}$$

and

$$\mathbf{P}_{1C} = \begin{bmatrix} \frac{5}{\sqrt{50.02}} & \frac{0.1}{\sqrt{50.02}} \\ \frac{5}{\sqrt{50.02}} & \frac{-0.1}{\sqrt{50.02}} \end{bmatrix};\ \mathbf{P}_{2C} = \begin{bmatrix} \frac{0.1}{\sqrt{50.02}} & \frac{5}{\sqrt{50.02}} \\ \frac{-0.1}{\sqrt{50.02}} & \frac{5}{\sqrt{50.02}} \end{bmatrix} \tag{4.26}$$

then it can be verified that Equation 4.24 holds. It is also clear that the first component is distinct for the first block and the second component is distinct for the second block given the sizes of the elements in $\mathbf{P}_{1C}$ and $\mathbf{P}_{2C}$. In fact, both blocks have little in common and the model is misspecified. The subsequent value for the GAC is 0.0404, which reflects this.
A completely different situation is when the loading matrices (assume the same scores as in the previous example) are given by

$$\mathbf{P}_{1C} = \begin{bmatrix} \frac{0.1}{\sqrt{1.5}} & \frac{0.2}{\sqrt{1.5}} \\ \frac{-0.2}{\sqrt{1.5}} & \frac{0.1}{\sqrt{1.5}} \end{bmatrix};\ \mathbf{P}_{2C} = \begin{bmatrix} \frac{0.8}{\sqrt{1.5}} & \frac{0.9}{\sqrt{1.5}} \\ \frac{-0.9}{\sqrt{1.5}} & \frac{0.8}{\sqrt{1.5}} \end{bmatrix} \tag{4.27}$$

where now the scale of $\mathbf{P}_{1C}$ is smaller than that of $\mathbf{P}_{2C}$ but the norms of the columns in $\mathbf{P}_{1C}$ are of similar sizes, as well as those of $\mathbf{P}_{2C}$. So, the components can be considered common. The subsequent GAC is now 1. This shows that the requirements set for $r(\mathbf{X}_1, \mathbf{X}_2) = 1$ are needed.

This small example also shows nicely how complicated the separation of common and distinct components can be in practice. In short, if common components happen to explain very different amounts of variation in the separate blocks then the question is how common they are. This will be taken up in subsequent chapters.

4.5 Partial Matrix Correlations

The interpretation of the RV-coefficient in terms of Pearson correlations of vectorised matrices opens up an interesting opportunity to define *partial matrix correlations* (Aben *et al.*, 2018). In general, a partial correlation measures the correlation between two variables corrected for (conditioned on) a third variable denoted as $r_p(\mathbf{x}_1, \mathbf{x}_2|\mathbf{x}_3)$. Hence, a partial correlation between two blocks $\mathbf{X}_1$ and $\mathbf{X}_2$ corrected for block $\mathbf{X}_3$ would be:

$$\text{RV}_p = corr(\text{Vec}(\mathbf{X}_1\mathbf{X}_1^t), \text{Vec}(\mathbf{X}_2\mathbf{X}_2^t)|\text{Vec}(\mathbf{X}_3\mathbf{X}_3^t)) \tag{4.28}$$

where we have simplified notation. This can, of course, also be used in the context of the modified-RV. Practically, this can be done by using residuals of regression steps (regressing both $\text{Vec}(\mathbf{X}_1\mathbf{X}_1^t)$ and $\text{Vec}(\mathbf{X}_2\mathbf{X}_2^t)$ on $\text{Vec}(\mathbf{X}_3\mathbf{X}_3^t)$ and using the residuals of those regressions). Note that similar ideas are used to distinguish direct from indirect effects in path modelling (Næs *et al.*, 2020). When multiple data blocks are available, it is a matter of choice whether to partialise two blocks on one other block or on multiple blocks. Also in ordinary partial correlations such choices have to be made and it is up to the user how to do this.

An application of partial matrix correlations with heterogeneous blocks of data is given in Example 4.2.

Example 4.2: iTOP: Inferring topology of genomics data

The data of this example is described in Iorio *et al.* (2016) and consists of 206 samples of cancer cell lines. Several genomics data were available which are summarised in Table 4.1.

Table 4.1 Overview of the data sets used in the genomics example.

	Dimension	Type
Mutation	300	Binary
CNA	425	Binary
Methylation	14429	Quantitative
Cancer type	31	Binary
Gene expression	17419	Quantitative
Proteomics	452	Quantitative
Drug response	265	Quantitative

In earlier examples in this book the nature of mutation, methylation, CNA and gene expression data has already been explained (see Section 1.4.2). Proteomics data consists of measured protein levels in the cells. The drug responses are IC50 values and these measure the amount of drug needed to kill 50% of the cells. Finally, there are cell lines from 31 different cancer cells available. The question now is how all these data sets pertaining to the same samples are related. To investigate this, partial RV-coefficients were constructed where for the quantitative data Euclidean distances were used and the Jaccard index for the binary data. The partialisation between two blocks was done in subsequent steps starting with one other block up to all other blocks of data (Colombo and Maathuis, 2014). When the partial matrix correlation was zero, then no more blocks were added. Permutation testing was used to check the significance of the found partial correlations. A high-level overview of the relationships is shown in Figure 4.2.

Figure 4.2 shows that all the information regarding drug response from DNA-derived properties (mutation, methylation and CNA) is channeled through gene-expression. There is a strong link between gene-expression and proteomics which is expected since genes encode for proteins. However, there is also a link between proteomics and drug response, hence, not all information about drug response can be explained by gene-expression. This also makes biological sense since some proteins which are translated by a gene undergo further modifications (so-called post-translational modifications) and a detailed analysis shows indeed that such proteins are involved in this relationship (Aben *et al.*, 2018).

For a more detailed look at what is happening in terms of the partial matrix correlations a zoom-in is provided in Figure 4.3. The upper part of the figure represents the (partial) correlations. The corresponding lower part shows between which data sets these correlations can be found. There is clearly a correlation between methylation and drug response, and between cancer type and drug response. However, if the methylation and cancer type is corrected for (i.e., partialised on) gene-expression, then the correlation disappears. The partialisation on the proteomics data set does not lower these correlations that much.

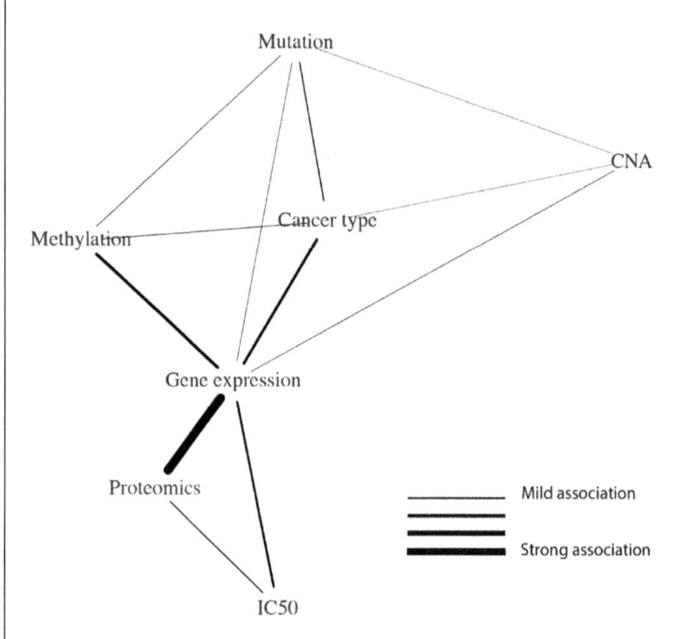

Figure 4.2 Topology of interactions between genomics data sets. Source: Aben *et al.* (2018). Reproduced with permission of Oxford University Press.

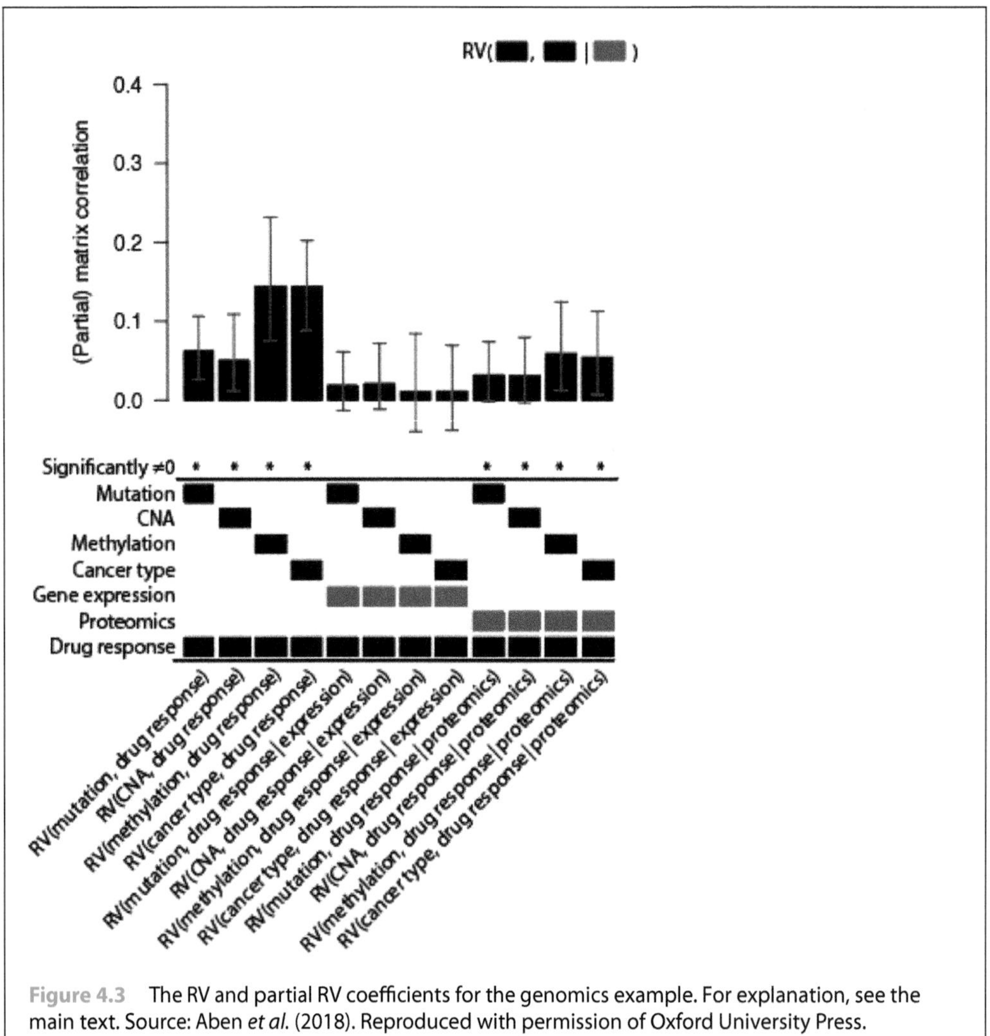

Figure 4.3 The RV and partial RV coefficients for the genomics example. For explanation, see the main text. Source: Aben *et al.* (2018). Reproduced with permission of Oxford University Press.

The generic vectorised framework means that also GCD and SMI easily lend themselves to the partial correlation formulation as defined for RV and RV_{mod}.

4.6 Conclusions and Recommendations

An overview of the matrix correlations which we discussed in this chapter is given in Figure 4.4. This figure shows that there is still room for development. The full matrix correlations for homogeneous data can conveniently be summarised in a generic framework. This is less clear for the partial matrix correlations and the matrix correlations for heterogeneous data.

A crucial aspect of matrix correlations is the choice about which aspects of the matrices should be compared. This goes back to rule three in Requirements 4.1 for which many choices are available. If only a comparison between subspaces is needed, then the GCD or SMI_{OP} can be used. If the configurations of the samples is important then the RV coefficient (or its

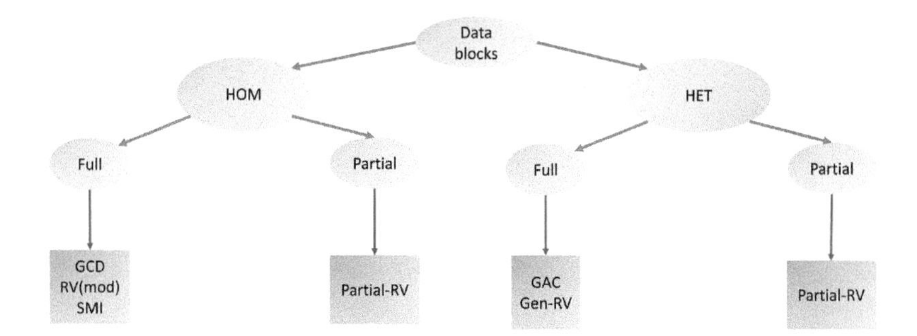

Figure 4.4 Decision tree for selecting a matrix correlation method. Abbreviations: HOM is homogeneous data, HET is heterogeneous data, Gen-RV is generalised RV, Full means full correlations, Partial means partial correlations. For more explanation, see text.

modified form for high-dimensional data) can be used. Finally, for heterogeneous data and partial correlations the choice is very limited.

4.7 Open Issues

Properties and practical use

The matrix correlation coefficients have different properties and these influence the choice of a particular matrix correlation coefficient for a specific problem. One example is exploring the practical importance of the effect of all components having the same impact, i.e., RV versus SMI. This is not obvious in practice and is often governed by the tradition in a field. It would be worthwhile to have some guidelines in this respect.

Framework for heterogeneous data

As mentioned above, extending the generic framework to also encompass heterogeneous data would be helpful. This may also give inspiration for developing new matrix correlations for heterogeneous data. Since heterogeneous data is getting more and more abundant this is needed.

Relationships between matrix correlation and common/distinct components

There are relationships between matrix correlations on the one hand and the methods for finding common and distinct components on the other hand. This is clearly true for partial correlations as evidenced by the GAC and its associated GAS model. It would be fruitful to see how different types of partial matrix correlations (e.g., partial-RV, partial-SMI) are related to certain models and methods. This may also shed light on the properties of those partial matrix correlations.

Another extension may be to quantify the commonality between more than two blocks of data in a summary statistic measuring such a commonality. This is clearly related to the concept of common and distinct since the commonality is in the common parts.

Part II

Selected Methods for Unsupervised and Supervised Topologies

5

Unsupervised Methods

5.i General Introduction

In this chapter we will discuss the most commonly used unsupervised methods with simple *undirected* linking structures such as shown in Figure 5.1. We will limit ourselves to skeletons with only shared sample modes or only shared variable modes; less used unsupervised methods will be discussed briefly in Chapter 9. In this chapter, we will also discuss a method that can deal with multiway arrays and a method that can handle blocks containing measurements of different scale-types.

Unsupervised multiblock data analysis methods are mostly used for exploring data sets. This process is visualised in Figure 5.2. The idea is that to study a complex system (upper part of Figure 5.2) a selection is made on parts of that system (A, B, and C) relevant for the research question. Then measurements are made using instrumental techniques sufficiently diverse to catch the relevant components of the parts A, B, and C. Subsequently, these data are analysed and 'glued together' to arrive at a hypothesised reconstruction of the system (bottom part of Figure 5.2). For biological systems, this may mean that for that system gene-expression, microbial composition and the metabolome are measured to arrive at a comprehensive description. For food related studies, this can entail sensory measurements, chemical measurements and consumer liking scores for the same set of food products.

There are many ways to explore a multiblock data set. As already explained in Chapter 1 there are many goals to pursue in multiblock data analysis, and one of the primary goals of unsupervised multiblock data analysis is to find consistent patterns in the data that may hint to underlying data generating mechanisms. Given the complexity of the data and types of systems studied, there are many unsupervised multiblock data analysis methods. Hence, this chapter will also discuss many methods. At the end of the chapter we will give some guidelines on when to use what method. In the empirical cycle of science, unsupervised data analysis (or exploratory data analysis) is typically the *inductive* part that should generate hypotheses (Tukey, 1980). In many cases it gives sufficient insight into a system to undertake further steps, e.g., in terms of a follow-up experiment or domain specific actions. Having said that, it remains the responsibility for the researcher to select a method which is appropriate for the particular data set, data structure and research questions asked from the data.

5.ii Relations to the General Framework

A brief summary of the methods described in this chapter is given in Table 5.1. This table shows that there is a variety of methods, some of which have similar characteristics. An

Multiblock Data Fusion in Statistics and Machine Learning: Applications in the Natural and Life Sciences,
First Edition. Age K. Smilde, Tormod Næs, and Kristian Hovde Liland.

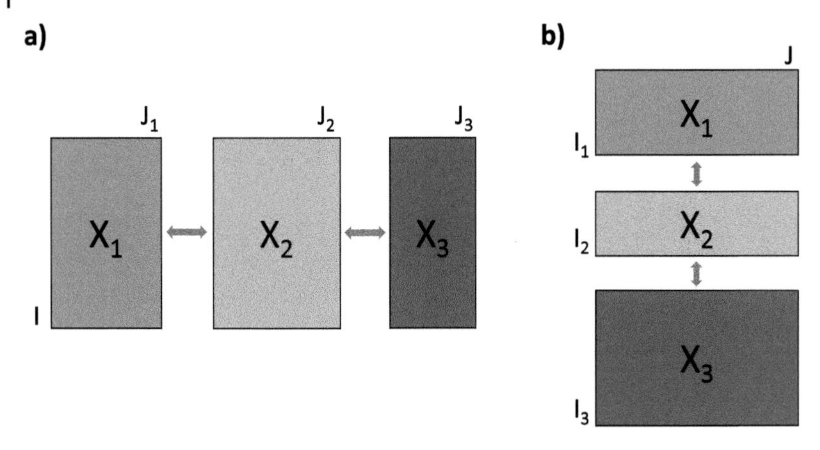

Figure 5.1 Unsupervised analysis as discussed in this chapter, (a) links between samples and (b) links between variables (simplified representations, see Chapter 3).

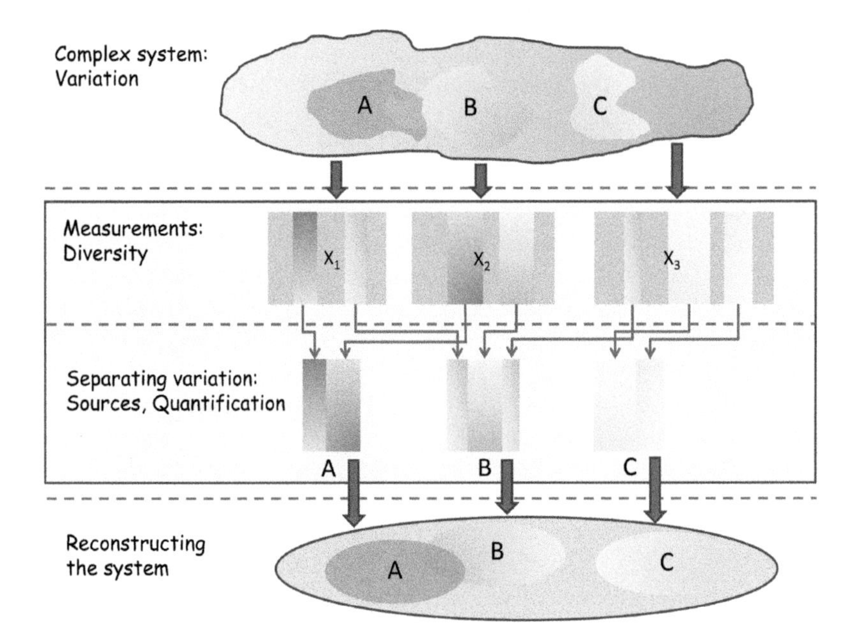

Figure 5.2 Illustration explaining the idea of exploring multiblock data. Source: Smilde *et al.* (2017). Reproduced with permission of John Wiley and Sons.

overarching theme in this chapter is the notion of common, local, and distinct components already explained briefly in Section 1.8. Table 5.1 (column E) shows that several such methods will be described in this chapter. Arriving at a decomposition into common, local and distinct components can be done in different ways: (i) by using explicit models, (ii) by using specific rotations and (iii) by imposing penalties or combinations thereof. Some methods are focused on finding only common components and these will be discussed first. Different types of optimisation criteria or loss-functions can be used, and these will be illustrated too (see Table 5.1 column F). Almost all methods in this chapter are simultaneous (see also Section 2.3.5 and Table 5.1 column C). Most of the methods can only handle homogeneous data and not data of different measurement scales (heterogeneous data; see Table 5.1 column B). We will start with

Table 5.1 Overview of methods. Legend: U=unsupervised, S=supervised, C=complex, HOM=homogeneous data, HET=heterogeneous data, SEQ=sequential, SIM=simultaneous, MOD=model-based, ALG= algorithm-based, C=common, CD=common/distinct, CLD=common/local/distinct, LS=least squares, ML=maximum likelihood, ED=eigendecomposition, MC=maximising correlations/covariances. For abbreviations of the methods, see Section 1.11.

		A			B		C		D		E			F			
	Section	U	S	C	HOM	HET	SEQ	SIM	MOD	ALG	C	CD	CLD	LS	ML	ED	MC
SCA	5.1.1.1	■			■			■	■		■			■		■	
Multigroup analysis	5.1.1.3	■			■		■			■	■						■
DISCO	5.1.2.1	■			■		■			■			■	■			
MCR	5.1.2.2	■			■			■	■				■	■			
MFA/STATIS	5.2.1.2	■			■			■	■		■			■			
GCA	5.2.1.3	■			■			■	■		■			■		■	
RGCCA	5.2.1.4	■			■		■			■	■						■
ESCA	5.2.1.5	■				■		■	■		■				■		
OS	5.2.1.6	■				■		■		■	■			■			
JIVE	5.2.2.1	■			■			■	■			■		■			
PCA-GCA	5.2.2.3	■			■		■		■				■	■			■
ACMTF	5.2.2.4	■			■			■	■				■	■			
PESCA	5.2.2.5	■				■		■	■				■		■		

methods for data sets that have a shared variable mode and in a later section we will discuss methods for a shared sample mode. The chapter ends by discussing a generic framework for defining unsupervised multiblock data analysis methods. This framework may help putting the methods of this chapter in perspective and may shed light on their (dis-)similarities.

5.1 Shared Variable Mode

Typical examples of a shared variable mode are multigroup studies in which different groups are compared using the same set of variables (see also Section 10.6). In chemistry, this may entail measuring different sets of samples using the same type of spectroscopy; in sensory science this may be measurements of the same sensory attributes across different sets of products and in biology this may encompass RNAseq measurements of the same system under different conditions.

5.1.1 Only Common Variation

5.1.1.1 Simultaneous Component Analysis

THE SCA MODEL

The notion of a shared variable mode was already explained in Chapter 3 (see also Figure 3.3). The method used mostly in multiblock data analysis for this situation will be explained for M blocks $\mathbf{X}_m$ $(I_m \times J)$; $m = 1, \ldots, M$ which are assumed to be mean-centred. The method is called Simultaneous Component Analysis (SCA). It was already discussed briefly in Chapter 2

Table 5.2 Different types of SCA, where $\mathbf{D}_m$ is a diagonal matrix and $\mathbf{\Phi}$ is a positive definite matrix (see Section 2.8). The correlations and variances pertain to the block-scores (see text).

Method	Constraint	Correlations	Variances
SCA-P	no	free	free
SCA-PF2	$(\frac{1}{I_m})\mathbf{T}_m^t\mathbf{T}_m = \mathbf{D}_m\mathbf{\Phi}\mathbf{D}_m$	equal across blocks	free
SCA-IND	$(\frac{1}{I_m})\mathbf{T}_m^t\mathbf{T}_m = \mathbf{D}_m^2$	equal to zero	free
SCA-ECP	$(\frac{1}{I_m})\mathbf{T}_m^t\mathbf{T}_m = \mathbf{\Phi}$	equal across blocks	equal across blocks

and originates from psychometrics (Levin, 1966; Kiers and Ten Berge, 1994; Timmerman and Kiers, 2003). The basic model is

$$\mathbf{X}_m = \mathbf{T}_m\mathbf{P}^t + \mathbf{E}_m;\ m = 1, \ldots, M \tag{5.1}$$

with block-scores $\mathbf{T}_m$ $(I_m \times R)$. The word 'simultaneous' derives from the fact that all decompositions of the data blocks in Equation 5.1 share the same loading matrix $\mathbf{P}(J \times R)$. There are four different versions of SCA as summarised in Table 5.2 with the different versions of SCA ordered according to increasingly constrained models. These constraints pertain to the scores in the different blocks and can put restrictions on their correlations and/or their variances. These correlations and variances are collected in the matrices $(\frac{1}{I_m})\mathbf{T}_m^t\mathbf{T}_m$ and thus these requirements are in terms of those matrices. The ordering of the methods according to increased constraints implies that, given an equal number of components R, the fit of the successive models is the same or worse. The gain of using a more constrained model is that if that model is a reasonable model for the data, it is less prone to error fitting and easier to interpret. An example of the constraints as shown in Table 5.2 is given in Elaboration 5.1.

The SCA model of Equation 5.1 can also be written as

$$\mathbf{X} = \begin{bmatrix} \mathbf{X}_1 \\ \cdot \\ \cdot \\ \cdot \\ \mathbf{X}_M \end{bmatrix} = \begin{bmatrix} \mathbf{T}_1 \\ \cdot \\ \cdot \\ \cdot \\ \mathbf{T}_M \end{bmatrix} \mathbf{P}^t + \begin{bmatrix} \mathbf{E}_1 \\ \cdot \\ \cdot \\ \cdot \\ \mathbf{E}_M \end{bmatrix} = \mathbf{T}\mathbf{P}^t + \mathbf{E} \tag{5.2}$$

and for the SCA-P version this is an (unrestricted) PCA model of the concatenated data. Thus it holds that the columns of $\mathbf{P}$ are in the row-space of $\mathbf{X}$ but not necessarily in the row-spaces of the individual matrices $\mathbf{X}_m$; $m = 1, \ldots, M$. Also, for SCA-P there are no restrictions on $\mathbf{T}$ and thus $\mathbf{T}$ is in the column-space of $\mathbf{X}$. Hence, it holds that $\mathbf{T} = \mathbf{X}\mathbf{W}$ for some $\mathbf{W}$ (for the SCA-P case it holds that $\mathbf{W} = \mathbf{P}$) and thus $\mathbf{T}_m = \mathbf{X}_m\mathbf{W}$ which means that the $\mathbf{T}_m$ are in the column-space of their respective blocks. An example of SCA-P in the `multiblock` R-package is found in Section 11.6.4. For other SCA versions than SCA-P, there are restrictions on $\mathbf{T}$ and, hence, these are not necessarily in the column-space of $\mathbf{X}$.

ELABORATION 5.1

Different constraints in SCA models

We will illustrate the different constraints in the SCA models using three blocks of data $\mathbf{X}_1$, $\mathbf{X}_2$ and $\mathbf{X}_3$ with, for simplicity, the same number of samples (hence, we will ignore the term $1/I_m$ here) and two SCA components per block. Note that the number of variables is equal since the blocks share the variables. Hence, block $\mathbf{X}_m$ has components $\mathbf{T}_m = [\mathbf{t}_{m1}|\mathbf{t}_{m2}]$. The inner product of these scores can now be written as:

$$\mathbf{T}_m^t\mathbf{T}_m = [\mathbf{t}_{m1}|\mathbf{t}_{m2}]^t[\mathbf{t}_{m1}|\mathbf{t}_{m2}] = \begin{bmatrix} \mathbf{t}_{m1}^t\mathbf{t}_{m1} & \mathbf{t}_{m1}^t\mathbf{t}_{m2} \\ \mathbf{t}_{m2}^t\mathbf{t}_{m1} & \mathbf{t}_{m2}^t\mathbf{t}_{m2} \end{bmatrix} = \begin{bmatrix} ||\mathbf{t}_{m1}||^2 & \mathbf{t}_{m1}^t\mathbf{t}_{m2} \\ \mathbf{t}_{m2}^t\mathbf{t}_{m1} & ||\mathbf{t}_{m2}||^2 \end{bmatrix} \quad (5.3)$$

and working out the different terms, it holds that:

$$\mathbf{T}_m^t\mathbf{T}_m = \mathbf{D}_m\mathbf{\Phi}_m\mathbf{D}_m \quad (5.4)$$

with

$$\mathbf{D}_m = \begin{bmatrix} ||\mathbf{t}_{m1}|| & 0 \\ 0 & ||\mathbf{t}_{m2}|| \end{bmatrix} \quad (5.5)$$

and

$$\mathbf{\Phi}_m = \begin{bmatrix} 1 & \frac{\mathbf{t}_{m1}^t\mathbf{t}_{m2}}{||\mathbf{t}_{m1}||\,||\mathbf{t}_{m2}||} \\ \frac{\mathbf{t}_{m2}^t\mathbf{t}_{m1}}{||\mathbf{t}_{m1}||\,||\mathbf{t}_{m2}||} & 1 \end{bmatrix} \quad (5.6)$$

for $m = 1, 2, 3$. So far, the solution is constraint-free thus resulting in SCA-P. The SCA-PF2 constraint now dictates that $\frac{\mathbf{t}_{m1}^t\mathbf{t}_{m2}}{||\mathbf{t}_{m1}||\,||\mathbf{t}_{m2}||}$ is the same for each m. Since the scores are centred (because the original data are centred) this comes down to requiring that the correlations between the scores in the same block remain the same between the blocks. The variances of the scores may change since there are no restrictions on $\mathbf{D}_m$. Such a restriction on the correlations may make the comparisons between groups easier.

The constraint of the SCA-IND model can be illustrated in a similar way. This constraint amounts to matrices $\mathbf{D}_m$ similar to those of Equation 5.5 and the matrix $\mathbf{\Phi}$ equals the identity matrix. In this case the variances of the scores are completely free and the correlations of the components within the blocks are zero. Hence, in each block two independent directions are chosen with unconstrained variances.

The first version (SCA-P) with no constraints on $\mathbf{T}_m$ is used most often and is also known as the SUMPCA for cross-product matrices in psychometrics (Kiers, 1991)[1]. The next version (SCA-PF2) resembles the three-way method PARAFAC2 (hence, the abbreviation PF2) (Kiers *et al.*, 1999; Smilde *et al.*, 2004). SCA-IND has a relationship with individual differences scaling (INDSCAL); a method for two-mode three-way data (e.g., crossproducts or distances) (Carroll and Arabie, 1980; Law *et al.*, 1984) (see Section 2.2.1). The relationship between SCA-IND and INDSCAL can be seen by taking the cross-product of the fitted part of Equation 5.1 and imposing the appropriate constraint:

$$\left(\frac{1}{I_m}\right)\widehat{\mathbf{X}}_m^t\widehat{\mathbf{X}}_m = \mathbf{P}\mathbf{D}_m^2\mathbf{P}^t;\ m = 1, \ldots, M \quad (5.7)$$

which equals the INDSCAL model (Carroll and Chang, 1970). The most restricted model is SCA-ECP, where ECP stands for equal average cross-products constraints. To our knowledge, this method has not been used in the natural sciences. Only SCA-P and SCA-ECP have

---

1 Not to be confused with SUM-PCA in chemometrics, see Section 5.2.1.1

rotational freedom; the other two versions provide unique solutions where uniqueness means that the solution cannot be rotated without losing fit[2].

The SCA-IND model has been used in plant science to study the chemical response of cabbage plants to a herbivore attack (Jansen *et al.*, 2012). A widely used model in chemistry is multivariate curve resolution (MCR) (Tauler *et al.*, 1995) with shared variables which is detailed later in this section.

The algorithms to fit the SCA models depend on the specific type of model. For the SCA-P model, the algorithm is simple because an SVD of $\mathbf{X}$ (see Equation 5.2) provides the $\mathbf{T}_1, \ldots, \mathbf{T}_M$ and $\mathbf{P}$. For the other versions of SCA, algorithms are available imposing the constraints which are all based on alternating least squares (ALS) minimising the sum of squared residuals (Timmerman and Kiers, 2003).

PREPROCESSING

As mentioned in Section 2.6, preprocessing concerns three aspects: centring and scaling within and between blocks. Regarding centring, what to do in which situation depends very much on the specific application and the type of model. However, some general recommendations can be given. For the general SCA models it is a good idea to centre the data per data block per variable across the samples in that block thereby removing level differences between blocks. Then constants are eliminated from the data without introducing artificial variation (Bro and Smilde, 2003; Harshman and Lundy, 1984) and the components are defined on the basis of the intra-block covariances instead of the inter-block covariances between variables. Also, the average component scores per block can be shown to be zero and this allows for the interpretation as given in Table 5.2 (Timmerman and Kiers, 2003). For interval-scaled data, this is also the most obvious choice of centring.

Within-block scaling – if needed – is recommended for each variable separately and basically there are two options. In option one, within-block scaling is done by using a scaling constant per variable calculated across the samples of all data blocks. This means that the within-block scaling is done with the same scaling constants within all the blocks (i.e., scaling across all samples in the row-concatenated blocks, see Equation 5.2). This type of scaling preserves between-block variability (Bro and Smilde, 2003; Harshman and Lundy, 1984). The second option is to scale the variables in each block using a scaling constant calculated per block. These options give different results (see Elaboration 5.2).

When centring and scaling based on the observed standard deviations across all samples is used (the first option), then the total sum-of-squares of each block $\mathbf{X}_m$ will be different if the numbers of samples are very different between the blocks. Hence, the blocks with a large number of samples will tend to dominate the solution (see Section 1.7). Between block scaling (i.e., multiplying each block with a constant which may be different for each block) may prevent this from happening but is not trivial since this will destroy the between-block differences.

ELABORATION 5.2

Within-block scaling in SCA models

Within-block scaling in SCA models is not trivial and the way to perform this depends on the application. The two options explained in the main text will be illustrated with a small example of two blocks of data sharing the same variables. The raw data in the two blocks are

2 This does not mean that the 'true' solution is obtained!

$$\mathbf{X}_1 = \begin{bmatrix} 1 & 20 \\ 5 & 14 \\ 3 & 65 \\ 2 & 34 \\ 4 & 57 \end{bmatrix}; \mathbf{X}_2 = \begin{bmatrix} 10 & 120 \\ 7 & 155 \\ 5 & 241 \\ 12 & 195 \\ 14 & 105 \\ 18 & 135 \end{bmatrix} \tag{5.8}$$

and it is clear that the two variables have very different ranges so a scaling might be considered. After centring both matrices around their own centres the result is:

$$\mathbf{X}_{1\text{Cent}} = \begin{bmatrix} -2 & -18 \\ 2 & -24 \\ 0 & 27 \\ -1 & -4 \\ 1 & 19 \end{bmatrix}; \mathbf{X}_{2\text{Cent}} = \begin{bmatrix} -1 & -34 \\ -4 & 1 \\ -6 & 60 \\ 1 & 41 \\ 3 & -49 \\ 7 & -19 \end{bmatrix} \tag{5.9}$$

and from this it becomes clear that some numbers are the same for the same variable in the two blocks (e.g., -1 and 1 in variable 1 in both blocks). In option one, the scaling constant is calculated across all samples in both blocks for the same variable and after this scaling the result is:

$$\mathbf{X}_{1\text{Scal1}} = \begin{bmatrix} -0.57 & -0.53 \\ 0.57 & -0.72 \\ 0 & 0.81 \\ -0.29 & -0.12 \\ 0.29 & 0.57 \end{bmatrix}; \mathbf{X}_{2\text{Scal1}} = \begin{bmatrix} -0.29 & -1.02 \\ -1.15 & 0.03 \\ -1.72 & 1.79 \\ 0.29 & 1.22 \\ 0.86 & -1.46 \\ 2.00 & -0.57 \end{bmatrix} \tag{5.10}$$

from which we can conclude that the scaling has worked since all values now have comparable magnitude. The original values of 1 and -1 for variable 1 in both blocks remain equal in this type of scaling (note also that the values 19 and -19 for variable 2 in both centred blocks keep the same magnitude and reversed sign).

The other type of scaling (option two) where scaling constants are calculated per variable per block results in

$$\mathbf{X}_{1\text{Scal2}} = \begin{bmatrix} -1.27 & -0.80 \\ 1.27 & -1.07 \\ 0 & 1.21 \\ -0.63 & -0.18 \\ 0.63 & 0.85 \end{bmatrix}; \mathbf{X}_{2\text{Scal2}} = \begin{bmatrix} -0.21 & -0.79 \\ -0.85 & 0.02 \\ -1.27 & 1.40 \\ 0.21 & 0.96 \\ 0.63 & -1.14 \\ 1.48 & -0.44 \end{bmatrix} \tag{5.11}$$

which is a completely different data set as the one of the first option (Equation 5.10). For instance, the values of 1 and -1 for variable 1 have now obtained different values in both blocks.

If values such as -1 and 1 are comparable between the two blocks, then the first scaling preserves that comparability. An example is in metabolomics where block $\mathbf{X}_1$ are measured concentrations for an individual at five different time points and block $\mathbf{X}_2$ are similar measurements for another individual. Then the centring operation means that we consider changes in metabolite concentrations relative to a baseline value. In that case a value of 1 for the first individual has as similar meaning as this value for the other individual. This comparability should not be destroyed and thus the first type of scaling is recommended in this case. Note that block-scaling this result (i.e., dividing both blocks of Equation 5.10 by their Frobenius norm) gives

$$\mathbf{X}_{1\text{Scal3}} = \begin{bmatrix} -0.35 & -0.33 \\ -0.35 & -0.44 \\ 0 & 0.50 \\ -0.18 & -0.07 \\ 0.18 & 0.35 \end{bmatrix}; \ \mathbf{X}_{2\text{Scal3}} = \begin{bmatrix} -0.07 & -0.24 \\ -0.27 & 0.01 \\ -0.41 & 0.43 \\ 0.07 & 0.29 \\ 0.21 & -0.35 \\ 0.48 & -0.14 \end{bmatrix} \tag{5.12}$$

which destroys the between-block structure.

A completely different example is from sensory science. Suppose that two assessors (blocks 1 and 2) judge the same products (samples) on the same attributes, It is known that assessors use sensory scales differently, so a value of 10 of assessor 1 is not the same as a value of 10 of assessor 2 for the same attribute and the same product. In this case, it would make sense to 'calibrate' both assessors by autoscaling (i.e., centring and within-block scaling) each block. After that, the sensory scales have been harmonised and the blocks can be analysed jointly.

These examples serve to illustrate the complexity of preprocessing multiblock data. The crucial question is what type of preprocessing makes the data comparable and relevant for the research question.

Validation

Validation of SCA models includes establishing model complexity, that is, how many components to choose (see also Chapter 2). There are no general rules for this but a method often used in practice is to calculate residual sum-of-squares and plotting those against model complexity (see also Chapter 2). The scree method is then utilised to select the number of components. A more advanced method also uses the number of free parameters in the model in combination with the fit (Ceulemans and Kiers, 2006). If an external estimate is available defining the residual sum-of-squares expected due to measurement noise, then this can be taken as a cut-off. Unfortunately, in many cases this is not available.

In deciding which SCA method to use, e.g., SCA-P versus SCA-PF2, comparing residual sum-of-squares is also helpful keeping in mind that there are hierarchical relationships between the methods. Choosing an appropriate model will always strike a balance between fit and interpretability (see, e.g., Smilde *et al.* (2015)). Unfortunately, there is no established theory yet about degrees of freedom for residual sum-of-squares for SCA models which would otherwise facilitate comparing such residual sum-of-squares across models. Cross-validation can also be used for model selection by leaving out elements in the different blocks and predicting those with the built models, much in the same way as performing cross-validation in PCA models to establish the number of principal components (see also Section 2.7.5).

The explained variance for the total model can be calculated using $||\mathbf{X}||^2$ and $||\mathbf{E}||^2$ and the property that given $\mathbf{T}$ the loadings $\mathbf{P}$ are obtained through a regression step invoking orthogonality of $\mathbf{TP}^t$ and $\mathbf{E}$ (see also Section 2.8). Calculating explained variances per block is more involved (see Elaboration 5.3).

The methods SCA-P and SCA-ECP have rotational freedom (Timmerman and Kiers, 2003), hence, $\mathbf{P}$ can then be chosen to be orthogonal without losing fit. This also means that for these SCA-versions, it is possible to calculate explained variances per component.

ELABORATION 5.3

Explained variances per block in SCA

Explained variances per block can be calculated for SCA-P. This can be seen as follows. Starting from the model in Equation 5.1 it is possible to work out the sum-of-squares of $\mathbf{X}_m$ using traces and cyclic permutation:

$$\|\mathbf{X}_m\|^2 = tr(\mathbf{X}_m^t\mathbf{X}_m) = tr(\mathbf{T}_m\mathbf{P}^t + \mathbf{E}_m)^t(\mathbf{T}_m\mathbf{P}^t + \mathbf{E}_m) = \quad (5.13)$$
$$tr(\mathbf{P}\mathbf{T}_m^t\mathbf{T}_m\mathbf{P}^t) + tr(\mathbf{E}_m^t\mathbf{E}_m) + 2tr(\mathbf{T}_m^t\mathbf{E}_m\mathbf{P})$$

and it is clear that a sufficient condition for a split-up of explained variation is $tr(\mathbf{T}_m^t\mathbf{E}_m\mathbf{P}) = 0$ (see also Section 2.8). For the SCA-P case, the scores $\mathbf{T}$ are unconstrained and then it holds that $\mathbf{EP} = 0$ as a consequence of the regression of $\mathbf{X}$ onto $\mathbf{P}$. Writing out $\mathbf{EP} = 0$ into its partitioned form shows that $\mathbf{E}_m\mathbf{P} = 0$; thus is holds that $tr(\mathbf{T}_m^t\mathbf{E}_m\mathbf{P}) = 0$ and explained variances can be calculated meaningfully per block. For the other SCA alternatives (SCA-IND, SCA-PF2 and SCA-ECP) this is more complicated.

The case for SCA-PF2 is illustrated by following the algorithm for estimating its parameters. The basic notion is that every matrix $\mathbf{T}_m$ that meets the constraint $(\frac{1}{I_m})\mathbf{T}_m^t\mathbf{T}_m = \mathbf{D}_m\mathbf{\Phi}\mathbf{D}_m$ can be written as $\mathbf{T}_m = \mathbf{Q}_m\mathbf{Z}\mathbf{D}_m$ with $\mathbf{Q}_m^t\mathbf{Q}_m = \mathbf{I}_R$, $\mathbf{Z}$ is an arbitrary $(R \times R)$ matrix and $\mathbf{D}_m$ is an $(R \times R)$ diagonal matrix (Kiers *et al.*, 1999). Then solving $\mathbf{D}_m$ for block m can be written as minimising:

$$\min_{\mathbf{D}_m} \|\mathbf{X}_m - \mathbf{Q}_m\mathbf{Z}\mathbf{D}_m\mathbf{P}^t\|^2 \quad (5.14)$$

and writing this problem in Vec(.) notation this becomes:

$$\min_{\mathbf{D}_m} \|\text{Vec}(\mathbf{X}_m) - [\mathbf{P} \otimes (\mathbf{Q}_m\mathbf{Z})]\text{Vec}(\mathbf{D}_m)\|^2 \quad (5.15)$$

where the equality $\text{Vec}(\mathbf{ABC}) = [\mathbf{C}^t \otimes \mathbf{A}]\text{Vec}(\mathbf{B})$ has been used. The term $[\mathbf{P} \otimes (\mathbf{Q}_m\mathbf{Z})]\text{Vec}(\mathbf{D}_m)$ can be written as $\mathbf{U}_m\mathbf{d}_m$ where $\mathbf{U}_m$ is the column-wise Kronecker product of $\mathbf{P}$ and $\mathbf{Q}_m\mathbf{Z}$ and $\mathbf{d}_m$ is the vector containing the diagonal elements of $\mathbf{D}_m$. Upon defining $\mathbf{x}_m = \text{Vec}(\mathbf{X}_m)$, the problem boils down to minimising the sum-of-squares of $[\mathbf{x}_m - \mathbf{U}_m\mathbf{d}_m]$ which is an ordinary regression problem. Hence, the residuals are orthogonal to the regressors and explained variances per block can be calculated[3].

A similar derivation can be made for SCA-IND and thus in that case there are also block-wise explained sums-of-squares. This does not hold for SCA-ECP. Thus, explained variances cannot always trivially be calculated per block or per component.

5.1.1.2 Clustering and SCA

When the number of blocks becomes very large, the SCA methods can be too restrictive because they all assume the same loadings across all blocks. This requirement can be relaxed in different ways: by clustering similar blocks based on a specific distance measure combined with fuzzy clustering (Dahl and Næs, 2009) or by explicitly modelling similarity (Clusterwise SCA, (De Roover *et al.*, 2012)).

FUZZY SCA CLUSTERING

Suppose that there are M blocks $\mathbf{X}_m$ $(I_m \times J)$ (column-centred), e.g., M assessors assessing products using the same sensory features. The fuzzy clustering approach is based on minimising the standard fuzzy clustering criterion

3 We thank Henk Kiers for pointing this out.

$$Q = \sum_{k=1}^{K} \sum_{m=1}^{M} u_{mk}^2 d_{mk}^2; \ 0 \leq u_{mk} \leq 1, \sum_{k=1}^{K} u_{mk} = 1; \ m = 1, \ldots M \tag{5.16}$$

where u_{mk} is the membership value of block m to cluster k and d_{mk} is the distance of block m to cluster k. The original fuzzy clustering algorithm was developed for Euclidean distances between object-cluster centres Bezdek (1981) but as shown in Næs and Isaksson (1991); Næs and Mevik (1999) the same idea and algorithm can be extended to many other measures of differences between objects and cluster models. These differences are defined implicitly since they are not based on an initial SCA model but based on weighted combinations from the different blocks and emerge through the algorithm. The method can be formulated within a context of a shared variable mode or a shared sample mode. The procedure for the shared variable mode is shown in Algorithm 5.1.

Algorithm 5.1

FUZZY SCA CLUSTERING

Here we follow the general algorithmic setup in Bezdek (1981) for fuzzy clustering which starts with random membership values and iterates between improving membership values and distances until convergence. Step 2 below concerns optimising the distances while step 3 concerns optimisation of membership values.

Step 1: Initialise $\mathbf{U}(M \times K)$ by using random nonnegative numbers with the restriction that the sum for each block is equal to 1 (see Equation 5.16). The value of K has to be chosen *a priori*.

Step 2: With the above definition of distances, the inner sum of Equation 5.16 can be written, e.g., for each cluster k, as

$$\sum_{m=1}^{M} ||u_{mk}\mathbf{X}_m - \mathbf{T}_{mk}\mathbf{P}_k^t||^2 \tag{5.17}$$

where u_{mk} is simply moved into the expression for the distance between data and model. The minimisation of this is achieved by applying SCA to the matrix

$$\begin{bmatrix} u_{1k}\mathbf{X}_1 \\ \cdot \\ \cdot \\ u_{Mk}\mathbf{X}_M \end{bmatrix} \tag{5.18}$$

which means that the blocks with the smallest membership values to cluster k are given the lowest weight in the SCA. Then the solution $||u_{mk}\mathbf{X}_m - \mathbf{T}_{mk}\mathbf{P}_k^t||$ is divided by u_{mk} (i.e., u_{mk} moved out of the Frobenius norm again) and the distance d_{mk} is obtained as $d_{mk} = ||\mathbf{X}_m - \frac{1}{u_{mk}}\mathbf{T}_{mk}\mathbf{P}_k^t||$. Note that a new SCA is calculated for each cluster k.

Step 3: Update the u_{mk} value for block m and cluster k using the standard updating scheme:

$$u_{mk} = [\sum_{l=1}^{K} (\frac{d_{mk}}{d_{ml}})^2]^{-1} \tag{5.19}$$

where the sum is over the clusters. It was proved in Bezdek (1981) that this optimises the Q for given distances.

Step 4: Go to 2 and continue until convergence.

In general, the numerical properties are quite good for these types of clustering methods (Rousseeuw *et al.*, 1995). The value of K can be chosen by assessing the quality of the clustering. The results of the algorithm are SCA-P models for each cluster and membership values for each block to each cluster. These membership values are useful in interpreting the strength of membership of the different blocks to the different clusters.

Clusterwise SCA

The model for clusterwise SCA is:

$$\mathbf{X}_m = \sum_{k=1}^{K} c_{mk}\mathbf{T}_{mk}\mathbf{P}_k^t + \mathbf{E}_m;\ m = 1, \ldots, M \tag{5.20}$$

and the SCA-ECP version is used to ensure that data blocks with a different within-block correlation structure will be allocated to different clusters. This is because the SCA-ECP version of SCA is the most restrictive one (see Table 5.2) and therefore gives the most crisp clustering looking for homogeneous clusters. The number of clusters is K and has to be chosen *a priori*. The values c_{mk} are elements of the binary partition matrix $\mathbf{C}(M \times K)$ which equals one if the data block belongs to the cluster and zero otherwise. The loading matrix $\mathbf{P}_k(J \times R)$ provides the common loadings for all blocks in cluster k. The estimation proceeds by a least squares algorithm that alternates between updating the matrix $\mathbf{C}$ and the matrices $\mathbf{P}_k$ and $\mathbf{T}_{mk}$ under the ECP-restrictions. This algorithm may get stuck in a local minimum so several restarts are recommended (De Roover *et al.*, 2012).

Both clustering methods are flexible ways of searching for groups of homogeneous data blocks. There are several differences in output of both methods. Clusterwise SCA gives a crisp clustering: each block is assigned to one cluster. This is not the case for fuzzy SCA clustering: the blocks obtain membership values between zero and one. These values can also be used to assess the quality of the clustering. The second difference is the type of SCA model within a cluster. For clusterwise SCA the method SCA-ECP is the best option as explained above and for fuzzy clustering SCA-P is chosen although SCA-ECP may also be an option for that method.

The fuzzy SCA clustering has been applied in sensory science (Dahl and Næs, 2009) and clusterwise SCA up to now only in psychometrics (De Roover *et al.*, 2012, 2013b,a). Yet, useful applications in life science data are easily foreseeable.

5.1.1.3 Multigroup Data Analysis

A sequential method for finding common components in multiblock data with a shared variable mode is called multigroup data analysis (MGA, Tenenhaus and Tenenhaus (2014)). This method solves the following problem:

$$\begin{aligned} &\max \sum_{m,m'=1}^{M} c_{m,m'} g[(\mathbf{X}_m^t\mathbf{X}_m\mathbf{w}_m, \mathbf{X}_{m'}^t\mathbf{X}_{m'}\mathbf{w}_{m'})] \\ &s.t.\, \mathbf{w}_m^t\mathbf{\Psi}_m\mathbf{w}_m = 1\,;\, m, m' = 1, \ldots, M \end{aligned} \tag{5.21}$$

where $c_{m,m'}$ (zero or one) encodes which blocks are connected in the skeleton of the problem and $\boldsymbol{\Psi}_m = (1-\tau_m)(\frac{1}{I})\mathbf{X}_m^t\mathbf{X}_m + \tau_m\mathbf{I}$ is a positive definite matrix with regularisation parameter $0 \leq \tau_m \leq 1$. The abbreviation *s.t.* means *subject to* and describes the constraints imposed on the solution. The matrix $\boldsymbol{\Psi}_m$ can be seen as a shrinkage estimator of the covariance matrix of $\mathbf{X}_m$ (Schäfer and Strimmer, 2005) and there are recommendations for the selection of τ_m (Ledoit and Wolf, 2004). The function g is identity, the absolute value or the square function and the argument within g (that is, $g[(,)]$) is the scalar or inner product. These different choices generate different types of solutions. Subsequent components are calculated orthogonal to the previous ones, e.g., using a deflation step. A worked out example is given in Example 5.1, which shows that multigroup data analysis (for some situations) is very related to SCA-P.

Example 5.1: Example of multigroup analysis

An example of multigroup data analysis is used to show the working of the method in a simple case. We assume two data blocks $\mathbf{X}_1(I_1 \times J)$ and $\mathbf{X}_2(I_2 \times J)$ both of full column rank. For simplicity we choose the function $g(.)$ to be the identity and $\tau_m = 0$. Hence, $\boldsymbol{\Psi}_1 = \mathbf{X}_1^t\mathbf{X}_1$ and $\boldsymbol{\Psi}_2 = \mathbf{X}_2^t\mathbf{X}_2$ and these are both positive definite. Then the multigroup method has to solve the problem:

$$\max_{\mathbf{w}_1,\mathbf{w}_2}(\mathbf{X}_1^t\mathbf{X}_1\mathbf{w}_1, \mathbf{X}_2^t\mathbf{X}_2\mathbf{w}_2);\ s.t.\ \mathbf{w}_m^t\boldsymbol{\Psi}_m\mathbf{w}_m = 1\ (m = 1,2) \tag{5.22}$$

and by writing out the inner product in Equation 5.22 this can also be written as

$$\max_{\mathbf{w}_1,\mathbf{w}_2}\mathbf{w}_1^t\mathbf{X}_1^t\mathbf{X}_1\mathbf{X}_2^t\mathbf{X}_2\mathbf{w}_2;\ s.t.\ \mathbf{w}_m^t\boldsymbol{\Psi}_m\mathbf{w}_m = 1\ (m = 1,2) \tag{5.23}$$

and upon writing out the term $\mathbf{w}_m^t\boldsymbol{\Psi}_m\mathbf{w}_m = \mathbf{w}_m^t\mathbf{X}_m^t\mathbf{X}_m\mathbf{w}_m$ and defining $\mathbf{t}_m = \mathbf{X}_m\mathbf{w}_m$ this becomes:

$$\max_{\mathbf{w}_1,\mathbf{w}_2}\mathbf{t}_1^t\mathbf{X}_1\mathbf{X}_2^t\mathbf{t}_2;\ s.t.\ \mathbf{t}_m^t\mathbf{t}_m = 1\ (m = 1,2) \tag{5.24}$$

Now the term $\mathbf{X}_m^t\mathbf{t}_m$ can be defined as loading $\mathbf{p}_m$. Due to the constraint $\mathbf{t}_m^t\mathbf{t}_m = 1$ it holds that $\mathbf{p}_m$ is the least squares solution of the regression $\mathbf{X}_m = \mathbf{t}_m\mathbf{p}_m^t + \mathbf{E}_m$ (see Elaboration 2.4). Thus, the multigroup problem can also be formulated as

$$\max_{\mathbf{w}_1,\mathbf{w}_2}\mathbf{p}_1^t\mathbf{p}_2;\ s.t.\ \mathbf{t}_m^t\mathbf{t}_m = 1\ (m = 1,2) \tag{5.25}$$

which shows that components are sought in the row-spaces of $\mathbf{X}_1$ and $\mathbf{X}_2$ that are maximally aligned and least squares for their respective blocks (note that the optimisation is over $\mathbf{w}'s$ which are hidden in the loss function). Hence, this is very similar to SCA-P but the loadings for the two blocks are not forced to be exactly the same (but very related). In terms of linking structure (see Section 3.4), multigroup analysis has a more flexible link than SCA-P.

5.1.2 Common, Local, and Distinct Variation

Apart from only finding common variation it is also useful to distinguish common from distinct variation. The idea of separating common and distinct variation is already visualised in Figure 1.9. The power of such an approach is that sources of variation can be separated and interpreted individually. This is especially powerful when exploring the relationships between multiple data blocks which contain measurements of a diverse structure. In that case, the data blocks may not 'pick up' the same type of variation of the studied system.

We can take things one step further in separating common from distinct variation in the case of more than two data blocks. Then there can also exist local components, that is, components shared between only some of the data blocks but not all. Whereas in Chapters 1 and 2 we have described common and distinct components in terms of column-spaces, the ideas can also be carried over to row-spaces which are relevant for the current shared variable case. When discussing also local components, there have to be at least three blocks. For the case of three blocks, this idea is visualised in Figure 5.3. Figure 5.3 is a representation of the sum of the row spaces of the matrices $\mathbf{X}_1(I_1 \times J)$, $\mathbf{X}_2(I_2 \times J)$ and $\mathbf{X}_3(I_3 \times J)$ which is a subspace of $\mathbb{R}^J$. The subspace which is in common for all blocks is called $\mathbf{X}^t_{123C}$. Next, the subspace in common for block one and block two is called $\mathbf{X}^t_{12L}$ with $\mathbf{X}^t_{123C} \cap \mathbf{X}^t_{12L} = \{0\}$. Similar definitions hold for $\mathbf{X}^t_{13L}$ and $\mathbf{X}^t_{23L}$. Finally, the distinct subspaces are indicated by $\mathbf{X}^t_{1D}$, $\mathbf{X}^t_{2D}$ and $\mathbf{X}^t_{3D}$ and these have only $\{0\}$ in common with all other spaces. Simply stated, the sum of the ranges of $\mathbf{X}^t_1$, $\mathbf{X}^t_2$ and $\mathbf{X}^t_3$ is divided in mutually different subspaces (see also Section 2.8).

5.1.2.1 Distinct and Common Components

The distinct and common components (DISCO) method can distinguish common from local and distinct components, and we will explain the DISCO method with three blocks. The DISCO method starts with the SCA solution of Equation 5.1 where the standard SCA-P methods can be used to find the number of components. To find common, local, and distinct components an extra step is required. Suppose that $\mathbf{T}_1, \mathbf{T}_2$, $\mathbf{T}_3$ and $\mathbf{P}$ (with $\mathbf{P}^t\mathbf{P} = \mathbf{I}$) give an optimal approximation of $\mathbf{X}$ according to this SCA model, then $[\mathbf{T}^t_1|\mathbf{T}^t_2|\mathbf{T}^t_3]^t\mathbf{Q}$ and $\mathbf{PQ}$ will do the same for an orthogonal rotation matrix $\mathbf{Q}$. Hence, this rotational freedom can be used to rotate towards a partially specified target that defines distinct and local components as components with all zero scores for the data block in which they are not involved (and keep the remaining parts unspecified) (Schouteden *et al.*, 2014). This target has to be chosen *a priori* and if $\mathbf{T}_{target}$ is this target then $\mathbf{Q}$ can be found by solving:

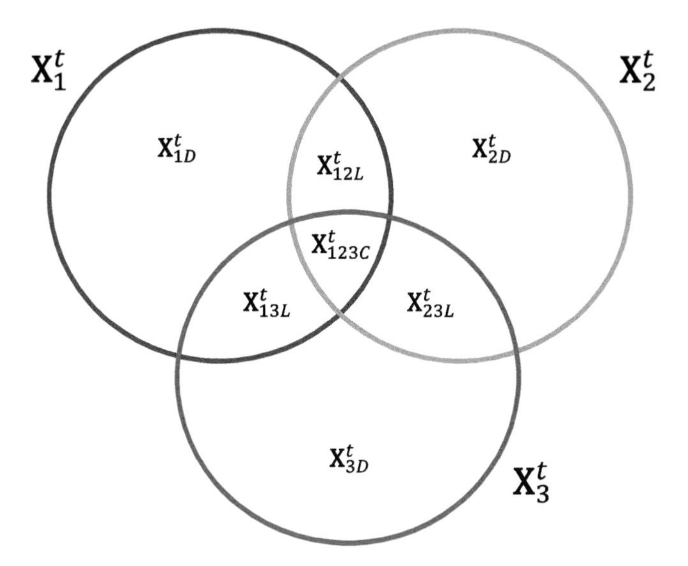

Figure 5.3 The idea of common (C), local (L) and distinct (D) parts of three data blocks. The symbols $\mathbf{X}^t$ denote row spaces; $\mathbf{X}^t_{13L}$, e.g., is the part of $\mathbf{X}^t_1$ and $\mathbf{X}^t_3$ which is in common but does not share a part with $\mathbf{X}^t_2$.

$$\min_{\mathbf{Q}} \|\mathbf{W}_{\text{target}} * (\mathbf{T}_{\text{target}} - [\mathbf{T}_1^t|\mathbf{T}_2^t|\mathbf{T}_3^t]^t\mathbf{Q})\|^2 \; s.t. \; \mathbf{Q}^t\mathbf{Q} = \mathbf{I} \tag{5.26}$$

where $\mathbf{W}_{\text{target}}$ is a matrix with ones on the positions corresponding to the zeros specified in the target and with zeros on all remaining positions (note that * denotes the element-wise or Hadamard product, see Section 1.10). This means that the matrix $\mathbf{Q}$ is found in such a way that the solution is as close as possible to the zeros (i.e., the rotated matrices $\mathbf{T}_m$ come as close as possible to the target). The resulting value of the loss function of Equation 5.26 is called the non-congruence value and indicates to what extent the defined target holds for the data. The choice of the number of common, local and distinct components can be facilitated by trying out different combinations and visualising the results in a plot as shown in Example 5.4. One of the targets to test for these three blocks with five, four, and six samples, respectively, and four overall components may then look like

$$\begin{bmatrix} \mathbf{T}_{1\text{target}} \\ \mathbf{T}_{2\text{target}} \\ \mathbf{T}_{3\text{target}} \end{bmatrix} = \begin{bmatrix} x & x & x & 0 \\ x & x & x & 0 \\ x & x & x & 0 \\ x & x & x & 0 \\ x & x & x & 0 \\ \hdashline x & 0 & 0 & x \\ x & 0 & 0 & x \\ x & 0 & 0 & x \\ x & 0 & 0 & x \\ \hdashline x & x & 0 & 0 \\ x & x & 0 & 0 \\ x & x & 0 & 0 \\ x & x & 0 & 0 \\ x & x & 0 & 0 \\ x & x & 0 & 0 \end{bmatrix} \tag{5.27}$$

where an 'x' is an arbitrary value and 0 is a forced zero. Now the first component is assumed to be in common, the second component is assumed to be local for block one and three, the third component is assumed to be distinct for block one, and the fourth component is assumed to be distinct for block two. Obviously, this is not a very scalable approach since for many blocks the possible number of targets to test increases rapidly. An example of DISCO in the `multiblock` R-package is found in Section 11.6.4. Note that DISCO is a sequential method as explained in Chapter 2: it consists of two standard steps in multivariate data analysis. An example of the use of DISCO in genomics for a two block case is presented in Example 5.2.

Example 5.2: Example of DISCO in genomics

The data pertains to measured gene-expression in humans vaccinated against influenza (Nakaya *et al.*, 2011). Two different types of vaccines (TIV and LAIV) were administered and a micro-array gene-expression experiment was performed for each participant on Day0, Day3 and Day7. We will focus on the data of the difference between the measurements of Day3 and Day0 (Van Deun *et al.*, 2013). This results in two data sets $\mathbf{X}_{TIV}(24 \times 54715)$

and $\mathbf{X}_{LAIV}(27 \times 54715)$ in which the same probe-sets are used for the micro-array measurements. Hence, the variables are shared. A first step in the analysis is to run an SCA model on the concatenated data set (hence, a PCA on the combined (51×54715) data set). The component-wise explained variances are shown in Figure 5.4 for each block separately and for the combination of the two blocks.

From Figure 5.4, we decided to take six overall components since those explain a reasonable amount of variance and make interpretation manageable. The method DISCO was applied on this data, and the resulting proportions of variance accounted for by SCA and DISCO are shown in Table 5.3.

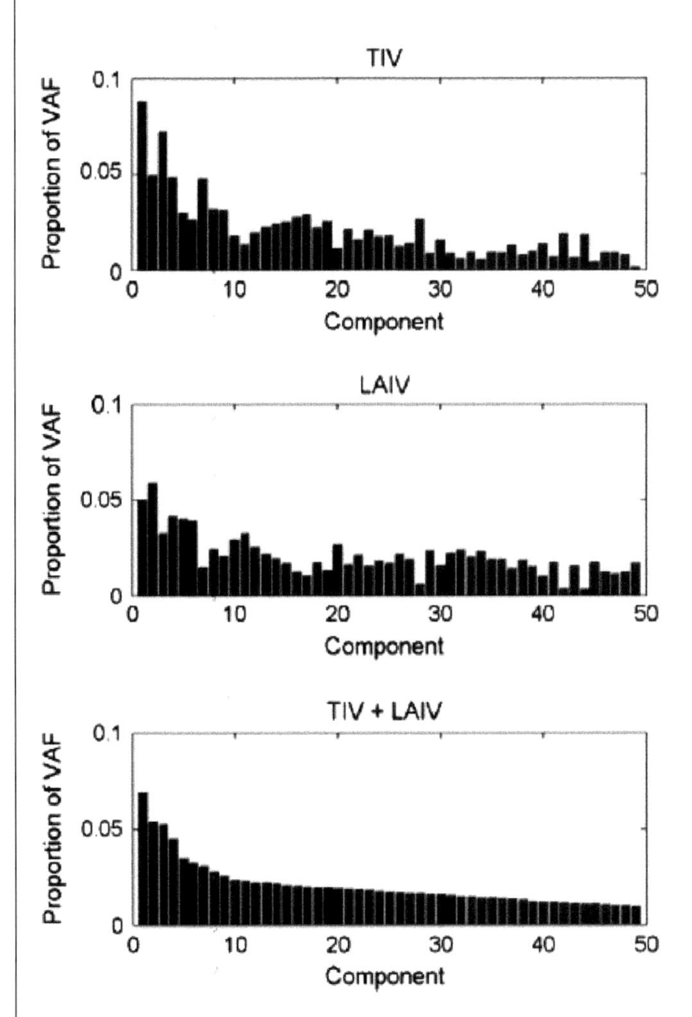

Figure 5.4 Proportion of explained variances (variances accounted for) for the TIV Block (upper part); the LAIV block (middle part) and the concatenated blocks (lower part). Source: Van Deun *et al.* (2013). Reproduced with permission of Elsevier.

Table 5.3 Proportions of explained variance per component (C1, C2,...) and total in each of the blocks for the two different methods. Legend: conc is the abbreviation of concatenated; yellow is distinct for TIV; red is distinct for LAIV; green is common (see text).

	SCA			DISCO		
	TIV	LAIV	Conc.	TIV	LAIV	Conc.
C1	0.09	0.05	0.07	0.02	0.06	0.04
C2	0.05	0.06	0.05	0.09	0.03	0.06
C3	0.07	0.03	0.05	0.08	0.02	0.05
C4	0.05	0.04	0.04	0.05	0.05	0.05
C5	0.03	0.04	0.03	0.04	0.05	0.05
C6	0.03	0.04	0.03	0.03	0.04	0.03
Total	0.31	0.26	0.29	0.31	0.26	0.29

In the SCA columns, the components are ordered as they appear in the SCA solution (largest component first; see Figure 5.4 bottom panel). In the columns for DISCO the order is arbitrary due to permutational freedom of the components. The total explained proportions of variance are the same for SCA and DISCO since the rotation as applied in DISCO does not change this result. The colour coding in Table 5.3 shows the distinct and common components which are chosen by trying different targets and selecting the one with the lowest non-congruence value (see Equation 5.26). This resulted in one distinct component for LAIV (first row in the DISCO part; a higher explained variance for the LAIV column); two distinct components for TIV and three common components. Note that the distinction between distinct and common is not crisp as the explained variances are not exactly zero. For more explanation, see elsewhere (Van Deun *et al.*, 2013).

The DISCO method can also be applied in the case of a shared sample mode, see Section 5.2.2.2.

5.1.2.2 Multivariate Curve Resolution

One of the most used methods in chemometrics in the case of a shared variable mode is multivariate curve resolution (MCR). The goal of MCR is to resolve mixtures of chemical constituents, and in the original form it was meant to analyse a single data block. This area of chemistry started with the seminal paper of Lawton and Sylvestre (1971), was later systematised and expanded (Tauler *et al.*, 1995) and currently has very many applications. In most applications there are spectroscopic measurements involved, and due to Beer's law (Christian and O'Reilly, 1988) such measurements can be expressed in such a way that it allows for bilinear models. Note that centring the matrices in this case is not advisable since it would destroy the structure.

Moving to the case of analysing multiple blocks simultaneously, advantages are an increasing resolving power and less rotational freedom (Tauler *et al.*, 1995). In its most basic form, we consider two blocks of data $\mathbf{X}_1(I_1 \times J)$ and $\mathbf{X}_2(I_2 \times J)$ containing measurements obtained with the same type of instrument, e.g., a UV-spectrometer using the same J wavelengths. Also suppose that these two data blocks have measurements of mixtures of the same R chemical constituents. A model of both data blocks would then be

$$\begin{aligned} \mathbf{X}_1 &= \mathbf{C}_1\mathbf{S}^t + \mathbf{E}_1 \\ \mathbf{X}_2 &= \mathbf{C}_2\mathbf{S}^t + \mathbf{E}_2 \end{aligned} \tag{5.28}$$

where $\mathbf{C}_1\,(I_1 \times R)$ and $\mathbf{C}_2(I_2 \times R)$ contain the concentrations; $\mathbf{S}(J \times R)$ contains the pure spectra of each of the R chemical constituents; $\mathbf{E}_1$ and $\mathbf{E}_2$ contain measurement noise. These models are instances of Beer's law and clearly, the MCR-model of Equation 5.28 is a special case of the SCA model of Equation 5.1. In most cases extra constraints are imposed on the MCR model such as (i) non-negativity of $\mathbf{C}_1$, $\mathbf{C}_2$ and $\mathbf{S}$; (ii) unimodality on $\mathbf{C}_1$ and/or $\mathbf{C}_2$ depending on the context or (iii) closure to reflect mass conservation or a combination of those. There are also MCR versions for a shared sample mode (see Section 5.2.2.6) and for cases with shared samples and shared variables (e.g., L-shaped data, see Section 8.5.3).

Equation 5.28 can be rewritten as

$$\mathbf{X} = \begin{bmatrix} \mathbf{X}_1 \\ \mathbf{X}_2 \end{bmatrix} = \begin{bmatrix} \mathbf{C}_1 \\ \mathbf{C}_2 \end{bmatrix} \mathbf{S}^t + \begin{bmatrix} \mathbf{E}_1 \\ \mathbf{E}_2 \end{bmatrix} = \mathbf{C}\mathbf{S}^t + \mathbf{E} \tag{5.29}$$

where now $\mathbf{C}$ and $\mathbf{S}$ are not least-squares (LS) estimates but are LS-estimates under restrictions as mentioned earlier, and these constraints can be different for $\mathbf{C}_1$ and $\mathbf{C}_2$. The residuals of Equation 5.29 are minimised subject to these constraints. Hence, contrary to the 'free' SCA-P case, the estimates $\mathbf{C}_1$, $\mathbf{C}_2$ and $\mathbf{S}$ are not necessarily in the column and row spaces of $\mathbf{X}$ (see Elaboration 5.4). For MCR, the properties of the estimated parameters are less clear than for ordinary SCA (Rajko, 2009; Tauler, 2010; Rajko, 2010). One of the problems of MCR is rotational freedom which means that the solutions can be rotated to some extent. This is an unwanted property since the solutions should provide chemical information. Some theory about uniqueness is available (Manne, 1995; Smilde *et al.*, 2001) and also methods to establish the 'degree of uniqueness' in terms of rotatability of the solution while still obeying the restrictions (Tauler, 2001; Olivieri and Tauler, 2021).

ELABORATION 5.4

Stay in the row-space or not?

The question whether or not to restrict solutions to be in certain spaces can be illustrated with the case of estimation of pure spectra in MCR. The measured spectra are collected in the rows of the respective matrices. Hence, it may be an option to look for the true pure spectra in the row-spaces of the respective matrices but these true pure spectra are not necessarily located in these row-spaces due to noise on the data. For a single matrix $\mathbf{X}$, Figure 5.5 shows the true pure spectra as blue arrows spanning the true (blue) row-space. For easy of visualisation, the row-spaces are shown but actually the solutions are in a cone due to the restriction of concentrations being non-negative (Lawton and Sylvestre, 1971). The estimated row-space of $\mathbf{X}$ is shown in green and due to the noise on the data this is slightly different from the true row-space. When the estimated spectra are restricted to be in the green space, this results in spectra indicated as green arrows which are clearly not the true ones. Abandoning the constraint of the estimated spectra to be in the row-space of $\mathbf{X}$ may give solutions (red arrows) that are closer to the true ones.

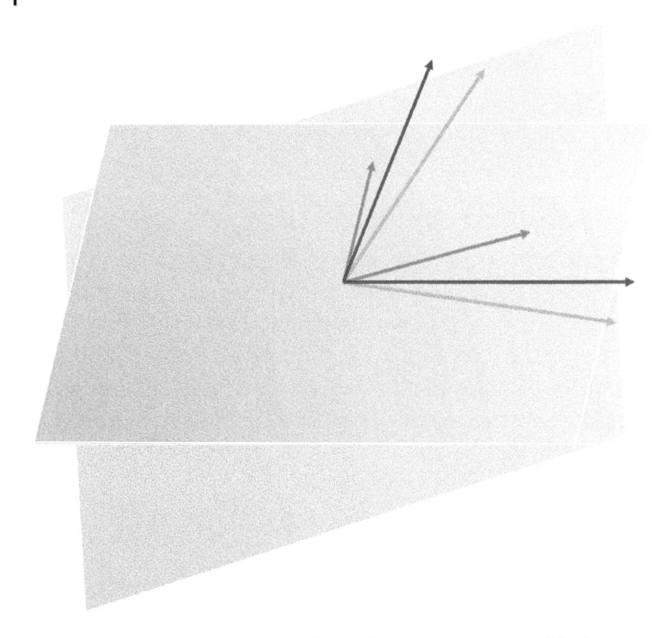

Figure 5.5 Row-spaces visualised. The true row space (blue) contains the pure spectra (blue arrows). The row-space of **X** is the green plane which contains the estimated spectra (green arrows). The red arrows are off the row-space and closer to the true pure spectra.

For homoscedastic noise at low levels, the green and blue subspaces will not differ that much. For heteroscedastic and/or large amounts of noise, these subspaces may differ considerably (de Juan and Tauler, 2020). Of course, in practice we do not know the pure spectra and moving far out of the row-space might be dangerous. One rule-of-thumb may be to calculate the distance of the red arrows to the green plane by projecting the red arrows on the row-space of the respective matrices and calculate the residuals of that projection. These residuals can then be compared to the measurement error. If these residuals are too large, then probably the solution is not reliable.

The method MCR can also be used to distinguish common from local and distinct variation. Starting from Equation 5.29 it is possible to include situations where some chemical constituents are not present in sets of samples. This may look like:

$$\begin{bmatrix} \mathbf{X}_1 \\ \mathbf{X}_2 \\ \mathbf{X}_3 \end{bmatrix} = \begin{bmatrix} \mathbf{c}_{11} & \mathbf{c}_{12} & 0 \\ \mathbf{c}_{21} & 0 & \mathbf{c}_{23} \\ \mathbf{c}_{31} & \mathbf{c}_{32} & 0 \end{bmatrix} \mathbf{S}^t + \begin{bmatrix} \mathbf{E}_1 \\ \mathbf{E}_2 \\ \mathbf{E}_3 \end{bmatrix} \tag{5.30}$$

where the vectors $\mathbf{c}_{ml}$ indicate concentration vectors of constituent l in block m. This shows that chemical constituent one is present in all data blocks, chemical constituent two is only present in data blocks one and three, and constituent three is only present in data block two. Such knowledge may be available beforehand or inferred from the data. This is phrased as selectivity in the MCR literature and special diagnostics and algorithms are available for this situation (Maeder and Zuberbuehler, 1986; de Juan *et al.*, 2014). The final MCR solution can then be calculated under these selectivity constraints.

MCR requires special algorithms depending on the type of constraints applied. Also in this case, the algorithms are (mostly) of the alternating least squares type (Jaumot *et al.*, 2015).

It is also possible to use weighted least squares estimation to account for heterogeneous measurement error (Dadashi *et al.*, 2013).

In general, centring is not recommended for MCR since it would destroy the model structure. Some pretreatment of spectral data may be required prior to data analysis to remove offsets and baselines. MCR has also been used in genomics (Wentzell *et al.*, 2006) which may require yet another type of preprocessing. Also scaling the data is usually not recommended. An exception is the use of maximum likelihood PCA as a preprocessing step to whiten heteroscedastic nose (Hoefsloot *et al.*, 2006; Dadashi *et al.*, 2013).

For MCR, selecting the model and number of components works differently than for general SCA. The model is selected based on chemical knowledge and the number of components is selected based on the quality of the extracted components. If these components get too noisy or their shape too erratic, then probably too many components are used. Also the residuals of the modelling can be used to diagnose model quality.

5.2 Shared Sample Mode

In most cases in multiblock data analysis, the sample mode is shared (see Figure 3.6). This happens often in practice where on the same set of samples different measurements are performed. There are many examples of this in the natural and life sciences; one obvious example being multi-omics studies in which different omics data are collected for the same set of samples. In this section we will describe the most frequently used methods and we will divide the methods in whether they are focused on common variation or trying to separate common from local and distinct variation. The idea of common and distinct (or unique) variation in the shared sample mode is visualised in Figure 1.9.

5.2.1 Only Common Variation

5.2.1.1 SUM-PCA

The most straightforward method for finding common variation is known as SUM-PCA in chemometrics[4] (Smilde *et al.*, 2003). The model is

$$\mathbf{X}_m = \mathbf{T}\mathbf{P}_m^t + \mathbf{E}_m; \; m = 1, \dots, M \tag{5.31}$$

with common components $\mathbf{T}$ $(I \times R)$ and block-specific loadings $\mathbf{P}_m$ $(J_m \times R)$. It has become customary to call this model also SCA (originally developed for the shared variable mode situation, see Section 5.1.1) which may be confusing[5]. To stay in line with common practice, we will call this method SCA from now on. Hence, we will call both variable-wise and sample-wise simultaneous component analysis simply SCA and the meaning will be clear from the context. Upon defining $\mathbf{X} = [\mathbf{X}_1| \dots |\mathbf{X}_M]$, model 5.31 can be fitted by solving

$$\min_{\mathbf{T},\mathbf{P}} \|\mathbf{X} - \mathbf{T}\mathbf{P}^t\|^2 \tag{5.32}$$

where $\mathbf{P}^t = [\mathbf{P}_1^t| \dots |\mathbf{P}_M^t]$ which can be understood as a PCA on $\mathbf{X}$. This problem can be rephrased as

$$\max_{\mathbf{T}} tr(\mathbf{T}^t\mathbf{X}\mathbf{X}^t\mathbf{T}) = \max_{\mathbf{T}} tr(\mathbf{T}^t[\sum_{m=1}^{M} \mathbf{X}_m\mathbf{X}_m^t]\mathbf{T}); s.t. \; \mathbf{T}^t\mathbf{T} = \mathbf{I} \tag{5.33}$$

---

4 Not to be confused with SUMPCA in psychometrics which is equivalent to variable-wise SCA-P (see Section 5.1.1)
5 Multiblock data analysis is rich in confusing terminology!

using the fact that the scores of PCA can be obtained as the eigenvectors of the cross-product $\mathbf{XX}^t$. For this case, this cross-product can be written as the summation in Equation 5.33 explaining the name SUM-PCA. Usually, $\mathbf{T}$ is taken to have orthogonal columns and $\mathbf{P}$ as a matrix with orthonormal columns. This does not mean that each $\mathbf{P}_m$ has orthonormal columns. Note that the SUM-PCA model has rotational freedom just like any PCA model.

In most cases, centring each variable in each block is useful to do since the interest is usually in the differences between samples. Depending on the specific situation, scaling each variable in each block is also an option. Even after auto-scaling per block, the total sum-of-squares in the blocks (which are IJ_m per block) can be very different thereby biasing the solution towards the large blocks. In most cases, it is recommended to give each block sum-of-squares one (after the within-block preprocessing). This is related to the concept of fairness (see Section 1.7).

Equation 5.31 can also be written as

$$\mathbf{X} = \mathbf{TP}^t + \mathbf{E} \tag{5.34}$$

using the definitions of $\mathbf{X}$ and $\mathbf{P}$ as above and defining $\mathbf{E}$ accordingly (see also Equation 5.32). Given the $\mathbf{P}$, this is a regression equation for $\mathbf{X}$ and $\mathbf{T}$ so that the columns of $\mathbf{T}$ are in the column-space of $\mathbf{X}$, but not necessarily in the column-space of the individual data sets. Hence, the columns of $\mathbf{T}$ are actually *consensus* components (see Section 1.8). Also, $\mathbf{P}$ is in the row-space of $\mathbf{X}$, so $\mathbf{P} = \mathbf{X}^t\mathbf{W}$ and thus $\mathbf{P}_m = \mathbf{X}_m^t\mathbf{W}$ which means that $\mathbf{P}_m$ is in the row-space of $\mathbf{X}_m$. It also holds that $\mathbf{T}^t\mathbf{E} = 0$; thus also $\mathbf{T}^t\mathbf{E}_m = 0; m = 1, \ldots, M$. Because of the latter property, explained variation can be calculated per block since

$$tr(\mathbf{X}_m^t\mathbf{X}_m) = tr(\mathbf{P}_m\mathbf{T}^t\mathbf{TP}_m^t) + tr(\mathbf{E}_m^t\mathbf{E}_m);\ m = 1, \ldots, M \tag{5.35}$$

and, likewise, the total sum-of-squares in $\mathbf{X}$ can be split between the parts described by the components and one part of the residuals. If the components $\mathbf{T}$ (and/or $\mathbf{P}$) are chosen to be orthogonal, then the amount of explained variation can also be calculated per component.

Sometimes it is convenient to have scores that are in the column-spaces of the blocks (see Elaboration 5.5)

ELABORATION 5.5

Block-scores

To find scores in the column-space of the associated matrices, we assume $\mathbf{P}_m$ fixed and solve:

$$\mathbf{X}_m = \mathbf{T}_m\mathbf{P}_m^t + \widetilde{\mathbf{E}}_m;\ m = 1, \ldots, M \tag{5.36}$$

for $\mathbf{T}_m$ and thus

$$\mathbf{T}_m = \mathbf{X}_m\mathbf{P}_m(\mathbf{P}_m^t\mathbf{P}_m)^{-1};\ m = 1, \ldots, M \tag{5.37}$$

where we used the tilde to indicate the difference with the previous residuals (from Equation 5.31) and the current ones. Due to this extra regression step, the $\mathbf{T}_m$ is now in the column-space of $\mathbf{X}_m$. To distinguish the two sets of scores, sometimes the $\mathbf{T}$ are called super-scores and $\mathbf{T}_m$ are called block-scores (Westerhuis *et al.*, 1998). Strictly speaking (in our terminology) these block-scores are actually the common scores. Note that in general $\mathbf{E}_m \neq \widetilde{\mathbf{E}}_m$ and also $\|\widetilde{\mathbf{E}}_m\|^2 \leq \|\mathbf{E}_m\|^2$. This change in residuals has repercussions for calculating explained sum-of-squares depending on whether super-scores $\mathbf{T}$ or block-scores $\mathbf{T}_m$ are used.

It is also possible to define a sparse version of SCA (Van Deun *et al.*, 2011). This builds upon the framework of sparse PCA (see Section 2.1.3). A short explanation is given in Elaboration 5.6.

ELABORATION 5.6

Sparse SCA

There are several possibilities to define sparse SCA models (Van Deun *et al.*, 2011) and we will show one version based on sparse loadings. To this end, the following problem can be solved:

$$\min_{\mathbf{T},\mathbf{P}} \|\mathbf{X} - \mathbf{T}\mathbf{P}^t\|^2 + \lambda G(\mathbf{P}) \tag{5.38}$$

where $\mathbf{X}$ is as defined before and $\mathbf{T}$(with $\mathbf{T}^t\mathbf{T} = \mathbf{I}$), are scores and loadings, respectively (see also sparse PCA in Section 2.1.3). The term $G(\mathbf{P})$ is generic for any type of sparsity penalty that can be imposed and $\lambda \geq 0$ is again a tuning parameter to be chosen *a priori*. Examples are the group-lasso or the elastic net (see also Elaboration 5.12); there are many options with different properties (Van Deun *et al.*, 2011).

5.2.1.2 Multiple Factor Analysis and STATIS

The multiple factor analysis (MFA) model resembles the one of SCA:

$$d_{m1}^{-1}\mathbf{X}_m = \mathbf{T}\mathbf{P}_m^t + \mathbf{E}_m;\ m = 1, \ldots, M \tag{5.39}$$

where d_{m1} is the largest singular value of $\mathbf{X}_m(I \times J_m)$ (Pagès, 2005). This amounts to a special kind of prior scaling of the data block whereby differences in number of variables and of redundancy of information is corrected for. This can be seen by realising that the 'size' of a matrix can be measured by the sum of the squared singular values $\sum_l d_{ml}^2$; where $l = 1, \ldots L$ is an index for the singular values (this is the squared Frobenius norm, see Section 2.8, and L is the rank of the matrix) which is J_m for auto-scaled data. To correct for size differences, we have to divide the matrix $\mathbf{X}_m$ by the term $\sqrt{\sum_l d_{ml}^2}$. Another aspect which relates to the fairness concept is redundancy. This can be measured by $d_{m1}^2 / \sum_l d_{ml}^2$ and if a matrix is very redundant (that is, almost of rank one), then the term $d_{m1}^2 / \sum_l d_{ml}^2$ is almost one. Such a matrix will not contribute a lot to the total solution of Equation 5.39 (maybe only in the first component). To correct for this redundancy, the matrix $\mathbf{X}_m$ can be divided by the term $\sqrt{d_{m1}^2 / \sum_l d_{ml}^2}$ thereby giving it less influence. Hence, in total this amounts to the matrix $\mathbf{X}_m$ being corrected (i.e., block-scaled) by

$$\frac{1}{\sqrt{\sum_l d_{ml}^2}} \frac{1}{\sqrt{\frac{d_{m1}^2}{\sum_l d_{ml}^2}}} \mathbf{X}_m = d_{m1}^{-1}\mathbf{X}_m;\ m = 1, \ldots, M \tag{5.40}$$

which is the correction used by MFA. It is customary to estimate MFA under the constraint that $\mathbf{T}^t\mathbf{T} = \mathbf{I}$ to identify the solution. The estimation procedure is then minimising the sum of squared residuals ($\mathbf{E}_m$) summed across all M blocks under the above orthogonality constraint of $\mathbf{T}$, e.g., by using an SVD on the concatenated corrected matrices $\mathbf{X}_m$. An example of MFA in the `multiblock` R-package is found in Section 11.6.3.

The STATIS (Structuration des Tableaux à Trois Indices de la Statistique) method is very similar to MFA and the basic equation is

$$a_m \mathbf{X}_m = \mathbf{T}\mathbf{P}_m^t + \mathbf{E}_m; \; m = 1, \ldots, M \tag{5.41}$$

where the weights a_m are derived in the following way (Lavit *et al.*, 1994). First, derive the cross-product matrices $\mathbf{S}_m = \mathbf{X}_m \mathbf{X}_m^t$ $(m = 1, \ldots, M)$ (these matrices contain the covariances between objects). Secondly, form the matrix $\mathbf{Z}$ by vectorising the matrices $\mathbf{S}_m$ and placing these M vectors as columns in this matrix. The weights a_m are now the loadings of the first principal component of $\mathbf{Z}$. Larger weights can be expected for (a) blocks with larger values, (b) larger blocks, (c) blocks with more covariance between the objects, and (d) blocks with more similar cross-product matrices to other matrices (Van Deun *et al.*, 2009). The latter property is the motivation for STATIS: it wants to find consensus between the cross-product matrices. Estimation proceeds by minimising the sum of squared errors in Equation 5.41. Note that the correlation between the vectorised matrices $\mathbf{S}_m$ is exactly the RV-coefficient between the associated matrices (see Section 4.2.3). Hence, the weights can also be regarded as being calculated using the RV-coefficients between all pairs of matrices. An example of STATIS in the `multiblock` R-package is found in Section 11.6.3.

Preprocessing for MFA and STATIS is similar to the preprocessing of SCA. Within-block autoscaling can be performed (prior to calculating the weights!) and – if needed – block-scaling can be applied (Stanimirova *et al.*, 2004) but this may interfere with the applied weights.

From the above, it is clear that SCA, MFA and STATIS can be presented in a unified framework (Van Deun *et al.*, 2009):

$$v_m \mathbf{X}_m = \mathbf{T}\mathbf{P}_m^t + \mathbf{E}_m; \; m = 1, \ldots, M \tag{5.42}$$

where the methods only differ regarding their choice of the weights v_m.

5.2.1.3 Generalised Canonical Analysis

Generalised canonical analysis (GCA) was already introduced in Section 2.1.10 but for completeness is treated here again more extensively. The basic idea is to identify linear combinations of the blocks, $\mathbf{X}_m \mathbf{W}_m$, which fit best to a set of orthonormal components collected in $\mathbf{T}$. This is done by minimising the criterion

$$\min_{(\mathbf{T}, \mathbf{W}_m)} \sum_{m=1}^{M} \|\mathbf{X}_m \mathbf{W}_m - \mathbf{T}\|^2 \tag{5.43}$$

with respect to $\mathbf{T}^T\mathbf{T} = \mathbf{I}$ and $\mathbf{W}_m$ $(m = 1, \ldots, M)$ (Van der Burg and Dijksterhuis, 1996).

In general, the columns in the respective data blocks will be centred prior to a GCA analysis. GCA is insensitive to within- and between-block scaling since such scaling weights can be compensated for by the $\mathbf{W}_m$s. This may or may not be a useful property.

The same solution as in Equation 5.43 can be obtained by maximising a sum of squared correlations between linear combinations of the X blocks (Van der Burg and Dijksterhuis, 1996), which is a direct generalisation of the case with only two X blocks (Hotelling, 1936b). In practice, the actual solution $\mathbf{T}$ is found as the eigenvectors (scaled to length one) of the matrix

$$\mathbf{Z}_{GCA} = \sum_{m=1}^{M} \mathbf{X}_m (\mathbf{X}_m^t \mathbf{X}_m)^+ \mathbf{X}_m^t \tag{5.44}$$

where the superscript + means the Moore–Penrose (pseudo-)inverse (see Section 2.8). The weights $\mathbf{W}_m$ can then be found by regressing $\mathbf{T}$ on $\mathbf{X}_m$: $\mathbf{W}_m = \mathbf{X}_m^+\mathbf{T}$. Note that the matrix in the middle $(\mathbf{X}_m^t\mathbf{X}_m)$ of Equation 5.44 may contain small eigenvalues of which the inverse is taken. Hence, GCA may be sensitive to noise. For a proof of Equation 5.44, see Elaboration 5.7.

ELABORATION 5.7

GCA as an eigenproblem

We will prove Equation 5.44 for the case of two matrices $\mathbf{X}_1$ and $\mathbf{X}_2$. The GCA problem is then:

$$\min_{(\mathbf{T},\mathbf{W}_1,\mathbf{W}_2)} \|\mathbf{X}_1\mathbf{W}_1 - \mathbf{T}\|^2 + \|\mathbf{X}_2\mathbf{W}_2 - \mathbf{T}\|^2 \tag{5.45}$$

under the constraint $\mathbf{T}^t\mathbf{T} = \mathbf{I}$. Given a solution for $\mathbf{T}$ the $\mathbf{W}'s$ can be found as $\mathbf{W}_1 = \mathbf{X}_1^+\mathbf{T}$ and $\mathbf{W}_2 = \mathbf{X}_2^+\mathbf{T}$. Filling in these solutions in Equation 5.45 and working out the sums-of-squares in traces using $\|\mathbf{X}\|^2 = tr(\mathbf{X}^t\mathbf{X})$, the GCA problem becomes

$$\min_{\mathbf{T}} tr(-\mathbf{T}^t\mathbf{X}_1\mathbf{X}_1^+\mathbf{T} + \mathbf{I} - \mathbf{T}^t\mathbf{X}_2\mathbf{X}_2^+\mathbf{T} + \mathbf{I}) \tag{5.46}$$

which has the same solution as the problem

$$\max_{\mathbf{T}} tr(\mathbf{T}^t\mathbf{X}_1\mathbf{X}_1^+\mathbf{T} + \mathbf{T}^t\mathbf{X}_2\mathbf{X}_2^+\mathbf{T}) \tag{5.47}$$

which can be solved by taking $\mathbf{T}$ as the leading eigenvectors (scaled to length one) of the matrix $\sum_{m=1}^{2}\mathbf{X}_m\mathbf{X}_m^+ = \sum_{m=1}^{2}\mathbf{X}_m(\mathbf{X}_m^t\mathbf{X}_m)^+\mathbf{X}_m^t$.

In our strict terminology for common and distinct, the components as collected in $\mathbf{T}$ are consensus components which are situated in $R([\mathbf{X}_1|\mathbf{X}_2])$ and the common components are $\mathbf{T}_m = \mathbf{X}_m\mathbf{W}_m$. Interpretation of GCA models can be done by studying the components $\mathbf{T}$ and the weights $\mathbf{W}_m$ making up these components. However, with interpreting $\mathbf{W}_m$ we have to be careful: these do not necessarily reflect the importance of the variables, see Elaboration 5.8.

ELABORATION 5.8

GCA Correlation loadings

Correlation loadings can be defined for PCA (see Section 2.1.2) but a similar concept can be used for GCA which is shown in the following. The weights $\mathbf{W}_m$ may not reflect the importance of variables (Van der Burg and Dijksterhuis, 1996). To see this, consider again calculating the weights $\mathbf{W}_m$ from $\mathbf{W}_m = \mathbf{X}_m^+\mathbf{T}$. This is actually the solution of the regression equation

$$\mathbf{T} = \mathbf{X}_m\mathbf{W}_m + \mathbf{E}_m \tag{5.48}$$

where $\mathbf{T}$ has R components; $\mathbf{W}_m$ has size $(J_m \times R)$ and $\mathbf{T}_m = \mathbf{X}_m\mathbf{W}_m = \mathbf{X}_m\mathbf{X}_m^+\mathbf{T}$ and thus the common components $\mathbf{T}_m$ are simply the orthogonal projections of the consensus components $\mathbf{T}$ onto the column-space of $\mathbf{X}_m$, i.e., as close as possible to $\mathbf{T}$.

Now, we can calculate the correlation loadings between the consensus components $\mathbf{T}$ and the variables in $\mathbf{X}_m$ which are $\mathbf{P}_m = \mathbf{X}_m^t\mathbf{T}$ $(J_m \times R)$, assuming that both $\mathbf{T}$ and $\mathbf{X}_m$ are standardised (without lack of generality). This equation can be seen as the solution of the regression problem

$$\mathbf{X}_m = \mathbf{T}\mathbf{P}_m^t + \mathbf{F}_m \tag{5.49}$$

for a given $\mathbf{T}$ since $\mathbf{T}^t\mathbf{T} = \mathbf{I}$. In general, the values in $\mathbf{P}_m$ are different from the ones in $\mathbf{W}_m$ and these correlation loadings can be interpreted as the importance of the variables.

It is also possible to express the variables' importance in terms of the common components. Then we have to solve the regression

$$\mathbf{X}_m = \mathbf{T}_m\mathbf{Q}_m^t + \widetilde{\mathbf{F}}_m \tag{5.50}$$

where $\mathbf{T}_m = \mathbf{X}_m\mathbf{W}_m$ and $\mathbf{Q}_m$ are the loadings (i.e., regression coefficients of Equation 5.50 for fixed $\mathbf{T}_m$). This equation is the basis for the method PCA-GCA (see Section 5.2.2.3) since a PCA on $\widetilde{\mathbf{F}}_m$ would give the distinct components (to be discussed later). The difference between the matrices $\mathbf{W}_m$, $\mathbf{P}_m$ and $\mathbf{Q}_m$ is visualised in Figure 5.6 where we assumed, for simplicity, a single consensus component $\mathbf{t}$.

In this figure, the column-space of $\mathbf{X}_m$ is shown where $\mathbf{X}_m$ has two variables, $\mathbf{x}_{m1}$ and $\mathbf{x}_{m2}$ (green arrows) which are not orthogonal. The consensus component $\mathbf{t}$ (red arrow) is outside this column-space and the common component $\mathbf{t}_m$ (blue arrow) is the projection of $\mathbf{t}$ onto that column-space. This $\mathbf{t}_m$ has weights w_1 and w_2 (same as the ones for $\mathbf{t}$) and clearly variable one contributes more to the consensus component. The loadings $\mathbf{p}$ from Equation 5.49 are shown in the figure as p_1

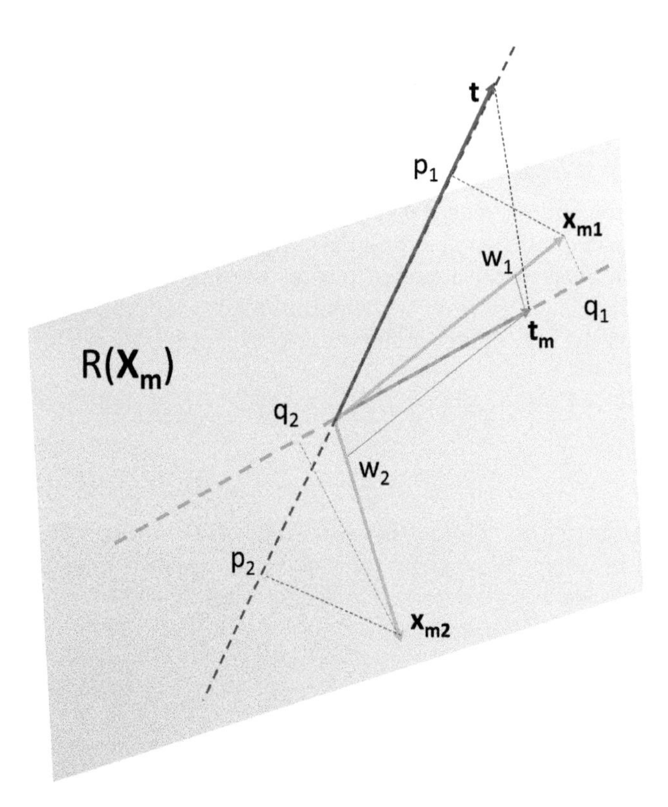

Figure 5.6 Difference between weights and correlation loadings explained. Green arrows are variables of $\mathbf{X}_m$; red arrow is the consensus component $\mathbf{t}$; blue arrow is the common component $\mathbf{t}_m$. Dotted lines represent projections.

and p_2 which are obtained by projecting the variables of $\mathbf{X}$ onto $\mathbf{t}$. The loadings of the variables on the common component (Equation 5.50) are shown as the values q_1 and q_2. For the first variable, the weight and loading are similar (although not the same), but for the second variable the weight and loading are very different. Hence, all three weights and loadings are different and so is their interpretation. The weights w show the contribution of the variables $\mathbf{x}_{m1}$ and $\mathbf{x}_{m2}$ to the consensus component; the correlation loadings p show how these variables are correlated with the consensus component and the loadings q show how the variables are related to the common component.

There is an interesting relationship between GCA and SCA, see, e.g., Dahl and Næs (2006). SCA finds the components as eigenvectors of the matrix $\mathbf{Z}_{SCA} = \sum_{m=1}^{M} \mathbf{X}_m \mathbf{X}_m^t$ and GCA finds its components as eigenvectors of $\mathbf{Z}_{GCA} = \sum_{m=1}^{M} \mathbf{X}_m(\mathbf{X}_m^t \mathbf{X}_m)^+ \mathbf{X}_m^t$. This suggest a hybrid method based on finding eigenvectors of $\mathbf{Z}_\alpha = \sum_{m=1}^{M} \mathbf{X}_m[(1-\alpha)\mathbf{X}_m^t \mathbf{X}_m) + \alpha \mathbf{I}]^+ \mathbf{X}_m^t$ with $0 \leq \alpha \leq 1$. If $\alpha = 0$, we obtain GCA and for $\alpha = 1$, SCA is obtained.

The differences between GCA and SCA also becomes clear when using the SVD of $\mathbf{X}_m = \mathbf{U}_m \mathbf{D}_m \mathbf{V}_m^t$ and writing out the $\mathbf{Z}$ matrices in both cases. Then $\mathbf{Z}_{SCA} = \sum_{m=1}^{M} \mathbf{U}_m \mathbf{D}_m^2 \mathbf{U}_m^t$ and $\mathbf{Z}_{GCA} = \sum_{m=1}^{M} \mathbf{U}_m \mathbf{U}_m^t$. The matrices $\mathbf{D}_m$ encode the within-block correlations strengths and thus GCA is not considering these within-block correlations whereas SCA does consider this. Again, this goes back to the fundamental choices (see Section 1.7). An example of GCA in the `multiblock` R-package is found in Section 11.6.3.

5.2.1.4 Regularised Generalised Canonical Correlation Analysis

GCA cannot be used when the number of variables exceeds the number of samples. Hence, in high-dimensional settings GCA does not work. To combat this problem, in the psychometrics tradition a path has been chosen that led to regularised generalised canonical correlation analysis (RGCCA) (Tenenhaus *et al.*, 2017). This method has a close resemblance with multigroup data analysis (see Section 5.1.1.3). The basic equation for RGCCA is:

$$\begin{aligned} &\max_{\mathbf{w}_m, \mathbf{w}'_m} \sum_{m,m'=1}^{M} c_{m,m'} g[\text{cov}(\mathbf{X}_m \mathbf{w}_m, \mathbf{X}_{m'} \mathbf{w}_{m'})] \\ &s.t.\ \mathbf{w}_m^t \boldsymbol{\Psi}_m \mathbf{w}_m = 1\ ; m, m' = 1, \ldots, M \end{aligned} \tag{5.51}$$

where $c_{m,m'}$ (zero or one) encodes which blocks are connected; $g[.]$ is a continuously differentiable convex function including $g[x] = |x|$ (the case of $x = 0$ does not occur in practical applications since that would mean that $\text{cov}(\mathbf{X}_m \mathbf{w}_m, \mathbf{X}_{m'} \mathbf{w}_{m'}) = 0$ or $\mathbf{X}_m^t \mathbf{X}_{m'} = 0$) and $\boldsymbol{\Psi}_m = (1 - \tau_m)(\frac{1}{I})\mathbf{X}_m^t \mathbf{X}_m + \tau_m \mathbf{I}$ is a positive definite matrix with regularisation parameter $0 \leq \tau_m \leq 1$. The regularisation comes into play by using the identity matrix $\mathbf{I}$. Problem (5.51) can be simplified by using the transformations $\mathbf{Z}_m = \mathbf{X}_m \boldsymbol{\Psi}_m^{-1/2}$ and $\mathbf{v}_m = \boldsymbol{\Psi}_m^{1/2} \mathbf{w}_m$. Then the problem becomes:

$$\begin{aligned} &\max_{\mathbf{v}_m, \mathbf{v}'_m} \sum_{m,m'=1}^{M} c_{m,m'} g[cov(\mathbf{Z}_m \mathbf{v}_m, \mathbf{Z}_{m'} \mathbf{v}_{m'})] \\ &s.t.\ \mathbf{v}_m^t \mathbf{v}_m = 1\ ; m, m' = 1, \ldots, M \end{aligned} \tag{5.52}$$

which is easier to solve. Subsequent components are calculated orthogonal to the previous ones, e.g., using a deflation step. A simple example for a three-block situation is given in Elaboration 5.9. As stated already in Section 5.1.1.3 there are ways to select the regularisation parameters. RGCCA is a very versatile method and encompasses many multiblock data analysis methods based on calculating subsequent components as special cases, such as PLS and redundancy analysis (see also Chapter 10). Hence, it can also serve as a framework for such multiblock methods. A generic algorithm for solving problem 5.51 is available (Tenenhaus *et al.*, 2017). RGCCA is available in the R-package RGCCA (Table 11.1).

ELABORATION 5.9

A simple example of RGCCA

A simple example may serve to understand the machinery of RGCCA better. We assume three blocks of data $\mathbf{X}_1(I \times J_1)$, $\mathbf{X}_2(I \times J_2)$ and $\mathbf{X}_3(I \times J_3)$; $\tau_m = 1(\mathbf{\Psi}_m = \mathbf{I}); \forall m$; $c_{m,m'} = 1$ for $m \neq m'$; $c_{m,m} = 0; \forall m$ and the function $g(.)$ is the identity. Then Equation 5.51 becomes

$$\max_{\mathbf{w}_m' s}[2\text{cov}(\mathbf{X}_1\mathbf{w}_1, \mathbf{X}_2\mathbf{w}_2) + 2\text{cov}(\mathbf{X}_1\mathbf{w}_1, \mathbf{X}_3\mathbf{w}_3) + 2\text{cov}(\mathbf{X}_2\mathbf{w}_2, \mathbf{X}_3\mathbf{w}_3)]$$
$$s.t.\, \mathbf{w}_m^t \mathbf{w}_m = 1\,; m, m' = 1, \ldots, M \tag{5.53}$$

and upon using $\mathbf{t}_m = \mathbf{X}_m\mathbf{w}_m$ (and dropping the scalar 2), this can also be written as

$$\max_{\mathbf{w}_m' s}[\mathbf{t}_1^t\mathbf{t}_2 + \mathbf{t}_1^t\mathbf{t}_3 + \mathbf{t}_2^t\mathbf{t}_3]$$
$$s.t.\, \mathbf{w}_m^t \mathbf{w}_m = 1\,; m, m' = 1, \ldots, M \tag{5.54}$$

Hence, the scores of the blocks are found in such a way that they covary as much as possible. Note that due to the length-one restrictions on the weights $\mathbf{w}_m$ the scores are also LS for the blocks.

5.2.1.5 Exponential Family SCA

Many of the heterogeneous data in the natural and life sciences can theoretically be modelled using a Gaussian, Bernoulli, Poisson, or negative binomial distribution. These distributions are important in the life sciences since quantitative (interval- or ratio-scaled) data can be modelled using the normal distribution; binary data can be modelled with the Bernoulli distribution and count data with the Poisson or negative Binomial distribution. Examples of these kinds of data will be given in the following sections. The ESCA framework uses these distributions and is thus very useful for modelling heterogeneous multiblock data.

We will explain the generalised-SCA (GSCA) and its extension exponential family-SCA (ESCA) method by using the example of modelling two blocks of data, namely a binary block $\mathbf{X}_1(I \times J_1)$ and a block $\mathbf{X}_2(I \times J_2)$ containing quantitative measurements (see Section 2.2.2 and Elaboration 2.7). The basic idea is to define a systematic underlying latent variable model for both blocks where the samples are shared and all information is in common and represented by the shared latent variables. Hence, we assume that there are no distinct systematic parts in the blocks. Later, in the PESCA model this assumption is relaxed, see Section 5.2.2.5. Then the model is made stochastic by assuming a distribution for the measurements conditional on the systematic part. Subsequently, the likelihood functions of both parts can be derived and combined.

For $\mathbf{X}_1$ we assume that there is a low-dimensional systematic structure $\mathbf{\Theta}_1(I \times J_1)$ underlying $\mathbf{X}_1$ and the elements of $\mathbf{X}_1$ (which are all zeros or ones) follow a Bernoulli distribution with parameters ϕ_{1ij}, thus $x_{1ij} \sim B(\phi_{1ij})$ (each element of $\mathbf{X}_1$ receives its own Bernoulli parameter). The logistic function $\phi = \eta(\theta) = (1+\exp(-\theta))^{-1}$ is then used and x_{1ij}, θ_{1ij} are the ij^{th} elements of $\mathbf{X}_1$ and $\mathbf{\Theta}_1$, respectively. The logistic function is explained in Elaboration 5.10. The $\mathbf{\Theta}_1$ is now assumed to be equal to $\mathbf{1}\boldsymbol{\mu}_1^t + \mathbf{T}\mathbf{P}_1^t$ where $\boldsymbol{\mu}_1$ represents the offset term, $\mathbf{T}$ the consensus scores and $\mathbf{P}_1$ the loadings for the binary data.

ELABORATION 5.10

The logistic function

The logistic function is often used as a function to go from the real domain $\theta \in \Re$ to a value in the interval $(0, 1)$ which may represent a probability. This function is shown in Figure 5.7.

This function is used much in generalised linear models and specifically in logistic regression (e.g., for classification purposes).

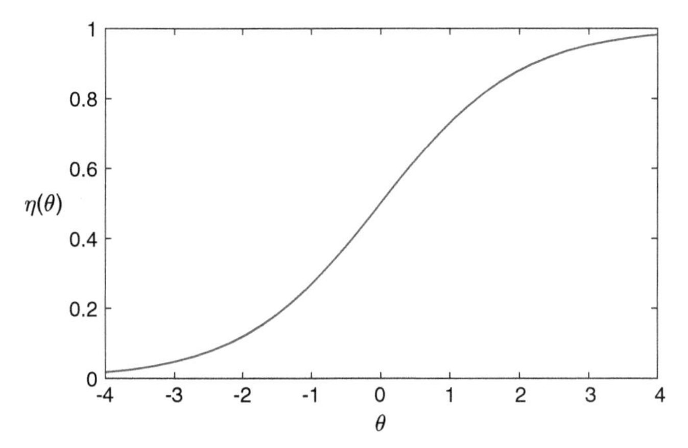

Figure 5.7 The logistic function $\eta(\theta) = (1 + exp(-\theta))^{-1}$ visualised. Only the part between [−4,4] is shown but the function goes from $-\infty$ to $+\infty$.

The quantitative measurements $\mathbf{X}_2$ are assumed to follow the model $\mathbf{X}_2 = \mathbf{1}\boldsymbol{\mu}_2^t + \mathbf{T}\mathbf{P}_2^t + \mathbf{E} = \boldsymbol{\Theta}_2 + \mathbf{E}$ where the elements e_{ij} of $\mathbf{E}$ are independently normally distributed with mean 0 and variance σ^2. The matrix $\mathbf{P}_2$ contains the loadings of the quantitative data; $\mathbf{T}$ are the consensus scores and the constraints $\mathbf{T}^t\mathbf{T} = \mathbf{I}_R$ and $\mathbf{1}^t\mathbf{T} = 0$ are imposed for identifiability (hence, the scores are orthonormal and centred). Note that for this model, we assume $\mathbf{X}_1$ and $\mathbf{X}_2$ not to be centred. For technical reasons, these centres are modelled explicitly through the parameters $\boldsymbol{\mu}_1$ and $\boldsymbol{\mu}_2$. The shared information between $\mathbf{X}_1$ and $\mathbf{X}_2$ is assumed to be represented by the consensus latent variables $\mathbf{T}$. Thus $\mathbf{X}_1$ and $\mathbf{X}_2$ are stochastically independent given these latent variables and the negative log-likelihoods of both parts can be summed:

$$f_1(\boldsymbol{\Theta}_1) = -\sum_i^I \sum_j^{J_1} [x_{1ij} log(\eta(\theta_{1ij})) + (1 - x_{1ij}) log(1 - \eta(\theta_{1ij}))] \tag{5.55}$$

$$f_2(\boldsymbol{\Theta}_2, \sigma^2) = \frac{1}{2\sigma^2} ||\mathbf{X}_2 - \boldsymbol{\Theta}_2||_F^2 + \frac{1}{2} log(2\pi\sigma^2)$$

$$f(\boldsymbol{\Theta}_1, \boldsymbol{\Theta}_2, \sigma^2) = f_1(\boldsymbol{\Theta}_1) + f_2(\boldsymbol{\Theta}_2, \sigma^2)$$

and minimised[6] simultaneously for the parameters $\mathbf{T}, \mathbf{P}_1, \mathbf{P}_2, \boldsymbol{\mu}_1, \boldsymbol{\mu}_2, \sigma^2$ under the constraints $\boldsymbol{\Theta}_1 = \mathbf{1}\boldsymbol{\mu}_1^t + \mathbf{T}\mathbf{P}_1^t; \boldsymbol{\Theta}_2 = \mathbf{1}\boldsymbol{\mu}_2^t + \mathbf{T}\mathbf{P}_2^t$ and some extra constraints; details are given elsewhere (Song *et al.*, 2018). This method is called generalised simultaneous component analysis (GSCA) and an example is given below (Example 5.3). This method can be extended to any member of the exponential family of distributions and that method is called exponential family SCA (ESCA) (Song *et al.*, 2019).

Example 5.3: GSCA example

The example for GSCA is from cancer research, specifically, from cancer cell lines (Iorio *et al.*, 2016). Cancer cells from breast, lung, and skin cancer were analysed in terms of gene-expression and copy number aberration (CNA) (see Example 1.2 for an explanation). This resulted in a binary CNA data block $\mathbf{X}_1(160 \times 410)$ and data block $\mathbf{X}_2(160 \times 1000)$ containing gene-expression of the same cells. The CNA data are extremely sparse and unbalanced, as can be seen in Figure 5.8.

For the sake of illustration, a GSCA model is fitted using only three components. Prior to fitting this model, we also fitted a logistic PCA (Song *et al.*, 2017) on the CNA data. The resulting score plot is shown in Figure 5.9(a) which shows no structure. The consensus scores ($\mathbf{T}$) plot of the GSCA model (Figure 5.9(b)) shows clear clustering, and even within the breast cancer the hormone-positive group and an MITF-high group within the skin cancer group form clusters. Hence, the gene-expression helps in finding clear patterns in the CNA data resulting in better interpretable CNA loading plots. A PCA on the gene-expression alone reveals a very similar score plot as for the GSCA model indicating that indeed the gene-expression 'steers' the CNA data in an interpretable direction (Smilde *et al.*, 2020).

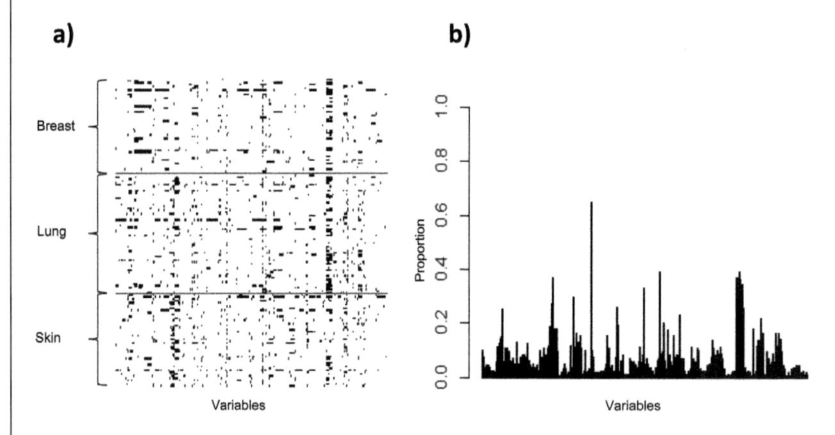

Figure 5.8 CNA data visualised. Legend: (a) each line is a sample (cell line), blanks are zeros and black dots are ones; (b) the proportion of ones per variable illustrating the unbalancedness. Source: Song *et al.* (2021). Reproduced with permission of Elsevier.

6 It is customary to minimise the negative log-likelihood instead of maximising the log-likelihood.

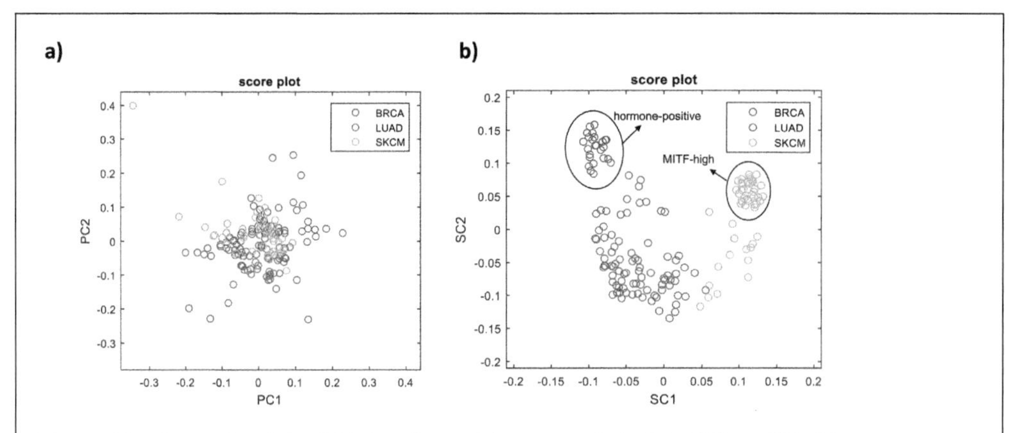

Figure 5.9 Score plot of the CNA data. Legend: (a) scores of a logistic PCA on CNA; (b) consensus scores of the first two GSCA components of a GSCA model (MITF is a special gene). Source: Smilde *et al.* (2020). Licensed under CC BY 4.0.

5.2.1.6 Optimal-scaling

An alternative to using GSCA or ESCA for heterogeneous data is optimal-scaling. The concept of optimal-scaling was already very briefly explained in Elaboration 2.9. A more detailed exposure is given here. Since optimal-scaling can handle data of different measurement scales, it is very suitable for heterogeneous data fusion. It gives, e.g., a straightforward way of generalising SCA to the case of heterogeneous data. This will be explained at the end of the section after introducing the concept of optimal-scaling. Optimal-scaling is a general methodology, and we will explain it starting with the example of PCA.

BASIC IDEA

As already briefly stated in Chapter 2 the basic equation of optimal-scaling for PCA is

$$\min_{H(.),\mathbf{T},\mathbf{P}} ||H(\mathbf{X}) - \mathbf{T}\mathbf{P}^t||^2 \tag{5.56}$$

where $\mathbf{X}$ is a matrix containing the original data (of any measurement scale) and $H(.)$ is a function that depends on the type of measurement scale of the data in $\mathbf{X}$. The basic idea is to make the data in $\mathbf{X}$ 'interval-scale-like' through the function $H(.)$ such that it is amenable to a PCA model with scores $\mathbf{T}$ and loadings $\mathbf{P}$.

To give an idea of the functions $H(.)$ and the machinery behind optimal-scaling we will follow the exposition given by Michailidis and de Leeuw (1998). Suppose that the matrix $\mathbf{X}(I \times J)$ contains J categorical variables not necessarily with the same number of categories. To arrive at an 'interval-scale' level of these categorical variables we need two ingredients: (i) numerical encodings for these categorical variables, and (ii) quantifications that transform these encoded categorical variables into 'interval-scale' level. The encoding for each variable $\mathbf{x}_j$ is done using an indicator matrix $\mathbf{G}_j(I \times L_j)$ where L_j is the number of categories for variable j and a one indicates membership of the associated category. The idea of optimal-scaling is to find category quantification matrices $\mathbf{Y}_j$ $(L_j \times R; j = 1, \ldots, J)$ such that these fit as good as possible to a set of scores $\mathbf{T}$ $(I \times R)$ by solving the following problem (Michailidis and de Leeuw, 1998):

$$\min_{\mathbf{T},\mathbf{Y}_j} \sum_{j=1}^{J} ||\mathbf{T} - \mathbf{G}_j\mathbf{Y}_j||^2 \tag{5.57}$$

under the constraints that $(1/I)\mathbf{T}^t\mathbf{T} = \mathbf{I}$ and that these scores are centred around zero (to avoid trivial solutions of Equation 5.57). This method – including the alternating optimisation method to solve Equation 5.57 – is called homogeneity analysis or HOMALS for short (Gifi, 1990). The rows of $\mathbf{T}$ give a low dimensional representation of the objects, and the matrices $\mathbf{Y}_j(j = 1, \ldots, J)$ give the optimal quantifications of the categorical variables. These optimal quantifications $\mathbf{Y}_j(j = 1, \ldots, J)$ are different for the R components, namely $\mathbf{y}_{jr}(L_j \times 1; r = 1, \ldots, R)$ where $\mathbf{y}_{jr}$ is the rth column of $\mathbf{Y}_j$. These quantifications transform the categorical variables into interval-scaled-like variables and thus cannot be interpreted as loadings.

The optimal quantifications as calculated above are different per component, and this may be difficult to interpret. It is convenient to restrict the rank of $\mathbf{Y}_j(j = 1, \ldots, J)$ to be one (by using $\mathbf{Y}_j = \mathbf{y}_j\mathbf{p}_j^t$) thereby making the quantifications across the r components proportional which simplifies interpretation. Then Equation 5.57 can be rewritten as

$$\min_{\mathbf{T},\mathbf{y}_j,\mathbf{p}_j} \sum_{j=1}^{J} ||\mathbf{T} - \mathbf{G}_j\mathbf{y}_j\mathbf{p}_j^t||^2 \tag{5.58}$$

with the same constraints on $\mathbf{T}$ as before. In order to identify the solution we impose $\mathbf{y}_j^t\mathbf{G}_j^t\mathbf{G}_j\mathbf{y}_j = I$. Now, the vectors $\mathbf{p}_j(R \times 1)$ are the loadings and $\mathbf{y}_j(L_j \times 1)$ contain the quantifications which are the same for all R dimensions of the solution. Finally, Equation 5.58 can be written as

$$\min_{\mathbf{T},\mathbf{y}_j,\mathbf{p}_j} ||H(\mathbf{X}) - \mathbf{T}\mathbf{P}^t||^2 \tag{5.59}$$

where $H(\mathbf{X})$ represents the optimal-scaled data. Hence, optimal-scaling performs two things: (i) it (optimally) transforms the data, and (ii) gives a low dimensional representation of this transformed data. This is done in an iterative way: optimal-scaling is not a preprocessing step which is performed once on the raw data but works as an iterative process in which the optimal quantifications are obtained and simultaneously a low rank approximation of the optimal-scaled data is obtained. Equation 5.59 resembles PCA with scores $\mathbf{T}$ and loadings $\mathbf{P}$; this solution is therefore called non-linear PCA (PRINCALS) (Gifi, 1990; Michailidis and de Leeuw, 1998). Elaboration 5.11 explains in more technical detail how to get from Equation 5.58 to Equation 5.59.

ELABORATION 5.11

Non-linear PCA

The relationship between (linear) PCA and non-linear PCA becomes clear when rewriting Equation 5.58 (following (Gifi, 1990), p.167–168) as

$$\min_{\mathbf{T},\mathbf{y}_j,\mathbf{p}_j} \sum_{j=1}^{J} ||\mathbf{T} - \mathbf{G}_j\mathbf{y}_j\mathbf{p}_j^t||^2 = \tag{5.60}$$

$$\min_{\mathbf{T},\mathbf{y}_j,\mathbf{p}_j} \sum_{j=1}^{J} tr(\mathbf{T}^t\mathbf{T}) - 2\sum_{j=1}^{J} tr(\mathbf{T}^t\mathbf{G}_j\mathbf{y}_j\mathbf{p}_j^t) + \sum_{j=1}^{J} tr(\mathbf{p}_j\mathbf{y}_j^t\mathbf{G}_j^t\mathbf{G}_j\mathbf{y}_j\mathbf{p}_j^t) =$$

$$\min_{\mathbf{T},\mathbf{y}_j,\mathbf{p}_j} IJtr\mathbf{I} - 2\sum_{j=1}^{J} tr(\mathbf{T}^t\mathbf{G}_j\mathbf{y}_j\mathbf{p}_j^t) + I\sum_{j=1}^{J} tr(\mathbf{p}_j\mathbf{p}_j^t)$$

using the constraints on $\mathbf{T}$ and $\mathbf{y}_j$. The function in Equation 5.60 differs only a constant from the function

$$g(\mathbf{T}, \mathbf{y}_j, \mathbf{p}_j) = \sum_{j=1}^{J} ||\mathbf{G}_j\mathbf{y}_j - \mathbf{T}\mathbf{p}_j||^2, \tag{5.61}$$

as follows from rewriting $g(\mathbf{T}, \mathbf{y}_j, \mathbf{p}_j)$ using the constraints on $\mathbf{T}$ and $\mathbf{y}_j$:

$$\begin{aligned} &g(\mathbf{T}, \mathbf{y}_j, \mathbf{p}_j) = \\ &\sum_{j=1}^{J} \mathbf{y}_j^t\mathbf{G}_j^t\mathbf{G}_j\mathbf{y}_j - 2\sum_{j=1}^{J} tr(\mathbf{p}_j^t\mathbf{T}^t\mathbf{G}_j\mathbf{y}_j) + \sum_{j=1}^{J} tr(\mathbf{p}_j^t\mathbf{T}^t\mathbf{T}\mathbf{p}_j) = \\ &IJ - 2\sum_{j=1}^{J} tr(\mathbf{T}^t\mathbf{G}_j\mathbf{y}_j\mathbf{p}_j^t) + I\sum_{j=1}^{J} tr(\mathbf{p}_j^t\mathbf{p}_j). \end{aligned} \tag{5.62}$$

Thus, it has been shown that minimising Equation 5.58 subject to the constraints $(1/I)\mathbf{T}^t\mathbf{T} = \mathbf{I}$ and $\mathbf{y}_j^t\mathbf{G}_j^t\mathbf{G}_j\mathbf{y}_j = I$ is equivalent to the problem

$$\begin{aligned} &\min_{\mathbf{T},\mathbf{y}_j,\mathbf{p}_j} \sum_{j=1}^{J} ||\mathbf{G}_j\mathbf{y}_j - \mathbf{T}\mathbf{p}_j||^2 = \\ &\min_{\mathbf{T},\mathbf{y}_j,\mathbf{p}_j} ||[\mathbf{G}_1\mathbf{y}_1| \ldots |\mathbf{G}_J\mathbf{y}_J] - \mathbf{T}\mathbf{P}^t||^2 = \\ &\min_{\mathbf{T},\mathbf{y}_j,\mathbf{p}_j} ||H(\mathbf{X}) - \mathbf{T}\mathbf{P}^t||^2 \end{aligned} \tag{5.63}$$

where $\mathbf{P}$ has rows $\mathbf{p}_j^t$ and $[\mathbf{G}_1\mathbf{y}_1| \ldots |\mathbf{G}_J\mathbf{y}_J]$ is written as $H(\mathbf{X})$. This is seen to be the (non-linear) analogue of ordinary PCA (Gifi, 1990).

The nature of the measurement scale can now be incorporated by allowing the quantifications to be free for nominal-scale data and monotonic for ordinal-scaled data. The latter quantification ensures the order in the ordinal-scaled data. Framed in terms of Equation 5.59 this becomes:

$$x^*_{ij} > x^*_{kj} \;\; if \;\; x_{ij} > x_{kj} \tag{5.64}$$

where x^*_{ij} and x^*_{kj} are elements of $H(\mathbf{X})$; x_{ij} and x_{kj} are elements of $\mathbf{X}$. Ties in the original data can be treated in different ways depending on whether the underlying measurements can be considered continuous or discrete (De Leeuw *et al.*, 1976; Takane *et al.*, 1977; Young *et al.*, 1978) but this is beyond the scope of this book. Binary data represents a special case. When considered as categorical data, non-linear PCA using optimal-scaling is the same as performing a (linear) PCA on the standardised binary data (Smilde *et al.*, 2020).

Multiblock optimal-scaling

There are different ways to use optimal-scaling for analysing multiblock data. One method generalises (generalised) canonical correlation analysis (OVERALS (Van der Burg and Dijksterhuis, 1996)) and the other method generalises simultaneous component analysis (SCA) (MORALS (Young, 1981)). An example is given for the SCA approach. Suppose there are four blocks of data, respectively ratio-scaled, ordinal-scaled, nominal-scaled, and binary data.

Upon writing $H(\mathbf{X}) = [H_1(\mathbf{X}_1)|H_2(\mathbf{X}_2)|H_3(\mathbf{X}_3)|H_4(\mathbf{X}_4)]$ the problem becomes

$$\min_{\text{Par}} ||H(\mathbf{X}) - \mathbf{T}\mathbf{P}^t||^2 = \qquad (5.65)$$
$$\min_{\text{Par}} ||[H_1(\mathbf{X}_1)|H_2(\mathbf{X}_2)|H_3(\mathbf{X}_3)|H_4(\mathbf{X}_4)] - \mathbf{T}[\mathbf{P}_1^t|\mathbf{P}_2^t|\mathbf{P}_3^t|\mathbf{P}_4^t]||^2$$

with an obvious partition of the loading matrix $\mathbf{P}$ and where the term 'Par' stands for all parameters. Apart from the scores $\mathbf{T}$ and loadings $\mathbf{P}$ these are the following. For the ratio-scaled block there are no extra parameters since the original scale is used (i.e., $H_1(\mathbf{X}_1) = \mathbf{X}_1$). The second (ordinal-scaled) block puts restrictions on $H_2(\mathbf{X}_2)$ following Equation 5.64 and estimation of the elements of $H_2(\mathbf{X}_2)$ can be done with monotonic regression methods (De Leeuw *et al.*, 1976). The third block (nominal-scaled) has underlying indicator matrices $\mathbf{G}$ and associated quantifications $\mathbf{y}$ obeying the rules of Equation 5.58. Finally, the binary block $H_4(\mathbf{X}_4)$ is simply the standardised version of $\mathbf{X}_4$ and this ensures an Optimal-Scaling as mentioned above for binary variables. An example of optimal-scaling is provided in Chapter 9 Example 9.2.

5.2.2 Common, Local, and Distinct Variation

Also for the shared sample case, there are methods for separating common, local, and distinct variation. There are many ways to separate common from local and distinct variation: methods based on least-squares or maximum likelihood with restrictions and methods based on rotations. These will be discussed with examples from each group. Methods based on extensions of the singular value decomposition will be discussed in Chapter 9.

5.2.2.1 Joint and Individual Variation Explained

The method of joint and individual variation explained (JIVE; (Lock *et al.*, 2013)) can only distinguish common from distinct variation. For two blocks, it derives directly a decomposition according to:

$$\begin{aligned} \mathbf{X}_1 &= \mathbf{T}_C\mathbf{P}_{1C}^t + \mathbf{T}_{1D}\mathbf{P}_{1D}^t + \mathbf{E}_1 = \mathbf{X}_{1C} + \mathbf{X}_{1D} + \mathbf{E}_1 \\ \mathbf{X}_2 &= \mathbf{T}_C\mathbf{P}_{2C}^t + \mathbf{T}_{2D}\mathbf{P}_{2D}^t + \mathbf{E}_2 = \mathbf{X}_{2C} + \mathbf{X}_{2D} + \mathbf{E}_2 \end{aligned} \qquad (5.66)$$

where $\mathbf{T}_C$ are the common scores (actually, consensus scores since they are not in the column-space of the respective matrices) and $\mathbf{T}_{1D}, \mathbf{T}_{2D}$ are the distinct scores. In estimating this decomposition, the following constraints are used:

$$\mathbf{X}_{1C}^t\mathbf{X}_{1D} = 0; \quad \mathbf{X}_{1C}^t\mathbf{X}_{2D} = 0; \quad \mathbf{X}_{2C}^t\mathbf{X}_{1D} = 0; \quad \mathbf{X}_{2C}^t\mathbf{X}_{2D} = 0 \qquad (5.67)$$

and, thus, the distinct part in a block is orthogonal to the common parts in all blocks but the distinct parts in different blocks are not necessarily mutually orthogonal (see the discussion on common and distinct in Chapter 2, Section 2.5). Estimation is done by minimising the sum-of-squared residuals $(\mathbf{E}_1, \mathbf{E}_2)$ under the restrictions of Equation 5.67. The (low) ranks of all common and distinct matrices involved are determined by permutation tests[7]. JIVE is within- and between-block scale dependent and has been applied in gene-expression analysis (Lock *et al.*, 2013). An example is given in Example 5.6. Also, an example of JIVE in the `multiblock` R-package is found in Section 11.6.3.

---

7 Which does not work very well in practice. It usually overestimates the number of components (Måge *et al.*, 2019).

5.2.2.2 Distinct and Common Components

The distinct and common components (DISCO) method goes along similar lines as in the case of a shared variable mode. This means that first an SCA is performed and subsequently a rotation. The only difference is that in the current case the target matrix is defined in terms of the loadings. We will explain this for the case of three blocks of data. The method starts again with an SCA of the three blocks:

$$\mathbf{X}_m = \mathbf{T}\mathbf{P}_m^t + \mathbf{E}_m;\ m = 1, 2, 3 \tag{5.68}$$

with $\mathbf{T}^t\mathbf{T} = \mathbf{I}$, and after choosing the total number of SCA components, target loading matrices $\mathbf{P}_{\text{target}} = [\mathbf{P}_{1target}^t|\mathbf{P}_{2target}^t|\mathbf{P}_{3target}^t]^t$ should be defined expressing the hypotheses about the common, local, and distinct structure. For three blocks with five, four, and nine variables, such a target may look like:

$$\begin{bmatrix} \mathbf{P}_{1\text{target}} \\ \mathbf{P}_{2\text{target}} \\ \mathbf{P}_{3\text{target}} \end{bmatrix} = \left[\begin{array}{ccccc} x & x & 0 & 0 & x \\ x & x & 0 & 0 & x \\ x & x & 0 & 0 & x \\ x & x & 0 & 0 & x \\ x & x & 0 & 0 & x \\ \hline x & 0 & x & x & 0 \\ x & 0 & x & x & 0 \\ x & 0 & x & x & 0 \\ x & 0 & x & x & 0 \\ \hline x & x & 0 & x & 0 \\ x & x & 0 & x & 0 \\ x & x & 0 & x & 0 \\ x & x & 0 & x & 0 \\ x & x & 0 & x & 0 \\ x & x & 0 & x & 0 \\ x & x & 0 & x & 0 \\ x & x & 0 & x & 0 \\ x & x & 0 & x & 0 \end{array}\right] \tag{5.69}$$

where it is assumed that the first component is common; the second component is local between blocks 1 and 3; the third component is distinct for block 2; the fourth component is local for blocks 2 and 3; and the last component is distinct for block 1. Note that the sorting of components as common (actually, consensus), local and distinct is arbitrary and any permutation will give the same solution. This is one of the possible targets which can be hypothesised but many others should also be tried (see Example 5.4). There are some recommendations on how to choose targets in practice (Schouteden *et al.*, 2014). Subsequently, the matrix $\mathbf{P} = [\mathbf{P}_1^t|\mathbf{P}_2^t|\mathbf{P}_3^t]^t$ from Equation 5.68 is rotated towards this target similarly to DISCO for the shared variable case:

$$\min_{\mathbf{Q}} \|\mathbf{W}_{\text{target}} * (\mathbf{P}_{\text{target}} - [\mathbf{P}_1^t|\mathbf{P}_2^t|\mathbf{P}_3^t]^t\mathbf{Q})\|^2\ s.t.\ \mathbf{Q}^t\mathbf{Q} = \mathbf{I} \tag{5.70}$$

where the matrix $\mathbf{W}_{\text{target}}$ encodes the positions of the zeros and non-zeros in the target as in the case of Equation 5.26. There are many different targets already for three blocks. For a five component model such targets may be 1 common, 1 local, and 3 distinct; 2 common, 1 local, and 2 distinct. In total, for a five component model there are already 462 possible targets. This method does not generalise easily to more than three blocks since then the number of targets grows rapidly and model selection becomes a serious problem. Moreover, the non-congruence values of Equation 5.70 can be very similar thereby hampering clear conclusions. This also holds for the shared variable mode case.

5.2.2.3 PCA-GCA

The PCA-GCA method combines two of the previously mentioned methods and resembles the method PO-PLS (see Section 7.4). First, a PCA is performed on the individual data blocks (using loading matrices $\mathbf{P}_m$ with orthonormal columns) and subsequently, a GCA is performed on the resulting score matrices $\mathbf{T}_m$ $(m = 1, \ldots, M)$. These score matrices do not necessarily have to be of the same size, i.e., they can have different numbers of principal components. Subsequently, the following problem is solved:

$$\min_{(\mathbf{A}, \mathbf{W}_m)} \sum_{m=1}^{M} \|\mathbf{T}_m \mathbf{W}_m - \mathbf{A}\|^2 \tag{5.71}$$

where $\mathbf{A}$ contains the orthogonal *consensus* components, and the *common* components $\mathbf{T}_m\mathbf{W}_m = \mathbf{X}_m\mathbf{P}_m\mathbf{W}_m$ are in the column-spaces of the respective matrices. Hence, these column-spaces can be orthogonalised (deflated) for the components $\mathbf{T}_m\mathbf{W}_m$ $(m = 1, \ldots, M)$, and a subsequent PCA on these deflated column-spaces then gives the distinct components. An alternative is to perform PCAs on the separate blocks under the restriction that these principal components are orthogonal to $\mathbf{A}$ (see Section 7.3.4). The reason to use a PCA step prior to GCA is that GCA tends to overfit and cannot be used for high-dimensional data.

In this procedure there are several metaparameters to choose:

1) The number of initial PCs per data block.
2) The number of consensus components in the GCA step.
3) The number of distinct components per data block.

This selection can be done with the regular tools such as cross-validation (for steps 1 and 3) and permutation testing (step 2) (van den Berg *et al.*, 2009). Also plotting the canonical variates in step 2 is helpful. In general, setting these metaparameters needs some experience and is not trivial. The PCA-GCA method itself is straightforward and easy to implement. Example 5.4 (Smilde *et al.*, 2017) shows how to select the common and distinct components in DISCO and PCA-GCA.

Example 5.4: Example of DISCO and PCA-GCA on sensory data

The sensory example concerns the smell and taste of flavoured water and how these can be optimised using a recipe. In particular, we are interested in knowing which aspects of smell and taste profiles are in common and what is distinct in the two sensory profiles. The data in this example consists of descriptive smell and taste sensory attributes of flavoured water samples and is a subset of a larger data set (Måge *et al.*, 2012). The 18 water samples were created according to a full factorial experimental design with two flavour types (A and B), three flavour doses (0.2, 0.5, and 0.8 g/l) and three sugar levels (20, 40, and 60 g/l). A trained sensory panel consisting of 11 assessors evaluated the samples first by smelling (9 descriptors) and then by tasting (14 descriptors), using an intensity scale from 1 to 9. Two data blocks (SMELL and TASTE) were constructed by averaging across assessors. The blocks were mean-centred and block-scaled to sum-of-squares one prior to analysis.

A crucial aspect of the decomposition is to decide the dimensions of the common and distinct subspaces. For DISCO, this is a two-step process: first, the number of SCA components is selected. Then, the most appropriate target matrix is sought by evaluating the non-congruence value (Equation 5.70) for all possible allocations of common and distinct components. Since there are three independent design factors in this experiment (flavour type, flavour dose, and sugar level), we choose to keep three SCA components even though the third component explains very little variation (Figure 5.10(a)). The lowest non-congruence value is approximately equal for models with one and two common components (Figure 5.10(b)), but after a closer inspection of the scores we choose the model with one common component and one distinct component per block. For real data, the non-congruence value is never zero, meaning that the zeros in the target matrix are not exactly zero in the rotated loadings. This means that the distinct component for one block also explains some variation in the other block (so-called spill-over).

For PCA-GCA, the dimension selection is also a stepwise procedure. First, an appropriate number of principal components is selected for each data block. Next, the correlation coefficients and explained variances from GCA are evaluated in order to decide the number of common components. The number of distinct components is then given as the difference between the total number of components per block minus the number of common components. In this example, we choose to keep three components for each block, following the same argument as for DISCO (three design factors). Figure 5.10(c) shows that the canonical correlation together with the explained variances clearly suggest one consensus component (correlation = 0.98), which means that the distinct subspaces are two-dimensional.

The subspaces found by PCA-GCA and DISCO are very similar. The correlation between the common DISCO component and the common PCA-GCA component is 0.98. The correlation between the distinct SMELL component from DISCO and the first distinct SMELL component

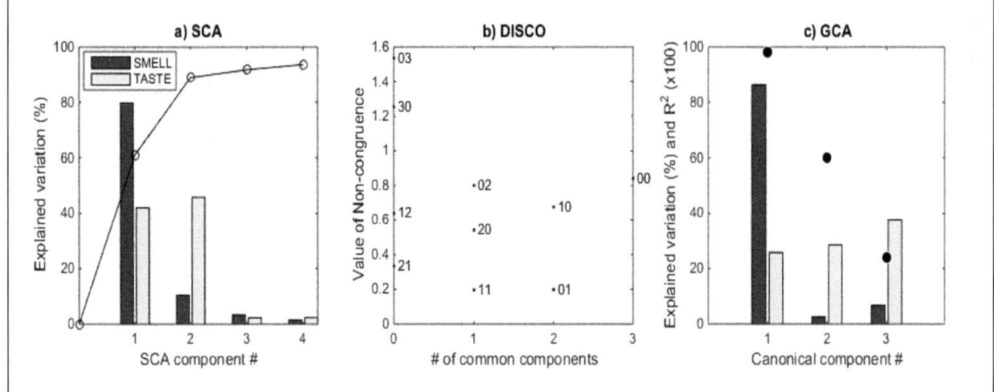

Figure 5.10 Plots for selecting numbers of components for the sensory example. (a) SCA: the curve represents cumulative explained variance for the concatenated data blocks. The bars show how much variance each component explains in the individual blocks. (b) DISCO: each point represents the non-congruence value for a given target (model). The plot includes all possible combinations of common and distinct components based on a total rank of three. The horizontal axis represents the number of common components and the numbers in the plot represent the number of distinct components for SMELL and TASTE, respectively. (c) PCA-GCA: black dots represent the canonical correlation coefficients between the PCA scores of the two blocks (x100) and the bars show how much variance the canonical components explain in each block. Source: Smilde *et al.* (2017). Reproduced with permission of John Wiley and Sons.

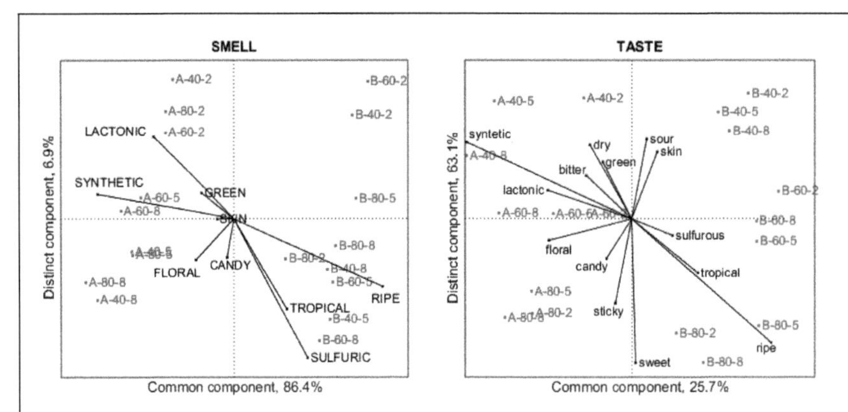

Figure 5.11 Biplots from PCA-GCA, showing the variables as vectors and the samples as points. The samples are labelled according to the design factors flavour type (A/B), sugar level (40,60,80) and flavour dose (2,5,8). The plots show the common component (horizontal) against the first distinct component for each of the two blocks. Source: Smilde *et al.* (2017). Reproduced with permission of John Wiley and Sons.

from PCA-GCA is 0.74. The corresponding number for the distinct TASTE components is 0.99. Figure 5.11 shows biplots from PCA-GCA for each of the two blocks. It is clear that the common component distinguishes between flavour type (A and B). This component explains 86% of the SMELL variation and 26% of the TASTE variation. As a validation of the commonness, note that the sensory attributes that span this subspace are the same both for smelling and tasting: synthetic/lactonic/oral for flavour A versus ripe/tropical/sulphurous for flavour type B. The first distinct SMELL component explains 7% of the variation and is related to the flavour dose, showing that the lowest dose tends to give a more lactonic smell. The first distinct TASTE component explains 63% of the variation and describes differences in sugar level. The attributes that span this component are sweet/ripe versus sour/synthetic/skin/dry.

This example shows that both methods are able to separate common and distinct subspaces in a similar way. The subspaces that explain a large proportion of the variation (common and distinct TASTE) are practically equal for both methods (correlations > 0.98), while there is less agreement regarding the weaker distinct SMELL component (correlation = 0.74).

For the PCA-GCA method with more than two blocks an ad-hoc procedure can be used to find common, local and distinct components. This is summarised in Algorithm 5.2

Algorithm 5.2

PCA-GCA FOR THREE BLOCKS OF DATA

The algorithm of a PCA-GCA model for finding common, local, and distinct components for three data blocks has the following generic steps:

1) Perform individual PCA models per block. The selection of the number of components per block can be made, e.g., with cross-validation.

2) Calculate consensus components from the scores of these models using Equation 5.71, and decide on the number of consensus components by calculating how much they explain in the respective blocks.
3) Subtract the common components associated with these found consensus components from each block and continue with pairwise consensus components between two blocks. These then constitute the local components.
4) Subtract these local components associated with the respective blocks (i.e., the projections of these local components on their respective blocks) and perform a PCA on the remaining parts to find the distinct components.

Obviously, for more than three blocks this method becomes a bit cumbersome. However, the appeal of the PCA-GCA method is that it contains established and well defined steps and can easily be calculated with standard software. An example of PCA-GCA in the `multiblock` R-package is found in Section 11.6.3. An illustration of PCA-GCA and DISCO with three blocks in medical biology is given in Example 5.5.

Example 5.5: Example of DISCO and PCA-GCA in medical biology

The examples concerned 14 obese patients with Diabetes Mellitus Type II (DM2) who underwent gastric bypass surgery (Lips *et al.*, 2014; Smilde *et al.*, 2017). It is relevant to study the metabolism of these patients before and after the surgery and also whether this surgery affects the difference in metabolism before and after taking a meal. Blood samples were taken four weeks before and three weeks after surgery, and on each occasion samples were taken both before and after a meal. The blood samples were then analysed on multiple analytical platforms for the determination of amines, lipids and oxylipins. The three data blocks amines (A), lipids (L), and oxylipins (O) consist of 14 subjects x 4 samples = 56 rows, and 34, 243 and 32 variables, respectively. To correct for skewness of the data, all data were square-root transformed. Individual differences between subjects were removed by subtracting each subjects' average profile per metabolite across the four conditions (before/after meal, before/after surgery). After that, all variables were then scaled to unit variance across the whole data set. The blocks were also scaled to unit norm prior to SCA, to normalise scale differences between blocks.

As the start for the DISCO modelling we choose five SCA components although there is no clear cut-off point seen in Figure 5.12a. Subsequently, different target loadings were investigated, and this resulted in one common component, two local components for the A and L blocks, and one distinct component for both the A and O blocks. The split-up of variation in the common, local and distinct parts is shown in Figure 5.13 (upper panel).

For PCA-GCA the starting point is PCA models of the different blocks separately (see Figure 5.12b). It is difficult to decide on how many PCs to include in the models for the different blocks, hence, different numbers were tried and subsequently canonical correlations were calculated. This resulted in two common components and one local component for the blocks A and L. The split-up in variances including the distinct parts are shown in Figure 5.13 (lower panel). Both methods agree that there is a common part, and also a local component between blocks A and L. Focusing on the DISCO results, there is clearly some 'spill-over': the

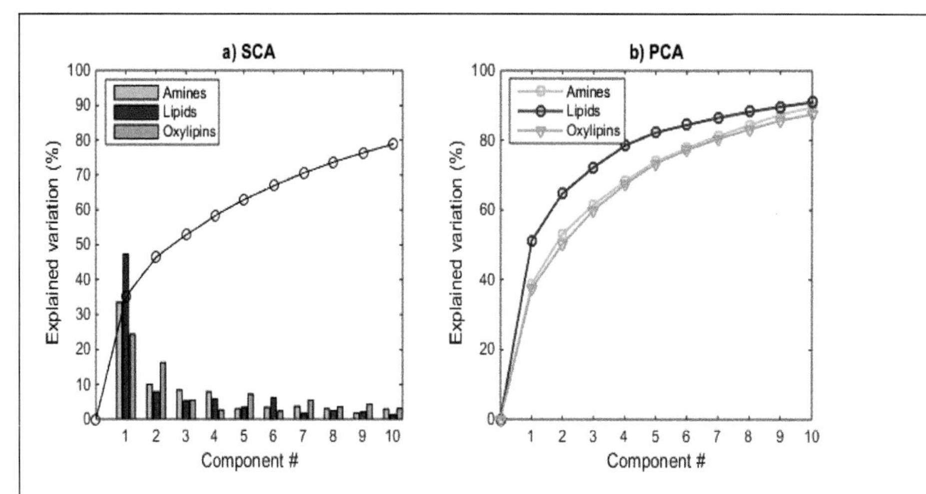

Figure 5.12 Amount of explained variation in the SCA model (a) and PCA models (b) of the medical biology metabolomics example. Source: Smilde *et al.* (2017). Reproduced with permission of John Wiley and Sons.

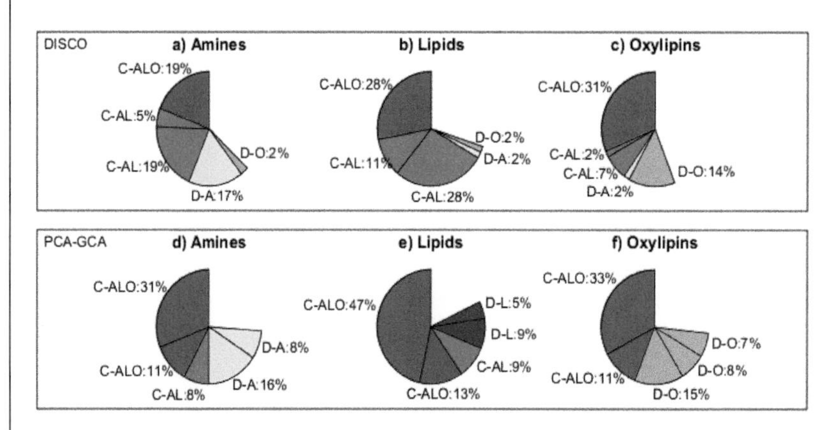

Figure 5.13 Amount of explained variation in the DISCO and PCA-GCA model. Legend: C-ALO is common across all blocks; C-AL is local between block A and L; D-A, D-O, D-L are distinct in the A, O and L blocks, respectively. Source: Smilde *et al.* (2017). Reproduced with permission of John Wiley and Sons.

D-O component also has small contributions in blocks A and L. Similarly, D-A has small contributions in blocks L and O. Hence, the distinctions are not clear cut. The scores and loadings of the large common component of DISCO is shown in Figure 5.14.

This figure clearly shows the effect of the meal and the surgery across a large part of the metabolism (A, L, and O). Moreover, it also shows the effect in terms of the loadings of the individual metabolites. The most striking observation is that the branched-chain amino acids leucine, valine, and L-2-aminoadipic acid (closely related to branched-chain amino acids) are down-regulated after surgery which confirms earlier findings (Lips *et al.*, 2014). A more detailed analysis can be found in Smilde *et al.* (2017).

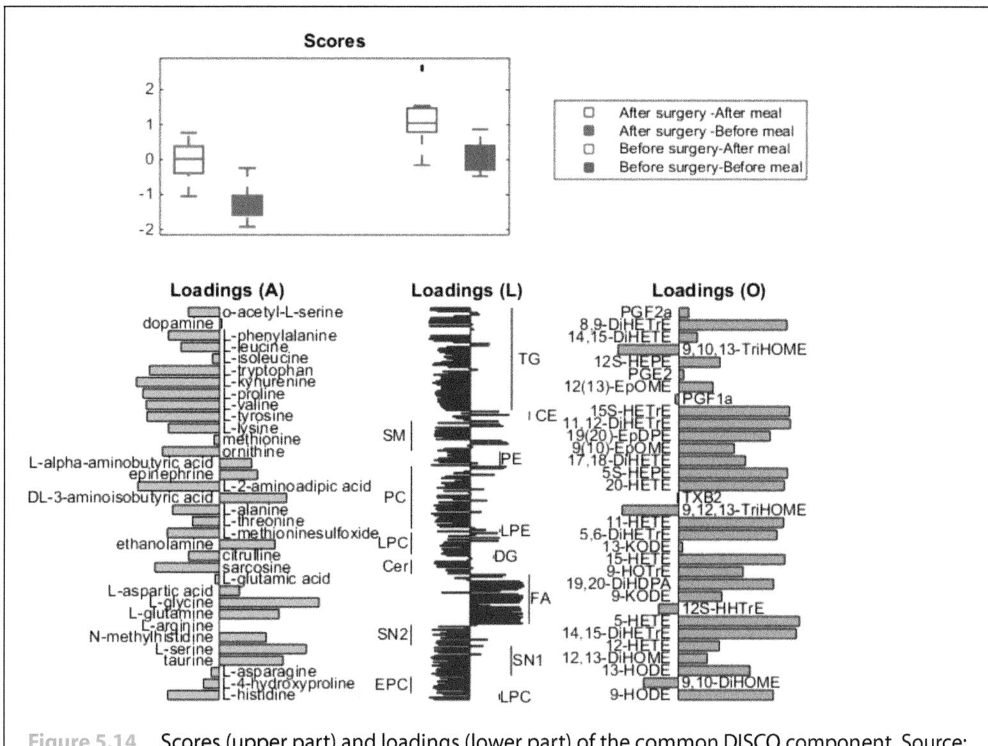

Figure 5.14 Scores (upper part) and loadings (lower part) of the common DISCO component. Source: Smilde *et al.* (2017). Reproduced with permission of John Wiley and Sons.

5.2.2.4 Advanced Coupled Matrix and Tensor Factorisation

A completely different way of finding common, local and distinct components is by using penalised matrix factorisations (Acar *et al.*, 2014, 2015). The idea is to force components in a block to zero by an appropriate penalty. For two matrices $\mathbf{X}_1$ and $\mathbf{X}_2$ with a shared sample mode the following problem is solved:

$$\min_{(\mathbf{T},\mathbf{P}_1,\mathbf{P}_2,\mathbf{D}_1,\mathbf{D}_2)} \|\mathbf{X}_1 - \mathbf{T}\mathbf{D}_1\mathbf{P}_1^t\|^2 + \|\mathbf{X}_2 - \mathbf{T}\mathbf{D}_2\mathbf{P}_2^t\|^2 + \lambda[\|\text{diag}(\mathbf{D}_1)\|_1 + \|\text{diag}(\mathbf{D}_2)\|_1] \quad (5.72)$$

where $\mathbf{T}, \mathbf{P}_1$ and $\mathbf{P}_2$ are full rank matrices containing the R scores and loadings, respectively; $\mathbf{D}_1$ and $\mathbf{D}_2$ are diagonal matrices with the positive elements on their diagonals collected in the vectors $\text{diag}(\mathbf{D}_1)(R \times 1)$ and $\text{diag}(\mathbf{D}_2)(R \times 1)$, respectively. The symbols $\|.\|_1$ are the 1-norms of vectors (sum of absolute values, see Section 2.8) and λ is a positive penalty parameter to be chosen *a priori*. The first part of the objective function in Equation 5.72 is simply an SCA-type model (and also called coupled matrix tensor factorisation (Acar *et al.*, 2013)); and the second part penalises the diagonal elements of $\mathbf{D}_1$ and $\mathbf{D}_2$ to induce sparseness. The resulting method is called advanced coupled matrix tensor factorisation or ACMTF for short. This sparseness has such a structure that from the pattern of the found solutions the common and distinct components (and thus the total number of components) can be read off.

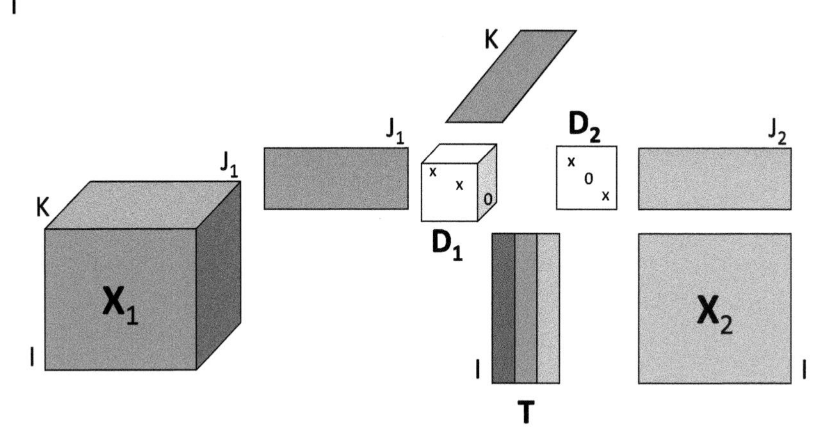

Figure 5.15 ACMTF as applied on the combination of a three-way and a two-way data block. Legend: an 'x' means a non-zero value on the superdiagonal (three-way block) or the diagonal (two-way block). The three-way block is decomposed by a PARAFAC model. The red part of $\mathbf{T}$ is the common component, the blue part is distinct for $\mathbf{X}_1$, and the yellow part is distinct for $\mathbf{X}_2$ (see also the x and 0 values).

If, for instance, diag$(\mathbf{D}_1) = (0, 2, 4)$ and diag$(\mathbf{D}_2) = (3, 3, 0)$ then the first component is distinct for $\mathbf{X}_2$, the second component is common, and the third component is distinct for $\mathbf{X}_1$. This sparseness is different from the one in sparse SCA (see Elaboration 5.6) since in that case the loadings of the SCA solution are made sparse. Note that the common components of ACMTF are actually consensus components since they are not necessarily located in the column-spaces of $\mathbf{X}_1$ or $\mathbf{X}_2$.

This method can easily be extended to more than two data blocks and also to analysing a combination of matrices and tensors (i.e., three-way arrays) (Acar *et al.*, 2015). This is one of the few methods that can analyse common, local, and distinct components including multi-way arrays. A graphical illustration of the latter is given in Figure 5.15. The three-way block is decomposed by imposing a PARAFAC-type model (see Chapter 9) and the two-way block by imposing a PCA-type model. A three-component solution is found in which the first component is common (red; the non-zero values on the (super)diagonals); the second component is distinct for the three-way block (blue) and the third component is distinct for the two-way block (yellow). Note that the zeros and non-zeros are found by the algorithm and not selected *a priori*. Applications are reported in analytical chemistry (Acar *et al.*, 2014) and metabolomics (Acar *et al.*, 2015) and an example of comparing JIVE with ACMTF is given in Example 5.6.

Example 5.6: Example of JIVE and ACMTF

The methods JIVE and ACMTF were tested in a real example from analytical chemistry (Acar *et al.*, 2014, 2015). Five chemicals were selected: two peptides (Val-Tyr-Val and Trp-Gly), a single amino acid (Phe), a sugar (Malto), and an alcohol (Propanol), and 29 samples were prepared with varying concentrations according to a predetermined design. Diffusion NMR-spectra of these samples were recorded using a gradient resulting in NMR-spectra at different time points, thus generating a three-way array with modes *samples, time* and *chemical shift.*

An LC-MS analysis was also performed on the same samples resulting in a two-way array with samples in the first mode and integrated peaks in the second mode. The method JIVE cannot handle three-way data, thus the NMR-spectra were matricised keeping the sample mode intact. From chemistry, it is known that propanol will not give a signal in LC-MS. Hence, we expect four common components and one distinct component for the NMR block.

The selection of the number of components is a crucial issue. For JIVE we had to give the correct number (five) of total components since the advised permutation test gave far too many components. For ACMTF we needed six components because there was structured noise in the LC-MS part which needed an extra component; a five component ACMTF model did not recover the profiles that well. Figure 5.16 shows the recovery of the sample concentrations for JIVE and ACMTF. The recovery of JIVE is poor and the recovery of the sample concentrations is good for ACMTF. For a more detailed comparison, see the original publication (Acar *et al.*, 2015).

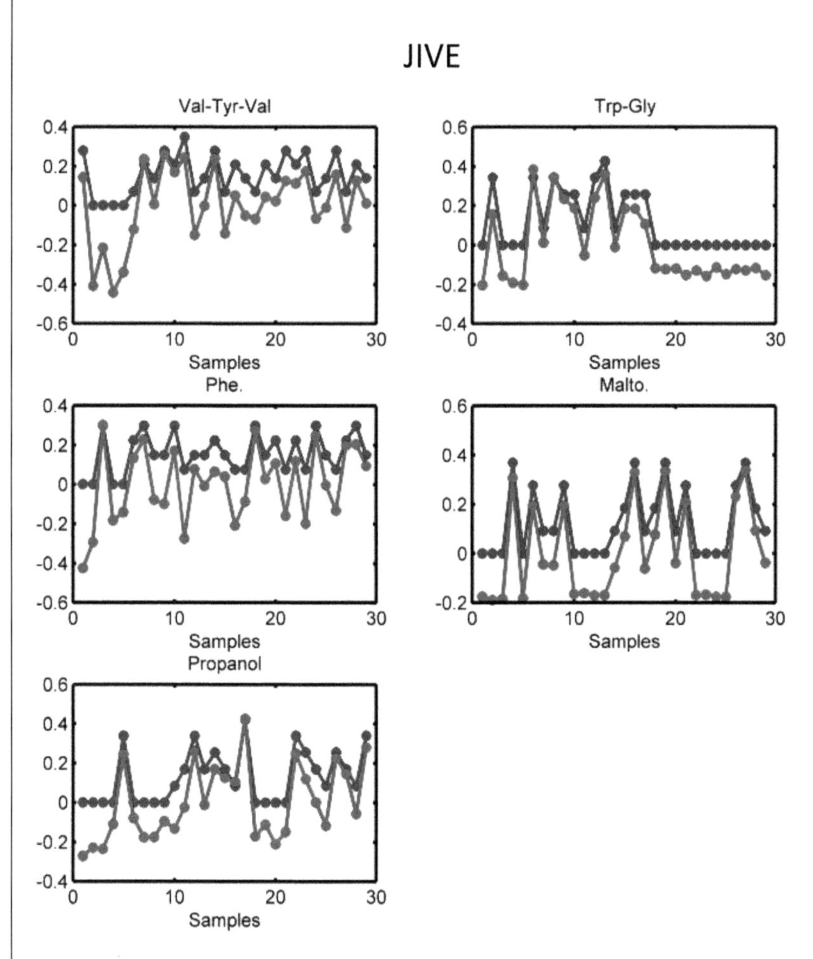

Figure 5.16 True design used in mixture preparation (blue) versus the columns of the associated factor matrix corresponding to the mixtures mode extracted by the JIVE model (red). Source: Acar *et al.* (2015). Reproduced with permission of IEEE.

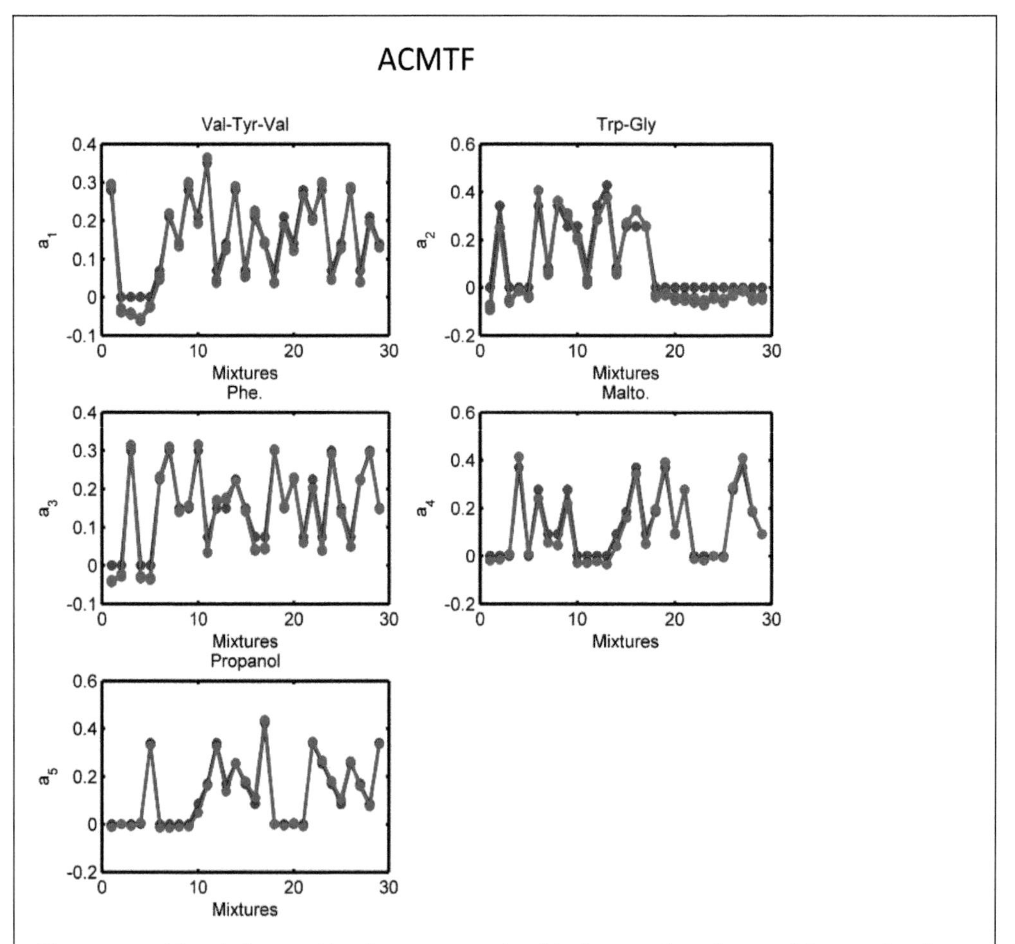

Figure 5.17 True design used in mixture preparation (blue) versus the columns of the associated factor matrix corresponding to the mixtures mode extracted by the ACMTF model (red). Source: Acar *et al.* (2015). Reproduced with permission of IEEE.

5.2.2.5 Penalised-ESCA

There is another way of using penalties as in ACMTF which is illustrated in the method penalised-ESCA (PESCA) that can be used in cases with heterogeneous data, i.e., data blocks of different measurement scales. This is one of the few methods that can combine finding common, local and distinct components in heterogeneous data sets. It does so by using penalties forcing parts of the loading matrices to zero thereby creating the common, local and distinct components. It builds on the method GSCA/ESCA (see Section 5.2.1.5).

The PESCA method will be explained by using three blocks of data: $\mathbf{X}_1(I \times J_1)$ with binary data (e.g., mutation data); $\mathbf{X}_2(I \times J_2)$ with binary data (e.g., CNA data) and $\mathbf{X}_3(I \times J_3)$ with quantitative data (e.g., gene-expression). These blocks are assumed to be non-centred and their centres are modelled explicitly as was also done in the GSCA model. As explained above in the GSCA model (see Equation 5.55), we can use the Bernoulli and normal distributions to write the models of these blocks in terms of their underlying parameters $\mathbf{\Theta}_1, \mathbf{\Theta}_2$ and $\mathbf{\Theta}_3$

with

$$\Theta_1 = \mathbf{1}\boldsymbol{\mu}_1^t + \mathbf{T}\mathbf{P}_1^t$$
$$\Theta_2 = \mathbf{1}\boldsymbol{\mu}_2^t + \mathbf{T}\mathbf{P}_2^t$$
$$\Theta_3 = \mathbf{1}\boldsymbol{\mu}_3^t + \mathbf{T}\mathbf{P}_3^t \tag{5.73}$$

which can also compactly be written as

$$[\Theta_1|\Theta_2|\Theta_3] = \mathbf{1}[\boldsymbol{\mu}_1^t|\ \boldsymbol{\mu}_2^t|\ \boldsymbol{\mu}_3^t] + \mathbf{T}[\mathbf{P}_1^t|\mathbf{P}_2^t|\mathbf{P}_3^t] \tag{5.74}$$

where the scores $\mathbf{T}(I \times R)$ hold all components: common, local, and distinct. In this model, the parameters Θ_1, Θ_2 and Θ_3 are the underlying estimated parameters of the three blocks using a Bernoulli distribution (block 1 and block 2) and a Gaussian distribution (block 3). Group-wise penalties can now be imposed on the loadings $\mathbf{P}$. When the loading matrices are partitioned as $\mathbf{P}_1 = [\mathbf{p}_{11}|\dots|\mathbf{p}_{1R}]$, $\mathbf{P}_2 = [\mathbf{p}_{21}|\dots|\mathbf{p}_{2R}]$ and $\mathbf{P}_3 = [\mathbf{p}_{31}|\dots|\mathbf{p}_{3R}]$ with R the total number of components and $\sigma_{mr} = ||\mathbf{p}_{mr}||^2$ then for the binary blocks this penalty is:

$$\lambda_{bin}[\sqrt{J_1}\sum_{r=1}^{R} g(\sigma_{1r}) + \sqrt{J_2}\sum_{r=1}^{R} g(\sigma_{21r})] \tag{5.75}$$

where $g(\sigma) = log(1 + \frac{\sigma}{\gamma})$ is the generalised double Pareto (GDP) penalty (see Elaboration 5.12). There are many alternatives for using penalties, but the GDP strikes a good balance between being continuous and not shrinking important groups of loadings too much (see Elaboration 5.12). For the quantitative data the penalty is similar:

$$\lambda_{quan}[\sqrt{J_3}\sum_{r=1}^{R} g(\sigma_{3r})] \tag{5.76}$$

and the $\sqrt{.}$ terms are normalisation constants. Estimation proceeds by maximum likelihood using a majorisation approach under the restrictions of Equation 5.73 and the group-penalties of Equations 5.75 and 5.76 (Song *et al.*, 2019). Elaboration 5.12 shows which penalties can be used. It is important to realise that the penalty is on the σ_{mr} values; this means that if such a σ_{mr} is forced to be zero then the whole r^{th} loading of block m becomes zero. Hence, by inspecting which parts of $\mathbf{P}_1 = [\mathbf{p}_{11}|\dots|\mathbf{p}_{1R}]$, $\mathbf{P}_2 = [\mathbf{p}_{21}|\dots|\mathbf{p}_{2R}]$ and $\mathbf{P}_3 = [\mathbf{p}_{31}|\dots|\mathbf{p}_{3R}]$ become zero, the common, local and distinct components can be found. For identifiability the restrictions $\mathbf{1}^t\mathbf{T} = 0$ and $\mathbf{T}^t\mathbf{T} = \mathbf{I}$ are imposed. This means that all components (common, local, and distinct) are orthogonal to each other. Note that also in this case the common components are actually consensus components since they are not guaranteed to be in the column-spaces of the respective matrices. For obvious reasons, this method is called penalised-ESCA or PESCA for short. An example is given in Chapter 9 (Example 9.4).

ELABORATION 5.12

Group-wise penalties

Penalties are well-known methods in machine and statistical learning to regularise parameters in complex estimation situations by inducing sparseness. A prime example is the ridge-type penalty in regression (i.e., ridge regression) which shrinks all regression coefficients towards zero. Another prime example is the lasso penalty which forces a set of regression coefficients to zero (Hastie *et al.*, 2009, 2015). These penalties can be imposed on individual parameters but also on groups

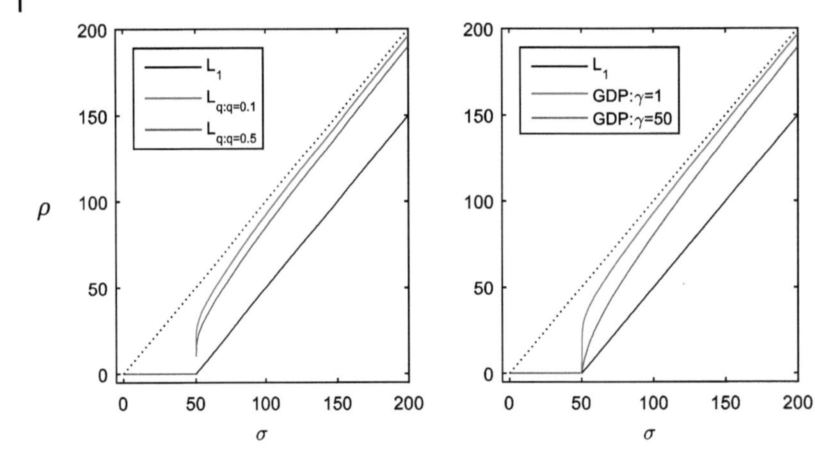

Figure 5.18 Example of the properties of group-wise penalties. Left panel: the family of group-wise L-penalties. Right panel: the GDP penalties. The x-axis shows the L_2 norm of the original group of elements to be penalised; the y-axis shows the value of this norm after applying the penalty. More explanation, see text. Source: Song *et al.* (2021). Reproduced with permission of John Wiley and Sons.

of parameters, such as a predefined group of variables. The most well-known method in the latter field is the group-lasso (Yuan, 2006), and also combinations of sparsity on group and individual level can be imposed (Simon *et al.*, 2013). There are many options for such penalties and they have different properties. Figure 5.18 shows different alternatives.

The group penalty is built up using two elements (Fu, 1998). The first element is the L_2-norm of the *k*th group denoted by the σ_k; which is the Frobenius norm of a vector $\mathbf{p}_k$. These σ_k values are subsequently combined into one penalty. Hence, this penalty forces some σ_k to become zero thereby ensuring that a whole group of coefficients becomes zero (i.e., the whole vector $\mathbf{p}_k$) since a vector with norm zero is the null-vector. For this group-wise penalty there are several choices. The L_1 penalty is the 1-norm of a vector (lasso penalty, see also Section 2.8). The L_q penalty is defined as $\sum_{k=1}^{K} \sigma_k^q;\ 0 \leq q \leq 1$ where $q = 0$ corresponds to no penalty (ordinary least squares); $q = 1$ corresponds to the lasso penalty (since $\sigma_k > 0$). The L_1 penalty drives all σ values below a certain threshold to 0 and values above this threshold are shrunken; see Figure 5.18 where a threshold of 50 was chosen as illustration, but this threshold is data dependent. This property is the reason for refitting regressions after an L_1 thresholding to restore the original sizes for the non-penalised values which is also standard practice in lasso-type regressions.

The L_q penalties are discontinuous (see the y-values corresponding to $\sigma = 50$ in Figure 5.18 left panel) and shrink the non-penalised values less than the L_1 penalty. The equation for the generalised double Pareto (GDP) shrinkage is $log(1 + \frac{\sigma}{\gamma})$ where γ is a tuning parameter (Armagan *et al.*, 2013) (see Figure 5.18 right panel). The GDP is continuous and has nice shrinkage properties. The continuity of the GDP penalty is favourable for applying certain types of optimisation algorithms.

5.2.2.6 Multivariate Curve Resolution

The multivariate curve resolution (MCR) method can also be used to analyse blocks of data sharing the sample mode. This may happen, e.g., when different spectroscopies are used for the same samples. For two such blocks the models may then look like:

$$\mathbf{X}_1 = \mathbf{C}\mathbf{S}_1^t + \mathbf{E}_1 \qquad (5.77)$$
$$\mathbf{X}_2 = \mathbf{C}\mathbf{S}_2^t + \mathbf{E}_2$$

where $\mathbf{X}_1$ and $\mathbf{X}_2$ are the original data; $\mathbf{C}$ $(I \times R)$ represents the shared concentrations of the R chemical constituents and $\mathbf{S}_1$ $(J_1 \times R)$, $\mathbf{S}_2$ $(J_2 \times R)$ contain the underlying spectra of these constituents in the different spectroscopies (Tauler *et al.*, 1995). The advantage of using both blocks simultaneously is enhanced selectivity by combining two spectroscopies thereby reducing rotational freedom. Also in this case, centring and scaling is not recommended since it would destroy the model structure.

MCR can also be formulated as a special method for finding common, local, and distinct components in chemical systems for the case of shared samples. This may be the case when different spectroscopies are used on the same set of samples containing chemical constituents of which some do not show up in one of the spectroscopies. This is again a form of selectivity and can be dealt with in the MCR context (Maeder and Zuberbuehler, 1986; de Juan *et al.*, 2014).

Since estimation and validation goes similarly as in the case of a shared variable mode (see Section 5.1.2.2) we do not discuss this further.

5.3 Generic Framework

As has become clear by now, there is a wide variety of unsupervised multiblock data analysis methods. Luckily, there are some generic frameworks which can be helpful in understanding the differences and commonalities between the methods. In fact, the linking structure between blocks as is used in this book as an organising principle, is an important part of one of these generic frameworks, namely the one for simultaneous multiblock data analysis methods with shared samples (Van Mechelen and Smilde, 2010). Another organising principle in this book is based on the difference between methods for finding only common variation and for methods that can distinguish between common, local and distinct components. The framework on which this is based was already discussed in Section 2.5 (Smilde *et al.*, 2017). Since most methods are now presented in this chapter, we can discuss both frameworks and discuss how the methods fit into those.

5.3.1 Framework for Simultaneous Unsupervised Methods

Most of the simultaneous unsupervised methods in this chapter can be summarised as special cases within a framework of possibilities of unsupervised multiblock data analysis methods. We will first describe the basics of this framework and later describe how the methods fit into this framework. One ingredient of the framework – linking structures – was already discussed in Chapter 3 as part of the organising principle of multiblock data analysis methods.

5.3.1.1 Description of the Framework

The framework for simultaneous unsupervised multiblock data analysis consists of three ingredients:

1) quantifications of data blocks
2) block-specific association rules
3) linking structure between the blocks

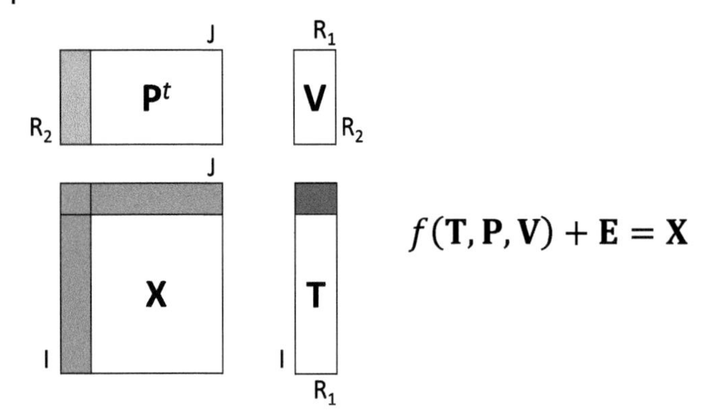

Figure 5.19 Quantification of modes and block-association rules. The matrix $\mathbf{V}$ 'glues together' the quantifications $\mathbf{T}$ and $\mathbf{P}$ using the function $f = (\mathbf{T}, \mathbf{P}, \mathbf{V})$ to approximate $\mathbf{X}$.

and these different ingredients will be explained briefly. The first ingredient is the idea of quantifying modes (see Figure 5.19) which was already introduced briefly in Chapter 2. The quantifiers as collected in $\mathbf{T}$ should describe the differences between the samples in the first mode of the data block $\mathbf{X}$. Thus, the first row of $\mathbf{X}$ (blue row) is summarised by the first row of $\mathbf{T}$ (red part). Likewise, the first column of $\mathbf{X}$ (blue column) is summarised as the first column of $\mathbf{P}$ (yellow part). To arrive at a more parsimonious description of the data the number of quantifiers (R_1 and R_2) should be smaller than the matrix size. Principal component scores and loadings are one instance of quantifiers. The next thing to consider is that these quantifiers should somehow represent the original matrix $\mathbf{X}$ as good as possible. This can be made quantitative by defining an association rule that uses and connects the quantifiers with a function to represent $\mathbf{X}$ (see the function f in Figure 5.19 and Elaboration 5.13).

ELABORATION 5.13

Association rules

A useful concept in component based methods is an approximation rule. Such a rule can be formulated as having a model for a block $\mathbf{X}_m$ of data that uses the quantifiers:

$$\mathbf{X}_m = f(\mathbf{T}_m, \mathbf{P}_m, \mathbf{V}_m) + \mathbf{E}_m \tag{5.78}$$

where the $\mathbf{T}_m$ and $\mathbf{P}_m$ are the quantifiers for the first and second mode of $\mathbf{X}_m$, respectively, and $\mathbf{V}_m$ is an auxiliary matrix, e.g., used to combine $\mathbf{T}_m$ and $\mathbf{P}_m$ to approximate $\mathbf{X}_m$. The most simple instance of this rule is by using $\mathbf{V}_m = \mathbf{I}$ to obtain:

$$\mathbf{X}_m = \mathbf{T}_m \mathbf{P}_m^t + \mathbf{E}_m \tag{5.79}$$

which is used a lot in component models, and the approximation is now defined in terms of a minimal sum of squared errors. Another example is the core-array $\mathbf{G}$ in Tucker3 models (see Chapter 9) thus the matrix $\mathbf{V}$ is also sometimes called a core-matrix. There are, however, very many different ways of defining approximation rules (Van Mechelen and Schepers, 2007).

Not all multiblock data analysis methods use approximation rules. Canonical correlation, e.g., does not define a model for a single block. Many covariance/correlation methods do not define such models. For the least squares based methods such models are explicitly used, and also in maximum likelihood models for blocks are used.

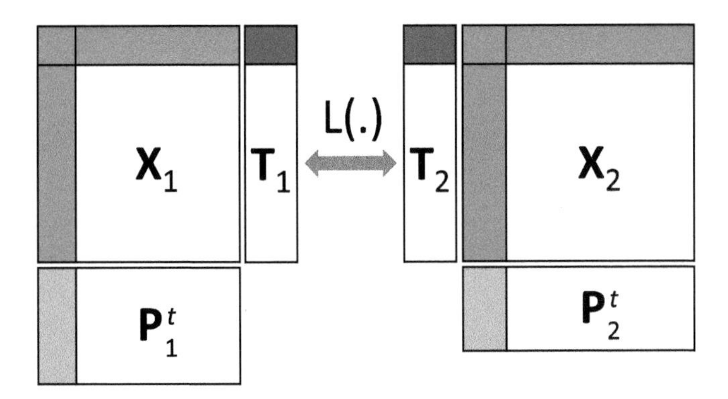

Figure 5.20 Linking the blocks through their quantifications.

The final ingredient is to define the relationships between the quantifiers of the shared mode of the blocks by selecting a linking structure (see also Chapter 3) between those quantifiers of the shared modes (see the symbol $L(.)$ in Figure 5.20). Also for such links there are many possibilities, some of them are displayed in Figure 3.14. linking structures can be seen as restrictions on the quantifiers of the blocks. The identity link (a) forces the quantifiers of both blocks to be equal. Hence, this can be seen as a structural asset of the global multiblock model. This link is by far the most used one. The partially vertical link (c) restricts some of the quantifiers ($\mathbf{T}_{12C}$) to be equal between the blocks. This is the linking structure used by methods that distinguish common, local and distinct components. The flexible link (b) encompasses many alternatives where the quantifiers of the blocks are forced to be related in a certain way, e.g., maximising the correlation between these quantifiers. Such a link is also sometimes called a soft coupling (Farias *et al.*, 2015). The quantifications, association rules and linking structures are not independent of each other but together form a multiblock data analysis model.

5.3.1.2 Framework Applied to Simultaneous Unsupervised Data Analysis Methods

Most of the methods as explained in Section 5.2.1 can be cast in this framework as follows. The majority of the methods use components as quantifiers for the different modes and the identity link. Although there is no defined association rule for GCA, this could be constructed by regressing the blocks $\mathbf{X}_m$ on their respective canonical variates, see Elaboration 5.8, Equation 5.48. ESCA and optimal-scaling are special cases; ESCA uses an identity link and an association rule in terms of likelihood. Optimal-scaling also uses an identity link, but the association rule is in terms of transformed variables. The partial linking structure (c) is discussed more extensively in Section 2.5. This framework shows that there are many alternatives for fusing multiblock data which have not been explored yet.

5.3.1.3 Framework of Common/Distinct Applied to Simultaneous Unsupervised Multiblock Data Analysis Methods

In Chapter 2, Section 2.5 we gave a framework for common and distinct components. The basic idea was to separate the sum of the column-spaces generated by the different blocks into common, local and distinct subspaces. There are many possibilities to do so, e.g., different choices can be made regarding orthogonality of such subspaces.

All methods separating common from distinct components can be cast in this framework and make different choices. Some methods are model-based, some methods are algorithmic-based; both approaches can include penalties or restrictions (see also Table 5.1).

Table 5.4 Properties of methods for common and distinct components. The matrix **D** indicates a diagonal matrix with all positive elements on its diagonal.

Methods	Subspace properties	Orthogonality
JIVE	$R(\mathbf{T}_C) \subseteq R[\mathbf{X}_1\vert\ldots\vert\mathbf{X}_M]$; $R(\mathbf{T}_{mD}) \subseteq R(\mathbf{X}_m)$	$\mathbf{X}^t_{mC}\mathbf{X}_{m'D} = 0; \forall m, m'$
	$R(\mathbf{T}_C) \nsubseteq R(\mathbf{X}_m); \forall m$	$(\mathbf{X}_{m'D})^t\mathbf{X}_{mD} \neq 0\ (m \neq m')$
DISCO	$R(\mathbf{T}) \subseteq R[\mathbf{X}_1\vert\ldots\vert\mathbf{X}_M]; R(\mathbf{T}) \nsubseteq R(\mathbf{X}_m); \forall m$	$\mathbf{T}^t_C\mathbf{T}_{mD} = \mathbf{D}; \mathbf{T}^t_{mD}\mathbf{T}_{m'D} = \mathbf{D}; \forall m, m'$
PCA-GCA	$\mathbf{T}_m\mathbf{W}_m \subseteq R(\mathbf{X}_m)$	$\mathbf{X}^t_{mC}\mathbf{X}_{mD} = 0;\ (\mathbf{X}_{mD})^t\mathbf{X}_{m'D} \neq \mathbf{D};\ (m \neq m')$
		$(\mathbf{X}_{mC})^t\mathbf{X}_{m'D} \neq \mathbf{D};\ (m \neq m')$
ACMTF	??	No orthogonality
PESCA	??	$\mathbf{T}^t\mathbf{T} = \mathbf{I}$
MCR	$R(\mathbf{C}) \subseteq R[\mathbf{X}_1\vert\ldots\vert\mathbf{X}_M]$	No orthogonality

As explained already in Section 2.5, there are properties to consider when discussing methods for separating common and distinct components. These properties have repercussions for issues regarding calculating explained variances and interpreting different plots (loadings and scores). In the latter case, it is of interest in what subspaces the scores of the common and distinct components are located. For calculating explained variances it is of importance what the orthogonality properties are. These properties are summarised in Table 5.4.

This table shows that the methods have very different properties depending on the choices they make, e.g., regarding orthogonality of the different components. One of the methods finds common components in the column-spaces of the respective blocks (PCA-GCA). Other methods find the common components in the summation of the column-spaces of the blocks (JIVE, DISCO, MCR) and thus are actually consensus components; and for some methods this is unclear (PESCA, ACMTF). Also the orthogonality properties differ. JIVE makes explicit assumptions regarding orthogonality: all distinct components are orthogonal to the common ones, but not amongst the distinct components themselves (see Figure 2.13a). DISCO and PESCA force all components (common and distinct) to be orthogonal. MCR and ACMTF do not impose any orthogonality, hence, explained variances cannot be calculated per component but only for the whole fitted model.

5.4 Conclusions and Recommendations

Properties of Some of the Methods

Some methods have very clear properties regarding the estimated parameters (i.e., weights, scores and loadings). Properties should be regarded in a broad sense: (i) in what spaces are the weights, scores and loadings, (ii) are they orthogonal to each other in the different component dimensions, (iii) how stable are they under resampling? Broadly speaking, the simultaneous methods are the most tractable in this sense and algorithmic approaches are the least tractable.

Which Method to Use?

Probably the most important question is which method to use in what situation? This depends very much on the goal of the analysis, on the knowledge domain and on the properties of the methods and the data. In case of shared variable modes, a decision tree is shown Figure 5.21.

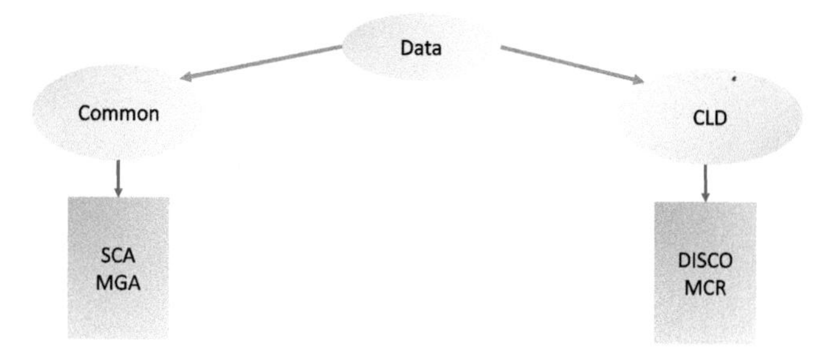

Figure 5.21 Decision tree for selecting an unsupervised method for the shared variable mode case. For abbreviations, see the legend of Table 5.1. For more explanation, see text.

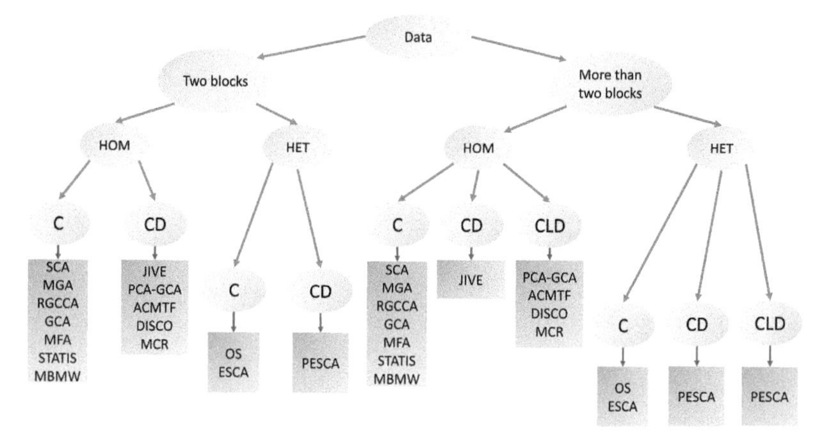

Figure 5.22 Decision tree for selecting an unsupervised method for the shared sample mode case. For abbreviations, see the legend of Table 5.1. For more explanation, see text.

This tree is rather simple and it can be used for an arbitrary number of blocks but only for homogeneous data. The MCR models are special since they require a very special type of data as explained in the respective sections. For the case of a shared sample mode, the tree is more involved, see Figure 5.22. The first decision point is whether there are only two blocks of data or more. The next decision is whether the data are homogeneous (i.e., of the same measurement scale) or heterogeneous. Of course, the methods for heterogeneous data can also be used for homogeneous data. These two decisions are given by the data. The next decisions are up to the user.

A crucial decision from the part of the user is whether or not they are interested in also discovering local or distinct components apart from the common ones. The methods that can separate common from local and distinct components can of course also be used to find common components. Likewise, the methods for heterogeneous data can also be used for homogeneous data. Following this decision tree, the user comes to the leaves which are showing which methods are available.

As an example, if a user has three data blocks all of a quantitative nature (homogeneous) and is interested to find, apart from common components, also local and distinct, then the methods PCA-GCA, DISCO, MCR and ACMTF are at their disposal. The next decision depends now on (i) the specifics of the application and (ii) the properties of the methods. If the multiblock data set is from chemistry and the purpose is to resolve mixtures, then MCR is an option. For three blocks of other types of data PCA-GCA, DISCO and ACMTF are options.

Which one to choose now depends on (i) availability of software, (ii) desired properties of the method, and (iii) personal taste. As an example of (ii) the user has to decide which orthogonality properties of common and distinct components they want. When there are more than three blocks then PCA-GCA and DISCO become unpractical and rather complicated; then the choice may be ACMTF. In this way, extra decisions need to be taken in the final leaf of the tree.

If one or more of the blocks are three-way data, then also a choice has to be made whether to analyse such data with a method that keeps the three-way structure intact such as ACMTF or whether to matricise the three-way data into a two-way data block and use the methods for this type of data. This choice is not explicitly shown in the tree to keep things simple.

It is also immediately clear from the tree in Figure 5.22 that the branches related to homogeneous data and only common variation are the most populated ones. The heterogeneous part is much less populated. Some other methods in this field are given in Chapter 9.

5.5 Open Issues

There are still many open issues in unsupervised multiblock data analysis. Below we will discuss some of them.

Meta-parameter or hyper-parameter selection

All methods in this chapter have meta-parameters or hyper-parameters as they are called in machine learning. These can be the number of components or the choice of a penalty parameter. Such meta-parameters have to be selected or tuned (see Section 2.7.8). A very popular technique to support this tuning is cross-validation. This should, however, be done with care, and it is not always the best approach. Alternatives are based on permutation testing, but experience shows that this does not work that well in all methods, as was shown in JIVE (van der Kloet *et al.*, 2016) where it suggested too many components which may be due to the fact that permutation testing is sensitive to the sample size (increasing the sample size will give more components as significant). Another possibly better alternative in some cases is stability selection (Meinshausen and Bühlmann, 2010) which has not been used a lot in multiblock data analysis.

The number of components (i.e., model complexity) can also be chosen based on sum of squared residuals and explained variance, e.g., using scree plots or other procedures (Wilderjans *et al.*, 2013). These can also be combined with plotting the scores and loadings and visually inspecting these.

Variable selection

Variable selection methods have not received a lot of attention in the unsupervised multiblock literature. One obvious way of performing variable selection is to start with selecting variables in the single blocks, e.g., based on a PCA of those blocks. Various methods exists for this (Jolliffe, 2010) but the obvious drawback is that this approach does not take the relationships between the blocks into account; it is basically an instance of mid-level fusion (see Section 1.3.2) and thus a form of local variable selection.

A global way of selecting variables is to use penalties. An example of this is sparse GCA which induces sparseness of the canonical weights $\mathbf{W}_m$ thereby implicitly selecting variables (Rohart *et al.*, 2017b). Note that the group penalty methods (SLIDE, ACMTF, GFA, and the like) impose a penalty on a whole block of loadings and thus do not select within such a block.

It is also possible to combine group penalties with penalties on the individual loadings but these methods tend to be complicated and not easy to use. One example of this is the MOFA method (Argelaguet *et al.*, 2018) which will be discussed briefly in Chapter 9.

Non-linearities

There can be non-linear relationships between blocks and within blocks. The between-block non-linear relationships would call for a non-linear linking structure. These are not explored yet in unsupervised multiblock analysis. Within-block non-linearities can to some extent be tackled with optimal-scaling methods.

Missing data

For many of the methods in this chapter there are ways to deal with missing data. For the methods based on least squares or maximum likelihood this can be accommodated by using weights in the loss function with weight zero for the missing data entry and weight one otherwise. How many missing data can be handled and to what extent this depends on the missingness patterns (missing completely at random, missing at random, and missing not at random (Little and Rubin, 2019)) is for most methods unknown.

Outliers and performance of the methods

There are always outliers in real data. Several diagnostic are available to detect outliers (see Section 2.7) but for many methods in this chapter it is not clear what the robustness of these methods is with respect to outliers. This touches on a broader aspect, namely the overall performance of the methods in terms of stability and statistical properties of the estimated parameters. Many of the methods are relatively new and complex. Hence, we may have to rely on resampling methods to assess the performance of the methods in a statistical sense.

6

ASCA and Extensions

6.i General Introduction

In many cases in the natural and life sciences we know more about our samples or variables than what we have collected in a multiblock data set. This calls for methods that can include such prior information. In this chapter we will discuss methods that incorporate a special kind of information, namely if the samples are collected according to an experimental design. There is an increasing number of applications in the natural and life sciences where data have such a structure. Although there exist classical ANOVA based methods for analysing multivariate output data – such as MANOVA – these methods are not always simple to use for interpretation. As long as least squares is used, the estimates are fine, and they can be used to test the effect of design variables on the ensemble of output variables if the number of samples is larger than the number of variables. It is, however, not always easy to interpret relations between the responses in cases with many highly correlated output variables.

A class of methods that can deal with (multiblock) high-dimensional data with an underlying experimental design is ANOVA-SCA (ASCA) which combines analysis of variance techniques with dimension reduction steps (Smilde *et al.*, 2012). This method has been extended to deal with unbalanced data (Thiel *et al.*, 2017) and mixed-effect models (Martin and Govaerts, 2021); to deal with common and distinct components (Song *et al.*, 2019; Alinaghi *et al.*, 2020); and also a sparse version is available (group-wise ASCA or GASCA) (Saccenti *et al.*, 2018). Moreover, some extra validation tools were developed for ASCA (Zwanenburg *et al.*, 2011; Liland *et al.*, 2018). A special case of ASCA is multilevel SCA (MSCA) (Timmerman, 2006; Jansen *et al.*, 2005a).

A closely related method to ASCA is ANOVA-PCA (APCA) (Harrington *et al.*, 2005) and also a comparison between ASCA and APCA is available (Zwanenburg *et al.*, 2011). Other alternatives are analysis-of-variance-PLS (El Ghaziri *et al.*, 2015) and PC-ANOVA (Luciano and Næs, 2009). Since the ASCA methodology is more widely used, we will mainly focus on that method in this chapter. There is also a relationship between ASCA and SO-PLS which will be explained in Chapter 7.

6.ii Relations to the General Framework

Like in the previous chapters, we will start again by giving an overview of the methods to be discussed and their high-level properties, see Table 6.1. The ASCA and MSCA methods can be described in different ways and we will illustrate those. One of these ways uses a sequential

Multiblock Data Fusion in Statistics and Machine Learning: Applications in the Natural and Life Sciences,
First Edition. Age K. Smilde, Tormod Næs, and Kristian Hovde Liland.

Table 6.1 Overview of methods. Legend: U=unsupervised, S=supervised, C=complex, HOM=homogeneous data, HET=heterogeneous data, SEQ=sequential, SIM=simultaneous, MOD=model-based, ALG= algorithm-based, C=common, CD=common/distinct, CLD=common/local/distinct, LS=least squares, ML=maximum likelihood, ED=eigendecomposition, MC=maximising correlations/covariances. For abbreviations of the methods, see Section 1.11.

		A			B		C		D		E			F			
	Section	U	S	C	HOM	HET	SEQ	SIM	MOD	ALG	C	CD	CLD	LS	ML	ED	MC
ASCA	6.1																
ASCA+	6.1.3																
LiMM-PCA	6.1.3																
MSCA	6.2																
PE-ASCA	6.3																

approach but in the end ASCA and MSCA can be written as a multivariate multiple regression problem. The ASCA+ method is defined from the onset as a multivariate multiple regression problem. LiMM-PCA is a sequential method in which principal components (which have least squares properties, see Chapter 2) are subjected to a linear mixed model (ML). PE-ASCA is defined as a restricted factor analysis model. Hence, we indicate ASCA and MSCA as both sequential and simultaneous, LiMM-PCA as sequential, and ASCA+ and PE-ASCA are clearly simultaneous. We regard all these methods as supervised since they take into account the experimental design underlying the samples and this can be made explicit for ASCA, ASCA+ and MSCA by writing those as a multivariate multiple regression problem.

ASCA, ASCA+, LiMM-PCA and MSCA are developed for one block of response data. When multiple blocks of response data are available, then concepts like common and distinct can be considered. The PE-ASCA method combines a common and distinct analysis with ASCA which is useful for multiple data sets collected according to the same underlying experimental design. Moreover, the PE-ASCA method is based on maximum likelihood and can also be used for heterogeneous data. We will also give an example of this method.

6.1 ANOVA-Simultaneous Component Analysis

ANOVA-simultaneous component analysis (ASCA) (Smilde *et al.*, 2005a) is an exploratory method for designed experiments with multivariate output. The method was developed for balanced factorial designs without complicating elements such as repeated measures, random effects and nesting structures but the idea can be extended to more complex cases (see, e.g., Thiel *et al.* (2017); Martin and Govaerts (2021)). Hence, ASCA is not a full statistical method yet including rigorous inference machinery, but can be regarded as an exploratory method with some validation based on resampling and multivariate statistics. In this chapter we will focus mainly on fixed-effect ANOVA models and its extensions to ASCA (Smilde *et al.*, 2012) since those are the most used implementations.

6.1.1 The ASCA Method

We can think of ASCA as a two-step procedure where the mean values for all responses for each level of a selected design factor are concatenated and subjected to data compression based on PCA. This is repeated for all design factors. Scores and loadings will give low-dimensional views into the responses' relation to the different design factors. The explanation below assumes balanced data.

The prototypical way of explaining ASCA is by starting with the ANOVA decomposition for the response of a single response variable y_j. Suppose we have two factors A and B at levels $a = 1, \ldots, A$ and $b = 1, \ldots, B$. When the usual constraints are applied to the ANOVA model (Searle, 1971), the variation of a single response is modelled as:

$$y_{i_{ab}abj} = y_{...j} + (y_{.a.j} - y_{...j}) + (y_{..bj} - y_{...j}) + \\ (y_{.abj} - y_{.a.j} - y_{..bj} + y_{...j}) + (y_{i_{ab}abj} - y_{.abj}) \tag{6.1}$$

where the index i_{ab} signifies the ith observation in the treatment group ab and j the variable. The dot-notation in the subscripts indicates over which index the mean is taken within the corresponding group. On the right-hand side of Equation (6.1), the mean over all I ($I = \sum_{i_{ab}}$) observations, the main effects of factor A, the main effects of factor B, the interactions between A and B and the individual contributions can be seen. In short, Equation (6.1) shows the break-up of a measured value as a sum of means (systematic variation) and an individual contribution, most often thought of as noise.

The next step in ASCA is to collect all the contributions to the variation in Equation (6.1) for all (j=1,...,J) responses in the proper matrices:

$$\mathbf{Y} = \mathbf{Y}_{\texttt{Mean}} + \mathbf{Y}_{\texttt{A}} + \mathbf{Y}_{\texttt{B}} + \mathbf{Y}_{\texttt{AB}} + \mathbf{Y}_{\texttt{Ind}}, \tag{6.2}$$

where the naming of the matrices of contributions is according to the terms in Equation (6.1) with obvious subscripts for the matrices, e.g., $\mathbf{Y}_{\texttt{Mean}}$ is the matrix containing the column-means of $\mathbf{Y}$. The dimensions of the different blocks of factor effects in the equation are the same as for the original block $\mathbf{Y}$ and these blocks represent the effects of the different factors in the design. A schematic representation of these matrices is shown in Elaboration 6.1. For each of the terms in this equation, a data compression using PCA is used. Hence, the full ASCA model becomes:

$$\mathbf{Y} = \mathbf{Y}_{\texttt{Mean}} + \mathbf{T}_{\texttt{A}}\mathbf{P}_{\texttt{A}}^t + \mathbf{T}_{\texttt{B}}\mathbf{P}_{\texttt{B}}^t + \mathbf{T}_{\texttt{AB}}\mathbf{P}_{\texttt{AB}}^t + \mathbf{Y}_{\texttt{Ind}} + \mathbf{F} \tag{6.3}$$

where the residuals $\mathbf{F}$ collect all residuals from the PCA-steps involved. If needed, also a PCA model of $\mathbf{Y}_{\texttt{Ind}}$ can be included. In other words, the ASCA method consists of first using ANOVA based estimation for splitting variability into contributions from the different design factors and then using PCA for each of the blocks.

ELABORATION 6.1

ASCA: setup up of the matrices involved

The factor effect matrices in ASCA are of a special form. An example of these matrices for two variables is given in Figure 6.1. This example has two factors (time at four levels and treatment at three levels) and two replicates per factor combination. The columns labelled 'Tyrosine' and 'Lysine' contain the centred data of these metabolites. The columns on the right labelled with 'Time' contain the estimates of the time effect for these two metabolites. Per time point these values are the same for all levels of the treatment and the replicates. Similarly, the 'Treatment' columns on the right contain the values for the treatment effect for the metabolites which are the same across the time points and the replicates. There is also a time × treatment interaction term in the model which will also result in highly structured matrices but these are not shown for simplicity. The rightmost two columns contain the residuals.

Sample	Time	Treatment	Tyrosine	Lysine		Time	Time		Treatment	Treatment			
1	1	1	-0.3966	-0.2017		-0.0753	0.1822		-0.4702	-0.5567		0.1489	0.1728
2	1	1	-0.3879	-0.4718		-0.0753	0.1822		-0.4702	-0.5567		0.1576	-0.0973
3	1	2	-0.2193	0.3284		-0.0753	0.1822		-0.0559	-0.088		-0.0881	0.2342
4	1	2	-0.0334	0.1096		-0.0753	0.1822		-0.0559	-0.088		0.0978	0.0154
5	1	3	0.1903	0.4425		-0.0753	0.1822		0.526	0.6447		-0.2604	-0.3844
6	1	3	0.3952	0.8862		-0.0753	0.1822		0.526	0.6447		-0.0555	0.0593
7	2	1	-0.4892	-0.7128		0.1923	0.062		-0.4702	-0.5567		-0.2113	-0.2181
8	2	1	-0.3101	-0.3931		0.1923	0.062		-0.4702	-0.5567		-0.0322	0.1016
9	2	2	-0.0788	0.3619		0.1923	0.062		-0.0559	-0.088		-0.2152	0.3879
10	2	2	0.3941	0.3118		0.1923	0.062		-0.0559	-0.088		0.2577	0.3378
11	2	3	-0.1609	0.2076		0.1923	0.062		0.526	0.6447		-0.8792	-0.4991
12	2	3	1.7989	0.5966	=	0.1923	0.062	+	0.526	0.6447	+	1.0806	-0.1101
13	3	1	-0.4608	-0.6939		-0.2169	-0.1015		-0.4702	-0.5567		0.2263	-0.0357
14	3	1	-0.6839	-0.7274		-0.2169	-0.1015		-0.4702	-0.5567		0.0032	-0.0692
15	3	2	-0.0912	-0.516		-0.2169	-0.1015		-0.0559	-0.088		0.1816	-0.3265
16	3	2	-0.1951	-0.4889		-0.2169	-0.1015		-0.0559	-0.088		0.0777	-0.2994
17	3	3	0.1603	1.0772		-0.2169	-0.1015		0.526	0.6447		-0.1488	0.534
18	3	3	-0.031	0.7402		-0.2169	-0.1015		0.526	0.6447		-0.3401	0.197
19	4	1	-0.5412	-0.7148		0.0999	-0.1427		-0.4702	-0.5567		-0.1709	-0.0154
20	4	1	-0.4916	-0.5382		0.0999	-0.1427		-0.4702	-0.5567		-0.1213	0.1612
21	4	2	0.0237	-0.3661		0.0999	-0.1427		-0.0559	-0.088		-0.0203	-0.1354
22	4	2	-0.247	-0.4446		0.0999	-0.1427		-0.0559	-0.088		-0.291	-0.2139
23	4	3	1.7142	0.1468		0.0999	-0.1427		0.526	0.6447		1.0883	-0.3552
24	4	3	0.1413	1.0603		0.0999	-0.1427		0.526	0.6447		-0.4846	0.5583

Figure 6.1 ASCA decomposition for two metabolites. The break-up of the original data into factor estimates due to the factors Time and Treatment is shown[1].

In Figure 6.1 we only show the ASCA matrices for two metabolites, but the full model contains 11 metabolites, and the resulting factor effect matrices for this full ASCA model are shown in Figure 6.2. The factor effect estimates are highly structured (see the colours in the figures) and also of low rank due to the repeats of rows with the same numbers. These ranks limit the number of principal components that can be selected for each factor effect matrix (Jansen *et al.*, 2005b). For a factor of two levels, the resulting factor effect matrix is only rank one and thus can be fully represented by one principal component.

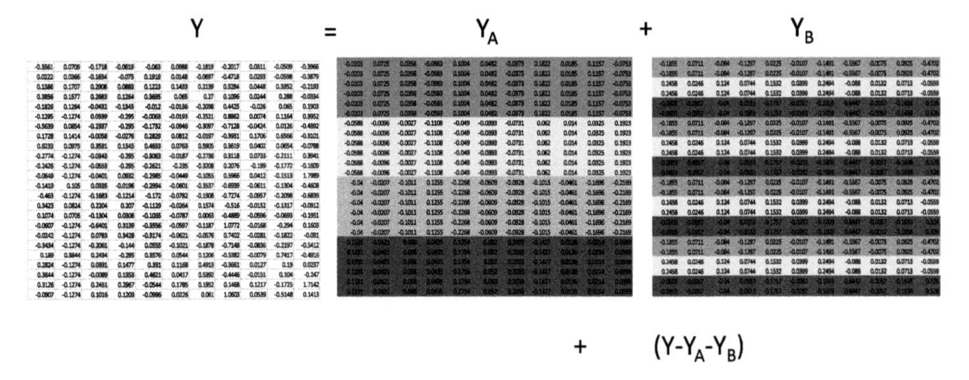

Figure 6.2 A part of the ASCA decomposition. Similar to Figure 6.1 but now for 11 metabolites.

1 We thank Frans van der Kloet for making these figures.

For the matrices on the right hand side of Equation 6.2 it can be shown that their column spaces are orthogonal, i.e., $\mathbf{Y}_{\mathbf{A}}^{t}\mathbf{Y}_{\mathbf{B}} = 0$ (Jansen *et al.*, 2005b) for balanced data. This is a desirable property for several reasons, one of them being the resulting partitioning of sums-of-squares:

$$\|\mathbf{Y}\|^2 = \|\mathbf{Y}_{\mathtt{Mean}}\|^2 + \|\mathbf{Y}_{\mathtt{A}}\|^2 + \|\mathbf{Y}_{\mathtt{B}}\|^2 + \|\mathbf{Y}_{\mathtt{AB}}\|^2 + \|\mathbf{Y}_{\mathtt{Ind}}\|^2, \quad (6.4)$$

where the symbol $\|\mathbf{Y}\|^2$ refers to the squared Frobenius norm of $\mathbf{Y}$ and equals the sum of squared entries of $\mathbf{Y}$. Note that calculating mean sum-of-squares (MS) as in ANOVA is not trivial for ASCA since that would require degrees of freedom in the denominator and such degrees of freedom are not yet available.

ASCA is used in many different areas, for instance in metabolomics (Jansen *et al.*, 2008; Xia *et al.*, 2011; Lemanska *et al.*, 2012; Ly-Verdú *et al.*, 2015; Thiel *et al.*, 2017), genomics (Nueda *et al.*, 2007; Tarazona *et al.*, 2012), food chemistry (Bevilacqua *et al.*, 2013), and sensory analysis (Luciano and Næs, 2009; Næs *et al.*, 2014). An example of ASCA in the `multiblock` R-package is found in Section 11.7.

We will start with an example of ASCA (Example 6.1) which is an extension of Example 1.1.

Example 6.1: Plant Metabolomics

The example was already introduced in Chapter 1. Briefly, plants are grown under different light conditions (factor 1) and multiple metabolites were measured in time (factor 2). An ASCA model can be made of these data, and the results for the factor *light* are shown in Figure 6.3(a). The first ASCA component shows an upwards trend in terms of increasing light, and the corresponding loading plot (Figure 6.3(b)) shows which metabolites are related to this behaviour. The second ASCA component differentiates between the light levels.

The results for the factor *time* are shown in Figure 6.4(a). The first ASCA-time component shows a strong increase and then it levels off. The second ASCA-time component is again a contrast. The loadings (Figure 6.4(b)) reveal that mostly metabolite 47 is responsible for the first component.

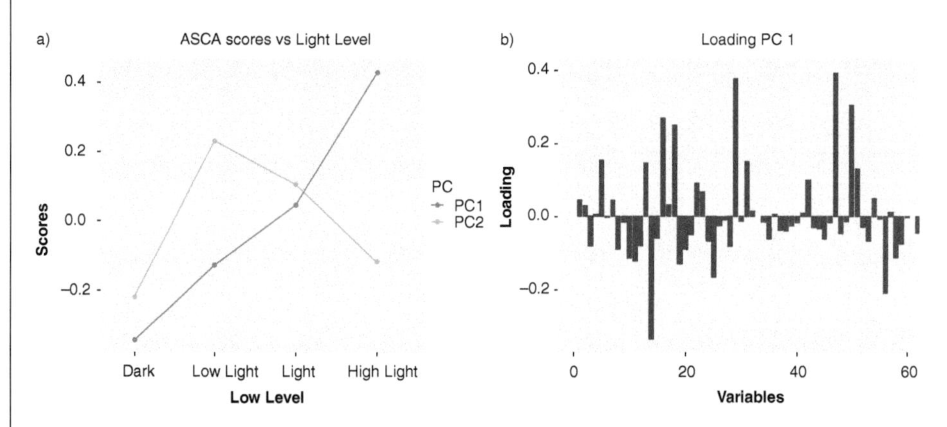

Figure 6.3 The ASCA scores on the factor *light* in the plant example (panel (a); expressed in terms of increasing amount of light) and the corresponding loading for the first ASCA component (panel (b)).

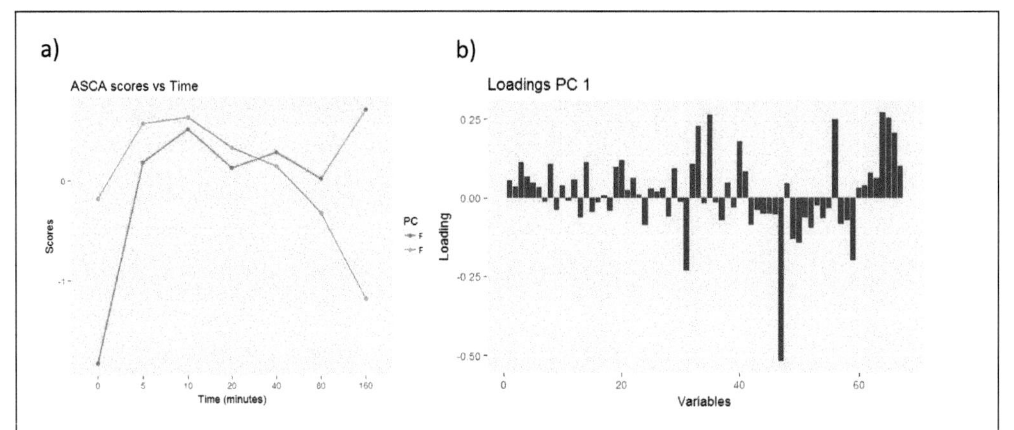

Figure 6.4 The ASCA scores on the factor *time* in the plant example (panel (a)) and the corresponding loading for the first ASCA component (panel (b)).

Finally, the results for the interaction between light and time are shown in Figure 6.5 where only the first ASCA component is shown. There is hardly an interaction between time and light/low light whereas there is clearly an interaction between time and dark versus high light. Again, metabolite 47 seems to be important.

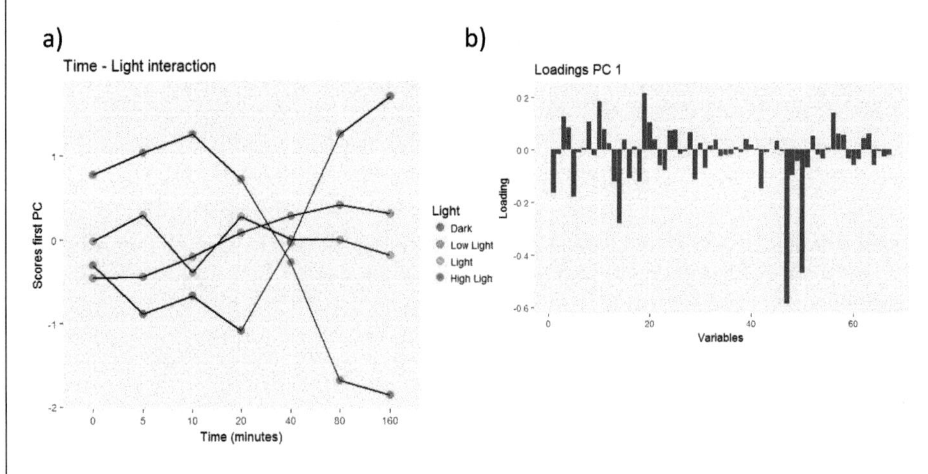

Figure 6.5 The ASCA scores on the interaction between light and time in the plant example (panel (a)) and the corresponding loading for the first ASCA component (panel (b)).

For models such as ASCA, preprocessing is not straightforward. In ASCA, several alternatives are possible depending on the situation. Centring is relatively straightforward but scaling the original variables can be done in several ways (just like when scaling in shared variable mode in Elaboration 5.2). This needs to be carefully considered since it can severely change the results (Timmerman *et al.*, 2015). The general idea is that scaling should reduce unwanted variation and enhance induced variation.

It is also possible to describe ASCA using a multiple multivariate regression framework:

$$\mathbf{Y} = \mathbf{X}_1\mathbf{B}_1 + \mathbf{X}_2\mathbf{B}_2 + \cdots + \mathbf{X}_m\mathbf{B}_m + \cdots + \mathbf{X}_M\mathbf{B}_M + \mathbf{F}, \tag{6.5}$$

where the design matrix $\mathbf{X}_{\mathrm{m}}$ for experimental factor m is represented by a binary (dummy) data matrix with one column for each level of the design factor and the matrix $\mathbf{Y}$ is assumed to be centred. The rows represent the observations and the columns the variables. Interactions are also allowed in this framework and can be obtained by multiplying together values from the individual blocks leading to extra blocks in the model. For this, it is important to have the proper way of coding the factors levels (see also (Liland *et al.*, 2018)).

The next step is to calculate and collect the LS means $\widehat{\mathbf{Y}}_{\mathrm{m}} = \mathbf{X}_{\mathrm{m}}\widehat{\mathbf{B}}_{\mathrm{m}}$ for each block in a matrix with the design factor levels as rows and response variables as columns. Then a PCA is performed on these LS means. The LS-means of the different factors correspond exactly to the matrices in the decomposition in Equation 6.2. This multivariate multiple regression framework lends itself for an extension of ASCA to cases of unbalanced designs, called ASCA+ (Thiel *et al.*, 2017).

Since the PCA is performed directly on LS-means for each factor, the maximum number of components that can be extracted for the main effects is limited by the number of factor levels minus one. For instance, an estimated LS means matrix for factor m, $\widehat{\mathbf{Y}}_{\mathrm{m}}$, based on a four-level factor will be completely 'emptied' after three principal components have been extracted. In other words, there will be no variation left after three components. For the interactions between design factor m and m′ the number of components is limited by the product of factor ranks for the two factors (Jansen *et al.*, 2005b). This is a restatement of the comments regarding the ranks of the factor effect matrices in Elaboration 6.1.

There are many variations possible using the setup of an ASCA model. One of the variations sometimes used is to combine effect and interactions. Sometimes this gives more interpretable models than expanding all factors and interactions fully. It is also possible for some designs to couple the ASCA model with three-way methods such as PARAFAC. An example of both approaches is given in Example 6.2.

Example 6.2: Toxicology example

This example of ASCA is from the field of toxicology (Jansen *et al.*, 2008). In this study, the response of rats to the administration of hydrazine hydrochloride, a hepato- and nephrotoxicant, is monitored. Of interest is the dose dependence of the toxicity. Therefore it is administered in a low and a high dose (30 and 90 mg kg^{-1} body weight respectively). Also a control group is included in the design. The treatments are indicated by $a = 1, \ldots, A(= 3)$ in this study. Each treatment group consists of five rats. The individual rats are nested within a treatment group, i.e., each rat belongs to only one treatment group, designated by indices. To monitor the dynamic response of the rats, their urine is sampled at eight time-points from 1 to 120 hours, indicated by $b = 1, \ldots, B(= 8)$, and urine is collected during the intervals between these time-points. For the ASCA example below, a simplified model with only main effects for 'time', 'treatment group' and their interaction is used without any attempt to model the repeated measurement structure for the rats. The latter aspect is primarily important for testing the effects, and since only factor estimates are used in ASCA, this simplified model was chosen here for illustration purposes.

All samples collected in this experiment have been measured using NMR spectroscopy. These spectra consist of 202 chemical shifts ($j = 1, \ldots, J(= 202)$) between 0 and 10 ppm. Denoting the measurement at chemical shift j, at time point b, of rat i in dose group a as y_{iabj} (where we have suppressed an indicator of the nesting of the rats in the index i for simplicity) and collecting these data in the matrix $\mathbf{Y}(IAB \times J)$ (where i is the number of rats in one group, five in this case), a PCA model of the data would be

$$\mathbf{Y} = \mathbf{Y}_{\mathtt{mean}} + \mathbf{T}\mathbf{P}^t + \mathbf{F} \tag{6.6}$$

The score of the first principal component of this PCA model arranged in the time order is shown in Figure 6.6 and shows grouping of the treatment groups.

Taking this one step further, the ASCA model of the data reads

$$\mathbf{Y} = \mathbf{Y}_{\mathtt{mean}} + \mathbf{Y}_{\mathtt{B}} + \mathbf{Y}_{(\mathtt{A+AB})} + \mathbf{Y}_{\mathtt{Ind}} \tag{6.7}$$

where $\mathbf{Y}_{\mathtt{B}}$ is the matrix of main effect time; $\mathbf{Y}_{(\mathtt{A+AB})}$ is the matrix of the combined main dose effect and dose–time interaction and $\mathbf{Y}_{\mathtt{Ind}}$ are the individual contributions. We choose to not have a main dose effect since that would be hard to interpret because the dose is only taking its effect gradually in time. This is a matter of choice; we can also single out the main dose effect if needed. The most interesting part of this decomposition is the matrix $\mathbf{Y}_{\mathtt{A+AB}}$, and the first two scores of a PCA on this matrix are shown in Figure 6.7.

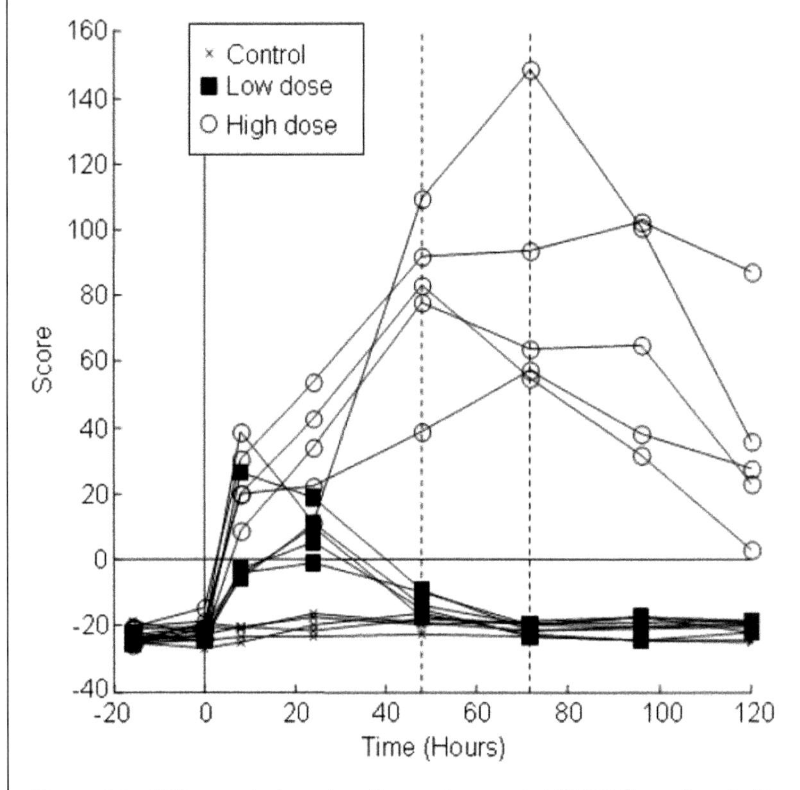

Figure 6.6 PCA on toxicology data. Source: Jansen *et al.* (2008). Reproduced with permission of John Wiley and Sons.

This analysis can be taken even one step further by re-arranging $\mathbf{Y}_{\mathrm{A+AB}}$ into a three-way array $\underline{\mathbf{Y}}_{\mathrm{A+AB}}(\mathrm{I} \times \mathrm{A} \times \mathrm{B})$ and subsequently using a PARAFAC model to decompose this three-way array giving the results of Figure 6.8.

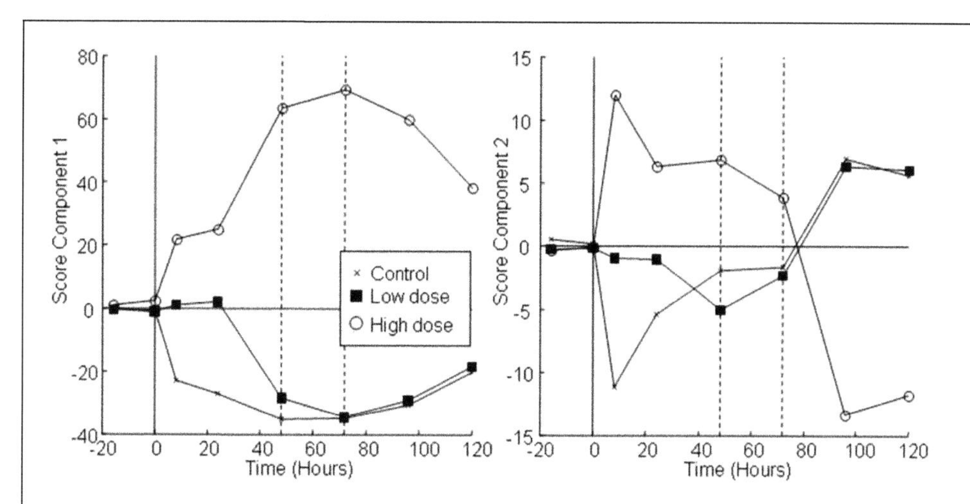

Figure 6.7 ASCA on toxicology data. Component 1: left; component 2: right. Source: Jansen *et al.* (2008). Reproduced with permission of John Wiley and Sons.

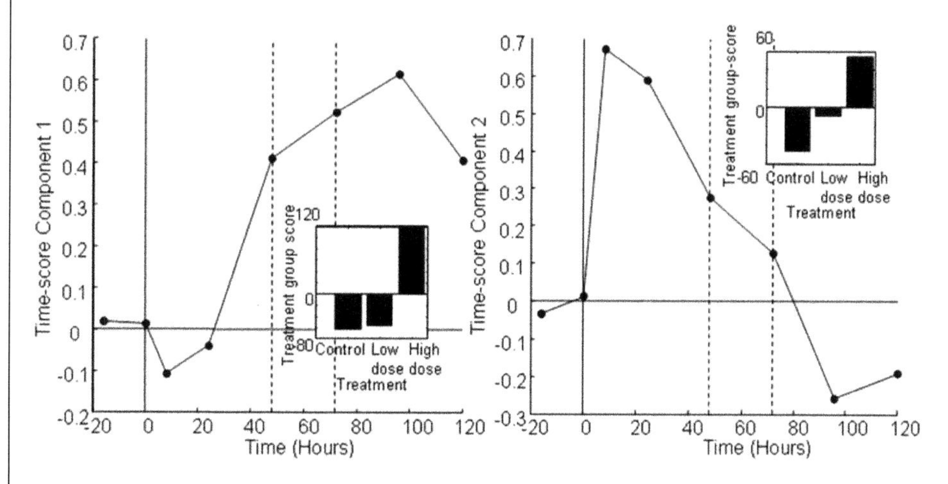

Figure 6.8 PARAFASCA on toxicology data. Component 1: left; component 2: right. The vertical dashed lines indicate the boundary between the early and late stages of the experiment. Source: Jansen *et al.* (2008). Reproduced with permission of John Wiley and Sons.

This method has been coined PARAFASCA, and Figure 6.8 shows two distinct stages in the toxicological intervention: an early transient phase (component 2) and a later persistent phase (component 1). For more details regarding the interpretation, see the original publication (Jansen *et al.*, 2008).

6.1.2 Validation of ASCA

Originally, the ASCA method was a method to visualise data with an underlying experimental design. Some methods are available to validate ASCA-models but a full statistical inference is not yet available. These methods are permutation testing, and data and confidence ellipsoids. These will be discussed briefly in the following sections.

6.1.2.1 Permutation Testing

The idea of permutation testing goes back a long time (Fisher, 1937) and is explained in Section 2.7.6. Permutation testing in ANOVA is also known for some time, see, e.g., Anderson and Ter Braak (2003) and this was the basis for developing a permutation strategy for ASCA models (Vis *et al.*, 2007). The idea of a permutation test for a simple case of one factor at two levels is shown in Example 6.3.

Example 6.3: Simple permutation test

Consider a single factor in ANOVA at two levels and three replicates per level. The data may look like:

$$\mathbf{y} = \begin{bmatrix} \mathbf{y}_1 \\ \mathbf{y}_2 \end{bmatrix} = \begin{bmatrix} 5 \\ 4 \\ 3 \\ -3 \\ -4 \\ -5 \end{bmatrix}, \quad \bar{\mathbf{y}} = \begin{bmatrix} 4 \\ 4 \\ 4 \\ -4 \\ -4 \\ -4 \end{bmatrix} \tag{6.8}$$

where the data are assumed to be centred and $\mathbf{y}_1$ and $\mathbf{y}_2$ contain the responses of the replicates at the two levels of the factor. The vector $\bar{\mathbf{y}}$ contains the mean responses per factor level. The effect size can now be calculated as $4 - (-4) = 8$. Under the null-hypothesis that the factor has no influence, the six values in $\mathbf{y}$ can be exchanged since these are simply replicates now. Hence, upon permuting the third and fourth value we can obtain:

$$\mathbf{y}_p = \begin{bmatrix} 5 \\ 4 \\ -3 \\ 3 \\ -4 \\ -5 \end{bmatrix}, \quad \bar{\mathbf{y}}_p = \begin{bmatrix} 2 \\ 2 \\ 2 \\ -2 \\ -2 \\ -2 \end{bmatrix} \tag{6.9}$$

where $\mathbf{y}_p$ and $\bar{\mathbf{y}}_p$ are now the permuted versions of the data and the means. This results in an effect size of $2 - (-2) = 4$ which is considerably lower than 8. This process can be repeated many times to create a null-distribution which can then be used to judge the significance of the found effect size of 8.

In the multivariate context this is illustrated with a small simulation. The data set consists of 20 samples where 10 variables are measured. There is an underlying design with one factor at two levels. The first 10 samples are replicates at the high level and the last 10 samples are replicates at the low level. In the first case, there is an effect of 1 for each variable, i.e., each measurement of the first 10 samples has an average of 1 and of the last 10 samples an average

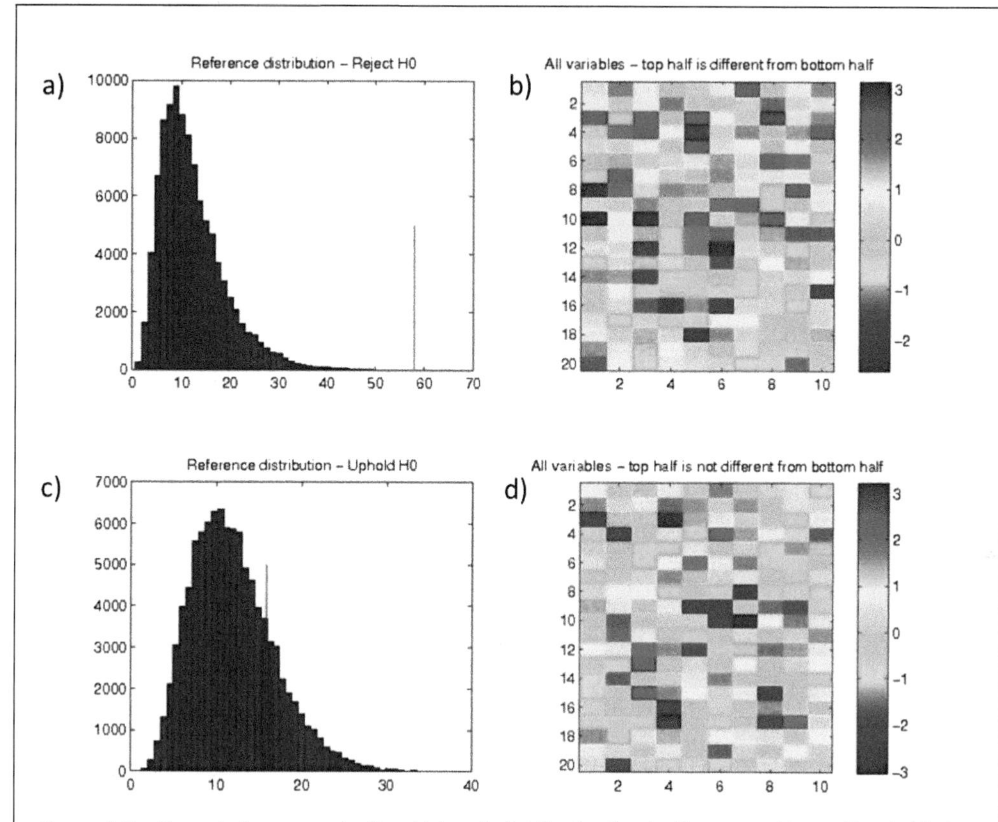

Figure 6.9 Permutation example. Panel (a): null-distribution for the first case with an effect (with size indicated with red vertical line). Panel (b): the data of the case with an effect. Panel (c): the null-distribution of the case without an effect and the size (red vertical line). Panel (d): the data of the case with no effect. Source: Vis *et al.* (2007). Licensed under CC BY 2.0.

of 0. In the second case, there is no effect, i.e., all samples and all variables have average 0. Normally distributed random noise is put on the data with mean 0 and variance 1 in all cases. The idea of permuting the numbers in the rows of $\mathbf{y}$ of Equation 6.8 can now be extended to permuting a whole row in the data matrix (see Figure 6.9(b) and (d)). Then sum-of-squares (SSQ) can be calculated for each permutation. For the original data the SSQ is calculated once and indicated with the vertical red lines of Figure 6.9(a) and (c). When many permutations are performed, a null-distribution can be constructed and used to judge the significance of the found effect sizes (see again Figure 6.9(a) and (c)). Clearly, for the first case there is a significant effect but not for the second case, as it should be.

In principle, the permutation testing in ASCA goes along the lines as shown in Example 6.3. For the main effects this is relatively straightforward, but for the interaction effects this is more involved. In the latter case, first the main effects have to be removed from the data and subsequently the residuals of that operation can be permuted (Anderson and Ter Braak, 2003). An example of permutation testing on the plant metabolomics example is found in Example 6.4.

Example 6.4: Plant metabolomics validation

The score and loading plots for the plant metabolomics example are already given in Figures 6.3, 6.4, and 6.5. The question is then of course to what extent these plots show significant trends. Permutation tests can answer that question and the result of such a test on the factor light and the interaction between light and time is shown in Figure 6.10.

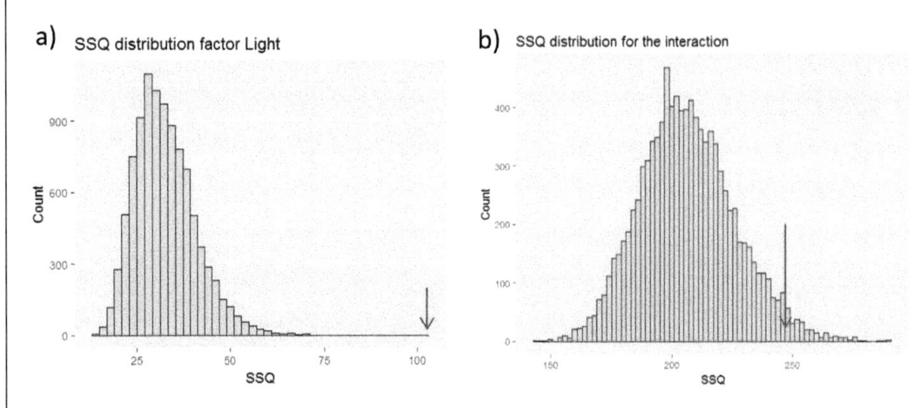

Figure 6.10 Permutation test for the factor light (panel (a)) and interaction between light and time (panel (b)). Legend: blue is the null-distribution and effect size is indicated by a red vertical arrow. SSQ is the abbreviation of sum-of-squares.

Panel (a) of Figure 6.10 shows that clearly the factor light is significant. The interaction between light and time (panel (b)) is borderline significant. When needed, also p-values can be calculated.

6.1.2.2 Back-projection

One way of visualising the variation of the replicated measurements is by using back-projection (Zwanenburg *et al.*, 2011). The idea is as follows. Suppose that an ASCA model is made, then the individual variation can be added to the effect matrices (e.g., for factor $\mathtt{A}$) and projected using the loadings of the corresponding PCA model:

$$\mathbf{Z} = (\widehat{\mathbf{Y}}_{\mathtt{A}} + \widehat{\mathbf{Y}}_{\mathtt{ind}})\mathbf{P}_{\mathtt{A}} \tag{6.10}$$

where $\mathbf{P}_{\mathtt{A}}$ are the loadings related to factor $\mathtt{A}$. The new variables $\mathbf{Z}$ can be plotted and show the spread of the points around the effect matrix $\widehat{\mathbf{Y}}_{\mathtt{A}}$. This can be done also for the other effect matrices, see Example 6.5, and ellipsoids can be included. This approach shows the variability in the data but not in the estimated factor effects (see below). Note that this approach differs from the one of ANOVA-PCA (APCA). In that case, the $\mathbf{Y}_{\mathtt{Ind}}$ is added first to $\mathbf{Y}_{\mathtt{A}}$ and then a PCA is performed on $\mathbf{Y}_{\mathtt{A}} + \mathbf{Y}_{\mathtt{Ind}}$. This gives a different loading matrix and different results (Zwanenburg *et al.*, 2011).

6.1.2.3 Confidence Ellipsoids

A method for estimating and visualising uncertainty was suggested in Liland *et al.* (2018) based on the multivariate regression model (Mardia *et al.*, 1979). In contrast to the residual based ellipsoids proposed in Friendly *et al.* (2013) and the back-projected individual variation (see Section 6.1.2.2), the model ellipsoids will grow and shrink with the number of samples. In this section we skip the subscripts, m, from $\mathbf{X}_m$, $\widehat{\mathbf{B}}_m$ and $\mathbf{P}_m$ for compactness, but the equations are, nonetheless, with respect to one factor at the time. We start from the Hotelling's T^2 statistic, which is defined as $t^2 = (\bar{\mathbf{x}} - \boldsymbol{\mu})^t \boldsymbol{\Sigma}_{\bar{\mathbf{x}}}^{-1} (\bar{\mathbf{x}} - \boldsymbol{\mu})$, where $\boldsymbol{\Sigma}_{\bar{\mathbf{x}}}$ is the sample covariance matrix. The model ellipsoids in projected space (see Figures 6.11 and 6.12) are built from the linear function $\mathbf{x}_{i,*}^t \widehat{\mathbf{B}}\mathbf{P}$ (effect level means in projected space), where $\mathbf{x}_{i,*}$ is a row of the design matrix $\mathbf{X}$, $\widehat{\mathbf{B}}$ represents the associated regression coefficients and $\mathbf{P}$ the PCA/SCA loadings:

$$\frac{I-q}{Ig_i}(\mathbf{x}_{i,*}^t \widehat{\mathbf{B}}\mathbf{P} - \mathbf{x}_{i,*}^t \mathbf{B}\mathbf{P})(\mathbf{P}^t \widehat{\boldsymbol{\Sigma}} \mathbf{P})^{-1}(\mathbf{x}_{i,*}^t \widehat{\mathbf{B}}\mathbf{P} - \mathbf{x}_{i,*}^t \mathbf{B}\mathbf{P})^t = T^2_{1-\alpha,d,I-q}. \tag{6.11}$$

Here, $\widehat{\boldsymbol{\Sigma}}$ is the estimated residual covariance matrix of the model, q is the total number of coefficients in the multivariate regression model in Equation 6.5, $g_i = \mathbf{x}_{i,*}^t (\mathbf{X}^t \mathbf{X})^{-1} \mathbf{x}_{i,*}$, and d is the number of dimensions of the ellipsoid ($d = 2$ in 2D scatter-plots). The α is adjusted to obtain ellipsoids with the desired confidence, as visualised in Example 6.5.

When confidence ellipsoids are calculated for a single experimental factor, as in Equation 6.11, a scaled projected model covariance matrix $\mathbf{P}^t \widehat{\boldsymbol{\Sigma}} \mathbf{P} I g_i / (I - q)$ is used. Here, $g_i = q_m / I$ is a scaling factor for the balanced case with sum-coding of design factors and g_i is also the leverage for the ith design point. Further, q_m is the number of levels of the current factor, m. Translating from Hotelling's T^2 to the F-distribution, these ellipsoids are finally scaled with the following factor:

$$c = \sqrt{\frac{(I - q_m)d}{I - q_m - d + 1} F(1 - \alpha, d, I - q_m - d + 1)}. \tag{6.12}$$

Through this we obtain common confidence ellipsoids for all levels of a factor, following the patterns of uncertainty. This is analogous to model error bars of effect levels often used together with ANOVA. Example 6.5 illustrates how these ellipsoids show the variability in the factor estimates.

The ellipsoids are easiest to visualise and interpret in two dimensions when superimposed on their corresponding score plots. If only two effect levels exist for a factor, the ellipsoids collapse into one dimension, i.e., error bars around effect level means. Also three-dimensional ellipsoids can be effective, though the viewing angle becomes important, and overlap and obstructions of view can pose problems. In Liland *et al.* (2018), there is also an example where the ellipsoids are used for statistical testing of effect level difference, where an effect level mean inside another effect level's 95% confidence ellipsoid is defined as not significantly different. An example of ASCA confidence ellipsoids in the `multiblock` R-package is found in Section 11.7.

Example 6.5: ASCA: Sensory assessment of candies

An example of the use of ASCA for a data set from sensory analysis of 5 candy products using a trained panel with 11 assessors (Luciano and Næs, 2009) is given in Figures 6.11, 6.12, and 6.13. This can be looked upon as a factorial design with 2 factors with 5 and 11 levels each. The assessors judged 9 attributes of the candies, both visually, physically and by taste, three times for each candy.

The ASCA plot of the two first components for the 5 candies is presented in Figure 6.11 both with the ellipsoids generated by the back-projection approach (left panel) and by the approach in Liland *et al.* (2018) (right panel). The data ellipsoids based on back-projection are calculated by fitting concentric ellipsoids to the points associated with a factor level. The process of obtaining the confidence ellipsoids involves estimating the ANOVA model covariance matrix, projecting this on the components using the associated loadings, and scaling based on the number of samples, factor levels, dimensions of the ellipsoids, and T^2 distribution. The ellipsoids represent 40%, 68%, and 95% confidence. A clear separation is observed between the candies on the left side and between those to the left and those to the right. The three candies to the right are less easy to distinguish from each other. The rest of the points represent the residuals from the ANOVA model.

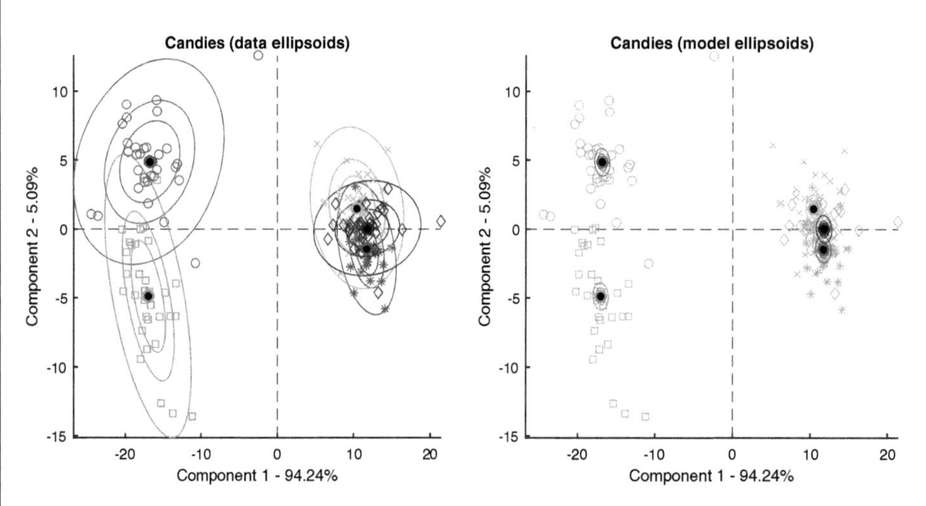

Figure 6.11 ASCA candy scores from candy experiment. The plot to the left is based on the ellipses from the residual approach in Friendly *et al.* (2013). The plot to the right is based on the method suggested in Liland *et al.* (2018). Source: Liland *et al.* (2018). Reproduced with permission of John Wiley and Sons.

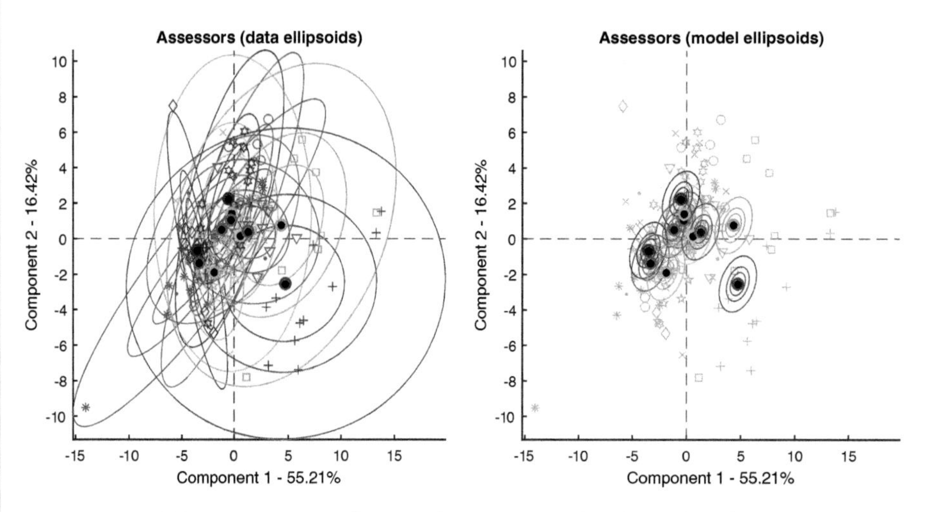

Figure 6.12 ASCA assessor scores from candy experiment. The plot to the left is based on the ellipses from the residual approach in Friendly *et al.* (2013). The plot to the right is based on the method suggested in Liland *et al.* (2018). Source: Liland *et al.* (2018). Reproduced with permission of John Wiley and Sons.

Corresponding plots for the assessors are given in Figure 6.12. Assessors should ideally not have much separation as they are calibrated to assess as similar as possible. A couple of the assessors are here seen to be non-overlapping with the others in the confidence ellipsoids, while most are clustered together.

As can be seen from the loading plot (Figure 6.13), the first axis is texture oriented and the second taste oriented. In the paper by Liland *et al.* (2018), the results of the confidence ellipsoids are taken further into tests for pair-wise differences between factor levels.

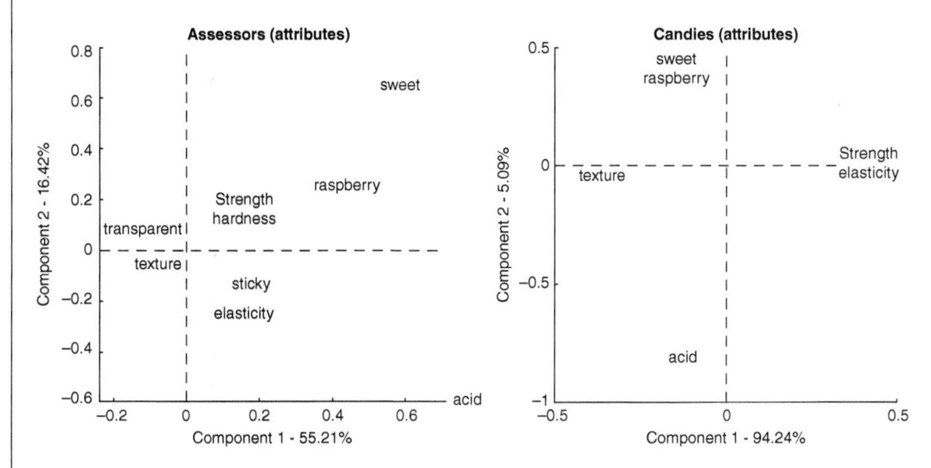

Figure 6.13 ASCA assessor and candy loadings from the candy experiment. Source: Liland *et al.* (2018). Reproduced with permission of John Wiley and Sons.

6.1.3 The ASCA+ and LiMM-PCA Methods

An extension of ASCA for the case of unbalanced data is based on the regression framework of ASCA as presented in Equation 6.5 and is called ASCA+ (Thiel *et al.*, 2017). For handling unbalanced designs, a generalised linear model framework is needed. This is somewhat more complicated than the standard and simple ASCA decomposition as presented in Equations 6.1 and 6.2. In such cases it is more natural to use Equation 6.5 as point of departure. This requires some careful coding of the design variables in the matrices $\mathbf{X}_m$. The nice split-up of sum-of-squares as in Equation 6.4 is not valid anymore. This is similar to ANOVA for the unbalanced case, and decisions have to be made regarding Type I, II, or III sum-of-squares (Hector *et al.*, 2010). Hence, things are not that straightforward anymore.

An interesting extension of ASCA is by also allowing random effects in the model, and there are two options. The first method is coined LiMM-PCA since it uses a linear mixed model framework (Martin and Govaerts, 2021). A recommended, but optional, step is to start with a PCA and then run linear mixed models on the scores of this PCA. The PCA step is taken to arrive at independent variables (i.e., scores) that can be subjected to linear mixed models and facilitates the subsequent statistical significance tests. The results can always be back-transformed to the original domain. After having obtained the estimated effect matrices (both for the fixed as well as random effects) these are subjected to a PCA as in ASCA. Note that this is an extension of ASCA+ (Thiel *et al.*, 2017).

The second method is a direct generalisation of ASCA and is called repeated-measures-ASCA since it accommodates the fact that the same measurements in time are performed on the same individuals. It first builds linear mixed models per variable and then performs PCAs on the effect matrices (Madssen *et al.*, 2020). The latter approach may be unpractical in high-dimensional applications.

There is a lot to say about the relative merits of these approaches. Having a PCA first and then performing ANOVA or LMM modelling on the scores has the advantage of scalability and being able to use advanced linear models on the PCs accommodating the design structure (e.g., split-plot, repeated measures). The advantage of performing ANOVAs or LMMs first (like in ASCA) is that the factors may have completely different underlying latent structures. ASCA would capture that but LiMM-PCA (and PC-ANOVA) would have more problems with that. The disadvantage of ASCA-type methods is that for high-dimensional data, many models have to be built.

6.2 Multilevel-SCA

Multilevel problems are abundant in data analysis. The key feature of multilevel data is that the data set has a hierarchical structure. Examples are subjects measured at different occasions as in a repeated measurement or cross-over design. Other examples are from chemical processes where measurements are available from different sites; from batch reactors within each site and from different time points of each batch. This calls for specialised data analysis methods taking into account the multilevel structure of the data.

Also in the natural sciences multilevel problems are occurring, and special multivariate analysis methods have been designed for this. Examples in metabolomics are (Jansen *et al.*, 2005a; van Velzen *et al.*, 2008; Westerhuis *et al.*, 2010; Rubingh *et al.*, 2011) where PLS and three-way methods have been generalised for the multilevel situation.

The SCA type models can also be accommodated to model multilevel structures (Timmerman, 2006). This originated in psychometrics but these methods also found their way into the natural and life sciences, especially in metabolomics (Jansen *et al.*, 2005a; Saccenti *et al.*, 2014) and in process chemometrics (de Noord and Theobald, 2005). Generally, these models are called multilevel simultaneous component analysis (MSCA) .

In its most basic form, e.g., for modelling groups of individuals where multiple measurements (e.g., at several time points) are available for each individual, the model is a simple one-way ANOVA model:

$$y_{i_a aj} = y_{..j} + (y_{.aj} - y_{..j}) + (y_{i_a aj} - y_{.aj}) \tag{6.13}$$

where the index a indicates the individual and the index i_a is the ith measurement of individual a. The first term of the right-hand side of Equation 6.13 is the grand mean, the second term is the between-subject variation and the third term represents the within-individual variation. Note that also in this case the within-individual terms are treated without a repeated measurement structure, so in this case this is simply an ANOVA with one factor (the individuals). Whereas in Example 6.1 interest is mainly in the effect of the factors, in MSCA the interest is in separating between- and within-individual effects and further studying these.

The next step is again collecting all terms (for $j = 1, \dots J$) in matrices and performing PCAs on those matrices. This MSCA model is thus seen as a special case of an ASCA model. In this simple case there are only two levels in the data, but more levels (e.g., sites, batch-reactors, and time points) can be accommodated by more involved ANOVA models. The MSCA concept remains the same: estimate separate ANOVAs per variable, collect all the terms in proper matrices and then perform PCAs on these effect matrices.

For MSCA, the between-block (i.e., the between individual) model collects the averages of the blocks and can be subjected to a centring and possibly a scaling operation. The within-block (i.e., within individual) parts of the MSCA model are already centred due to the subtraction of the between-block parts and can be scaled according to the general SCA model, if needed.

6.3 Penalised-ASCA

When multiple data blocks are collected on the same samples according to an experimental design, then ASCA can be combined with methods separating common from distinct variation (Alinaghi *et al.*, 2020). We will explain this using two blocks of data $\mathbf{Y}_1(I \times J_1)$ and $\mathbf{Y}_2(I \times J_2)$ with a shared sample mode, and this shared sample mode is organised with respect to an underlying experimental design. Hence, we now have the option to study whether the design is affecting the common and distinct components of the blocks differently.

Having these multiple blocks and the wish to study common and distinct components requires a method that can separate common from distinct variation and we will use the penalised-ASCA (PE-ASCA) method (see Section 5.2.2.5). This method is made for heterogeneous data but can also be used for homogeneous data as in the current case. Now there are two options: i) first perform PESCA on the combined data $\mathbf{Y} = [\mathbf{Y}_1|\mathbf{Y}_2]$ and then ASCA on the common and distinct parts or ii) first perform ASCA on each data block separately and then PESCA on the factor effect matrices. For noiseless data, the order does not matter, yet, since PESCA needs to estimate error variances for real data the first option is advisable since ASCA has a noise averaging effect thereby hampering the subsequent PESCA step (Alinaghi *et al.*, 2020). Suppose that the PESCA model gives the following decomposition of the matrices $\mathbf{Y}_1$ and $\mathbf{Y}_2$:

$$\begin{aligned}\mathbf{Y}_1 &= \mathbf{Y}_{1C} + \mathbf{Y}_{1D} + \mathbf{F}_1 = \mathbf{T}_C\mathbf{P}_{1C}^t + \mathbf{T}_{1D}\mathbf{P}_{1D}^t + \mathbf{F}_1 \\ \mathbf{Y}_2 &= \mathbf{Y}_{2C} + \mathbf{Y}_{2D} + \mathbf{F}_2 = \mathbf{T}_C\mathbf{P}_{2C}^t + \mathbf{T}_{2D}\mathbf{P}_{2D}^t + \mathbf{F}_2\end{aligned} \tag{6.14}$$

where the subscripts 'C' and 'D' refer to common and distinct. Next, an ASCA model is imposed, assuming two factors $\mathtt{A}$ and $\mathtt{B}$, time and treatment, respectively, on the matrices $\mathbf{Y}_{1C}$, $\mathbf{Y}_{2C}$, $\mathbf{Y}_{1D}$ and $\mathbf{Y}_{2D}$. These matrices are thus decomposed according to the ASCA model into factor effect matrices corresponding to the factors in the design. Next, the factor effect matrices are again subjected to a PCA for which the common part comes down to:

$$\begin{aligned}\mathbf{Y}_{1C_\mathtt{A}} &= \mathbf{T}_{C_\mathtt{A}}\mathbf{P}_{1C_\mathtt{A}}^t + \mathbf{F}_{1C_\mathtt{A}} \\ \mathbf{Y}_{2C_\mathtt{A}} &= \mathbf{T}_{C_\mathtt{A}}\mathbf{P}_{2C_\mathtt{A}}^t + \mathbf{F}_{2C_\mathtt{A}}\end{aligned} \tag{6.15}$$

where we only show the PCA-decompositions for the factor $\mathtt{A}$. Note that the scores for these common parts are the same for both blocks which is an assumption of the PESCA model. This is accommodated in the method by writing out the full model and maximising the likelihood under that restriction. The complete model also contains the PCA decompositions of all other matrices in Equation 6.14, and all error terms are collected and aggregated into one error term. This model is called PE-ASCA for obvious reasons (Alinaghi *et al.*, 2020). An example of this model is given in Example 6.6.

Example 6.6: PE-ASCA: NMR metabolomics of pig brains

The data from the example is a part of a previously published data set (Alinaghi *et al.*, 2019). Twenty-two piglets were delivered at 90% gestation from two sows (litters 1 and 2) and each delivered piglet was randomly assigned to one of the following groups: (i) a control group receiving anti-arterial saline and total parenteral nutrition (CON+TPN, n=7), (ii) an infected group receiving *Staphylococcus epidermidis* (SE) inoculation and total parenteral nutrition (SE+TPN, n=5), or (iii) a colostrum supplementation group receiving the SE inoculation and bovine colostrum and supplementary parenteral nutrition (SE+COL, n=10). The brain regions of the hypothalamus and midbrain were analysed by HR-MAS-NMR spectroscopy. The design was balanced by both leaving out and imputing samples (Alinaghi *et al.*, 2020). This resulted in the design as shown in the left part of Figure 6.14.

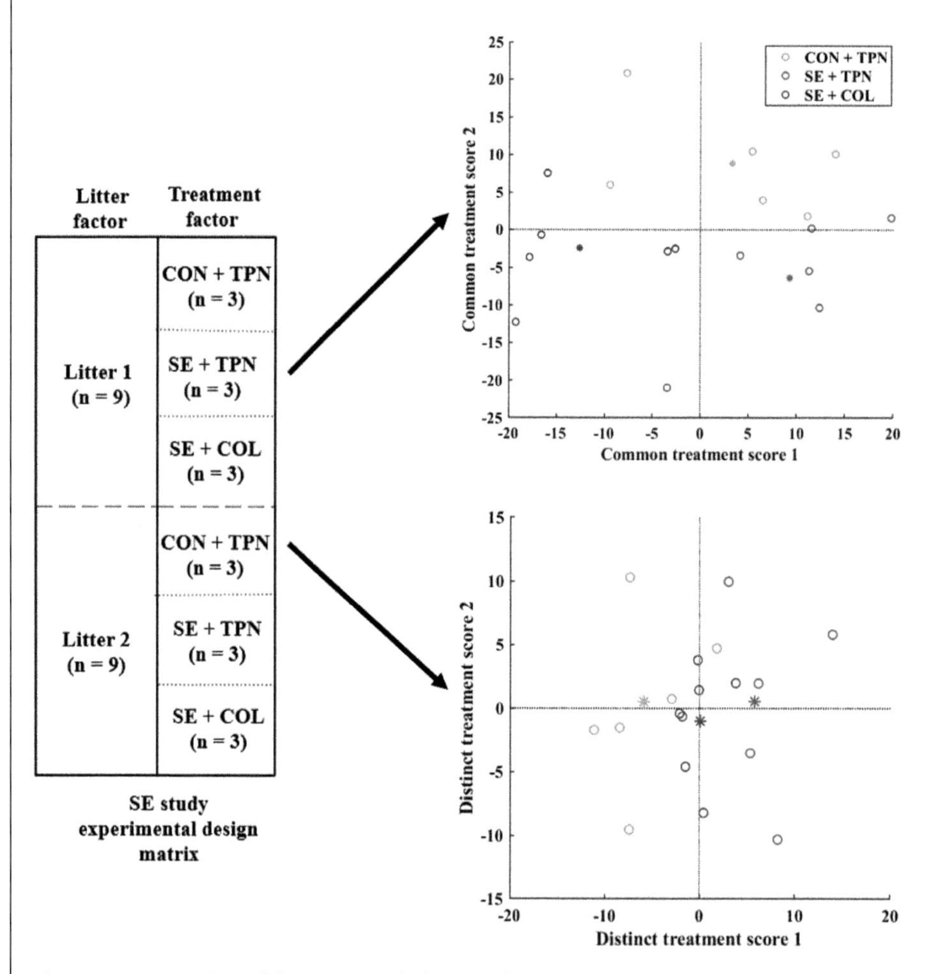

Figure 6.14 PE-ASCA of the NMR metabolomics of pig brains. Stars in the score plots are the factor estimates and circles are the back-projected individual measurements (Zwanenburg *et al.*, 2011). Source: Alinaghi *et al.* (2020). Licensed under CC BY 4.0.

The scores for the common treatment effect is shown in the upper part of Figure 6.14. This score plot clearly shows the treatment effect. For the hypothalamus, there is also distinct variation present in the treatments (lower part of Figure 6.14). The score plot shows again a separation between treatments. For a more detailed analysis, we refer to the original paper (Alinaghi *et al.*, 2020).

6.4 Conclusions and Recommendations

It is also important in this chapter to give recommendations for which method to use. This depends again on the properties of the methods and the data (see also Table 6.1). In all cases, the goal is to analyse the data with the experimental design in mind. Considerations for selecting a proper method are the following:

Design: What is the design of the study?
Balancedness: Is the design balanced or unbalanced?
Measurement-scale: Are the data quantitative or of a different measurement type?
Number of blocks: Is there only one independent block or multiple independent blocks?

and this is visualised in Figure 6.15. This tree should facilitate making decisions and should not be used as a strict decision tree. The design of the experiment from which the data are obtained dictates to a large extent which method can be used. This involves aspects such as fixed and/or random effects, nestedness, and balancedness. For balanced data and only crossed fixed effects, ASCA is the most straightforward method to use. ASCA can handle a certain amount of unbalancedness, but if this gets too large then ASCA+ is the preferred method. If the factors are nested (e.g., time series within individuals) then MSCA is the option.

Rigorous statistical significance testing is not yet available for ASCA. Good alternatives are the permutation tests and the confidence ellipsoids. The first can be used to get an idea of the overall significance for main effects and to some extent for interactions. The latter is a nice visual analytics tools which show differences between factor levels.

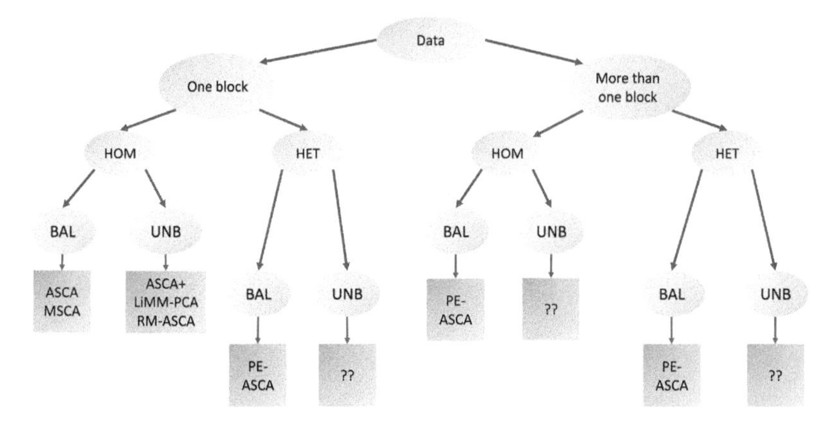

Figure 6.15 Tree for selecting an ASCA-based method. For abbreviations, see the legend of Table 6.1; BAL=Balanced data, UNB=Unbalanced data. For more explanation, see text.

If the design contains fixed and random effects, LiMM-PCA is an option. If there are multiple independent blocks then it may also be of interest to study common and distinct components within those blocks in relation to the design. There is only one method available then, which is PE-ASCA. This method is currently the only one that can be used for heterogeneous data.

6.5 Open Issues

It is clear from Figure 6.15 that there are many open spots. Hence, there is clearly room for method development, especially for heterogeneous data fusion. Even for the populated leaves of the tree, there are only a few methods. These methods are also rather new and their relative merits have to be established. Validation of the results generated by the methods rest mostly on resampling tools, and the behaviour of these tools in this context should also be investigated, e.g., how to perform a proper permutation test for complicated designs is not trivial. Also the extension of confidence ellipsoids to unbalanced data and mixed effects models could be explored.

Variable selection is also an issue in ASCA-related modelling. One option is to use a sparse version of ASCA (Saccenti *et al.*, 2018) but there are certainly other options based on inspecting the ASCA-loadings.

7

Supervised Methods

7.i General Introduction

The focus of this chapter is on supervised multiblock data analysis. This means methods which handle data blocks with a predictive relationships, i.e., some blocks are used for prediction (here denoted $\mathbf{X}_1$, $\mathbf{X}_2$, etc.) and some blocks are to be predicted ($\mathbf{Y}_1$, $\mathbf{Y}_2$, etc., usually only one block denoted $\mathbf{Y}$ is considered). In this chapter the focus is on methods which link along the sample mode, i.e., it is assumed that the rows in the blocks refer to the same objects/samples (see Chapter 3).

A typical situation where supervised methods are important is process modelling where measurements are made at different places and stages and the goal is to predict the end-product properties and to understand how the different input data influence the output. Another typical example is spectroscopy where we are often interested in combining information from different instruments for predicting the level of concentration of a chemical substance. In general, both good prediction ability and interpretation are important goals of a supervised analysis. Interpretation will here typically be based on the same types of score and loading plots as were considered in Chapter 2. Prediction ability is normally measured by standard tools such as cross-validation and prediction testing as also described in Section 2.7.

7.ii Relations to the General Framework

An overview of the relations to the general framework in Chapter 1 is presented in Table 7.1.

In this chapter only the supervised topology is treated, but an unsupervised method (GCA) is used for one of the methods (PO-PLS) as a part of the algorithm. All methods discussed are sequential; some methods combine different criteria sequentially while others are sequential only in the sense that the components are calculated one at a time. Heterogeneous data sets will not be covered.

Since all methods use PLS-type regressions, there are aspects of ED and MC in all the methods. Some methods also include LS steps in their algorithm. Common and distinct components play an important role, but here in a predictive sense. This concept has not yet got a precise mathematical definition and, therefore, a more intuitive approach is taken. ROSA is difficult to fit into this concept, but has an aspect of common/distinct as it is a block selecting method. We refer to Figure 7.1 for a more thorough discussion.

Multiblock Data Fusion in Statistics and Machine Learning: Applications in the Natural and Life Sciences,
First Edition. Age K. Smilde, Tormod Næs, and Kristian Hovde Liland.

Table 7.1 Overview of methods. Legend: U=unsupervised, S=supervised, C=complex, HOM=homogeneous data, HET=heterogeneous data, SEQ=sequential, SIM=simultaneous, MOD=model-based, ALG= algorithm-based, C=common, CD=common/distinct, CLD=common/local/distinct, LS=least squares, ML=maximum likelihood, ED=eigendecomposition, MC=maximising correlations/covariances. For abbreviations of the methods, see Section 1.11.

		A			B		C		D		E			F			
	Section	U	S	C	HOM	HET	SEQ	SIM	MOD	ALG	C	CD	CLD	LS	ML	ED	MC
MB-PLS	7.2																
SO-PLS	7.3																
PO-PLS	7.4																
ROSA	7.5																

7.1 Multiblock Regression: General Perspectives

7.1.1 Model and Assumptions

The methods discussed here are primarily linear in the coefficients for the X-variables or components extracted and based on a model like

$$\mathbf{Y} = \mathbf{X}_1\mathbf{B}_1 + \mathbf{X}_2\mathbf{B}_2 + \ldots + \mathbf{X}_M\mathbf{B}_M + \mathbf{F}, \tag{7.1}$$

where the $\mathbf{B}$ are the regression coefficients and $\mathbf{F}$ contains the (random) errors. It will be described here explicitly how the methods MB-PLS, SO-PLS, and ROSA can be formulated in this way in terms of original input variables in $\mathbf{X}_m$. For PO-PLS on the other hand, the situation is more complicated and a linear prediction equation will be presented only in terms of the components extracted.

The number of samples/rows in all data blocks is equal to I, the number of columns in $\mathbf{Y}$ is equal to J and the number of columns in the input blocks are $J_1, J_2, \ldots, J_M$. It will be assumed throughout this chapter that all blocks are centred and therefore the intercept is omitted from the equations. We refer to Chapter 1 for a description of how 'hats' over estimated parameters are used in this book.

7.1.2 Different Challenges and Aims

Collinearity among input variables occurs in most applications discussed in this book. If not taken into account, models become unstable and unreliable (see e.g., Næs *et al.* (2001)). All methods treated in this chapter are able to handle this issue.

Another desirable property is the ability to handle blocks of different complexity and type, both with respect to the number of variables in each of the blocks and the underlying dimensionality in the blocks (i.e., the number of components needed per block). Some of the methods discussed here handle this explicitly (for instance SO-PLS).

Invariance with respect to scaling of variables has two different aspects in multiblock analysis; within-block scaling and between-block scaling (see Section 2.6). The first means that methods give the same results, regardless of whether the variables are scaled or not within the same block. The methods treated here are not scale invariant within blocks, which means that we have to make a choice about the scaling. Invariance between blocks means that we can multiply a block with a constant c, do the analysis and multiply the solution with 1/c, and then get the same outcome as without the multiplication. This concept is related to, but is not the same as the fairness concept discussed in Chapter 1. Between block invariance usually makes analysis simpler as fewer choices need to be made, but it also means that alternative actions must be taken if emphasis on a particular block is needed. Some of the methods below are invariant to between-block scaling (SO-PLS, PO-PLS, and ROSA) while MB-PLS is not.

When regressing **Y**-variables onto several blocks of **X**-variables, there are a number of ways that the **X**-blocks can be combined (see, e.g., Næs *et al.* (2013)). This opens up an opportunity to use methods which highlight aspects that we are particularly interested in. Important concepts here are distinct and common variability (components) among blocks (see also Chapter 2) in a predictive sense. For all methods in this chapter, common and distinct variability among **X**-blocks will be either implicitly or explicitly involved, but the concepts are only explicit for the SO-PLS and PO-PLS methods in Sections 7.3 and 7.4 and to a certain extent for ROSA (Section 7.5). From a prediction ability point of view the methods are quite similar, the main difference lies in how the different parts of the variability are extracted and interpreted. Figure 7.1 shows how three of the methods in this chapter link to these concepts.

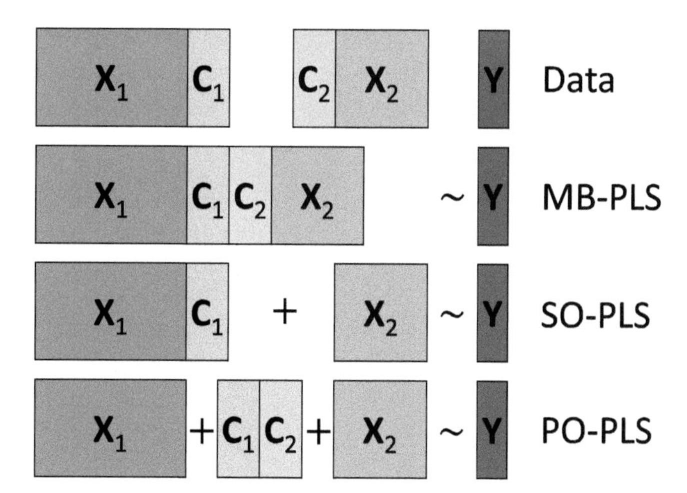

Figure 7.1 Conceptual illustration of the handling of common and distinct predictive information for three of the methods covered. The upper figure illustrates that the two input blocks share some information ($\mathbf{C}_1$ and $\mathbf{C}_2$), but also have substantial distinct components and noise (see Chapter 2), here contained in the $\mathbf{X}$ (as the darker blue and darker yellow). The lower three figures show how different methods handle the common information. For MB-PLS, no initial separation is attempted since the data blocks are concatenated before analysis starts. For SO-PLS, the common predictive information is handled as part of the $\mathbf{X}_1$ block before the distinct part of the $\mathbf{X}_2$ block is modelled. The extra predictive information in $\mathbf{X}_2$ corresponds to the additional variability as will be discussed in the SO-PLS section. For PO-PLS, the common information is explicitly separated from the distinct parts before regression.

7.2 Multiblock PLS Regression

One of the most well-known multiblock regression methods is multiblock PLS (MB-PLS). This method has been used in chemometrics for a long time, mostly in process chemometrics applications. Recently, it has also found its way into, for instance, bioinformatics, especially in the context of sparse methods. This section will describe both the standard and the sparse approaches.

7.2.1 Standard Multiblock PLS Regression

There exist a number of standard multiblock PLS regression (MB-PLS) proposals in the literature (see, e.g., Wangen and Kowalski (1988)). The works of Westerhuis *et al.* (1998) and Westerhuis and Smilde (2001) are important for clarifying the differences between those. The variant of the method considered here (Algorithm 7.1) is essentially a standard PLS regression based on concatenating the **X**-blocks (Figure 7.2) and postprocessing the scores and loadings of this analysis to obtain block-scores, block-loadings, and block-weights (Westerhuis *et al.*, 1998). For an alternative more true to the original formulations of MB-PLS, see Algorithm 7.2 (for sparse MB-PLS) where super-scores (or global-scores) are formed from individual block-scores and deflation is done for both **X** and **Y** using the super-scores.

In the following we will use the terms super-scores and super-weights for denoting quantities obtained from the concatenated data, not only for individual blocks. Likewise, block-scores, block-weights, and block-loadings are used to denote quantities obtained from or related to the individual blocks.

A major difference between many of the MB-PLS methods proposed is how the blocks are deflated before estimating a new component. As for regular PLS, there are three options, deflating **Y**, **X** or both (see Section 2.8). In addition we can do the deflation block-wise or based on the super-scores. Super-score and block-score deflation give different results; the block-score deflation can lead to reduced prediction ability (Westerhuis and Smilde (2001)).

When super-scores are used, all variants of deflation give the same predictions, but different interpretation tools. It was shown by Westerhuis and Smilde (2001) that for the super-score deflation based on both **X** and **Y**, as described above, the interpretation of variability within each of the blocks is influenced also by variability within the other blocks. The solution is, according to Westerhuis and Smilde (2001), to deflate based only on **Y** using super-scores. In this way, there is no carry over of information between the blocks. This same solution can alternatively be obtained as described in Algorithm 7.1.

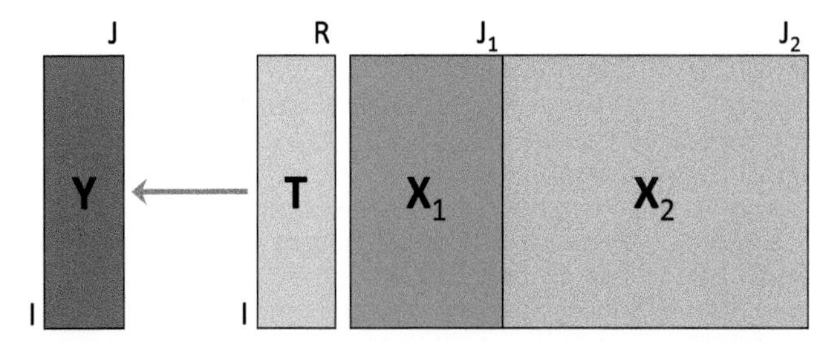

Figure 7.2 Illustration of link between concatenated **X** blocks and the response, **Y**, through the MB-PLS super-scores, **T**.

Algorithm 7.1

MB-PLS

Multiblock PLS with super-score deflation (of $\mathbf{Y}$) can be formulated as an ordinary PLS regression on concatenated blocks (blocks scaled by the square root of the number of variables) followed by a block-wise post-computation to extract block-weights, block-scores, and block-loadings for each block. The $\mathbf{Y}$-block deflation has the advantage that all local scores, weights and loadings are obtained from the original $\mathbf{X}_m$-blocks and not from blocks after deflation.

The MB-PLS predictor is obtained from the PLS model in the second line of the algorithm description where super-weights ($\mathbf{W}$), super-scores ($\mathbf{T}$), super-loadings ($\mathbf{P}$), Y-scores ($\mathbf{U}$), and Y-loadings ($\mathbf{Q}$) are produced. The prediction equation can, if wanted, be formulated in terms of original $\mathbf{X}$ variables giving an equation of the same shape as Equation 7.1 (see also results on regression coefficient calculations for PLS in Section 2.8). In the following algorithm the component index, r, is for simplicity omitted from the explanations of the individual block calculations.

1: $\mathbf{X} = [\mathbf{X}_1/\sqrt{J_1}|\ldots|\mathbf{X}_M/\sqrt{J_M}]$ – concatenate scaled blocks

2: $PLS(\mathbf{X},\mathbf{Y}) \Rightarrow \mathbf{W},\mathbf{T},\mathbf{P},\mathbf{U},\mathbf{Q}$ – partial least squares, PLS2

Loop over components – $r = 1, \ldots, R$

Loop over blocks – $m = 1, \ldots, M$

3: $\mathbf{w}_m = \mathbf{X}_m^t\mathbf{u}/(\mathbf{u}^t\mathbf{u})$ – block-weights

4: $\mathbf{w}_m^\star = \mathbf{w}_m/||\mathbf{w}_m||$ – normalise

5: $\mathbf{t}_m = (\mathbf{X}_m/\sqrt{J_m})\mathbf{w}_m^\star$ – block-scores

6: $\mathbf{p}_m = (\mathbf{X}_m^t/\sqrt{J_m})\mathbf{t}_m/(\mathbf{t}_m^t\mathbf{t}_m)$ – block-loadings

End block loop

End component loop

Since MB-PLS is neither scale invariant within nor between blocks, we here need to carefully consider the need for both within- and between-block standardisation. In Algorithm 7.1, we have applied scaling with the square root of the number of variables for each block, since this corresponds to the original formulation of the method (see Westerhuis *et al.* (1998)). Another option is to divide each block by the first singular value of each individual matrix, as done for the much used multiple factor analysis for unsupervised analysis (MFA, Escofier and Pagès (1994), see also Section 5.2.1.2). The block-wise standardisation has the advantage that it makes the blocks more comparable in the sense that their weight will be similar in the analysis (see discussion on fairness in Section 1.7).

A possible drawback with the MB-PLS method, in its basic version, is that it extracts the same number of components for all blocks. In cases with very different underlying dimensionality this may complicate interpretation (Jørgensen *et al.* (2007)). A major advantage of the MB-PLS method is its simplicity and the fact that for the predictions only one set of components is to be determined. The latter can make it less vulnerable to overfitting than other methods to be discussed below.

The predictions are, as in other PLS models, validated by empirical cross-validation or prediction testing (see Section 2.7). Scores and loadings as well as the links between them can be interpreted using the standard PLS tools as described in Chapter 2. An example of MB-PLS in the `multiblock` R-package is found in Section 11.8.2. Examples of applications of MB-PLS are available in many areas of the natural and life sciences. We give an example from chemistry (Example 7.1) and one from metabolomics (Example 7.2)

Example 7.1: MB-PLS: Raman on PUFA containing emulsions

This data set was introduced in Chapter 1. The baseline frame of reference for this analysis is the corresponding analyses using PLS on the concatenated data blocks in Chapter 2, i.e., the full Raman spectra before splitting into three consecutive blocks. As described in Chapter 1, there are two responses: (1) $PUFA_{sample}$ – PUFA measured as a percentage of the total sample – and (2) $PUFA_{fat}$ – PUFA measured as percentage of all fat types in the sample. When both responses are modelled simultaneously, they are standardised first.

We can follow the development of the cross-validated explained variance in Figure 7.3. For MB-PLS the explained variances were 88.1% and 97.9% after 7 components for the models predicting $PUFA_{sample}$ and $PUFA_{fat}$, respectively, and 92.3% after 7 components for the model predicting both PUFA measurements (see Chapter 1). The changes from using the reference models applying standard PLS regression and PLS2 were minimal ($PUFA_{sample}$: 88.4%, $PUFA_{fat}$: 97.9% and both PUFAs: 92.3% explained variance, respectively). The difference between standard PLS2 modelling on the Raman data and MB-PLS modelling is that each block is scaled by the square root of its number of variables before concatenation in MB-PLS (see Algorithm 7.1), while the blocks are directly concatenated when using PLS2. In addition, MB-PLS enables block-wise scores and loadings.

In Figures 7.4 and 7.5 we can observe how the weights (see Algorithm 7.1) are close to continuous across blocks. For this data set the super-weights ($\mathbf{w}$) and block-weights ($\mathbf{w}_m$) are very similar, and the largest contributions are found on the same wavelengths.

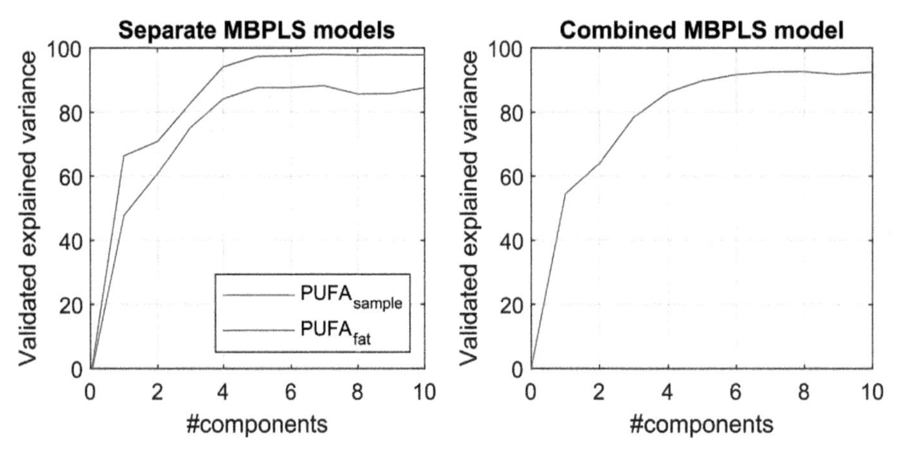

Figure 7.3 Cross-validated explained variance for various choices of number of components for single- and two-response modelling with MB-PLS.

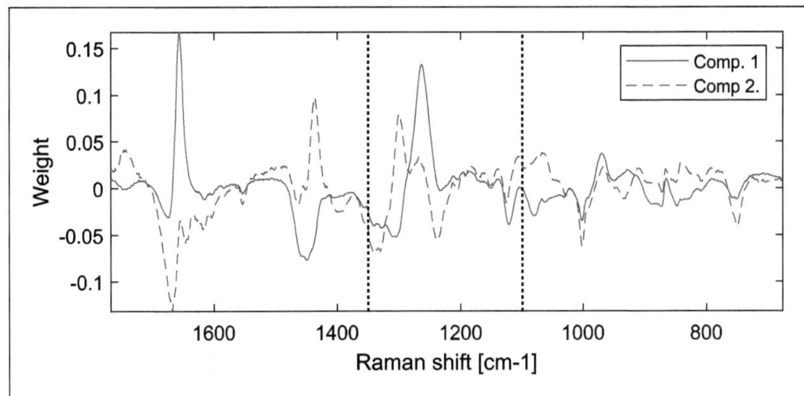

Figure 7.4 Super-weights ($\mathbf{w}$) for the first and second component from MB-PLS on Raman data predicting the $PUFA_{sample}$ response. Block-splitting indicated by vertical dotted lines.

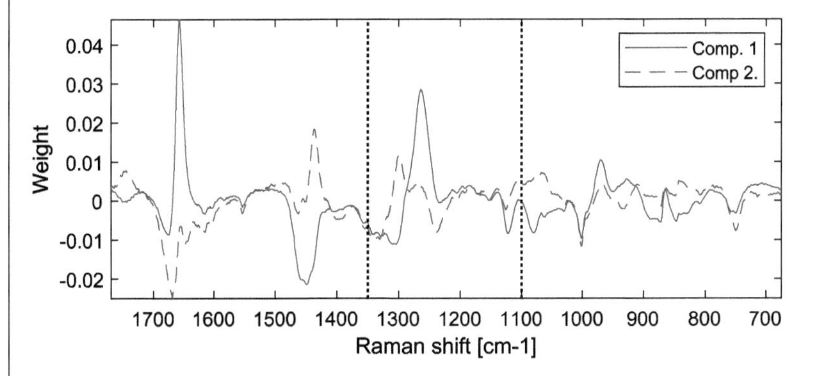

Figure 7.5 Block-weights ($\mathbf{w}_m$) for first and second component from MB-PLS on Raman data predicting the $PUFA_{sample}$ response. Block-splitting indicated by vertical dotted lines.

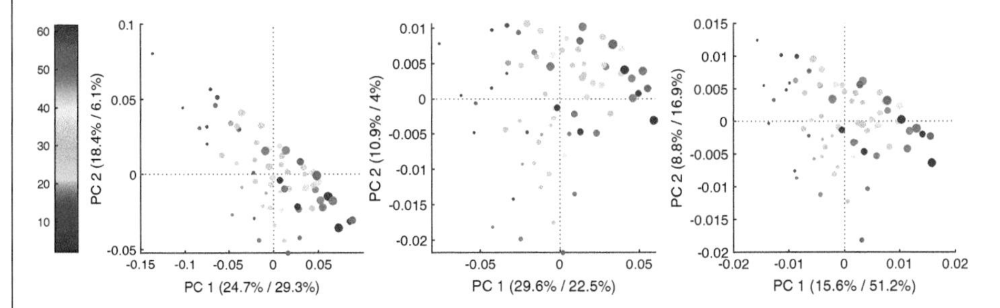

Figure 7.6 Block-scores ($\mathbf{t}_m$, for left, middle, and right Raman block, respectively) for first and second component from MB-PLS on Raman data predicting the $PUFA_{sample}$ response. Colours of the samples indicate the PUFA concentration as % in fat ($PUFA_{fat}$) and size indicates % in sample ($PUFA_{sample}$). The two percentages given in each axis label are cross-validated explained variance for $PUFA_{sample}$ weighted by relative block contributions and calibrated explained variance for the block ($\mathbf{X}_m$), respectively.

It is also evident from block-score plots in Figure 7.6 that the three blocks have different relations to the response. This shows that individual score plots can add to the interpretation of multiblock data. Here for instance, we see that the second block contributes most to the explanation of the response in the first component with large differences in point sizes, while the trends are less evident for the first and third block.

MB-PLS vs PLS2

Since MB-PLS is based on PLS2 on concatenated data blocks, the only reason that the explained variances in Figure 7.3 differ from those of PLS2 is because of the scaling applied to the individual blocks before concatenation, typically $\sqrt{J_m}$ for MB-PLS and no scaling for standard PLS2. The underlying equivalence with the concatenated analysis, means that the benefit of MB-PLS comes from its ability to show how different parts of the data (blocks) contribute to the global model through block-wise scores, loadings, weights, and explained variances. Using MB-PLS, all blocks are decomposed jointly using the same number of components for each block, regardless of the true underlying dimensionality of the blocks. This might not be a large issue for spectroscopic data, but can have a larger impact for more varied block data sources.

7.2.2 MB-PLS Used for Classification

It is common practice to use regression analysis also for classification problems (discriminant analysis, MB-PLS-DA) using a dummy response block having as many columns as there are categories/classes/groups (see Figure 7.7). Absence is represented by zeros and presence by ones in a category. It is well understood that using this matrix as the response matrix, **Y**, in a regression equation can form the basis for a classification rule (see e.g., Ripley (2007)). The idea has also been extended to handle multiblock situations using, for instance, the MB-PLS regression approach. The same idea can, however, also be used for other multiblock methods (see e.g., Biancolillo *et al.* (2015)). When only one input block is used, the method is often referred to simply as PLS-DA.

The regression model is first fitted, and the predicted values are used for classification. Different rules can be used for this; the most common is to classify directly according to which category gets the largest predicted value. Another possibility is to use either the predicted scores or the predicted response values as input to, for instance, linear discriminant analysis (LDA).

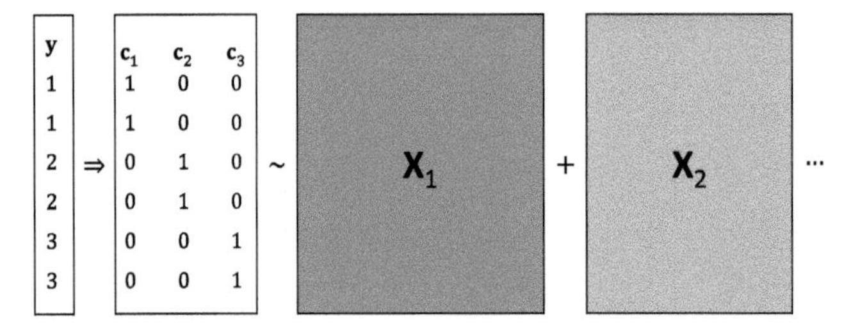

Figure 7.7 Classification by regression. A dummy matrix (here with three classes, c for class) is constructed according to which group the different objects belong to. Then this dummy matrix is related to the input blocks in the standard way described above.

Validation can be done as for standard regression procedures, but for classification it may sometimes be more natural to use a quality measure based on percentage of correct classifications in each of the groups. For selecting the number of components it was indicated in Biancolillo *et al.* (2015) that the RMSEP measure gave a more smooth development of the error curve. In that case it was, therefore, easier to use RMSEP rather than classification success for component selection (see also Westerhuis *et al.* (2008)), but this is not a general rule.

So-called canonical variates analyses (obtained from Fisher's linear discriminant analysis in more than one dimension, see Mardia *et al.* (1979)) is also useful for visualising the classification power. To avoid an overoptimistic impression of classification ability based on this method, it was proposed in Biancolillo *et al.* (2015) to calculate canonical variates on predicted values from cross-validated predictions. This is a general method and can be used for any classification approach based on regression analysis.

Example 7.2: MB-PLS for classification: Metabolomics in colorectal cancer

The study in Deng *et al.* (2020) is about colorectal cancer using metabolomics measurements on 234 blood samples from three groups of people: colorectal cancer (66 samples), colonic polyps (76 samples), and healthy controls (92 samples). In a targeted LC-MS approach 113 metabolites were detected of which 89 can be found in the *Homo sapiens* pathways of KEGG (Kanehisa and Goto, 2000). Using some exclusion criteria, the end-result includes 30 metabolic pathways with in total 81 metabolites. These pathways were considered as separate blocks in an MB-PLS-DA and thus an analysis on pathway level is possible instead of on metabolite level. Multiple comparisons were made between two groups, and in each classification the $\mathbf{y}$ vector contained zeros and ones to indicate the groups. The classification performance was measured using the receiver operating curve (ROC) which is plotting the true positive rate against the false positive rate at various threshold settings. From this, the area under the receiver operating curve (AUROC) can be calculated. Ideally the AUROC should be one and a value of 0.5 means the classification is no better than random guessing. These

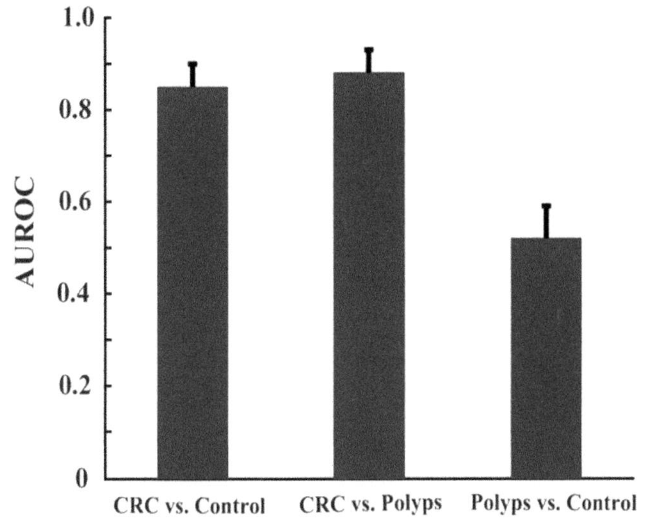

Figure 7.8 AUROC values of different classification tasks. Source: (Deng *et al.*, 2020). Reproduced with permission from ACS Publications.

AUROC values were calculated using cross-validation and they are shown in Figure 7.8. It is evident that no clear separation is possible between the polyps versus control group.

In ordinary PLS, variable importance can for instance be measured with the so-called variable influence on projection (VIP) statistic (Wold *et al.*, 1993; Eriksson *et al.*, 2013; Galindo-Prieto *et al.*, 2014). A generalisation of the VIP is developed which is called the PIP (pathway importance in projection), and this shows that for the discrimination of colorectal versus control and colorectal versus polyps three pathways are significant. For more details, we refer to the original publication (Deng *et al.*, 2020).

7.2.3 Sparse Multiblock PLS Regression (sMB-PLS)

The sparse version of MB-PLS (sMB-PLS) is based on sparse PLS (see Section 2.1.6). We will explain this method using a single $\mathbf{y}$ but there are also versions for multivariate $\mathbf{Y}$ (Lê Cao *et al.* (2008); Karaman *et al.* (2015)). The generalisation of sparseness to MB-PLS is more elaborate than sparse PLS since choices have to be made regarding crucial parts of the algorithm, such as the definition of super-scores and super-weights, and the specific deflation process. An instantiation of a sparse MB-PLS model for a single $\mathbf{y}$ is shown in Algorithm 7.2 which is one of the implementations of Karaman *et al.* (2015).

Algorithm 7.2

SPARSE MB-PLS

1: $\mathbf{X} = [\mathbf{X}_1 \vert \dots \vert \mathbf{X}_M]$	– define concatenated matrix $\mathbf{X}$
Loop over components	– $r = 1, \dots, R$
2: $\mathbf{w} = ST_\lambda(\mathbf{X}^t\mathbf{y})$	– calculate sparse weights for $\mathbf{X}^t\mathbf{y}$
3: $\Vert\mathbf{w}\Vert = 1$	– normalise $\mathbf{w}$
Loop over blocks	– $m = 1, \dots, M$
4: $\mathbf{w}(m)$	– split $\mathbf{w}$ according to the M blocks
5: $\widetilde{\mathbf{w}_m} = \mathbf{w}(m)/\Vert\mathbf{w}(m)\Vert$	– normalise block-weights
6: $\mathbf{t}_m = \mathbf{X}_m\widetilde{\mathbf{w}_m}$	– calculate block-scores
End block loop	
7: $\mathbf{T} = [\mathbf{t}_1 \vert \dots \vert \mathbf{t}_M]$	– collect block-scores
8: $\mathbf{w}_s = [\Vert\mathbf{w}(1)\Vert \vert \dots \vert \Vert\mathbf{w}(M)\Vert]^t$	– define super-weights of $\mathbf{X}$
9: $\mathbf{t} = \mathbf{T}\mathbf{w}_s$	– calculate super-scores of $\mathbf{X}$
10: $\mathbf{p} = \mathbf{X}^t\mathbf{t}/\mathbf{t}^t\mathbf{t}; q = \mathbf{y}^t\mathbf{t}/\mathbf{t}^t\mathbf{t}$	– calculate X and y-loadings
11: $\mathbf{X}_{\text{new}} = \mathbf{X} - \mathbf{t}\mathbf{p}^t$	– deflate X-block
12: $\mathbf{y}_{\text{new}} = \mathbf{y} - \mathbf{t}q$	– deflate y-block
End component loop	

As for regular MB-PLS, the number of components, r, is for simplicity omitted from notation. The crucial step in this algorithm is $\mathbf{w} = ST_{\lambda}(\mathbf{X}^t\mathbf{y})$ (ST is soft-thresholding again, see also Section 2.1.6). There are choices to make regarding calculating the super-weights and how to deflate. Standard MB-PLS deflation was performed using the super-scores (the $\mathbf{T}$ in the second line of Algorithm 7.1) since it would avoid artefacts (Westerhuis and Smilde, 2001) and (ii) a single PLS model and subsequent postprocessing can be used to obtain the block-related scores and loadings.

There seems to be a discussion on how to perform sparse MB-PLS. At least three different versions exist (Lê Cao *et al.*, 2008; Karaman *et al.*, 2015) and more can be envisaged. A drawback of the one presented in Algorithm 7.2 is that sparseness is imposed on the $\mathbf{w}$ which is a concatenation of weights across blocks. Hence, there is no guarantee that all block will have some non-zero weights; in the most extreme case this sparse MB-PLS version can be a block-selector! Moreover, it is not clear what the exact properties are of the solution found with Algorithm 7.2. It is claimed that property (ii) in Algorithm 7.2 is retained by this version, that is, a single sparse MB-PLS can be calculated using the super-weights as in this algorithm and subsequently all block-weights and block-scores can be recovered (Karaman *et al.*, 2015).

The sparse MB-PLS method has been used in a metabolomics application for biomarker discovery using different sets of metabolomics data (LC-MS and NMR) (Karaman *et al.*, 2015) (see Example 7.3). An example of sMB-PLS in the `multiblock` R-package is found in Section 11.8.3. Interesting extensions of sparse MB-PLS are reported where also sparseness on the weights for the Y-block (for more than one response) and on the X-block-scores are imposed. This leads to discovering gene-regulatory modules in genomics data (Li *et al.*, 2012). Another interesting extension is sparse PLS path-modelling with applications in genomics (Csala *et al.*, 2020).

Example 7.3: Sparse MB-PLS in metabolomics

In the study of Karaman *et al.* (2015), six piglets were subjected to three diets consisting of different types and amounts of wheat and rye indicated by the abbreviations WWG, WAF, and RAF. The same piglets were fed with these diets during some days and then a wash-out period was interspersed between the diets. Hence, this constitutes a cross-over design. Fasting blood samples were collected during the four days of treatment (the diets) and these were analysed on three different metabolomics platforms: LC-MS(+), LC-MS(-), and NMR. Cleaning the metabolomics data resulted in 1 016 variables for the LC-MS(+), 790 variables for the LC-MS(-), and 596 variables for the NMR. For some samples there were missing values, so finally the data set comprised 69 samples.

For such a study it is important to carefully consider the preprocessing steps. One of the things to think about is the cross-over nature of the study design allowing for removing between-individual variation. This was performed by centring across the samples pertaining to each individual pig, thereby centring all pigs around zero. Many more steps were involved including scaling within blocks and block-scaling, for more details see Karaman *et al.* (2015). There are now three X-blocks pertaining to the three metabolomics platforms. The Y-matrix of size (69×3) consists of dummy variables encoding the design; there were three treatments thus requiring three dummy variables. In the estimation part, the number of PLS components and the penalty parameter λ were chosen with cross-validation. This resulted in three components and 64 selected variables; which is a huge reduction in terms of variable selection. A part of the result is shown in Figure 7.9.

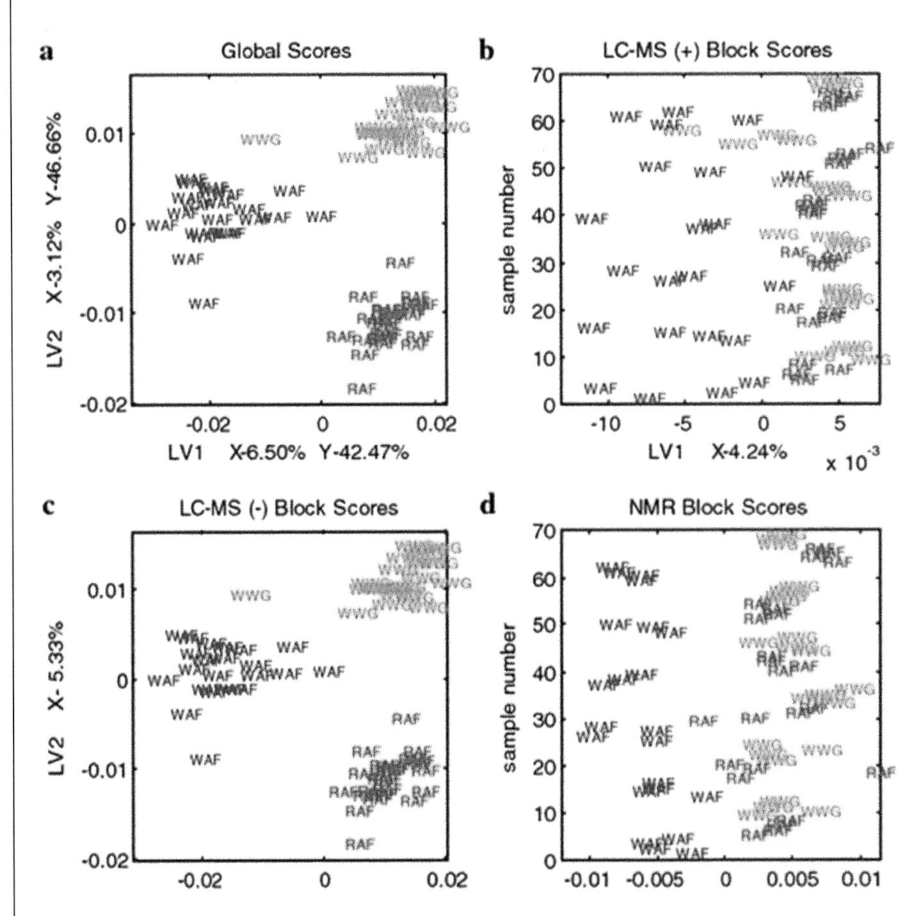

Figure 7.9 Super-scores (called global scores here) and block-scores for the sparse MB-PLS model of the piglet metabolomics data. Source: (Karaman *et al.*, 2015). Reproduced with permission from Springer.

The super-scores clearly show a separation between the treatment groups (see Figure 7.9a). Going down to the block-scores, the WAF group is separated from the others in the LC-MS(+) (Figure 7.9(b)) and the NMR (Figure 7.9(d)) but both platforms failed to separate the RAF and WWG groups (also not in the higher components). The LC-MS(-) platform, however, clearly separated all three groups and this was reflected in a high super-weight for that block in the sparse MB-PLS model. Correlation loadings were also calculated (not shown) to indicate which variables are responsible for the grouping, but unfortunately many of the features could not be identified as metabolites; a common set-back of this type of metabolomics.

Note that this study could also have been analysed using an ASCA model (see Chapter 6) which would also have allowed for studying the factor time and the interaction between time and treatment. Sparseness could then have been introduced by using the sparse version of ASCA called group-wise ASCA (GASCA) (Saccenti *et al.*, 2018).

7.3 The Family of SO-PLS Regression Methods (Sequential and Orthogonalised PLS Regression)

The SO-PLS method was originally developed for a situation in which one of the blocks is a design matrix and the second is a standard quantitative matrix (Jørgensen and Næs (2004) and Jørgensen and Næs (2008)). This situation is similar to the one often called covariance analysis in the ANOVA literature, with the exception that the covariates could be many and collinear. The method was called LS-PLS since the first step was based on LS regression and the second on PLS regression. LS-PLS was later extended to handle split-plot designs (Måge and Næs (2005)) and situations with several blocks that can be both design matrices and matrices based on regular quantitative measurements. The name SO-PLS was invented to comprise all these situations. The LS-PLS is then a special case of SO-PLS with the maximum possible number of PLS components for the design block. A number of extensions and modifications have been developed as will be shown below and in Chapter 10.

The basic SO-PLS method essentially consists of two different steps used in sequence, (i) PLS regression and (ii) orthogonalisation with respect to previously fitted blocks. In the following we will present the original version first before a slight modification is suggested for improved interpretation.

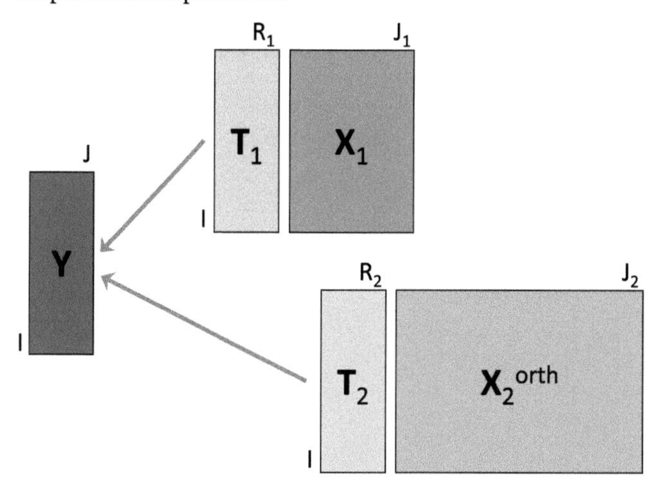

Figure 7.10 Linking structure of SO-PLS. Scores for both $\mathbf{X}_1$ and the orthogonalised version of $\mathbf{X}_2$ are combined in a standard LS regression model with $\mathbf{Y}$ as the dependent block.

7.3.1 The SO-PLS Method

The method will here first be described for two blocks, $\mathbf{X}_1$ and $\mathbf{X}_2$.

Algorithm 7.3

SO-PLS

The first step is to use regular PLS regression for $\mathbf{Y}$ versus $\mathbf{X}_1$. This gives scores $\mathbf{T}_1$ and loadings $\mathbf{P}_1$ as shown in Section 2.1.5. Then the $\mathbf{X}_2$ is orthogonalised with respect to the PLS components of $\mathbf{X}_1$ and a new PLS regression is run on the $\mathbf{Y}$-residuals. Finally the two sets of scores are combined in an LS regression as shown in the last line of the algorithm.

First block

1: $PLS(\mathbf{X}_1, \mathbf{Y}) => \mathbf{T}_1, \mathbf{P}_1$ – PLS regression for $\mathbf{X}_1$ and $\mathbf{Y}$

2: $\mathbf{X}_2^{\text{orth}} = (\mathbf{I} - \mathbf{T}_1(\mathbf{T}_1^t\mathbf{T}_1)^{-1}\mathbf{T}_1^t)\mathbf{X}_2$ – orthogonalised $\mathbf{X}_2$ wrt. $\mathbf{T}_1$

3: $\mathbf{Y}^{\text{orth}} = (\mathbf{I} - \mathbf{T}_1(\mathbf{T}_1^t\mathbf{T}_1)^{-1}\mathbf{T}_1^t)\mathbf{Y}$ – orthogonalised $\mathbf{Y}$ wrt. $\mathbf{T}_1$

Second block

4: $PLS(\mathbf{X}_2^{\text{orth}}, \mathbf{Y}^{\text{orth}}) => \mathbf{T}_2, \mathbf{P}_2$ – PLS regression for $\mathbf{X}_2^{\text{orth}}$ and $\mathbf{Y}^{\text{orth}}$

5: $\mathbf{Y} = \mathbf{T}_1\mathbf{Q}_1 + \mathbf{T}_2\mathbf{Q}_2 + \mathbf{F}$ – estimate prediction equation using LS

The linking structure of SO-PLS is illustrated in Figure 7.10.

The two sets of scores $\mathbf{T}_1$ and $\mathbf{T}_2$ are orthogonal and, therefore, the LS solution in the last line of the algorithm can be obtained fitting the two scores matrices separately. The second PLS regression in the algorithm can be performed without orthogonalising $\mathbf{Y}$ and, therefore, the LS solutions for $\mathbf{Q}_1$ and $\mathbf{Q}_2$ are exactly equal to the Y-loadings from PLS($\mathbf{X}_1, \mathbf{Y}$) and PLS($\mathbf{X}_2^{\text{orth}}, \mathbf{Y}$).

For more than two blocks, the same procedure is repeated for the additional blocks. First $\mathbf{X}_3$ is orthogonalised with respect to the scores of $\mathbf{T}_1$ and $\mathbf{T}_2$. Then the $\mathbf{Y}$-residuals from the previous predictions based on $\mathbf{T}_1$ and $\mathbf{T}_2$ are regressed onto the orthogonalised $\mathbf{X}_3$ (Figure 7.11). The procedure continues until all blocks have been fitted, switching between orthogonalisation and PLS regression. An example of SO-PLS in the `multiblock` R-package is found in Section 11.8.4.

Since the two sets of scores $\mathbf{T}_1$ and $\mathbf{T}_2$ in the algorithm are linear functions of $\mathbf{X}_1$ and the orthogonalised $\mathbf{X}_2$, the two terms in the linear LS equation at the end of the algorithm can, if wanted, be back-transformed to original block units multiplied by corresponding regression coefficients. Letting $\mathbf{R}_1$ and $\mathbf{R}_2$ be the matrices used to obtain the scores $\mathbf{T}_1$ and $\mathbf{T}_2$ from the blocks $\mathbf{X}_1$ and $\mathbf{X}_2^{\text{orth}}$ (i.e., $\mathbf{T}_1 = \mathbf{X}_1\mathbf{R}_1$ and $\mathbf{T}_2 = \mathbf{X}_2^{\text{orth}}\mathbf{R}_2$, see Equation 2.17 in Chapter 2), the estimate of the systematic part of the LS equation at the end of the algorithm can be written as:

$$\begin{aligned}
&\mathbf{X}_1\mathbf{R}_1\mathbf{Q}_1 + \mathbf{X}_2^{\text{orth}}\mathbf{R}_2\mathbf{Q}_2 \\
&= \mathbf{X}_1\mathbf{R}_1\mathbf{Q}_1 + (\mathbf{I} - \mathbf{T}_1(\mathbf{T}_1^t\mathbf{T}_1)^{-1}\mathbf{T}_1^t)\mathbf{X}_2\mathbf{R}_2\mathbf{Q}_2 \\
&= \mathbf{X}_1\mathbf{R}_1\mathbf{Q}_1 - \mathbf{T}_1(\mathbf{T}_1^t\mathbf{T}_1)^{-1}\mathbf{T}_1^t\mathbf{X}_2\mathbf{R}_2\mathbf{Q}_2 + \mathbf{X}_2\mathbf{R}_2\mathbf{Q}_2 \\
&= \mathbf{X}_1\mathbf{R}_1(\mathbf{Q}_1 - (\mathbf{T}_1^t\mathbf{T}_1)^{-1}\mathbf{T}_1^t\mathbf{X}_2\mathbf{R}_2\mathbf{Q}_2) + \mathbf{X}_2\mathbf{R}_2\mathbf{Q}_2 \\
&= \mathbf{X}_1\mathbf{B}_1 + \mathbf{X}_2\mathbf{B}_2
\end{aligned} \tag{7.2}$$

With reference to the conceptual illustration in Figure 7.1, line two of this equation shows that the common part of the two blocks (the term before the + sign) is handled in the first step and the distinct predictive part of $\mathbf{X}_2$ (the term after the + sign), i.e., the part that is not common (in the predictive sense) with $\mathbf{X}_1$, in the second step of the algorithm (see Figure 7.11).

The orthogonalisation and individual treatment of the blocks in SO-PLS has some important consequences. First of all the method allows for different underlying dimensionalities (i.e., different number of components) of the blocks, and the method becomes invariant to the relative scaling of the blocks. Another property is that we can calculate the additional predictive contribution of new blocks as they are incorporated. This is called additional explained

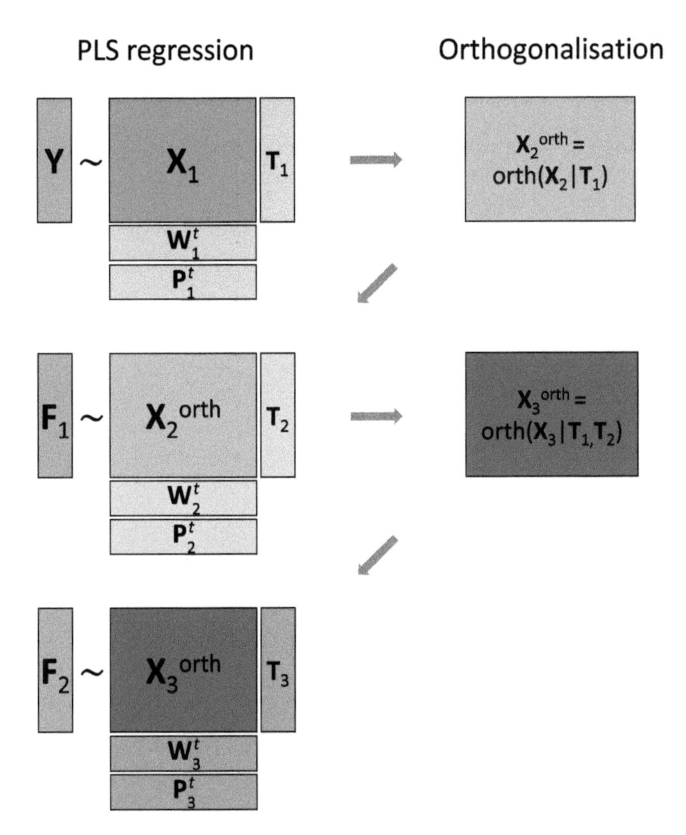

Figure 7.11 The SO-PLS iterates between PLS regression and orthogonalisation, deflating the input block and responses in every cycle. This is illustrated using three input blocks $\mathbf{X}_1$, $\mathbf{X}_2$, and $\mathbf{X}_3$. The upper figure represents the first PLS regression of $\mathbf{Y}$ onto $\mathbf{X}_1$. Then the residuals from this step, obtained by orthogonalisation, goes to the next (figure in the middle) where the same PLS procedure is repeated. The same continues for the last block $\mathbf{X}_3$ in the lower part of the figure. In each step, loadings, scores, and weights are available.

variance and is obtained by subtracting the explained variances for the two models (with and without the new block) from each other.

For the SO-PLS method the scores for blocks after the first block $\mathbf{X}_1$ will not lie in the column spaces of the corresponding original blocks (i.e., the columns of $\mathbf{T}_2$ are not in the column-space of $\mathbf{X}_2$, but rather in the space spanned by $[\mathbf{X}_1, \mathbf{X}_2]$). This means that these scores cannot be expressed as linear combinations of the columns of the original block $\mathbf{X}_2$ (see Section 2.8 for a discussion of this issue). This could possibly in some cases make interpretation of plots and explained variances less clear in terms of the original measurements. A new variant of the method, where the scores are forced to be within the spaces spanned by the blocks, is described below (see Algorithm 7.4). In the Raman example to follow the two methods give very similar interpretations, so in that case this phenomenon did not have any negative effect.

With the maximum number of components extracted in a block (= rank of the block), the PLS solution is the same as the LS solution. This is the reason why the original LS-PLS method fits into the general SO-PLS framework.

A method which will not be discussed further here, but which has some similarities with SO-PLS, is the TANDEM method published in Aben *et al.* (2016). In this method, the response is first fitted to some of the blocks using elastic net regression (Zou and Hastie,

2003). Next, the residuals of that regression are used as a new response to fit to other blocks of the data, again using elastic net regression.

7.3.2 Order of Blocks

In the case when there are many blocks to incorporate, it may not be obvious which blocks to incorporate first. This can have an impact on the solution. In the case of a design block and a covariate block, it is often natural to fit the design block first as done for LS-PLS in process modelling (see, e.g., Jørgensen and Næs (2004)), but in other cases we have to choose. The order is generally not so important for prediction ability, but interpretation will be affected. In such cases the user can choose the most natural order, or possibly try different orders for obtaining a better understanding of the data set. An example where the order is quite obvious is if we have an instrument in the laboratory and we are interested whether another and possibly new instrument will add to its prediction ability; the old instrument will then be used as $\mathbf{X}_1$ and the second one as $\mathbf{X}_2$. The order is also quite obvious if the blocks are measurements taken at different stages in an industrial production line.

Strategies have also been developed based on prediction ability for choosing block order (Niimi *et al.* (2018), Campos *et al.* (2018) and Niimi *et al.* (2020)). The idea behind these procedures is to first select the block with the best prediction ability and then choose the next keeping the first in the model. In the third step, the two first are used as $\mathbf{X}_1$ and $\mathbf{X}_2$ and we search for the third $\mathbf{X}_3$ that has the best additional prediction ability. This method may, however, be sensitive to overfitting in small data sets because of the many decisions made and must be used with care. It is clearly best suited for situations with not too many blocks and with a rather large number of samples. Developing multiple testing procedures for this purpose is clearly of interest, based on for instance ideas similar to the Bonferroni correction method. We refer to Figure 7.20 for an example of multiple testing of contributions of each block after the block sequence is determined.

7.3.3 Interpretation Tools

One of the aims of SO-PLS is to investigate the additional effect of new blocks added to the model. Therefore, the differences in prediction ability with and without the $\mathbf{X}_2$ in the model will play an important role when interpreting the full model. This will be called the additional explained variance. If the $\mathbf{T}_1$ and $\mathbf{T}_2$ are normalised (optional), the relative effect will also be visible in the $\mathbf{Q}$-values in the prediction equation in Algorithm 7.3.

For interpretation, the scores and loadings ($\mathbf{T}_1, \mathbf{P}_1, \mathbf{T}_2, \mathbf{P}_2$ etc.) for the models are always useful (alternatively weights can be used). A possible drawback with using the $\mathbf{T}_2$ scores is that because of orthogonalisation they are not in the column space of the original measurements $\mathbf{X}_2$. In Section 7.3.4 a recently developed modification of SO-PLS which gives scores in the space of $\mathbf{X}_2$, will be described.

Another tool that is useful in this context is the PCP method (principal components of prediction, Langsrud and Næs (2003)). This is a method which can be used for any regression situation (also multiblock regression) where the response matrix is multivariate. A discussion of the situation with univariate $\mathbf{y}$ is given in the same paper, but that case will not be pursued here. The PCP is particularly useful if there are many blocks since such situations lead to many PLS models to be interpreted. The PCP is based on first calculating the PCA scores and loadings of the predicted responses $\widehat{\mathbf{Y}}$. Then each of the variables in the input blocks, $\mathbf{x}_{mj}$ are regressed onto the principal components $\mathbf{T}$, using the model

$$\{\mathbf{T},\mathbf{P}\} = \text{PCA}(\widehat{\mathbf{Y}})$$
$$\mathbf{x}_{mj} = \mathbf{T}\boldsymbol{\alpha}_{mj} + \mathbf{e}_{mj}. \quad (7.3)$$

The PCA notation in the first line means that a standard PCA is used on the predicted value of $\mathbf{Y}$. This procedure leads to just one set of plots for the whole model, PCA scores of $\widehat{\mathbf{Y}}$, PCA loadings of $\widehat{\mathbf{Y}}$ and then a 'loading' plot based on the estimated coefficients $\boldsymbol{\alpha}$, indicating how the input variables relate to the scores $\mathbf{T}$. They play the same role as loadings in PLS and should be interpreted the same way. We refer to Example 7.5 for an illustration. In SO-PLS applications we can choose between using the original $\mathbf{X}$-measurements or the deflated matrices obtained in the SO-PLS sequence. In the latter, the importance of the variables is more explicitly expressed, but has the possible disadvantage that the deflated $\mathbf{X}$-blocks are not in the column space of the original measurements.

In Jørgensen *et al.* (2007)) a comparison between the interpretation of SO-PLS and MB-PLS was conducted when one of the blocks was a design block and the other a block of multivariate spectra. Some possible interpretation advantages of SO-PLS, related to very different size and characteristics of the blocks, were identified.

7.3.4 Restricted PLS Components and their Application in SO-PLS

As stated in the previous section, in the SO-PLS method the scores of the blocks beyond the first cannot be written as linear combinations of the columns of the original blocks. For instance, the $\mathbf{T}_2$ cannot be written as a linear combination of columns of $\mathbf{X}_2$. A new and alternative formulation of SO-PLS presented in Algorithm 7.4 solves this.

Algorithm 7.4

RESTRICTED PLS COMPONENTS AND THEIR USE IN SO-PLS

A PLS weight vector $\mathbf{w}$ ($||\mathbf{w}|| = 1$) can be calculated as the first right singular vector of $\mathbf{Y}^t\mathbf{X} = \mathbf{U}\mathbf{D}\mathbf{V}^t$ (see Section 2.1.5). This also means that it can be found as the first eigenvector $\mathbf{v}_1$ of $\mathbf{X}^t\mathbf{Y}\mathbf{Y}^t\mathbf{X} = \mathbf{V}\mathbf{D}^2\mathbf{V}^t$ ($= \mathbf{S}$). An equivalent formulation is that $\mathbf{w}$ is the vector with the restriction $||\mathbf{w}|| = 1$ that maximises $\mathbf{w}^t\mathbf{S}\mathbf{w}$.

We now propose to maximise $\mathbf{w}^t\mathbf{S}\mathbf{w}$ ($||\mathbf{w}|| = 1$) under the restriction that $\mathbf{w}^t\mathbf{C} = 0$, where $\mathbf{C}$ is a matrix. The solution to this can according to Rao (1964) be found as the first eigenvector (i.e., the one with the largest eigenvalue) of

$$(\mathbf{I} - \mathbf{C}(\mathbf{C}^t\mathbf{C})^{-1}\mathbf{C}^t)\mathbf{S} \quad (7.4)$$

assuming that the inverse exists. It was shown in Rao (1964) that the eigenvectors can also be obtained from the eigenvectors of a symmetric matrix. More precisely, the first eigenvector $\mathbf{v}_1$ of Equation 7.4 can be written as

$$\mathbf{v}_1 = (\mathbf{u}_1^t\mathbf{S}^{-1}\mathbf{u}_1)\mathbf{S}^{-1/2}\mathbf{u}_1 \quad (7.5)$$

where $\mathbf{u}_1$ is the first eigenvector of

$$\mathbf{S}^{1/2}(\mathbf{I} - \mathbf{C}(\mathbf{C}^t\mathbf{C})^{-1}\mathbf{C}^t)\mathbf{S}^{1/2}. \quad (7.6)$$

Since scores computed using $\mathbf{w}$ from the restricted model, $\mathbf{t} = \mathbf{X}\mathbf{w}$, are linear combinations of columns of $\mathbf{X}$, they will be embedded in the column space of $\mathbf{X}$. This method will now be called restricted PLS regression.

If we let $\mathbf{C}$ be the matrix $\mathbf{X}_2^t\mathbf{T}_1$ where $\mathbf{T}_1$ is the set of scores from block 1 in the SO-PLS sequence, the above restriction can be written as $\mathbf{w}^t\mathbf{C} = \mathbf{w}^t\mathbf{X}_2^t\mathbf{T}_1 = 0$. This means that the component $\mathbf{X}_2\mathbf{w}$, in addition to being embedded in $\mathbf{X}_2$, is forced to be uncorrelated with $\mathbf{T}_1$. It is assumed here that the matrix product $\mathbf{X}_2^t\mathbf{T}_1$ is non-zero, i.e., the matrices are not orthogonal to each other. As a consequence, the block orthogonalisation of SO-PLS is not used in the restricted case. Since the number of columns of $\mathbf{X}_2^t\mathbf{T}_1$ is equal to R (the number of columns in $\mathbf{T}_1$), the $\mathbf{C}$ will usually have full column rank, ensuring that $\mathbf{C}^t\mathbf{C}$ can be inverted.

The loadings matrix $\mathbf{P}_2$ for the obtained scores matrix $\mathbf{T}_2$ can be found by regressing $\mathbf{X}_2$ onto $\mathbf{T}_2$. This matrix will be embedded in the space spanned by the rows of $\mathbf{X}_2$.

Deflation goes as for standard PLS regression. When using restricted PLS components in SO-PLS, the process of prediction is exactly as with the ordinary SO-PLS, only basing regression coefficients on the loadings and weights obtained by the restricted PLS.

The advantage of this method as compared to standard SO-PLS is an interpretation that is directly related to the space of $\mathbf{X}_2$. Additional explained variances will for this method also be explicitly related to the additional explained variances of the measured variables (we refer to Section 2.8 for further discussion of the issue of explained variances and subspaces). A possible drawback is that the original proposal optimises prediction ability of the contributions from $\mathbf{X}_1$ and $\mathbf{X}_2$, while for the restricted this is not necessarily the case. In Example 7.5, the two versions give, however, the same predictive ability and the interpretation is also similar for the two. This is probably because of the high number of variables, as we have in simulations observed quite large differences for data with few variables, $J << I$.

7.3.5 Validation and Component Selection

Determining the number of components is an important issue also for the SO-PLS method. Incorporating too few or too many components will generally lead to poor predictions. As in most other cases considered in this book, cross-validation (CV) and prediction testing is used in connection with a criterion such as RMSECV/RMSEP or explained variance (Section 2.7). Since the number of components has to be decided for more than one block, it is, however, not obvious how to do it. There are basically two different strategies, so-called global and sequential selection. For the latter, we determine the number of components for each block in the sequential order they are incorporated, i.e., the number of components in $\mathbf{X}_1$ is decided before $\mathbf{X}_2$ is fitted and thereafter kept fixed. The other strategy is based on calculating the prediction ability for all combinations of components in the blocks and afterwards comparing the prediction ability for all of them. This last option may lead to better predictions, but is also more prone to overfitting since very many choices are tested. Therefore, prediction testing of the final solution on new data is important for practical use of the method. The two approaches correspond conceptually to forward selection and best subset selection in regression analysis.

In all cases regardless of the selection procedure, we typically use the so-called Måge plot (Næs *et al.* (2013)), which is a plot with the total (sum) number of components on the horizontal axis and the prediction ability on the vertical axis. The plot is useful for finding the best combination of components (see also DISCO in Section 5.1.2.1). For an application we refer to Figure 7.13 in Example 7.4.

It is of interest to test the effect of the additional contributions of the blocks using some type of significance testing, comparing for instance the contribution of the second block $\mathbf{X}_2$

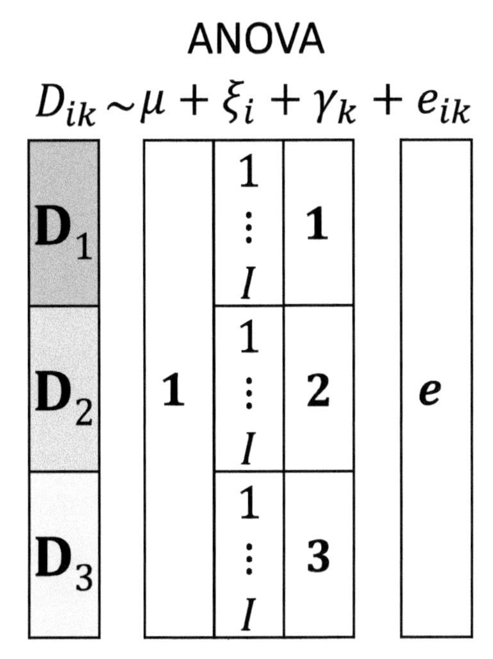

Figure 7.12 The CVANOVA is used for comparing cross-validated residuals $\mathbf{F}$ for different prediction methods/models or for different numbers of blocks in the models (in for instance SO-PLS). The squares or the absolute values of the cross-validated prediction residuals, D_{ik}, are compared using a two-way ANOVA model. The figure below the model represents the data set used. The indices i and k denote the two effects: sample and method. The I samples for each method/model (equal to three in the example) are the same, so a standard two-way ANOVA is used. Note that the error variance in the ANOVA model for the three methods is not necessarily the same, so this must be considered a pragmatic approach.

with the error sum-of-squares. Calculating degrees of freedom to be used in tests is more difficult for PLS than for least squares regression, although some suggestions have been put forwards (see e.g., Krämer and Sugiyama (2011)) and Rubingh *et al.* (2013)). An alternative and more pragmatic strategy for comparing prediction errors in general was proposed in Indahl and Næs (1998). The method was called CVANOVA (see also Section 2.7.5) and is based on comparing cross-validated residuals using ANOVA or t-tests. We simply calculate the squared (or absolute) prediction residuals, $D_{ik} = |y_{ik} - \hat{y}_{ik}|$, for each sample, i, and method, k, (here 'method' refers to a new component or block incorporated) and consider the residuals for the two methods as two different data columns (Cederkvist *et al.* (2005)). The paired t-test or Tukey's pair-wise test (or ANOVA if more than two methods are compared) is then used for significance testing (see Figure 7.12). A non-parametric ANOVA is an alternative here, but this is not tested.

7.3.6 Relations to ANOVA

Because of the orthogonality between blocks incorporated in SO-PLS, the sum-of-squares for each of the contributions can be added. In other words, the total sum-of-squares can be split into contributions for each block

$$\begin{aligned} tr(\mathbf{Y}^t\mathbf{Y}) &= SS(\mathbf{X}_1\hat{\mathbf{B}}_1) + SS(\mathbf{X}_2^{\text{orth}}\hat{\mathbf{B}}_2^{\text{orth}}) + SS(\hat{\mathbf{F}}) \\ &= tr(\hat{\mathbf{B}}_1^t\mathbf{X}_1^t\mathbf{X}_1\hat{\mathbf{B}}_1) + tr(\hat{\mathbf{B}}_2^{\text{orth},t}\mathbf{X}_2^{\text{orth},t}\mathbf{X}_2^{\text{orth}}\hat{\mathbf{B}}_2^{\text{orth}}) \\ &+ tr(\hat{\mathbf{F}}^t\hat{\mathbf{F}}) \end{aligned} \tag{7.7}$$

This extends to several blocks and also to the interactions case which will be discussed below. The decomposition is very similar to the Type I sum-of-squares decomposition in the ANOVA literature (see for instance Driscoll and Borror (2000)). We start with one block, estimate how much extra the next block contributes before going to the third and so on.

Example 7.4: SO-PLS: Sensory assessment of wines

This example is based on the Val de Loire Wine data presented in (Pagès (2005)). The data consist of sensory assessments of 21 wines where the attributes are stored in four blocks A–D (representing data/variables related to 'smell at rest', 'view', 'smell after shaking', and 'taste' respectively) and a response block E containing the variables 'overall quality' and 'typical'. We refer to the original publication for further description of the data. Fitting SO-PLS with up to six components results in the Måge plot seen in Figure 7.13. The order of the blocks in the SO-PLS model is given by the order in which the attribute types (smell at rest, view etc.) were measured (A, B, C, and D, having 5, 3, 10, and 9 columns, respectively).

In Figure 7.13, models using only a single block are connected with coloured lines. Maximum explained variance is found already after three components (see magnified view in Figure 7.13), one from each of the blocks, B, C, and D, excluding the first block, A. This corresponds to global selection of components. The results show that when the other blocks are in the model, the smell at rest (A) has no effect on the quality and typicality. After the three first components, the models start overfitting and a gradual reduction in explained variance is observed. Among the best candidates, the differences are very small leading to a large degree of overlap in the Måge plot. It can also be seen, in this case the block D used alone explains almost the same variance as the multiblock model (see purple line with maximum at 0.0.0.2 components in Figure 7.13). The multiblock model, on the other hand, describes also the sequential and additional contribution of the blocks B, C, and D. None of the blocks A, B, or C are even close to explaining as much variation as block D on their own.

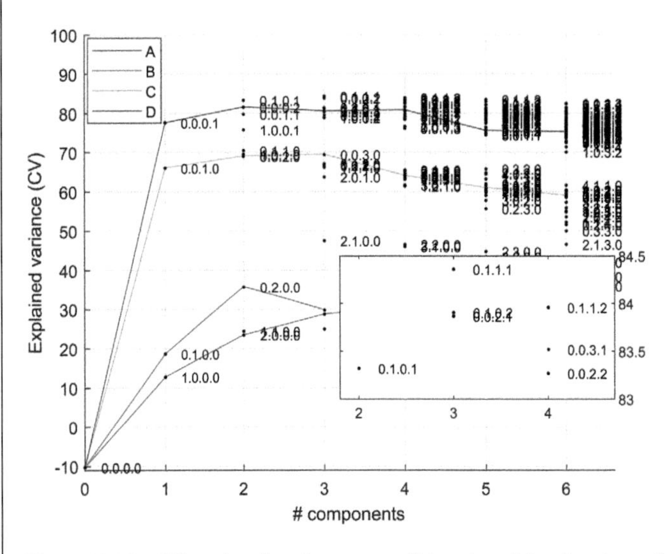

Figure 7.13 Måge plot showing cross-validated explained variance for all combinations of components for the four input blocks (up to six components in total) for the wine data (the digits for each combination correspond to the order A, B, C, D, as described above). The different combinations of components are visualised by four numbers separated by a dot. The panel to the lower right is a magnified view of the most important region (2, 3, and 4 components) for selecting the number of components. Coloured lines show prediction ability ($\mathbf{Q}^2$, see cross-validation in Section 2.7.5) for the different input blocks, A, B, C, and D, used independently.

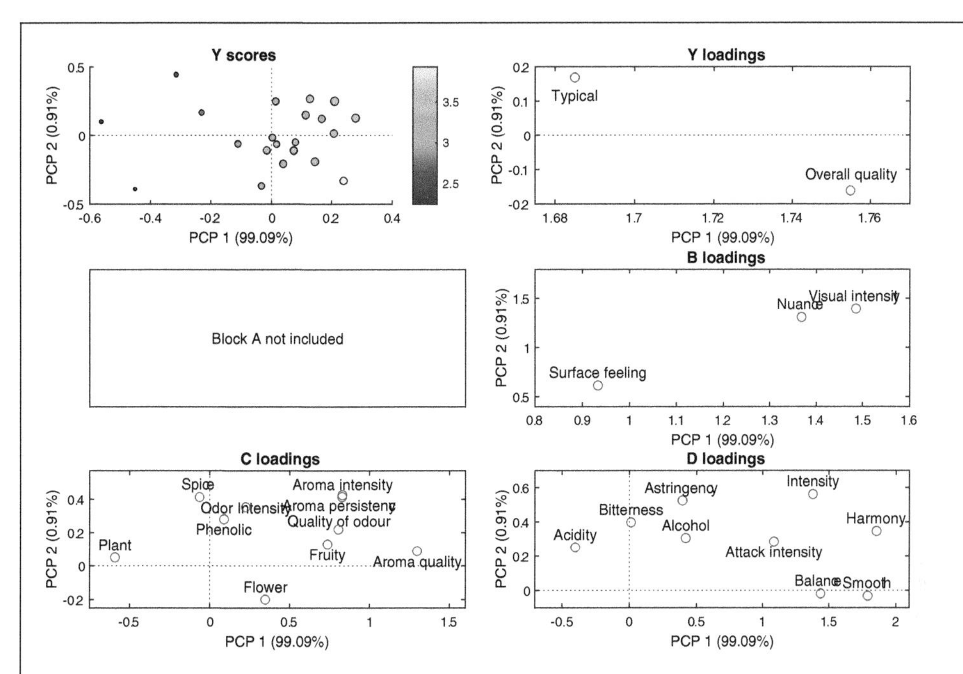

Figure 7.14 PCP plots for wine data. The upper two plots are the score and loading plots for the predicted Y, the other three are the projected input X-variables from the blocks B, C, and D. Block A is not present since it is not needed for prediction. The sizes of the points for the $\mathbf{Y}$ scores follow the scale of the 'overall quality' (small to large) while colour follows the scale of 'typical' (blue, through green to yellow).

After selection of the globally optimal model, with respect to explained variance, we generated PCP plots in Figure 7.14 using 0.1.1.1 components for blocks B through D. As can be seen, there is essentially only one component in the Y-matrix, mainly because of high correlation (0.85) between the two responses 'overall quality' and 'typical'. In the first component both responses are highly positive in the Y-loadings; 'overall quality' having a 4% higher loading than 'typical', while the second component is a contrast between the two. The other plots show how the different attributes in blocks B, C, and D are related to these components. All the 'view' attributes in block B have positive values in the first dimension, showing correlation to the responses. In blocks C and D there seems to be a tendency that positively loaded descriptors like 'quality', 'smooth', and 'harmony' correlate positively with the responses, while the more harsh descriptions like 'plant', 'acidity', 'bitterness', and 'astringency' have low or negative correlations.

Example 7.5: SO-PLS: Raman on PUFA containing emulsions

A single response SO-PLS model with $PUFA_{sample}$ is fitted to the three-block Raman data introduced in Chapter 1. The order of the blocks has been chosen to be (1) left, (2) middle, (3) right. In Figure 7.15 we see that the single block models (coloured lines) behave quite similarly, though the left block explains a little less variance than the other two blocks. Even after 10 components there is still a slight increase in explained variance. The global maximum cross-validated explained variance of 90.4% is achieved using 4+3+3 components from the left, middle, and right blocks, respectively. This is a little higher than MB-PLS and PLS (on concatenated blocks) having 88.1% and 88.4% explained variance, respectively (the difference between the two is due to different weights given to the blocks, see Algorithm 7.3). Another interesting choice is 1+1+2 components, i.e., four components in total, which could be easier to interpret, while still retaining around 85% explained variance. In general there are many choices of numbers of components that lead to almost identical explained variances, which in turn leads to a high degree of overlap in the Måge plot. In such cases it is generally natural to consider both prediction ability and simplicity of the interpretation.

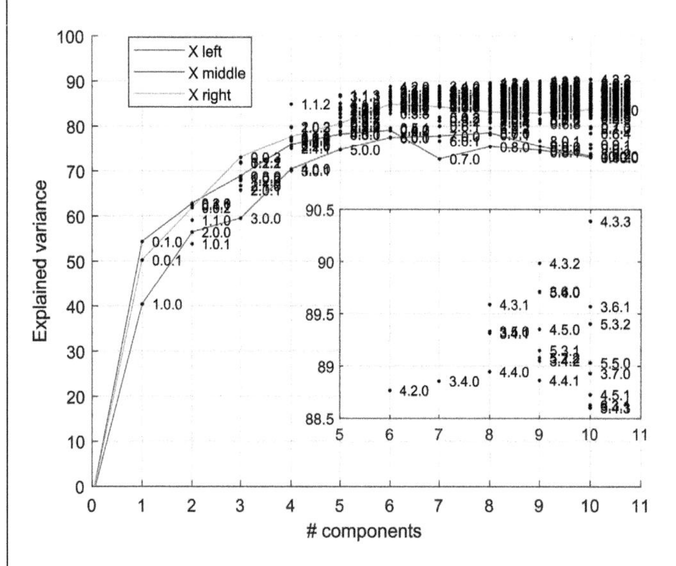

Figure 7.15 Måge plot showing cross-validated explained variance for all combinations of components from the three blocks with a maximum of 10 components in total. The three coloured lines indicate pure block models, and the inset is a magnified view around maximum explained variance.

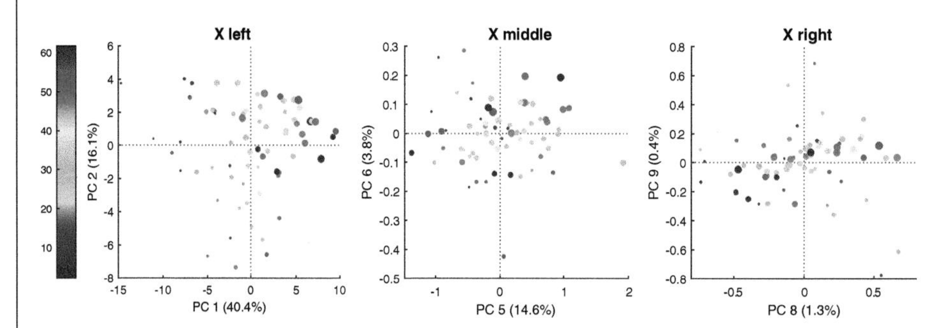

Figure 7.16 Block-wise scores ($\mathbf{T}_m$) with 4+3+3 components for left, middle, and right block, respectively (two first components for each block shown). Dot sizes show the percentage PUFA in sample (small = 0%, large = 12%), while colour shows the percentage PUFA in fat (see colour-bar on the left).

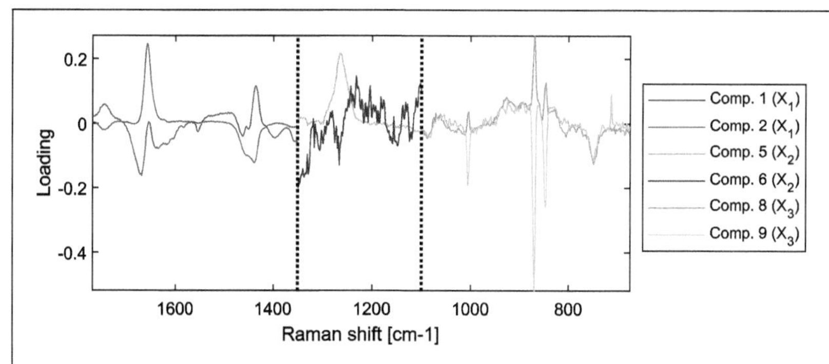

Figure 7.17 Block-wise (projected) loadings with 4+3+3 components for left, middle, and right block, respectively (two first for each block shown). Dotted vertical lines indicate transition between blocks. Note the larger noise level for components six and nine.

In Figure 7.16 there are three panels showing the two first scores of each block. In contrast to the MB-PLS solution in Example 7.1, the second and third block shows less trends in response values, as most of these have been caught by the first block.

In Figure 7.17 we see the two first vectors of loadings for each component. The left block-loadings look very similar to the MB-PLS solution (Figure 7.5), while the sequential orthogonalisation leads to different middle and right block-loadings. The similarity may indicate that the first two components in MB-PLS are dominated by the first block, $\mathbf{X}_1$, such that the two components for the other two blocks in MB-PLS are more related to variability in $\mathbf{X}_1$ than to their own local (distinct) variability.

Results from the new SO-PLS variant in Elaboration 7.4, where scores are restricted to their respective block-spaces, can be found in Figure 7.18. We observe that the components for the first block are exactly the same as for the original SO-PLS, while the subsequent blocks show minor deviations, especially so for the second block.

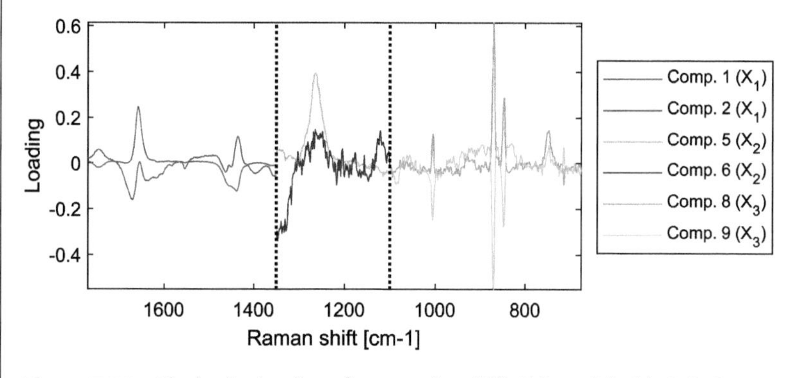

Figure 7.18 Block-wise loadings from restricted SO-PLS model with 4+3+3 components for left, middle, and right block, respectively (two first for each block shown). Dotted vertical lines indicate transition between blocks.

In Figure 7.19 we see a Måge plot for the restricted SO-PLS. The general trend is much the same, but the globally best model now uses 3+6+1 components (90.2% explained variance) instead of 4+3+3 (90.2% explained variance) using the unrestricted model. This means that the predictive ability of the two models is the same in this case. Fixing the restricted model at 4+3+3 components, reduces the explained variance to 87.8%.

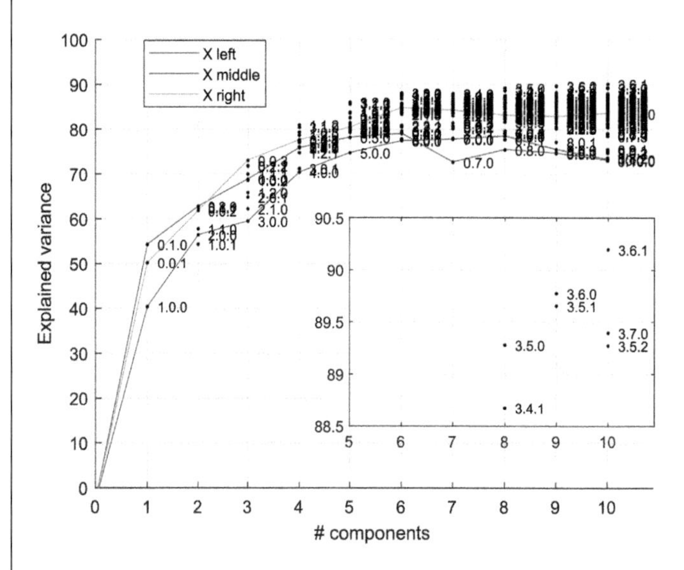

Figure 7.19 Måge plot for restricted SO-PLS showing cross-validated explained variance for all combinations of components from the three blocks with a maximum of 10 components in total. The three coloured lines indicate pure block models, and the inset is a magnified view around maximum explained variance.

Using CV-ANOVA (see Section 7.3.5 on validation) and Tukey's pair-wise test for significant differences of the block contributions (original SO-PLS method), we obtain Figure 7.20, indicating that there are significant contributions from the left and middle block, while the right block does not add much to the reduction in model error.

To illustrate principal components of predictions, again using the original SO-PLS version, we apply SO-PLS to the same data, but also include a second PUFA response, i.e., $PUFA_{fat}$. The optimal model uses 5+4+0 components to achieve 93.7% explained variance (compared to 92.2% for PLS and 92.4% for MB-PLS). Block-wise PCP loadings are shown in Figure 7.21.

Most of the modelled variation of the two PUFA responses is captured in the first PCP score, while the second component is more specific for $PUFA_{sample}$ (score plot not shown).

In the left part of the PCP loading plot (Figure 7.21) we observe similar trends as in the SO-PLS loading plots, especially for the first PCP component. The second component deviates more around the peak around 1660 cm^{-1} and around 1440 cm^{-1}. As the PCP components are calculated globally across blocks, while the SO-PLS components are calculated sequentially per block, it is natural that the right part of the PCP loading plot (1350–1100 cm^{-1}) is more different to the SO-PLS loadings (middle part of Figure 7.17). The components spanning the same wavelengths for SO-PLS start from component number five, leading to more noisy SO-PLS loadings in the region (1350–1100 cm^{-1}).

We emphasise that since the number of components is chosen for three blocks using cross-validation, some degree of overfitting may occur. It is, therefore, important for practical use of the model to validate it also by prediction testing.

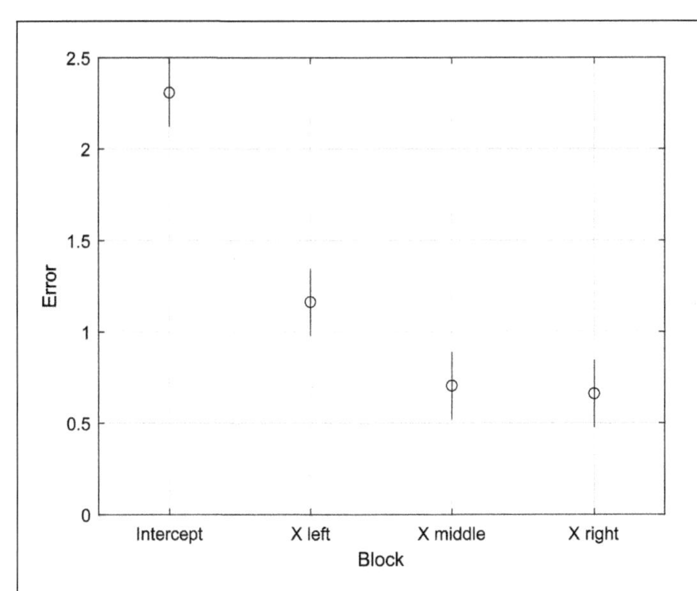

Figure 7.20 CV-ANOVA results based on the cross-validated SO-PLS models fitted on the Raman data. The circles represent the average absolute values of the difference between measured and predicted response, $D_{ik} = |y_{ik} - \hat{y}_{ik}|$, (from cross-validation) obtained as new blocks are incorporated. The four ticks on the x-axis represent the different models from the simplest (intercept, predict using average response value) to the most complex containing all the three blocks ('X left', 'X middle' and 'X right'). The vertical lines indicate (random) error regions for the models obtained. Overlap of lines means no significant difference according to Tukey's pair-wise test (Studentised range) obtained from the CV-ANOVA model. This shows that the 'X middle' adds significantly to predictive ability, while 'X right' has a negligible contribution.

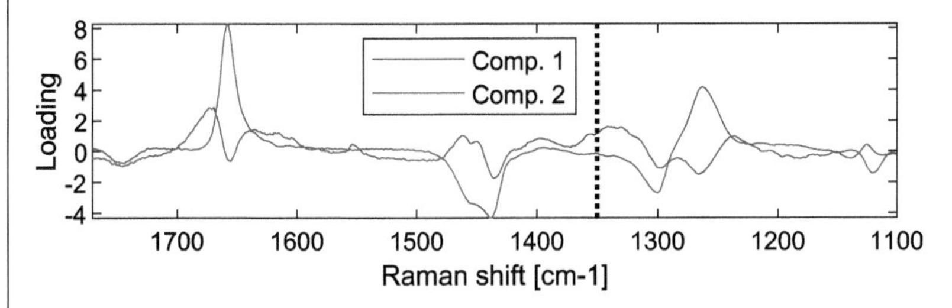

Figure 7.21 Loadings from Principal Components of Predictions applied to the 5+4+0 component solutions of SO-PLS on Raman data.

SO-PLS vs MB-PLS

For the Raman data, we observe that MB-PLS and SO-PLS find similar components despite their different approaches to modelling. This can both be attributed to the stable extraction of subspaces done by the underlying PLS methodology and that Raman spectra lend themselves easily to discovering these latent spaces. Different from MB-PLS, SO-PLS is invariant to block-scaling which can be an advantage if blocks have different underlying dimensionality or are otherwise more distinct. Though this can be beneficial in some analyses, it is not revealed using the spectroscopic data set, as the scaling and dimensionality in different spectral regions is quite homogeneous.

7.3.7 Extensions of SO-PLS to Handle Interactions Between Blocks

For polynomial regression models, it is common to add products of variables to represent the interaction between them. In the multiblock case with several variables in each block, however, it is not obvious how to define interactions.

In Næs *et al.* (2011b) a definition is suggested that generalises the standard polynomial solution and which comprises a number of important cases. It is based on first constructing linear combinations (based on selected $\mathbf{L}$ matrices, see below) of each of the blocks, $\mathbf{X}_1\mathbf{L}_1$ and $\mathbf{X}_2\mathbf{L}_2$, and then multiplying all variables in $\mathbf{X}_1\mathbf{L}_1$ with all variables in $\mathbf{X}_2\mathbf{L}_2$. The method is so far only developed for the interaction between two blocks. The number of columns in the interaction will be the same as the product of the number of columns of the two linear combinations involved. The interaction block will be denoted by $\mathbf{X}_3 = \mathbf{X}_1\mathbf{L}_1 \circ \mathbf{X}_2\mathbf{L}_2$, where $\circ$ means column-wise multiplication, i.e., each element in each column in the first linear combination is multiplied with the corresponding element in each column in the second. The interaction block $\mathbf{X}_3$ is incorporated in the SO-PLS the same way as described in Section 7.3.1. The SO-PLS procedure recommended is to fit $\mathbf{X}_1$ and $\mathbf{X}_2$ before the interactions in $\mathbf{X}_3$ (as for Type I ANOVA).

This definition comprises the use of the original variables ($\mathbf{L}$ is identity matrix) and also a subset of them, after variable selection ($\mathbf{L}$ is now a diagonal matrix with 1s and 0s on the diagonal, corresponding to which variables have been kept and eliminated). Another important case is when the linear combinations, one or both, represent principal components of the blocks ($\mathbf{L}$ then corresponds to principal component loadings). For situations with many variables in the blocks, this is perhaps the most interesting one. Note that constructing interactions in ANOVA, which is essentially based on regression onto categorical variables, follows the same idea, the interaction matrix is established by multiplying together the binary variables for the different factors. A simple example using one quantitative block and one design block is found in Example 7.6.

Example 7.6: Interactions through linear combinations

Given an arbitrary quantitative block $\mathbf{X}_1$ and a design block $\mathbf{X}_2$:

$$\mathbf{X}_1 = \begin{bmatrix} -0.4326 & -0.5298 \\ 0.6970 & -0.5851 \\ -0.9848 & 2.4080 \\ 1.3891 & -1.1712 \\ -0.9196 & -0.5880 \\ 0.2583 & 2.1419 \\ -0.5690 & -1.6878 \\ 0.5616 & 0.0119 \end{bmatrix} \quad \text{and} \quad \mathbf{X}_2 = \begin{bmatrix} 1 & 1 \\ 1 & 1 \\ 1 & -1 \\ 1 & -1 \\ -1 & 1 \\ -1 & 1 \\ -1 & -1 \\ -1 & -1 \end{bmatrix},$$

we can for instance create an interaction block using the following linear combinations:

$$\mathbf{L}_1 = \begin{bmatrix} 0.5 \\ 0.5 \end{bmatrix} \quad \text{and} \quad \mathbf{L}_2 = \begin{bmatrix} 1 & 0 \\ 0 & 1 \end{bmatrix}.$$

In other words, the effect is that the mean of each row in $\mathbf{X}_1$ is combined with the full design in $\mathbf{X}_2$ to produce:

$$\mathbf{X}_3 = \mathbf{X}_1\mathbf{L}_1 \circ \mathbf{X}_2\mathbf{L}_2 = \begin{bmatrix} -0.4812 & -0.4812 \\ 0.0560 & 0.0560 \\ 0.7116 & -0.7116 \\ 0.1090 & -0.1090 \\ 0.7538 & -0.7538 \\ -1.2001 & 1.2001 \\ 1.1284 & 1.1284 \\ -0.2867 & -0.2867 \end{bmatrix}.$$

Note that one of the individual blocks in the interaction block can be multiplied by a constant without changing the solutions, i.e., a multiplicative constant will be absorbed in the regression coefficients for the block. Therefore, invariance to between-block scaling is maintained with this definition of interaction.

Example 7.7: SO-PLS: Incorporating interactions

This example is a study of salt in fish fillets and is taken from Næs *et al.* (2011b). The input data consist of two blocks, $\mathbf{X}_1$ representing three levels of two design factors (salting levels and fish size, i.e., weight) and $\mathbf{X}_2$ representing six NIR variables (containing water and fat signals). The design matrix for each factor, which in its standard form has three linearly dependent columns, was reduced to two linearly independent columns each by simply eliminating one of them. An interaction block $\mathbf{X}_3 = \mathbf{X}_1 \circ \mathbf{X}_2$ was constructed using the column wise products as described above. The $\mathbf{X}_3$ then has 4×6 columns, i.e., 24 in total. The response $\mathbf{y}$ is the measured salt content of the fillets after salting.

Figure 7.22 shows the RMSEP values for varying numbers of interaction components (0, 1, and 2). It is clear from the plot that the interaction block adds to prediction ability.
The RMSEP (based on cross-validation) improves block-wise as follows: (only mean values) 1.21, (mean values plus design block, max = 4 components) 1.04, (mean values, design block, and NIR block, two components) 0.57, (all blocks, interaction block with two components) 0.49. This shows that all three blocks are important for prediction ability.

Figure 7.23 shows regression coefficients for the interaction block. It is evident that the highest level of salt (as compared to the other levels) is the driver of the improved prediction from including the interaction block. As can be seen, the contrast between the effect of the upper three NIR wavelengths and the lower three is clear for this level of salt. This means that going from the other combinations of design factors to this one changes the effect of the chemistry of the fish on the salt level after salting.

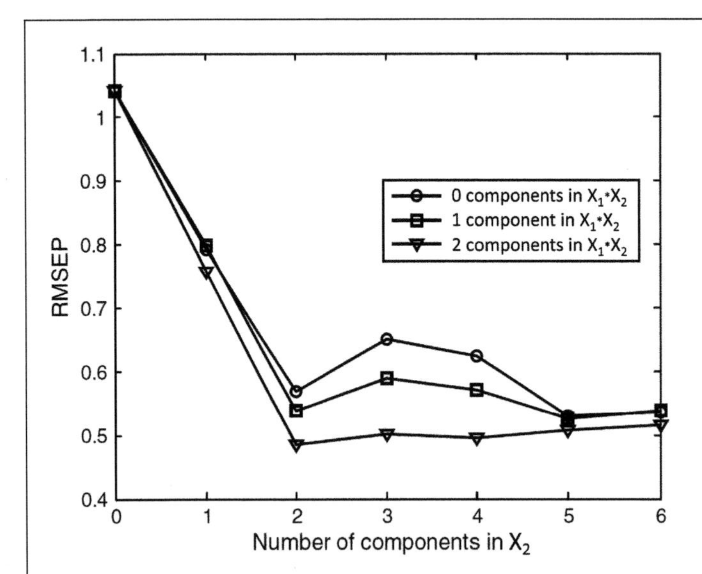

Figure 7.22 RMSEP for fish data with interactions. The standard SO-PLS procedure is used with the order of blocks described in the text. The three curves correspond to different numbers of components for the interaction part. The symbol * in the original figure (see reference) between the blocks is the same interaction operator as described by the ∘ above. Source: (Næs *et al.*, 2011b). Reproduced with permission from John Wiley and Sons.

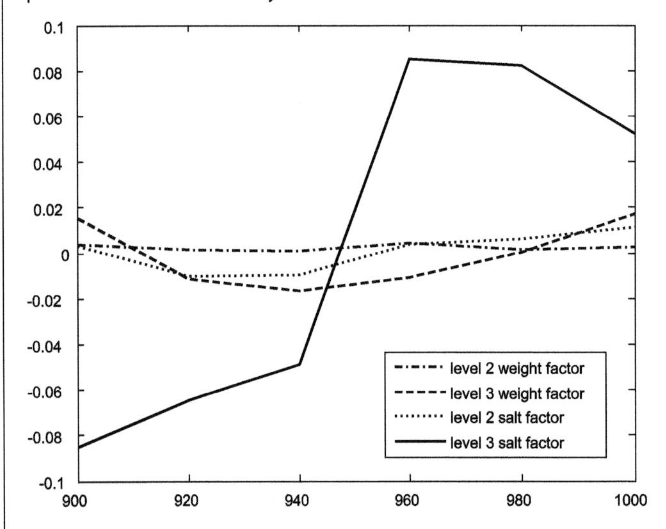

Figure 7.23 Regression coefficients for the interactions for the fish data with 4+2+2 components for blocks $\mathbf{X}_1$, $\mathbf{X}_2$ and the interaction block $\mathbf{X}_3$. Regression coefficients are obtained by back-transforming the components in the interaction block to original units in a similar way as shown right after Algorithm 7.3. The full regression vector for the interaction block (with 24 terms, see above) is split into four parts according to the four levels of the two design factors (see description of coding above). Each of the levels of the design factor has its own line in the figure. As can be seen, there are only two lines for each design factor, corresponding to the way the design matrix was handled (see explanation at the beginning of the example). The number on the x-axis represent wavelengths in the NIR region. Lines close to 0 are factor combinations which do not contribute to interaction. Source: Næs *et al.* (2011a). Reproduced with permission from Wiley.

7.3.8 Further Applications of SO-PLS

In addition to the initial examples related to process modelling in Jørgensen and Næs (2004), the original LS-PLS method has also been used in decision making in process monitoring (see Lepore *et al.* (2019)). The application is from modelling and monitoring of fuel consumption. The method was shown to be superior to established methods.

Studies of wine quality using the SO-PLS method can be found in Niimi *et al.* (2018) and Campos *et al.* (2017). In Niimi *et al.* (2018) the interest is in understanding the development of wines made from the grape variety Cabernet Sauvignon. The results of the paper focus on which chemical components that influence the sensory descriptors of the wines. The main methodological contribution of the paper is a procedure for how to use SO-PLS in situations with very many blocks. The first step of the method is to identify the input block with the best predictive power. The next step is to keep this block in the model and test the additional contribution of extra blocks. The best block is then chosen and the process continues. This procedure is quite similar to the ROSA method to be discussed below except that for the method discussed here the optimal number of components per blocks is estimated and as soon as a block is in the model, it is not considered again. Essentially the same procedure was proposed by Campos *et al.* (2017). The method is simple and seems to work well, but care should be taken to avoid overfitting.

A newly published paper by Roger *et al.* (2020) proposed a completely different use of the SO-PLS method in spectroscopy. Instead of incorporating new data blocks sequentially, different preprocessing methods are used to represent the blocks. If for instance three types of preprocessing are possible, all three are individually applied to the original data to create three new blocks. These three blocks are then incorporated sequentially in the SO-PLS. This means that the method has some resemblance with boosting strategies in machine learning (Freund (1995)), i.e., new challenges are handled along the way as new preprocessing methods are used. In other words, each preprocessing method is given a chance to improve prediction ability after the previous ones in the sequence have been applied.

7.3.9 Relations Between SO-PLS and ASCA

If the SO-PLS approach is used for data blocks that are orthogonal to each other, orthogonalisation is not needed and the method therefore corresponds exactly to using PLS regression for each of the input blocks separately. In a factorial design context with orthogonal factors, each of the input blocks represents a factor in the same way as described in Equation 6.5. The SO-PLS method will give separate PLS plots of scores and loadings for each of the experimental factors.

In the standard situation for ASCA with balanced factorial designs without repeated measures, and with N replicates per factor combination, the blocks in Equation 7.1 will be orthogonal to each other. Since the estimation of effects and the PCA part of ASCA in such cases is run for each factor separately without considering the rest, the primary difference between the two methods is that SO-PLS will give PLS plots for each block, while ASCA will give PCA plots of the LS means. For SO-PLS, the two steps of estimating effects and calculating components are combined while for ASCA they are separated. A demonstration of using SO-PLS in place of ASCA is shown in Example 7.8.

Example 7.8: SO-PLS: Sensory assessment of candies

The example used here is the same as the one used for ASCA in Example 6.5. The example is from sensory analysis of candies and there are two design variables: assessor (represented in $\mathbf{X}_1$) and sample/candy (represented in $\mathbf{X}_2$). There will therefore be two score plots and two corresponding loading plot to consider for the SO-PLS, one set for each design variable. The coding of variables is the same as for standard ASCA.

Figure 7.24 shows the two sets of scores and loadings, assessors to the left and candies to the right. The SO-PLS based score and loading plots are almost identical to the ASCA plots in Example 6.5. The only difference between the two approaches is that the former is based on PLS for each block and the latter on PCA.

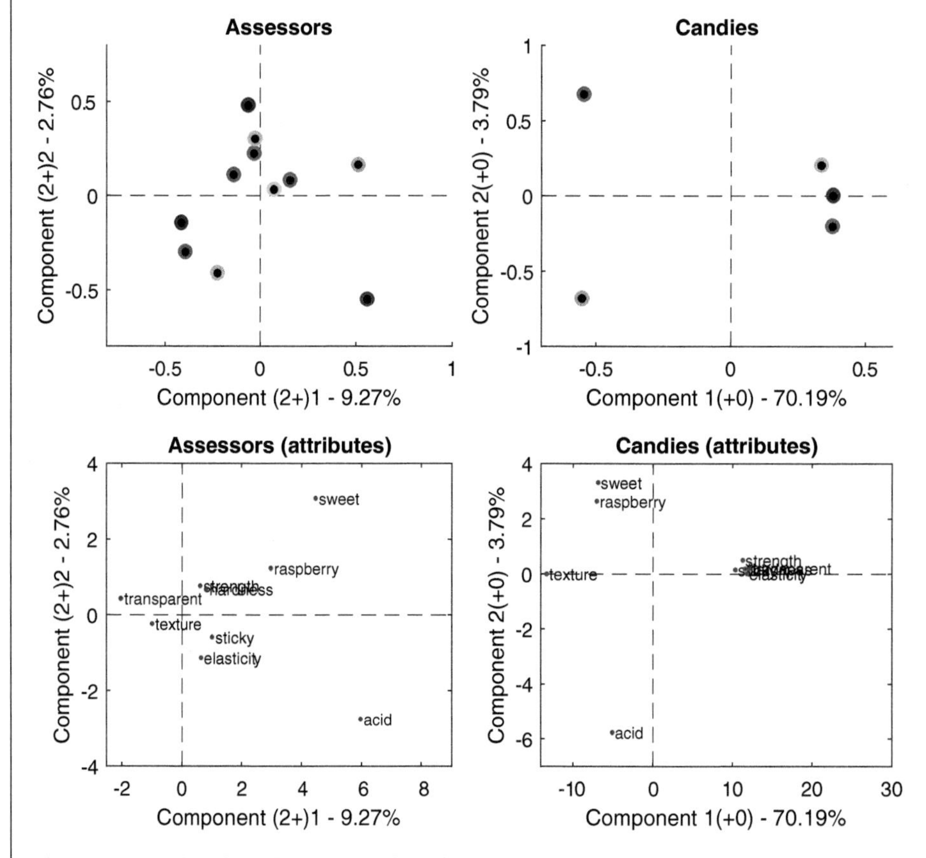

Figure 7.24 SO-PLS results using candy and assessor variables (dummy variables) as $\mathbf{X}$ and candy attribute assessments as $\mathbf{Y}$. Component numbers in parentheses indicate how many components were extracted in the other block before the current block.

7.4 Parallel and Orthogonalised PLS (PO-PLS) Regression

The parallel and orthogonalised PLS (PO-PLS) regression (Måge *et al.* (2008)) is strongly related to the concepts of common and distinct variability defined in Chapter 2. As such it can be seen as a prediction analogue to methodology such as, for instance, PCA-GCA. A major difference is that instead of starting with a PCA for reducing dimensionality before GCA as done for PCA-GCA, PLS is used for data compression in PO-PLS. For the rest, the methodology rests on using GCA for establishing common and local components before a final regression step is used for relating the obtained scores to the response variable. The methodology has a heuristic character, but builds on important principles that link strongly to the rest of the book. Here we present the method as it was originally proposed.

The goal is still to predict **Y** from all the **X** as accurately as possible, but in addition the method seeks information about how common, local, and distinct predictive variability in the **X** blocks are related to **Y**.

Algorithm 7.5

PO-PLS

1. First **Y** is modelled from each of the input blocks separately using PLS regression ($\mathbf{Y} \sim \mathbf{X}_m$), see Figure 7.25. This results in score matrices $\mathbf{T}_m$ and is done for the purpose of reducing dimensionality before canonical correlation analysis.
2. Then GCA (see Section 2.1.10) is used on the block-scores in order to obtain consensus scores $\mathbf{T}_C$, representing the common predictive variation of the blocks. If more than two blocks are used, consensus scores can be calculated for all relevant combinations of blocks, e.g., for three blocks: ($\mathbf{T}_{C12} = \text{GCA}(\mathbf{X}_1, \mathbf{X}_2)$, $\mathbf{T}_{C13} = \text{GCA}(\mathbf{X}_1, \mathbf{X}_3)$, $\mathbf{T}_{C23} = \text{GCA}(\mathbf{X}_2, \mathbf{X}_3)$, and $\mathbf{T}_{C123} = \text{GCA}(\mathbf{X}_1, \mathbf{X}_2, \mathbf{X}_3)$). The GCA consensus is in each case obtained in the same way as described in Chapter 2. We refer to Chapter 5 for further discussion on GCA and common components.
3. In order to establish distinct scores, each block-score matrix (from the first PLS step) and the response are thereafter first orthogonalised with respect to all the consensus scores ($\mathbf{T}_m^{\text{orth}} = \text{orth}(\mathbf{T}_m|\mathbf{T}_{C12}, \mathbf{T}_{C13}, \mathbf{T}_{C23}, \mathbf{T}_{C123})$, $\mathbf{F} = \text{orth}(\mathbf{Y}|\mathbf{T}_{C12}, \mathbf{T}_{C13}, \mathbf{T}_{C23}, \mathbf{T}_{C123})$). Then PLS is used again to obtain the individual distinct scores $\mathbf{T}_{Dm}$ for each block ($\mathbf{F} \sim \mathbf{T}_m^{\text{orth}}$). The $\mathbf{T}_{Dm}$ are not mutually orthogonal as $\mathbf{T}_m$ are only orthogonalised against the consensus scores. The scores in this original version of PO-PLS fall outside the column spaces of the blocks. The consequences of this have not been explored.
4. The final step is to regress **Y** onto the consensus scores $\mathbf{T}_C$ and all the individual distinct scores $\mathbf{T}_{Dm}$ using standard LS regression. For two input blocks the model can be written as

$$\mathbf{Y} = \mathbf{T}_C\mathbf{Q}_1^t + \mathbf{T}_{D1}\mathbf{Q}_2^t + \mathbf{T}_{D2}\mathbf{Q}_3^t + \mathbf{F}. \tag{7.8}$$

For more than two input blocks, all the consensus scores in point 3. must also be added. Since the scores in this model can be correlated, a multiple regression is needed. If the correlation is very high, a PLS regression can be used.

Figure 7.25 illustrates the idea. We also refer to Figure 7.1 for an illustration of how the PO-PLS method is based on singling out the common information and using that part separately

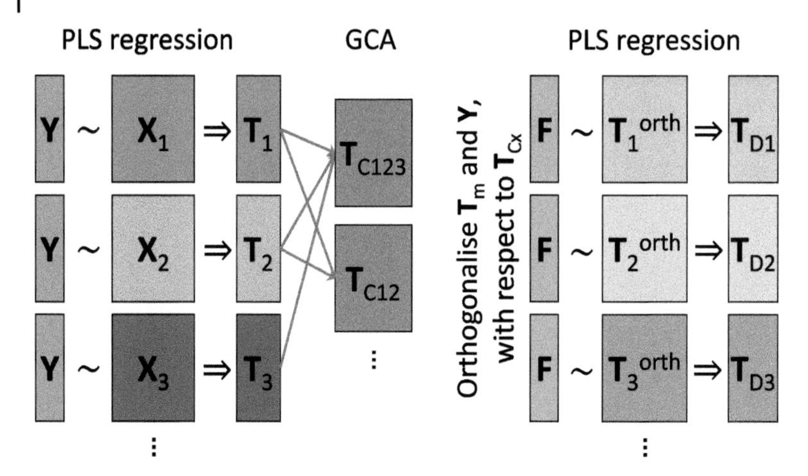

Figure 7.25 Illustration of the idea behind PO-PLS for three input blocks, to be read from left to right. The first step is data compression of each block separately (giving scores $\mathbf{T}_1$, $\mathbf{T}_2$ and $\mathbf{T}_3$) before a GCA is run to obtain common components. Then each block is orthogonalised (both the $\mathbf{X}_m$ and $\mathbf{Y}$) with respect to the common components, and PLS regression is used for each of the blocks separately to obtain block-wise distinct scores. The $\mathbf{F}$ in the figure is the orthogonalised $\mathbf{Y}$. The common and block wise-scores are finally combined in a joint regression model. Note that the different $\mathbf{T}$ blocks can have different numbers of columns.

in the model. Example 7.9 shows results from the same PUFA data set as used above. An example of PO-PLS in the `multiblock` R-package is found in Section 11.8.5.

In point 2 of Algorithm 7.5 the combinations of blocks can be all pairs of blocks, all triplets and so on, depending on how many blocks are used in the modelling. In Example 7.9, we have three blocks, resulting in one set of global consensus scores for blocks (1,2,3) and three sets of local consensus scores for blocks (1,2), (1,3) and (2,3). From point 4 above and Equation 7.8 we obtain a prediction equation and information about how the common and distinct parts of the variability in the $\mathbf{X}$ contribute to the variability in $\mathbf{Y}$. The latter is most easily obtained by comparing the $\mathbf{Q}$ values after standardisation of the $\mathbf{T}$, where larger values will indicate a stronger impact on the predictions.

When using PO-PLS for prediction, there is no simple vector of regression coefficients available to be applied directly to the raw $\mathbf{X}_m$ blocks. Instead the data blocks of the new data must be subjected to each step in the PO-PLS algorithm; first calculating predicted PLS scores, then projecting this onto the GCA solution to form global and local predicted scores, and finally orthogonalising the predicted scores on these to form distinct predicted scores. Then the relationship established in Equation 7.8 can be used to perform the predictions.

Applications of the PO-PLS method can be found in Måge *et al.* (2008), Næs *et al.* (2013) and in Kreutzmann *et al.* (2008). The method provides good interpretation opportunities and similar prediction results as an MB-PLS of all the blocks.

For PO-PLS, the number of components have to be selected both for the original PLS models, for the GCA analyses, and for the distinct models. The first of these is usually of minor importance as long as we do not incorporate components with too small variability. In the case of two blocks, we then end up with the selection of three sets of components, the common ones and the two distinct ones. The selection of components for the original block-wise models and the final distinct models is as usual done by the use of cross-validation. The components for the GCAs, are chosen by judging the component-wise squared canonical correlations and the calibrated explained variance (i.e., the explained variance obtained from the fitting to $\mathbf{Y}$) associated with these. A rule of thumb (may be data dependent) is to include components having more than 90% squared canonical correlations and at least a few percent explained variance. This rather subjective selection process may be regarded as one of the caveats of PO-PLS.

The idea behind PO-PLS can be modified in various ways. A simple alternative is to use PCA-GCA first on the input blocks and then relate the scores directly to the response variable in a regression. Another possibility is to combine it with SO-PLS in the following way. For two blocks, we first calculate the common variability, $\mathbf{T}_C$, as described in Algorithm 7.5 and then investigate how much extra block $\mathbf{X}_1$ explains after orthogonalisation, i.e., $\mathbf{X}_1^{\text{orth}}$ where orthogonalisation is done with respect to $\mathbf{T}_C$, before the same is done for $\mathbf{X}_2$ after orthogonalisation with respect to both the common variability and the distinct part of $\mathbf{X}_1$. This makes PO-PLS into a sequential procedure starting with the common variability, and then searching for distinct $\mathbf{X}_1$ variability and finally distinct $\mathbf{X}_2$ variability. See Zhao and Gao (2012) and also Kreutzmann *et al.* (2008) for similar ideas.

Example 7.9: PO-PLS: Raman on PUFA containing emulsions

The data set used here is the same as the one used for SO-PLS and MB-PLS. Predicting both responses of the PUFA data simultaneously, a model with four sets of common components for blocks; one global (1,2,3), and three local (1,2), (1,3), and (2,3), were created. The last combination (2,3) did not contribute and was later removed. Block-numbers are as in the previous examples, i.e., left, middle and right. In the following, the numbered list corresponds to the numbered list in Algorithm 7.5.

1) From each of the three blocks, PLSR models with 6, 5, and 5 components, respectively, were deemed optimal by manually assessing plots of cross-validated explained variance.
2) The corresponding scores ($\mathbf{T}_m$) formed the basis for the four GCA models used to extract the common components. A subjective assessment of squared canonical correlations and calibrated explained variance led to the choice of 1, 1, 1, and 0 common components, respectively, for the global (1,2,3), and three local (1,2), (1,3), and (2,3) block combinations.
3) After extracting common components the block-scores were orthogonalised against the common components using point 3 in the method description. A new set of PLSR models using the orthogonalised block-scores ($\mathbf{T}_m^{\text{orth}}$) resulted in 1, 3, and 2 distinct components, respectively, for the three blocks.
4) The resulting distinct block-scores ($\mathbf{T}_{Dm}$) were combined with the common scores in a model similar to Equation 7.8, where almost all variation was explained by the calibration (92.2% and 99.0%), while there was a slight drop of 3.5% and 1.3% from the calibrated/fitted to the validated explained variances, see Figure 7.26. The prediction ability is comparable to that of MB-PLS.

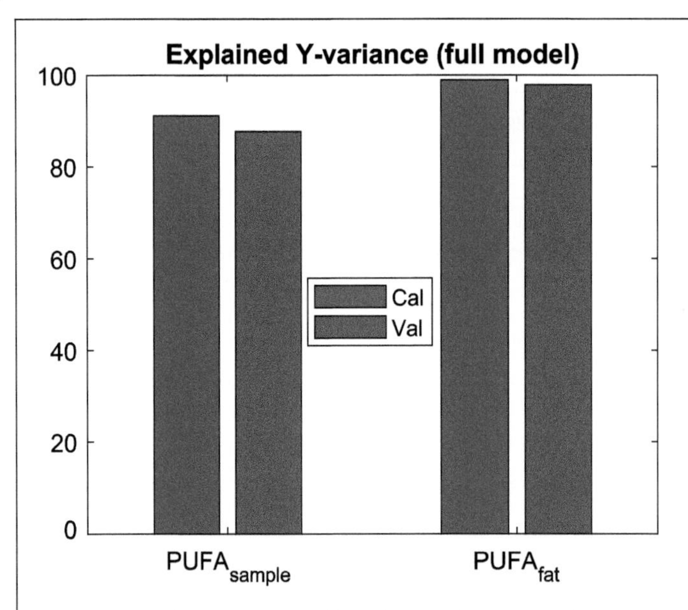

Figure 7.26 PO-PLS calibrated/fitted and validated explained variance when applied to three-block Raman with PUFA responses.

In Figure 7.27 the calibrated/fitted explained variance is shown for each common and distinct block (pairs of bars) and for each response (blue and red). On the right, the total calibrated explained variances are repeated from Figure 7.26 (92.2% and 99%). It is especially interesting to see how the responses get contributions from different distinct and common blocks. $PUFA_{sample}$ has its largest contributions from the common components of blocks 1 and 2 and the distinct components of block 3. $PUFA_{fat}$ has its largest contributions from local common components of block 1 and 3 and the distinct components of block 2. The global common component contributes almost equally to both responses. Disregarding the global common and the minimal distinct components for block 1, the responses are mainly explained by one block each and the local common components from the rest of the blocks.

In Figure 7.28 we have plotted the first global common component against the first local common component for the block combinations (1,2) and (1,3). There are clear trends in the

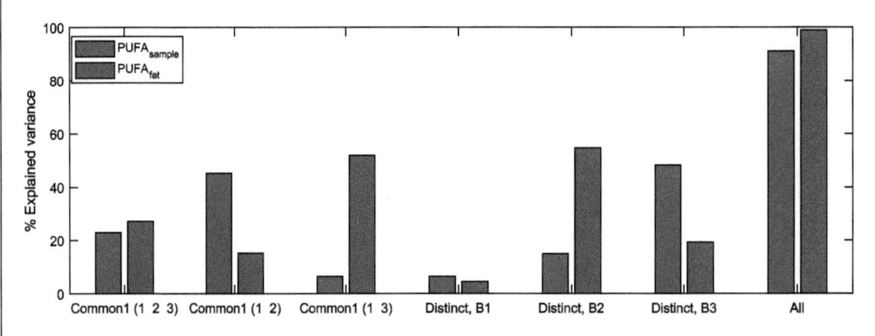

Figure 7.27 PO-PLS calibrated explained variance when applied to three-block Raman with PUFA responses.

distributions of PUFA both with regard to $PUFA_{sample}$ (size of points) and $PUFA_{fat}$ (colour of points). Especially the right plot shows this trend for $PUFA_{fat}$, which is in accordance with the explained variance for this response observed in Figure 7.27 being higher in combination (1,3) than in combination (1,2).

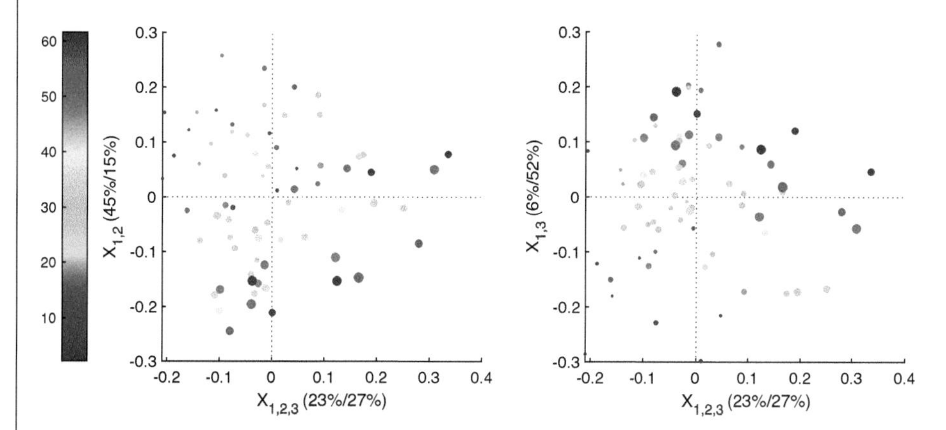

Figure 7.28 PO-PLS common scores when applied to three-block Raman with PUFA responses. The plot to the left is for the first component from $\mathbf{X}_{1,2,3}$ versus $\mathbf{X}_{1,2}$ and the one to the right is for first component from $\mathbf{X}_{1,2,3}$ versus $\mathbf{X}_{1,3}$. Size and colour of the points follow the amount of PUFA % in sample and PUFA % in fat, respectively (see also the numbers presented in the text for the axes). The percentages reported in the axis labels are calibrated explained variance for the two responses, corresponding to the numbers in Figure 7.26.

Finally, we have plotted both the common loadings and distinct loadings in Figures 7.29 and 7.30. As expected, distinct loadings are more erratic, since these are extracted after the common loadings. Both sets of loadings resemble the MB-PLS weights to a high degree. This might be surprising, given that the MB-PLS weights are simply PLS on weighted concatenated blocks, while the PO-PLS is a much more involved process that separates information into several categories. In other words, the processes to obtain the sets of loadings are very different. The similarity can be seen as a confirmation that different approaches find the same underlying patterns, and that PO-PLS is separating this into meaningful contributions.

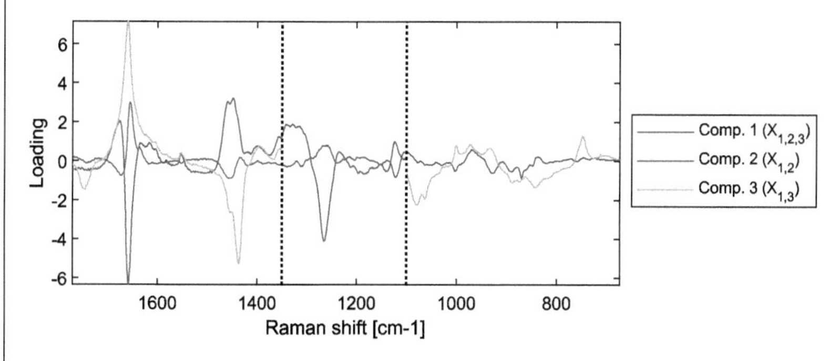

Figure 7.29 PO-PLS common loadings when applied to three-block Raman with PUFA responses.

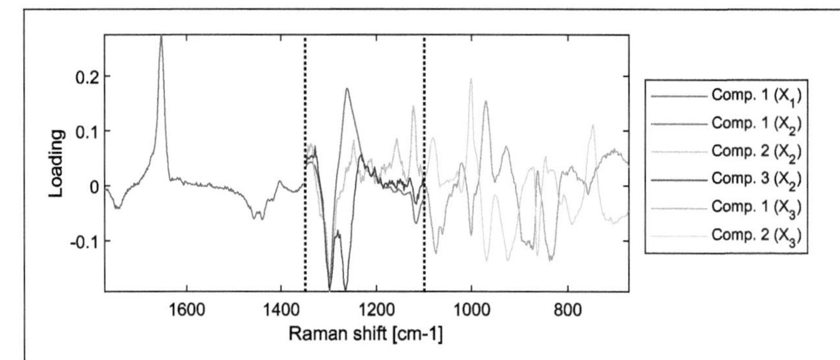

Figure 7.30 PO-PLS distinct loadings when applied to three-block Raman with PUFA responses.

It is important to emphasise that due to the many choices made regarding the number of components, a proper test set validation is needed in order to assess the 'true' prediction ability.

PO-PLS versus MB-PLS and SO-PLS

Applying PO-PLS to the Raman data revealed that for this particular data set, the most important subspace structures underlying the spectroscopic measurements were captured quite similarly to MB-PLS. We can also spot similarities to SO-PLS, but not to such a high degree. This can be explained by the different sequences of analysis where MB-PLS performs global modelling, then computes block-wise contributions from this (using the procedure described in Algorithm 7.1), SO-PLS performs sequential modelling, while PO-PLS uses different models as a basis for finding global and local models, before remodelling distinct components after orthogonalising on these.

7.5 Response Oriented Sequential Alternation

The response oriented sequential alternation method (ROSA, Liland *et al.* (2016)) is a close relative of SO-PLS with fewer user choices and which allows for an arbitrarily large number of blocks. Instead of using a fixed block order, ROSA extracts from any of the blocks each time a new components is computed. In this sense, ROSA makes all blocks compete equally for each component extracted. ROSA is also an alternative to the SO-PLS based method published by Niimi *et al.* (2018) for block selection in situations with many blocks.

7.5.1 The ROSA Method

The ROSA method was originally developed for one response variable only, but the procedure is essentially the same for a multivariate response. In the general case $SVD(\mathbf{X}_m^t\mathbf{Y})$ is used for calculating the weights in the same way as for standard PLS2 (see Chapter 2). Extending this with canonical PLS (Indahl *et al.*, 2009) further increases flexibility.

The basic idea of ROSA is to constrain the NIPALS algorithm to create components that each contain contributions from only one of the **X**-blocks. This is achieved through a component-wise competition where candidate scores are created per **X**-block, and the winning block is the one that reduces the residual **Y**-variance the most. The remaining candidate components are discarded while the winner is retained as block number r.

The consequence of the block selection is that for each new iteration, each block has a new chance and can possibly outperform a block which has been the most important at an earlier stage/iteration. The selection of components can switch between the blocks as the number of components increases. Regression coefficients for concatenated data-blocks can be calculated from scores, loadings and weights, just like with single block PLS.

An efficient and numerically stable way of calculating the ROSA is given in Algorithm 7.6.

Algorithm 7.6

ROSA

The ROSA algorithm can be efficiently implemented using PLS scores with unit norm and only deflating **Y**. When not deflating $\mathbf{X}_m$, orthogonalisation of extracted scores on previous components must be done (see Indahl (2014); Björck and Indahl (2017) and deflation in Section 2.8). The orthogonalisation is associated with a very low computational cost. The norm used is, as in most other places in the book, the standard Frobenius norm.

For each component, r, to be extracted, candidate PLS score vectors are computed for all M **X**-blocks, and these are individually orthogonalised with respect to previously extracted components $(1, 2, \ldots, r - 1)$. The candidate score vector which reduces the residual **Y**-variance the most is chosen as component r, and the remaining candidates are discarded. The response is deflated using the winning score vector before the procedure is repeated for the next component, $r + 1$.

Loop over components, r=1, ..., R – main loop, similar to PLS

Loop over blocks, m=1, ..., M – competition for current component

1: $PLS2(\mathbf{X}_m, \mathbf{Y})$ – one candidate component per block

2: $=> \mathbf{t}_m, \mathbf{w}_m$ – scores, weights; both scaled to unit norm

3: $\mathbf{t}_m^\star = \mathbf{t}_m - \mathbf{T}\mathbf{T}^t\mathbf{t}_m$ – orthogonalise on previous scores

End block loop

4: $\mathbf{t}^\dagger = \text{argmin}_m\{||\mathbf{Y} - \mathbf{t}_m^\star\mathbf{t}_m^{t,\star}\mathbf{Y}||\}$ – select block that minimises residuals

5: $\mathbf{t} = \mathbf{t}^\dagger/||\mathbf{t}^\dagger||$ – normalise scores

6: $\mathbf{w} = [0, \mathbf{w}_r^t, 0]^t$ – global weights (0 except winner)

7: $\mathbf{Y}_{\text{new}} = \mathbf{Y} - \mathbf{t}\mathbf{t}^t\mathbf{Y}$ – orthogonalise on winning block

End component loop

8: $\mathbf{P} = [\mathbf{X}_1, \ldots, \mathbf{X}_M]^t\mathbf{T}$ – global loadings for concatenated **X**

9: $\mathbf{Q} = \mathbf{Y}^t\mathbf{T}$ – global Y loadings

Here, $\mathbf{w}_r$ are the weights corresponding to the winning block for component r. Each time a winning block has been selected, all other candidate score vectors (blocks) are discarded. Choosing unit norm on scores and only deflating the response, we can postpone calculation of **X** and **Y** loadings until after all scores have been created. If loadings, **P**, are computed using the concatenated **X** matrix (as above), it will contain projections of all blocks onto the scores, not only the winning block. In Figure 7.31, this is indicated by shaded loadings for losing blocks. These shaded parts of the loadings show how the losing blocks relate to the scores of the winning block. If we choose to use this non-sparse **P** to calculate regression coefficients, the loser loadings will cancel out due to the corresponding zeros in the weights **W**, i.e., producing the same regression coefficients as using a sparse **P**.

Since only candidate scores and their associated prediction residuals are considered for block selection, the procedure does not depend on the units of the different input blocks. The ROSA is therefore invariant to between-block scaling, i.e., multiplying a whole block with a constant will not affect the block selection process or subsequent predictions.

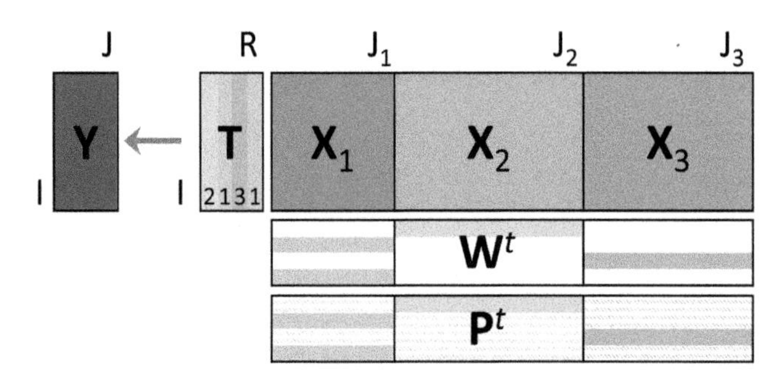

Figure 7.31 ROSA component selection searches among candidate scores ($\mathbf{t}_m$) from all blocks for the one that minimises the distance to the residual response **Y**. After deflation with the winning score ($\mathbf{Y}_{\text{new}} = \mathbf{Y} - \mathbf{t}_r\mathbf{q}_r' = \mathbf{Y} - \mathbf{t}_r\mathbf{t}_r^t\mathbf{Y}$) the process is repeated until a desired number of components has been extracted. Zeros in weights are shown in white for an arbitrary selection of blocks, here blocks 2,1,3,1. Loadings, **P**, and weights, **W** (see text), span all blocks.

Prediction is performed using concatenated **X** data, meaning that regression coefficients will take the same form as for PLS regression: $\mathbf{B} = \mathbf{W}(\mathbf{P}^t\mathbf{W})^{-1}\mathbf{Q}^t$ (see Section 2.8), where **W** and **P** span all blocks. To be able to use the weights, $\mathbf{w}_r$, across all blocks, zeros are added in all positions of **W** corresponding to non-winning blocks. These sparse weights relate correctly to the concatenated **X** and still have unit norms. This is visualised in Figure 7.31 for an arbitrary choice of blocks, emphasising the patterns of block selections in scores and zeros in weights.

One way of interpreting ROSA is as a PLS applied to the concatenated **X**, where a block-wise variable set selection is enforced, i.e., only allowing non-zero weights for one block per component. This means that the scores and loadings will stay in the column- and rowspace, respectively, of the concatenated **X** (not necessarily in the columnspace of each block). The method thus bears resemblance to how groups are handled in PESCA (Section 5.2.2.5) and group lasso (Section 5.12).

A weakness in the original ROSA formulation was its dependence on unvalidated block-scores. To mitigate this and make the procedure more robust, the implementation of the method used in this book uses cross-validated residuals to select the candidate block-scores for each component. This sometimes changes the block selection as was the case for the Raman data in Example 7.10, where prediction accuracy was also improved compared to the non-validated block selection. An example of ROSA in the `multiblock` R-package is found in Section 11.8.6.

The order of the blocks chosen for each component is solely based on how well the blocks predict the response. If the user has domain knowledge indicating a block order, this can be incorporated into the selection. The user can also intervene if two candidate score vectors yield similar predictions or are highly correlated, to nudge the solution in a desired direction.

We can mimic SO-PLS through the use of ROSA by forcing a sequence of block selections where all contributions from a specific block follow each other, e.g., block selections 1, 1, 1, 2, 2, 3, 3.

An application of ROSA closely related to interval PLS (Nørgaard *et al.* (2000)) is to split a predictor block into two or more variable ranges and treat this as a multiblock problem. Applying ROSA to this scenario serves two purposes; creating a robust prediction model and finding variable ranges that can be included or excluded in measuring and/or modelling. Selection of wavelength regions for low-cost spectroscopic measurement equipment is a practical example of usage.

7.5.2 Validation

Being a greedy algorithm with the possibility that at each step a block wins (is incorporated) with small margins to the next best, validation is important. As in ordinary PLS, cross-validation and test set validation are the primary options. However, because of the possible variation in block selections due to various subsets of samples, the block order should be determined as above before a final cross-validation (or prediction testing) is performed for the given block order. If cross-validation is used for both purposes, the final result may be over optimistic. To assess variability and indicate future performance, we can as an alternative adopt a double cross-validation strategy, letting block order change freely in the inner loop and validating the final model in the outer loop. This can be used to see if the model for the whole training data is stable with regard to block selection and also gives an indication of the predictive power we can assume for new data. In general the aim with double cross-validation is not to estimate the error rate for a fixed model, but rather to estimate future performance of the model given variation in data set and corresponding freedom in parameter choices when fitting the model.

7.5.3 Interpretation

Scores and loadings and the whole portfolio of plots associated with PLS can be applied and interpreted with ROSA. In addition, it is natural to plot some other quantities related to the sequence of candidate scores, for instance correlations to the winning score vector and/or the residual variation after each of the candidates have been applied. These two plots will reveal how close the blocks are in their prediction ability and give the user a basis for possible subjective changes in block orders. An example of such plotting can be found in Example 7.10, Figure 7.34.

Example 7.10: ROSA: Raman on PUFA containing emulsions

When modelling the PUFA responses separately, ROSA reaches its optimum cross-validated explained variance after five components. Block selections are different for the two responses. For $PUFA_{sample}$, ROSA selects as follows: middle, left, left, right, left (2,1,1,3,1), see Figure 7.32. For $PUFA_{fat}$, only the left block is used for the first nine components. Maximum cross-validated explained variances are 87.5 % and 97.7 %, respectively. Modelling both responses, ROSA achieves a combined explained variance of 93.0%, with selection: middle, middle, left, left, left, right, right, right, middle (2,2,1,1,1,3,3,3,2).

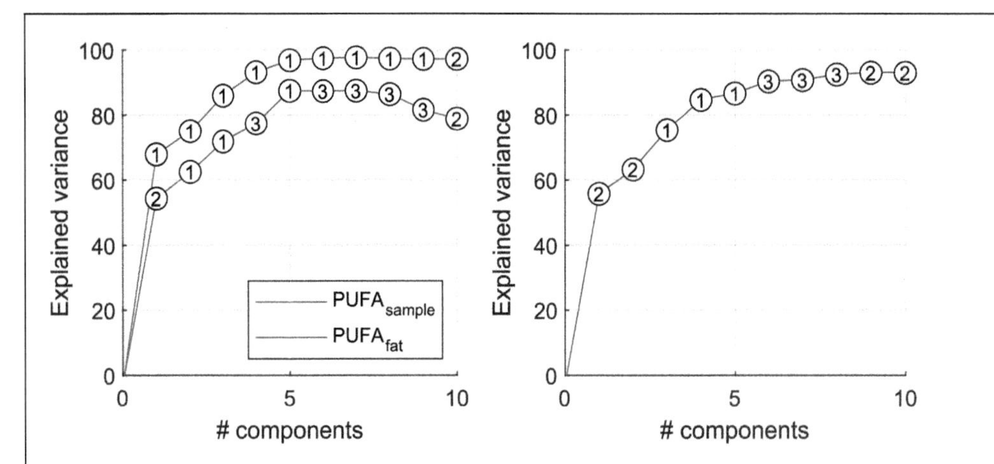

Figure 7.32 Cross-validated explained variance when ROSA is applied to three-block Raman with PUFA in sample and in fat on the left and both PUFA responses simultaneously on the right.

It is interesting to note how ROSA on PUFA$_{sample}$ selects first from the second block, then from the first block and finally from the third block. This is different from SO-PLS which is forced to select components according to a predefined order of the blocks. However, SO-PLS is able to explain 90.4% of the variation in the response for PUFA$_{sample}$, beating ROSA with almost 3%. It is, however, unclear whether this is significant or due to chance. Further validation on a test set is needed to verify this. Also MB-PLS has a slightly higher maximum explained variance at 88.1% and 97.9% for the single response models and similar explained variance at 92.3% when modelling both responses.

From the weights, for the PUFA$_{sample}$ response, in Figure 7.33 we can observe the progressively more jagged weight vectors. The focus and shape of the weights are quite different from both MB-PLS and SO-PLS applied to the same data. The most similar part is the first component using the middle block. The MB-PLS weights are heavily influenced by the first block for the first two components. The right block components of SO-PLS are more noisy than for ROSA since they come in later in the decomposition (component 8, 9, 10).

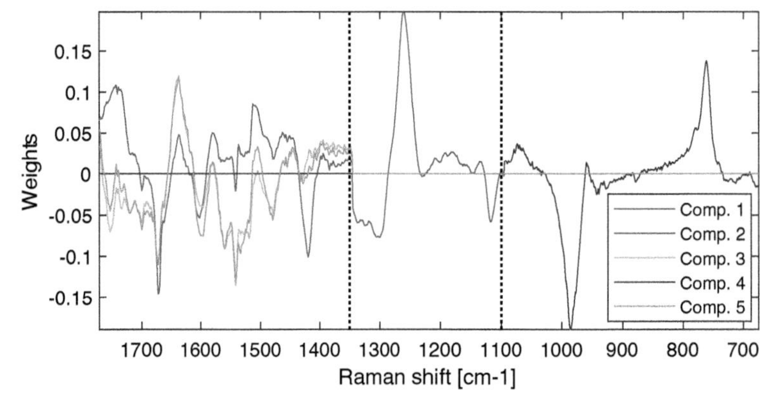

Figure 7.33 ROSA weights (five first components) when applied to three-block Raman with the PUFA$_{sample}$ response.

In Figure 7.34 we have plotted the progression of candidate scores for the model of $PUFA_{sample}$ in two ways. On the top are the RMSECV values that would result from applying the corresponding block score (both the winner and other candidates). In the lower plot are the correlations between each candidate block score and the selected block score. We note that the difference in fits applying different blocks can sometimes be very small, making the choice of block less important. This is because the spectroscopic data have response relevant information spread along the spectra, with some redundancy.

ROSA versus MB-PLS, SO-PLS and PO-PLS

For this particular data set, it seems that the sparsity gained by having components represent one block at a time comes at the expense of a tiny drop in explained variance. The observation that the explained variance is lower for ROSA than for MB-PLS may be explained by the fact that the MB-PLS components are linear combinations across all blocks, while ROSA is limited to one block at a time. As MB-PLS can be formulated as first performing a global PLS on the concatenated blocks, combinations of variables in different block are directly available for it, which can be useful. For the spectroscopic data, ROSA did not gain an advantage over MB-PLS from its block-scale invariance. Block-scale invariance is more useful for data sets with

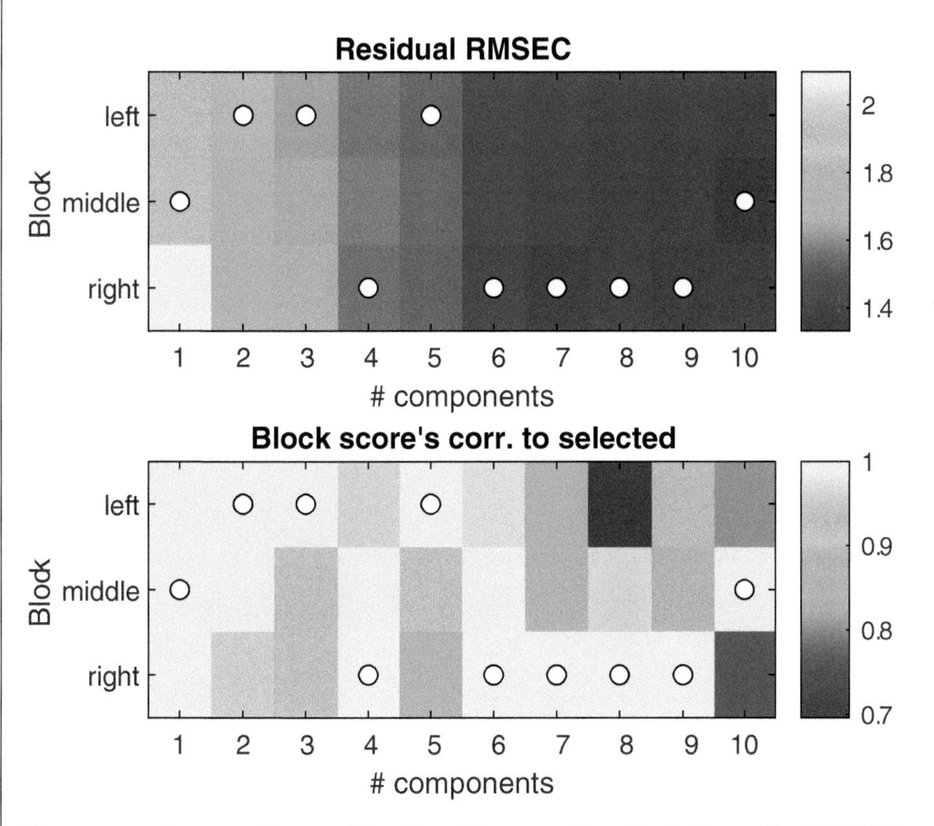

Figure 7.34 Summary of cross-validated candidate scores from blocks. Top: residual RMSECV (root mean square error of cross-validation) for each candidate component. Bottom: correlation between candidate scores and the score from the block that was selected. White dots show which block was selected for each component.

blocks of different scaling or underlying dimensionality. The Raman data are not large enough to properly validate if the advantages in explained variance using SO-PLS are significant, or if ROSA has been too greedy in its block selection. A comparison to PO-PLS could also be envisioned by including local and global blocks consisting of various concatenations of the data blocks.

7.6 Conclusions and Recommendations

In this chapter we have considered four different methods based on different ways of using the PLS regression method. All methods have their merit and, in our experience, from a prediction point of view it does not matter so much which one is used. From an interpretation point of view, however, they lend themselves to different situations depending on what type of information we are most interested in and a number of aspects related to the nature and shape of the blocks. We can say that the different methods look at the data set from different angles for different purposes. Most interpretation tools are based on various ways of presenting scores and loadings in scatter plots. Below we discuss some of the most important issues to consider when choosing a method, and then summarise this in the flowchart in Figure 7.35.

Invariance to between-block scaling

None of the methods here are invariant to within-block scaling. But all of the methods, except MB-PLS, are invariant to between-block scaling. This means that the solution will not be influenced by the relative scale of the blocks. For MB-PLS, on the other hand a decision must be made regarding the relative scale of input blocks, using for instance the Frobenius norm or the number of variables. If one of the blocks is considered more important or less important than the rest, it is then possible to impose this on the solution by using a different weight for that particular block (not possible for SO-PLS, PO-PLS and ROSA).

Choosing the number of components

Since MB-PLS is based on only one single concatenated regression, selecting the number of components is simpler for MB-PLS than for the others. This means that validation must be taken extra seriously for the other three methods in order to avoid overfitting. This makes MB-PLS simpler to use in situation with many input blocks. PO-PLS is the most difficult to use in the case of many blocks because of all the local relations between subsets of blocks to analyse.

Number of underlying dimensions

The methods SO-PLS, PO-PLS, and ROSA explicitly handle situations with different underlying dimensionality in the blocks. This is not the case for MB-PLS where the number of components extracted will be the same for all blocks. Subsequent analysis of the block-wise score and loading plots must be undertaken to investigate these issues.

Common versus distinct components

The only method which addresses this aspect explicitly is PO-PLS. The SO-PLS on the other hand focuses on additional effects which is close to distinct components in certain cases. For MB-PLS and ROSA, the relation to these concepts is more difficult.

MODIFICATIONS AND EXTENSIONS OF ORIGINAL VERSIONS

For MB-PLS a sparse version is developed (see Section 7.2.3 and sparse methods in Chapter 9). A similar approach could be taken for the other methods. The SO-PLS has been developed for handling interactions and, as will be shown in Chapter 10, it can also handle combinations of two-way and three-way blocks as well as more complex error structures (for instance split plot models with covariates).

7.7 Open Issues

An important issue to consider is the ability of the different methods to handle different underlying dimensionalities of the blocks. This may typically happen when two different instruments need very different numbers of important components to be explained properly or when a design matrix is combined with a spectrum. This is handled by all methods, but explicitly only for the sequential methods, for instance SO-PLS and ROSA. The question is how, for instance, MB-PLS handles this. A typical situation where this is relevant is when one of the blocks is a design matrix, but many other examples can be envisioned both in the omics area and in spectroscopy. For SO-PLS a design matrix is handled by using the maximum number of PLS components (as in LS-PLS). Another method that could be considered at this point is the TANDEM method published by Aben *et al.* (2016) which has some similarities with SO-PLS. The major differences are that the elastic net (Zou and Hastie (2005)) is used for estimation and the X-blocks are not orthogonalised.

A modification of the SO-PLS was proposed in Section 7.3.4, and tested on one single data set. This method is new and may have some advantages that deserve further attention in practice. The new version guarantees that all models provide scores and loadings which are embedded in the original block spaces. This is generally not the case for the original version. Similar modifications of PO-PLS could be envisioned.

Incorporating interactions between the blocks was considered in connection with SO-PLS. Only one definition of interaction was tested, namely column-wise multiplication of variables. An important question is then whether it is possible to make other definitions which are better and simpler to interpret. Another important question is how to incorporate definitions

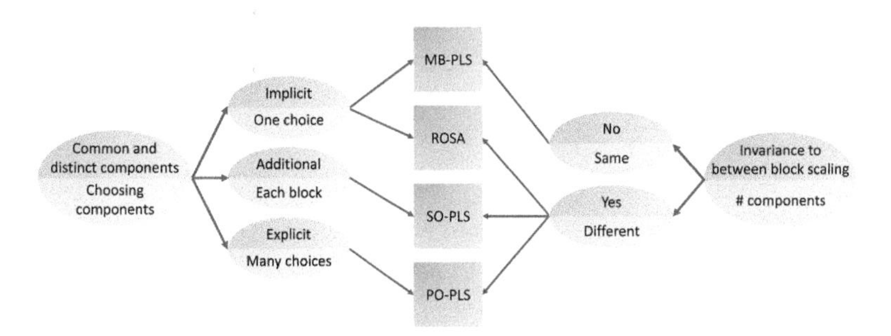

Figure 7.35 The decision paths for 'Common and distinct components; (implicitly handled, additional contribution from block or explicitly handled) and 'Choosing components' (single choice, for each block or more complex) coincide, as do 'Invariance to block scaling' (block scaling affects decomposition or not) and '# components' (same number for all blocks or individual choice). When traversing the tree from left or right, we therefore need to follow either a green or a blue path through the ellipsoids, e.g., starting from '# components' leads to choices 'Different' or 'Same'. More in depth explanations of the concepts are found in the text above.

of interaction for other methods than the SO-PLS. A third question is whether similar ideas can be extended to incorporate polynomial terms for handling non-linearities.

The noise level can potentially be very different for different input blocks. How this influences the different multiblock regression methods is still an open question.

A number of different methods have been discussed in this chapter. They are all useful and some of them are strongly related. The methods relate to concepts such as common, distinct and additional variance, which for prediction purposes are accounted for in different ways. The importance of these concepts in practice and how the different methods are able to handle them should be studied and evaluated in more practical applications.

Another important challenge is to incorporate data at different measurement scales (see Chapter 2), however, very little work has been done in this area.

Part III

Methods for Complex Multiblock Structures

8

Complex Block Structures; with Focus on L-Shape Relations

8.i General Introduction

This chapter focuses on data blocks that are related in more complex ways than those discussed in Chapter 5 and Chapter 7. The most important among the skeletons here is the L-structure (Figure 8.1(a)), but we will also briefly discuss the so-called domino structure (Figure 8.1(d)). The concepts 'L-structure' and 'domino structure' come from the resemblance of the data structure to the letter L and the position of the pieces in the game dominoes. The concrete methods to be discussed will be based on different topologies, both supervised and unsupervised ones. It should be mentioned that analysis of L-shape data is less developed than methodology for supervised and unsupervised situations.

We will discuss both sequential methods that handle either the shared sample mode (horizontal) or the shared variable mode (vertical) first, as illustrated by the grey colour in Figures 8.1(c) and (b), respectively, and methods that treat all blocks simultaneously.

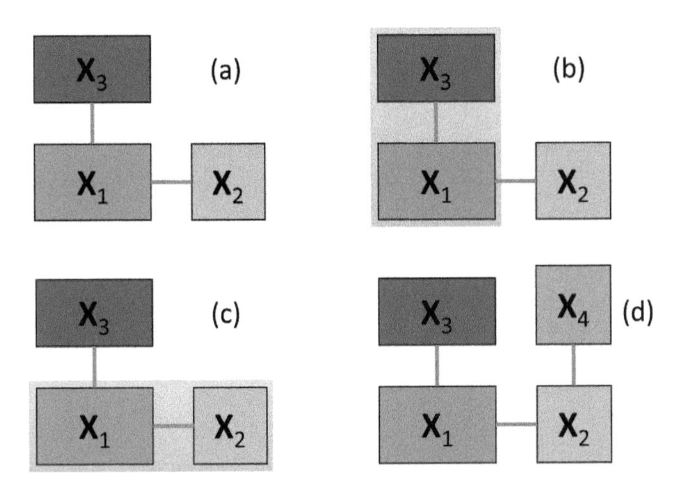

Figure 8.1 Figure (a)–(c) represent an L-structure/skeleton and Figure (d) a domino structure. See also notation and discussion of skeletons in Chapter 3. The grey background in (b) and (c) indicates that some methods analyse the two modes sequentially. Different topologies, i.e., different ways of linking the blocks, associated with this skeleton will be discussed for each particular method.

Multiblock Data Fusion in Statistics and Machine Learning: Applications in the Natural and Life Sciences,
First Edition. Age K. Smilde, Tormod Næs, and Kristian Hovde Liland.

Although relevant in other applications like omics analyses and spectroscopy (see Section 8.4.1), the main examples used in this chapter are from sensory and consumer science, where this structure is frequently occurring. Interest in this case often lies in interpreting relations between consumer liking of a series of products ($\mathbf{X}_1$), descriptive data for the products ($\mathbf{X}_2$), and information about the consumers ($\mathbf{X}_3$) as also depicted in Example 1.4. It is of interest in, for instance, product development to identify which product characteristics are important for liking and also to know the characteristics of the consumers that like the different products (for instance age and gender). This example will also be used throughout when describing methodology.

In this chapter the dimensions of the three matrices in the L-shape are, $\mathbf{X}_1$ $(I \times N)$, $\mathbf{X}_2$ $(I \times J)$, and $\mathbf{X}_3$ $(K \times N)$. In the sensory example, N is the number of consumers, K the number of consumer attributes (demographics, habits, attitudes etc.), I the number of samples, and J the number of sensory descriptors. Except for a few situations where one of the data sets ($\mathbf{X}_2$) represents an experimental design, only quantitative data will be treated. Unless stated otherwise, the data sets sharing the sample mode ($\mathbf{X}_1$ and $\mathbf{X}_2$) will be column centred. For the $\mathbf{X}_3$ block the centring will be described for each particular case. Standardisation is optional in all approaches treated (see Section 2.6). As in most other parts of the book, 'hats' over estimates are omitted (see also Chapter 1).

8.ii Relations to the General Framework

Table 8.1 gives a brief overview of how this chapter links to the general framework in Chapter 1. Some of the methods are tailor-made for handling L-shape data while others are based on treating the sample and variable modes in sequence using the standard supervised and unsupervised methods already discussed in Chapters 5 and 7. Only one of the methods is fully simultaneous (MCR, Section 8.5.3), the rest are either based on sequential component estimation after deflation or the use of two different methodologies in sequence.

Table 8.1 Overview of methods. Legend: U=unsupervised, S=supervised, C=complex, HOM=homogeneous data, HET=heterogeneous data, SEQ=sequential, SIM=simultaneous, MOD=model-based, ALG= algorithm-based, C=common, CD=common/distinct, CLD=common/local/distinct, LS=least squares, ML=maximum likelihood, ED=eigendecomposition, MC=maximising correlations/covariances. The green colour indicates that this method is discussed extensively in this chapter. The abbreviations for the methods represent the different sections and follow the same order. For abbreviations of the methods, see Section 1.11.

		A			B		C		D		E			F			
	Section	U	S	C	HOM	HET	SEQ	SIM	MOD	ALG	C	CD	CLD	LS	ML	ED	MC
PLS/PCR + ANOVA	8.2			■		■	■		■	■	■			■			■
L-PLS	8.3			■	■		■			■	■			■		■	
3BIF-PLS	8.4.2			■	■		■			■	■			■		■	
PCA external info	8.5.1			■	■		■		■	■		■		■		■	
PCA unlabelled	8.5.2			■	■		■		■	■	■			■		■	
MCR incomplete	8.5.3			■	■			■	■				■	■			
Corr. and PLS	8.5.4			■	■		■			■	■						■
Domino PLS	8.6			■	■		■			■	■			■		■	

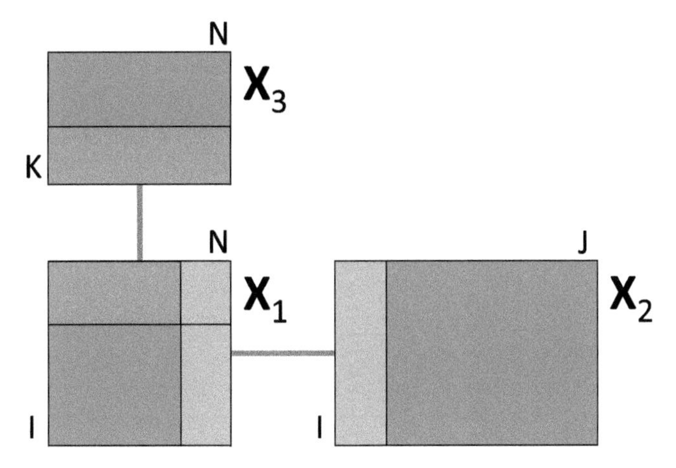

Figure 8.2 Conceptual illustration of common information shared by the three blocks. The green colour represents the common column space of $\mathbf{X}_1$ and $\mathbf{X}_2$ and the red the common row space of $\mathbf{X}_1$ and $\mathbf{X}_3$. The orange in the upper corner of $\mathbf{X}_1$ represents the joint commonness of the two spaces. The blue is the distinct parts of the blocks. This illustration is conceptual, there is no mathematical definition available yet about the commonness between row spaces and column spaces simultaneously.

The simultaneous method is based on model assumptions, the rest are standard component based strategies as discussed in for instance Chapters 5 and 7. Least squares regression is involved as a part of the algorithm for several of the methods. For L-shape methods, a special interest lies in the information that the three blocks have in common, i.e., how the estimated relations in the horizontal direction combine with the relations in the vertical direction. In other words, we are interested in how $\mathbf{X}_1$ links simultaneously to both $\mathbf{X}_2$ and $\mathbf{X}_3$. No precise mathematical definition of commonness in this case is available and, therefore, a more intuitive approach is taken. We refer to Figure 8.2 for a conceptual illustration of the idea. The 'HET' for the first method in the table refers to a combination of a design matrix with standard quantitative measurements (see also the setup for ASCA described in Chapter 6).

8.1 Analysis of L-shape Data: General Perspectives

Except for special cases, there is no specific model structure underlying L-shape data, other than the assumption that the blocks are linked through components of the blocks along both the sample and variable mode. A major goal for all situations is to identify as much common variability as possible between the three blocks, i.e., between the components representing the blocks.

As for the supervised and unsupervised methods discussed in previous chapters, an important challenge is to be able to handle collinear data in the presence of relatively few samples. Therefore, the component methods discussed in Chapter 2, which are developed for this purpose, will be important building blocks also in this chapter.

We will distinguish between two fundamentally different approaches. One of them is based on sequential handling of the two modes (sample mode and variable mode) using the information from the first analysis as basis for the second (Figures 8.1(b) and 8.1(c)). A possible advantage of this is that well-established methods such as PLS regression can be used for the two analyses. Interpretation can then be done sequentially. In other approaches both modes are analysed together through, for instance, an SVD of the products of all three matrices. The advantage of this is that we do not need to do more than one single analysis. Interpretation must then be done afterwards, but still using plots of the same style as for PLS. There is little research available that compares the two approaches. Within each of the two broad categories, we will discuss different concrete methods.

For most methods treated here the focus is interpretation, i.e., to gain insight into how well and in which way the three blocks relate to each other. This is typical for the applications in sensory science. For a few methods treated towards the end of this chapter, the focus is to use information in one of the blocks, for instance $\mathbf{X}_3$ to improve prediction ability among the other two. This is particularly important in spectroscopy.

The same methods of outlier detection (plots, residuals, and leverage) and validation as discussed in Chapter 2 can also be applied here. When the number of samples is very low, however, the use of cross-validation is questionable. In such cases, we will have to rely on other and more intuitive tools such as relations to the design of the study and/or other prior knowledge (see also Section 8.8).

8.2 Sequential Procedures for L-shape Data Based on PLS/PCR and ANOVA

8.2.1 Interpretation of $\mathbf{X}_1$, Quantitative $\mathbf{X}_2$-data, Horizontal Axis First

For this method, the horizontal axis in Figure 8.1(c) is analysed first before using the results for analysing the vertical relation. There exist two possible supervised topologies for the first step, either to predict $\mathbf{X}_1$ from $\mathbf{X}_2$ or $\mathbf{X}_2$ from $\mathbf{X}_1$, but here we will focus on the former (Figure 8.3(a)). In the consumer science literature (see also Example 8.1) this is referred to as external preference mapping. The other topologies in Figure 8.3 will be discussed in later subsections.

For the first step, typically either PCR or PLS will be used. The standard model used (Chapter 2) can be written as:

$$\begin{aligned} \mathbf{X}_2 &= \mathbf{T}\mathbf{P}^t + \mathbf{E} \\ \mathbf{X}_1 &= \mathbf{T}\mathbf{Q}^t + \mathbf{F} \end{aligned} \tag{8.1}$$

The score matrix, $\mathbf{T}$ $(I \times R_1)$, is obtained as $\mathbf{T} = \mathbf{X}_2\mathbf{W}$, where the $\mathbf{W}$ depends on whether PCR or PLS is used for estimation (see Chapter 2). The $\mathbf{T}$ represents the scores of the low-dimensional decomposition of the data and the loadings, $\mathbf{P}$ $(J \times R_1)$ and $\mathbf{Q}$ $(N \times R_1)$, describe how these scores are related to the original variables. For the consumer science example, the loadings $\mathbf{Q}$ represent the differences in liking among the N consumers along the different PLS/PCR axes. Interpreting these axes in light of both $\mathbf{P}$, $\mathbf{Q}$, and $\mathbf{T}$ is important before the next step. We refer to Example 8.1 for an illustration of external preference mapping using this model.

For step two (variable mode), a possible way to proceed is to relate the loadings for the $\mathbf{X}_1$ block (here the $\mathbf{Q}$) to the $\mathbf{X}_3$ block. Again, it is natural/common to use PCR or PLS regression, but now with the estimated $\mathbf{Q}$ (note again our convention of not using hats for estimates unless absolutely necessary, Section 1.10) as output and $\mathbf{X}_3^t$ as input block, i.e.,

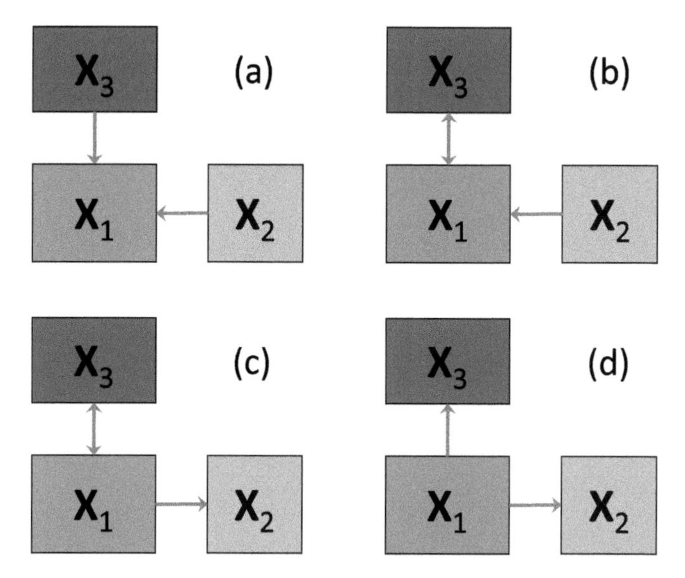

Figure 8.3 Topologies for four different methods. The three first ((a), (b), (c)) are based on analysing the two modes in sequence. (a) PLS used for both modes (this section). (b) Correlation first approach (Section 8.5.4). (c) Using unlabelled data in calibration (Section 8.5.2). The topology in (d) will be discussed in Section 8.3. We refer to the main text for more detailed descriptions. The dimensions of blocks are $\mathbf{X}_1$ $(I \times N)$, $\mathbf{X}_2$ $(I \times J)$, and $\mathbf{X}_3$ $(K \times N)$. The topology in (a) corresponds to external preference mapping which will be given main attention here.

$$\begin{aligned}\mathbf{X}_3^t &= \mathbf{U}\mathbf{V}^t + \mathbf{E}^* \\ \mathbf{Q} &= \mathbf{U}\mathbf{Z}^t + \mathbf{F}^*\end{aligned} \tag{8.2}$$

The columns in both $\mathbf{X}_3^t$ and $\mathbf{Q}$ are centred. The score matrix $\mathbf{U}$ (dimension $(N \times R_2)$) is here a linear combination of the columns of $\mathbf{X}_3^t$ (the linear combination depends on whether PLS or PCR is used for estimation, see Chapter 2), and represents the underlying variability in $\mathbf{X}_3$ which is relevant for predicting the consumer loadings $\mathbf{Q}$.

In the consumer science example, the rows of $\mathbf{X}_3^t$ and $\mathbf{Q}$ represent the N consumers, the columns of $\mathbf{X}_3^t$ the consumer attributes and the columns of $\mathbf{Q}$ the R_1 consumer loadings. The number of components in Model 8.2, R_2, can be different from the number of components R_1 for Model 8.1. Plotting of the loadings $\mathbf{V}$ $(K \times R_2)$ and $\mathbf{Z}$ $(R_1 \times R_2)$ in scatter plots gives information about which consumer attributes are related to the consumer loadings. We refer to Example 8.2 for an illustration of this step.

The rationale behind Model 8.2 is that the low-dimensional $\mathbf{Q}$ in Model 8.1 is considered a valid representation of the main variability in $\mathbf{X}_1$ that can be predicted from $\mathbf{X}_3$. Therefore, a direct regression equation is developed for the two matrices. Referring to the concept of common variability in Figure 8.2, the $\mathbf{Q}$ from step one is then considered the carrier of potential commonness with the $\mathbf{X}_3$ in step two of the method.

It is also possible to analyse one component (column) in $\mathbf{Q}$ at a time in order to highlight relations between $\mathbf{X}_3$ and the different components of $\mathbf{X}_1$. This is done in for instance Asioli *et al.* (2016).

The $\mathbf{Q}$ in Equation 8.2 can be approximated by $\mathbf{U}\mathbf{Z}^t$. Putting this into Equation 8.1 we get that $\mathbf{X}_1$ can be approximated by $\mathbf{T}\mathbf{Z}\mathbf{U}^t$. The estimate of $\mathbf{Z}$ from Equation 8.2 can also be written as $\mathbf{Q}^t\mathbf{U}(\mathbf{U}^t\mathbf{U})^{-1}$ leading to $\mathbf{T}\mathbf{Q}^t\mathbf{U}(\mathbf{U}^t\mathbf{U})^{-1}\mathbf{U}^t$ as an approximation for $\mathbf{X}_1$. If we further substitute $\mathbf{Q}^t$ by its estimate from Equation 8.1 which is equal to $(\mathbf{T}^t\mathbf{T})^{-1}\mathbf{T}^t\mathbf{X}_1$, we see that we can approximate $\mathbf{X}_1$ by

$$\mathbf{T}(\mathbf{T}^t\mathbf{T})^{-1}\mathbf{T}^t\mathbf{X}_1\mathbf{U}(\mathbf{U}^t\mathbf{U})^{-1}\mathbf{U}^t \tag{8.3}$$

In other words, $\mathbf{X}_1$ is approximated by projecting the columns of $\mathbf{X}_1$ onto the PLS/PCR scores $\mathbf{T}$ for Model 8.1 and the rows onto the PLS/PCR scores $\mathbf{U}$ in Model 8.2. This way of presenting the results creates a link to the L-PLS method (see for instance Equation 8.16) and also the method by Takane and Shibayama (1991) discussed in Section 8.5. A major difference is that in Takane and Shibayama (1991) the projections are done onto full spaces of $\mathbf{X}_2$ and $\mathbf{X}_3$, while here they are done onto low-dimensional sub-spaces defined by PLS components. We refer to Vinzi *et al.* (2007) for essentially the same result.

A frequently used approach for the second (vertical direction) step, at least in consumer science, is to segment/cluster the $\mathbf{Q}$ matrix row-wise (i.e., the consumers) and analyse these segments of consumers instead of the quantitative matrix $\mathbf{Q}$. The information in $\mathbf{Q}$ will then be transformed into a dummy matrix (with 0/1 values) representing the segments before relating the segments to $\mathbf{X}_3$. This more common approach will be discussed in detail in Section 8.2.3 and used in the Examples 8.1 and 8.2.

Similar procedures to those described here are discussed in Næs *et al.* (2018) and in Vinzi *et al.* (2007).

ELABORATION 8.1

Missing data and validation

Since PLS and PCR can handle some degree of missing values, the same holds also here. The residuals, both the $\mathbf{Y}$-residuals and $\mathbf{X}$-residuals, can as usual be used for outlier detection as shown in Chapter 2. Since often very few samples are used for testing in consumer science, validations using cross-validation is difficult in that case. Permutation tools, as suggested in Endrizzi *et al.* (2014) and in Vitale *et al.* (2017), can then be useful in combination with prior information and information about the design of the study.

8.2.2 Interpretation of $\mathbf{X}_1$, Categorical $\mathbf{X}_2$-data, Horizontal Axis First

The topology is here the same as in the previous section, the only difference is that $\mathbf{X}_2$ now represents an experimental design. A well-known example in this category is conjoint analysis (Gustafsson *et al.* (2007)) in consumer science, which will be used in the description below. In this case, the $\mathbf{X}_2$ represents the experimental factors, $\mathbf{X}_1$ the consumer liking of the products and $\mathbf{X}_3$ information about the individual consumers (demographics, habits, attitudes etc.).

A natural process here is to first use ANOVA for analysing the relation between the design matrix $\mathbf{X}_2$ and $\mathbf{X}_1$ before the results from the ANOVA are used for studying the relations to $\mathbf{X}_3$ (see, e.g., Næs *et al.* (2018)). This is the only difference from the methodology for quantitative data. In conjoint analysis the analysis has two scopes, estimating the effect on liking of the different factors and creating the relevant information for the second step. A possible alternative to classical ANOVA here is to use the PLS variant of ANOVA presented in for instance Martens and Martens (2001) (see also Chapter 6 for other ASCA based methods for analysing designed experiments with multivariate output).

In the case of two product factors (indexed by l and p) in the design ($\mathbf{X}_2$), the natural model (formulated in terms of population parameters) to use for liking is

$$x_{1,lpn} = \mu + \beta_l + \gamma_p + \beta\gamma_{lp} + \delta_n + \beta\delta_{ln} + \gamma\delta_{pn} + e_{lpn} \tag{8.4}$$

where $x_{1,lpn}$ represents the liking value (in block $\mathbf{X}_1$) for consumer n for the product representing levels l and p of the two product factors. The β and γ are the two product factor effects and δ is the consumer effect. The products of letters represent the interactions between the corresponding factors. In situations outside consumer science other models may be more appropriate, for instance a model with only the 'product effects' and their interactions. In Model 8.4, the consumer effect and all interactions with consumers are considered random effects (Næs *et al.* (2018)). This means that the model used is a linear mixed model (LMM).

The ANOVA gives estimates of main effects, interactions, and corresponding plots and p-values (see Example 8.2). Since estimated fixed main effects and their interactions ($\beta_l + \gamma_p + \beta\gamma_{lp}$) only represent average effects for the products in the population, they do not contain information about individual differences in liking among the consumers. But since $\mathbf{X}_3$ primarily contains information about the individual consumers, further analysis of these estimated fixed effects is not relevant in this context. The same can be said for the consumer effect (δ_n), which only refers to average liking for each consumer. These values can be of some interest, but interpretation is difficult due to their confounding with different use of the liking scale (consumers may use the liking scale differently). In addition they do not say anything about the relative liking of the products. Only the estimates of the interactions ($\beta\delta_{ln} + \gamma\delta_{pn}$) between consumers and products are, therefore, usually applied in relation to $\mathbf{X}_3$ (see Næs *et al.* (2018)). If wanted, these can also be estimated using BLUP (best linear unbiased predictor) taking the random factors into account, but this will not be pursued here since the standard procedure published is based on LS estimates of all parameters.

Reorganising the elements in Equation 8.4 we obtain

$$\beta\delta_{ln} + \gamma\delta_{pn} + e_{lpn} = x_{1,lpn} - \mu - \beta_l - \gamma_p - \beta\gamma_{lp} - \delta_n \tag{8.5}$$

From this we see that if we replace the parameters to the right with estimates, we obtain an estimate of the sum of the two relevant interactions and the unavoidable noise.

If we organise the right hand side of Equation 8.5 (with parameters substituted by estimates) as products$\times$consumers, or as the transpose), the resulting matrix can be analysed and interpreted by for instance PCA (see also Figure 8.4). For balanced designs, this matrix (products$\times$consumers) will be double centred (see Endrizzi *et al.* (2011)), i.e., across products and across consumers. Double centring means that both consumer and product average differences are eliminated and further focus is only on how the different consumers relate to the average consumer for each product. In other words, the PCA plots will illustrate how the consumers relate to each other for the different products without considering the average liking of the products. In Endrizzi *et al.* (2014) it is deemed advantageous for visual segmentation procedures to use double centred data in connection with PCA.

As for the methodology for quantitative $\mathbf{X}_2$, it is most common to use segmentation, i.e., clustering of the PCA loadings, when exploring the relation to consumer attributes in $\mathbf{X}_3$ (see Example 8.2).

ELABORATION 8.2

Hybrid approaches of methods in Sections 8.2.1 and 8.2.2

An alternative to using ANOVA in the first step of the method, is to use PCA directly on the consumer liking data, $\mathbf{X}_1$, and label the scores according to the design factors (see Næs *et al.* (2018)). Then a full interpretation of the scores, $\mathbf{T}$, and loadings, $\mathbf{Q}$, is made before the $\mathbf{Q}$ is related to $\mathbf{X}_3$ as above or by the use of cluster analysis and PLS-DA as will be discussed in Section 8.2.3.

In Endrizzi *et al.* (2014), a similar approach to the ANOVA method was suggested also for situations without an underlying experimental design for the products. The different samples were then simply used to represent a single experimental factor with the number of levels corresponding to the number of samples. This methodology can also be used in a standard preference mapping situation as described in Section 8.2.1 for eliminating product averages. A discussion of pros and cons of eliminating averages is given in the same paper.

8.2.3 Analysis of Segments/Clusters of $\mathbf{X}_1$ Data

For both the situations with quantitative and categorical $\mathbf{X}_2$ (Sections 8.2.1 and 8.2.2), most applications in the consumer science literature use segmentation/clustering of the **Q**-matrix before analysing relations to $\mathbf{X}_3$. These clusters can be defined by an external factor, so-called *a priori* segmentation (Næs *et al.* (2018)), or by the use of some sort of cluster analysis of the **Q** data (*a posteriori* segmentation, see Examples 8.1 and 8.2 below).

If a segmentation approach is used, using either an *a priori* or *a posteriori* approach, we can use dummy (0,1) variables to set up the matrix

$$\mathbf{D} = \begin{bmatrix} 1 & 0 & \cdots & 0 \\ 1 & 0 & \cdots & 0 \\ 0 & 1 & \cdots & 0 \\ \vdots & \vdots & \ddots & \vdots \\ 0 & 0 & \cdots & 1 \\ 0 & 0 & \cdots & 1 \end{bmatrix}. \tag{8.6}$$

Here each column corresponds to a segment and each row to a consumer (N in total). The 1s indicate the presence of a consumer in the actual segment. In the case of three consumer segments, the **D** will contain 3 columns and the row for consumer n will be 0, 0, 1 if the consumer belongs to segment number three.

The next step is to interpret the segments using information from $\mathbf{X}_3$ which contains information about the consumers. The idea is to try to understand who are the consumers in the different segments, which is of crucial importance in, for instance, marketing. One possible approach here is to use standard PLS-DA (for multi-class discrimination) as described in Chapter 7, i.e., using the same model as in Equation 8.2

$$\begin{aligned} \mathbf{X}_3^t &= \mathbf{U}\mathbf{V}^t + \mathbf{E}^* \\ \mathbf{D} &= \mathbf{U}\mathbf{Z}^t + \mathbf{F}^* \end{aligned} \tag{8.7}$$

but now the **Q** in Equation 8.2 is replaced by the dummy matrix **D**. This results in a classification rule as well as interpretation through standard PLS scatter plots of scores and loadings (see Example 8.2).

An alternative way of analysing clusters was discussed in Helgesen *et al.* (1997) for categorical consumer characteristics variables. The method analyses each variable (row k) in $\mathbf{X}_3$ by the use of tabulation. The percentages

$$100\,(\textit{Number of hits for level } q \textit{ of variable } k \textit{ in } \mathbf{X}_3)/N \tag{8.8}$$

for each of the segments obtained were calculated (see also Example 8.1). The N is as above the number of consumers. The phrase in parentheses means that we count how many consumers are present for the q different categorical levels of the actual variable (row k) in $\mathbf{X}_3$ (if the variable is quantitative it has to be split into categories before analysis). Note that the number of levels can be different for the different consumer variables as will be shown in Example 8.1. This gives a frequency table for each segment and each of the K variables in $\mathbf{X}_3$. If for instance the consumer characteristics variable, k, has two levels (younger than 30 years old ($q = 1$) and older than 30 ($q = 2$)), the table for that variable k will have two entries for each segment, one containing the percentage of consumers younger than 30 and the other the number of consumers older than 30. For two segments this gives a 2×2 table.

The segments can then be compared for each $\mathbf{X}_3$ variable separately using Chi-square tests, so-called homogeneity tests. In other words, these tests can be used to test whether the distribution of a variable in $\mathbf{X}_3$ is uniform across segments.

In Example 8.1 it is demonstrated how the tabulation approach can be used, and Example 8.2 shows the use of PLS-DA using Equation 8.7. In the first example, the $\mathbf{X}_2$ data are quantitative, in the other the $\mathbf{X}_2$ represents an experimental design (conjoint analysis).

Example 8.1: Preference mapping and segmentation of consumers

Preference mapping is an important concept in consumer science and refers to analysing the relation between consumer liking for a number of samples and the sensory characteristics of the same samples. This type of analysis provides information about which samples are the most or least liked and a sensory characterisation of these samples.

This example is taken from Helgesen *et al.* (1997) and is based on sensory analysis ($\mathbf{X}_2$) and consumer liking ($\mathbf{X}_1$) of dry fermented lamb sausages. A number of samples covering almost the whole market in Norway were analysed by standard descriptive sensory analysis. A subset of six different sausages were selected (evenly spread) from the initial PCA plot based on the sensory data. The same six samples were tested for liking by an untrained consumer panel. This is a low, but quite typical number of samples in consumer tests of food. A questionnaire was given to each consumer to be filled in ($\mathbf{X}_3$). The questionnaire contained questions about habits, attitudes, and demographics. The information flow for the study is shown in Figure 8.4.

The consumer liking values were first regressed onto the principal components of the sensory data, $\mathbf{X}_2$ (PCR, sensory data are in $\mathbf{X}_2$, liking data are in $\mathbf{X}_1$). The sensory PCA plot (scores

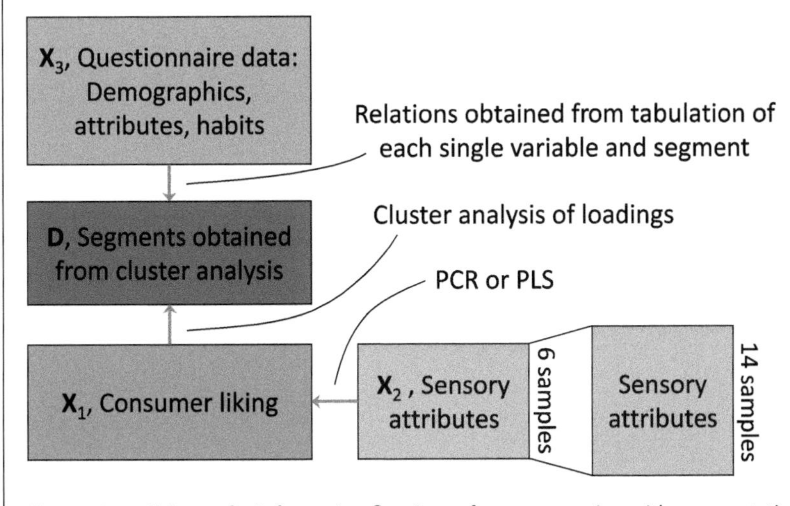

Figure 8.4 Scheme for information flow in preference mapping with segmentation of consumers.

T and loadings **P**) and the consumer loadings (**Q**) (from PCR in this case, see Equation 8.1) are presented in Figure 8.5(a) and (b), respectively. The PCA plot in Figure 8.5(a) can be used for interpreting for instance which samples are characterised by which attributes.

In the lower part of Figure 8.5, representing the consumer loadings **Q**, there are consumers in all directions of the plot, but there are relatively few to the left. This means that the samples to the left in Figure 8.5(a) (samples 1 and 3) are generally the least liked. From the sensory loadings we can see that they are characterised by among others the attributes hardness and dry outer rim.

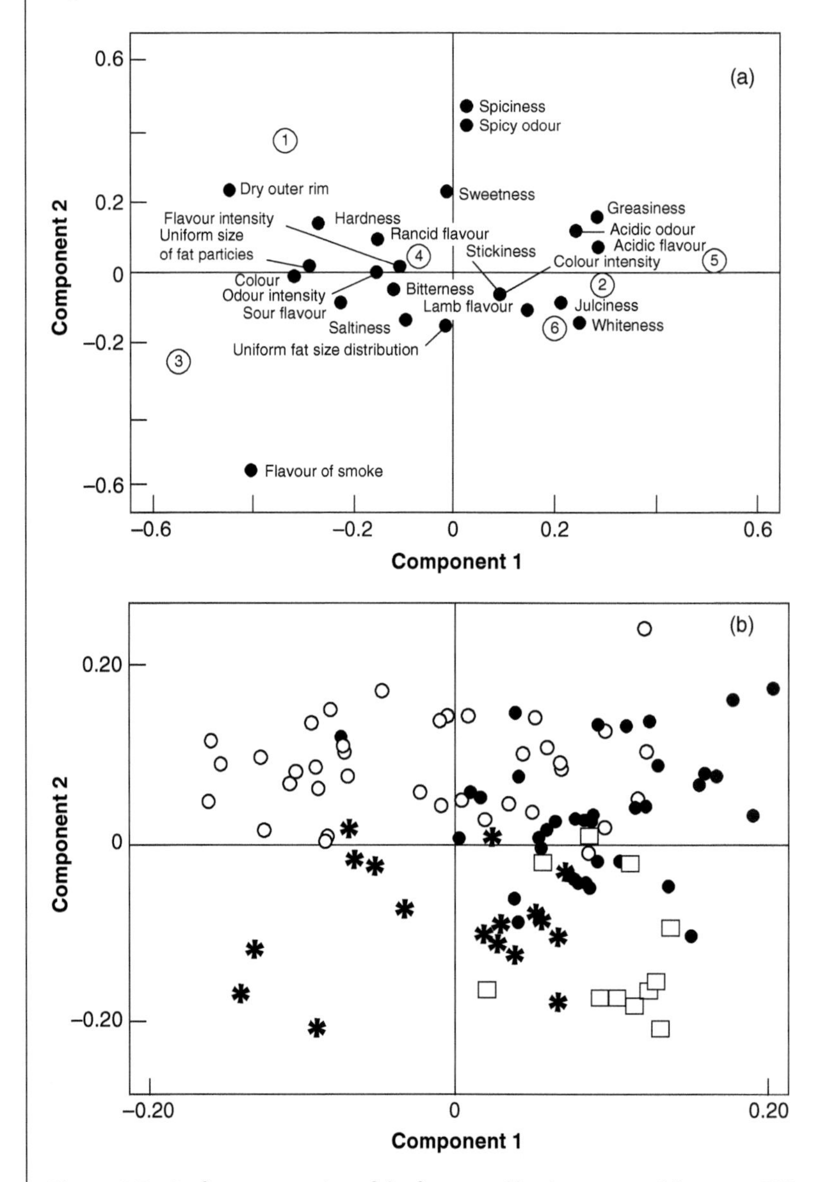

Figure 8.5 Preference mapping of dry fermented lamb sausages: (a) sensory PCA scores and loadings (from $\mathbf{X}_2$), and (b) consumer loadings presented for four segments determined by cluster analysis. Source: (Helgesen *et al.*, 1997). Reproduced with permission from Elsevier.

A hierarchical cluster analysis on consumer loadings using four clusters was performed for the liking data. The clusters were, however, not very clear in the analysis since there is no sharp border between the consumers in the plot. The four clusters are marked with different symbols in Figure 8.5b.

In order to link the liking, i.e., the four segments, to the $\mathbf{X}_3$ matrix of consumer characteristics, a tabulation approach was used in Helgesen *et al.* (1997). A small part of the table containing only two characteristics is shown in Table 8.2. For the 'lunch variable' (five levels) in the table, the groups are significantly different according to a chi-square test. For the gender variable (two levels), the differences are not significant, but there is a tendency of more men in cluster two as compared to cluster one. In the same way, it is possible to consider all consumer characteristics used in the questionnaire (see Helgesen *et al.* (1997) for more details).

Table 8.2 Tabulation of consumer characteristics. A selection of two consumer attributes/characteristics, gender, and lunch habits is given. The numbers represent percentages in each of the categories for each of the segments (subgroups). The sums in each column for each consumer characteristic variable is equal to 100. The lunch variable reflects the frequency of use with 1 representing the highest frequency and 5 'no answer'. Source: (Helgesen *et al.*, 1997). Reproduced with permission from Elsevier.

Demographic variables	Total sample %	Total sample number	*p*	Preference subgroups (%) Subgroup 1 (10.1%)	Subgroup 2 (16.5%)	Subgroup 3 (39.4%)	Subgroup 4 (33.9%)
Gender			0.317				
Male	56	61		36	**67**	51	**62**
Female	44	48		**64**	33	49	38
Lunch			0.003				
1	69	75		45	56	72	**79**
2	4	4		0	0	9	0
3	9	10		0	11	14	5
4	17	19		**55**	28	5	16
5	1	1		0	5	0	0

Example 8.2: Conjoint analysis, $\mathbf{X}_2$-matrix based on categorical variables

Conjoint analysis is the name of another important method in consumer science. The main difference from preference mapping is that for conjoint analysis the $\mathbf{X}_2$ contains design variables representing the design underlying the making of the samples.

An example of conjoint analysis based on consumer liking of cheese was published by Almli *et al.* (2011). The results presented here are from French consumers. In that case a fractional factorial design was established in the factors of interest. The design factors and their combinations ($\mathbf{X}_2$) are presented in Table 8.3. There are six factors in total (pasteurisation, packaging, organic, omega-3, price, and appropriateness) and 16 products to be tested by the consumers. The products were rated by 120 consumers, and the liking values were used as entries in $\mathbf{X}_1$ (dimensions: $I \times N = 16 \times 120$). The test was a so-called extrinsic test meaning that the consumers did not taste the products, all liking values were obtained based on pictures and information.

The first step was to run an ANOVA on $\mathbf{X}_2$ versus $\mathbf{X}_1$ using the model structure in Equation 8.4. The main effects plots from the ANOVA are given in Figure 8.6. Pasteurisation, price, and appropriateness are the dominating factors, followed by organic and packaging. The residuals (here organised with consumers as rows) were then calculated and subjected to PCA. The score and loading plots are given in Figure 8.7. It was quite clear when comparing the two plots to each other, that one group of consumers had a clear tendency towards preferring cheeses based on raw milk (upper right in the plot). Two segments were then established based on this criterion. Two dummy variables, one for each cluster, were created and then related to the consumer characteristics ($\mathbf{X}_3$ data) using standard PLS-DA (see Equation 8.7 and Chapter 7). The results of this analysis are given in Figure 8.8. It is evident that some of the consumer characteristics link clearly to the two segments indicated by points in the plot. For instance, it can be seen that age is related to preference for cheeses based on raw milk.

Table 8.3 Consumer liking of cheese. Design of the conjoint experiment based on six design factors. Source: (Almli *et al.*, 2011). Reproduced with permission from Elsevier.

Experimental conjoint design of the cheeses.

	Pasteurisation	Packaging	Organic	Omega-3	Price	Appropriateness
Trial 1	No	Original	No	No	Standard	Ordinary day
Trial 2	Yes	Original	No	No	Higher	Ordinary day
Trial 3	No	New	No	No	Higher	Special occasion
Trial 4	Yes	New	No	No	Standard	Special occasion
Trial 5	No	Original	Yes	No	Higher	Special occasion
Trial 6	Yes	Original	Yes	No	Standard	Special occasion
Trial 7	No	New	Yes	No	Standard	Ordinary day
Trial 8	Yes	New	Yes	No	Higher	Ordinary day
Trial 9	No	Original	No	Yes	Standard	Ordinary day
Trial 10	Yes	Original	No	Yes	Higher	Special occasion
Trial 11	No	New	No	Yes	Higher	Ordinary day
Trial 12	Yes	New	No	Yes	Standard	Ordinary day
Trial 13	No	Original	Yes	Yes	Higher	Ordinary day
Trial 14	Yes	Original	Yes	Yes	Standard	Ordinary day
Trial 15	No	New	Yes	Yes	Standard	Special occasion
Trial 16	Yes	New	Yes	Yes	Higher	Special occasion

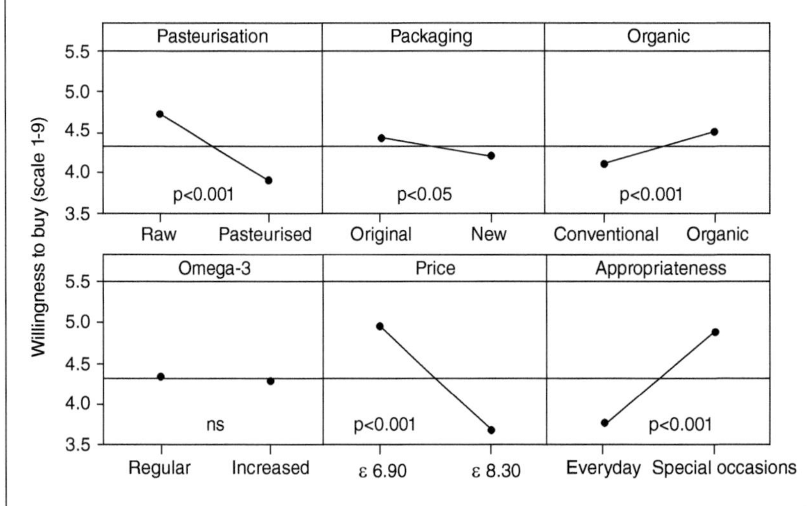

Figure 8.6 Results from consumer liking of cheese. Estimated effects of the design factors in Table 8.3. Source: Almli *et al.* (2011). Reproduced with permission from Elsevier.

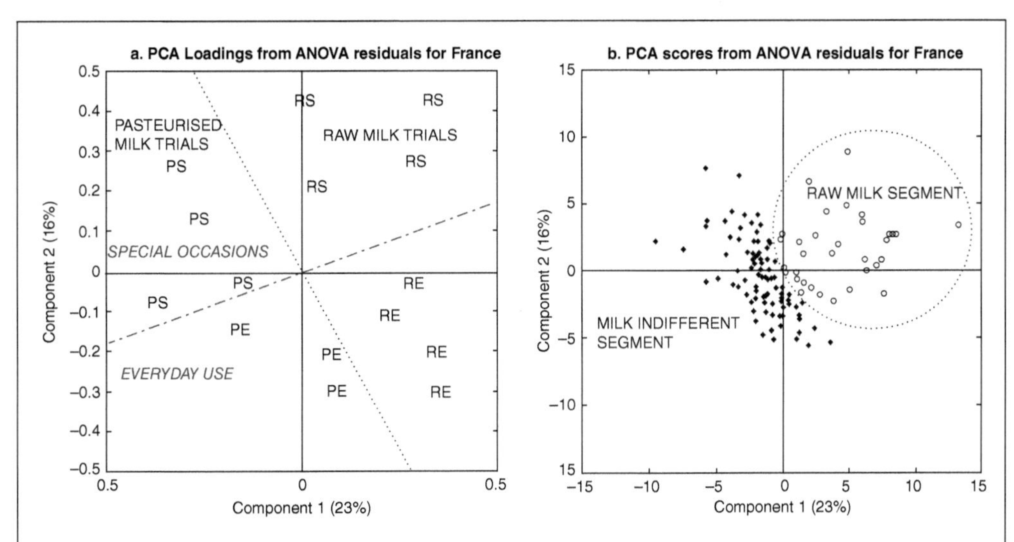

Figure 8.7 Results from consumer liking of cheese. (a) loadings from PCA of the residuals from ANOVA (using consumers as rows). Letters R/P in the loading plot refer to raw/pasteurised milk, and E/S refer to everyday/special occasions. (b) PCA scores from the same analysis with indication of the two consumer segments. Source: Almli *et al.* (2011). Reproduced with permission from Elsevier.

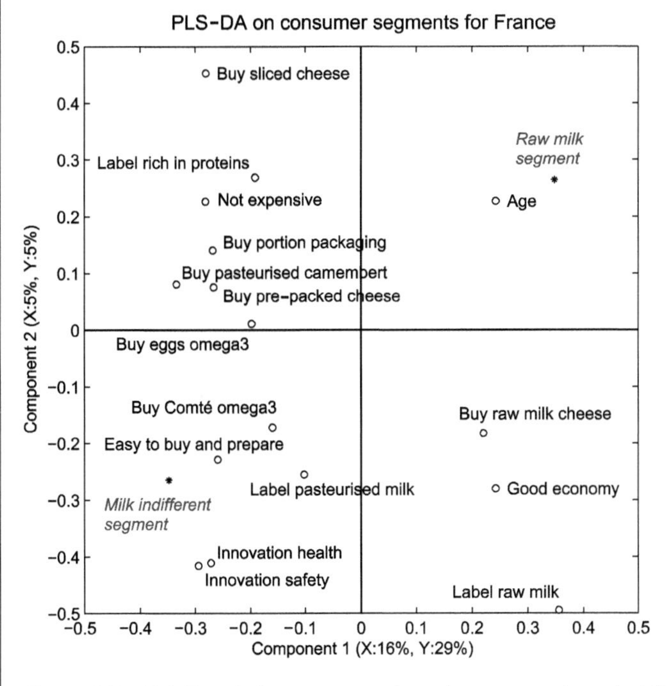

Figure 8.8 Relations between segments and consumer characteristics. Source: (Almli *et al.*, 2011). Reproduced with permission from Elsevier.

ELABORATION 8.3

Possible extensions

As was discussed in Måge *et al.* (2012), finding relations between descriptive sensory data and consumer liking data calls for multiblock methodology when the sensory profile block, $\mathbf{X}_2$, can be split into sub-blocks for instance according to modality (taste, smell, vision), i.e.,

$$\mathbf{X}_2 = [\mathbf{X}_{12}, \ldots, \mathbf{X}_{M2}] \tag{8.9}$$

This means that there is more than one $\mathbf{X}_2$-block along the horizontal axis in Figure 8.1(a). This structure can be analysed by using for instance the PO-PLS method for the horizontal link as done in Måge *et al.* (2012) (see Chapter 7) before the vertical axis in the L-shape is analysed based on scores and loadings from the PO-PLS (using for instance Model 8.2). This approach is useful for understanding more explicitly how the different parts of the sensory profile combine and link to liking.

In other cases it is natural to split the $\mathbf{X}_3$ block into sub-blocks with a structural link or path between them (Menichelli *et al.* (2014b)). This full skeleton is then a combination of a path diagram (SEM, see e.g., Bollen (1989b) and Section 10.5.4) and an L-shape structure (see Figure 8.9). This extension sheds light not only on how the different variables in $\mathbf{X}_3$ relate to the liking, but also how they relate to each other.

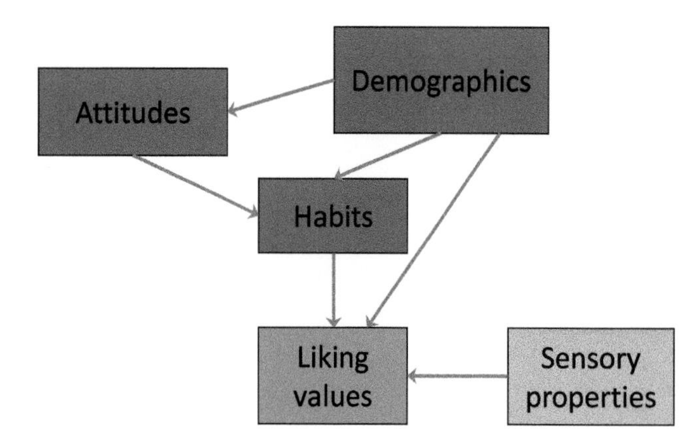

Figure 8.9 Topology for the extension. This is a combination of a regression situation along the horizontal axis and a path model situation along the vertical axis.

Combining categorical and quantitative variables in $\mathbf{X}_2$ (combining Sections 8.2.1 and 8.2.2) using the same topology as here is also possible as was discussed in for instance Asioli *et al.* (2017) and Menichelli *et al.* (2012).

8.3 The L-PLS Method for Joint Estimation of Blocks in L-shape Data

This section presents and discusses various L-PLS methods for joint analysis of all three blocks of data. The naming comes from the L-shape of the data structure and that it is based on principles/ideas very close to the original PLS algorithm. Note that the joint treatment of all three blocks is different from the methods in the previous sections where the shared sample mode and shared variable mode are analysed separately. The L-PLS is a sequential method in the sense that components are calculated in sequence after deflation.

An important difference between the L-PLS methods and the L-shape methods discussed in Section 8.2 is that for L-PLS the $\mathbf{X}_2$ and $\mathbf{X}_3$ play equivalent roles in the estimation while in the other approach main emphasis is first put on the relation between $\mathbf{X}_1$ and $\mathbf{X}_2$. The $\mathbf{X}_3$ comes in only at the interpretation stage of the first analysis. The consequences of the two different approaches have not been explored.

There exist different versions of the L-PLS method. We will first present the original version in Martens *et al.* (2005) which was later named Endo-L-PLS. This is based on the topology in 8.3a. Then we present the Exo-L-PLS as an alternative based on the topology in 8.3d. As can be seen, the two deviate in the way the predictive direction is defined either inwards towards $\mathbf{X}_1$ or outwards from $\mathbf{X}_1$ towards $\mathbf{X}_2$ and $\mathbf{X}_3$. Finally a modification of the original algorithm is presented which is developed for cases with focus on prediction of $\mathbf{X}_2$ from $\mathbf{X}_1$ and when $\mathbf{X}_3$ represents additional information about $\mathbf{X}_1$ to be used in the prediction (topology presented in Figure 8.3(c)). The main difference between the methods is the deflation step between each component. An example of L-PLS in the `multiblock` R-package is found in Section 11.9.1.

8.3.1 The Original L-PLS Method, Endo-L-PLS

The original L-PLS method (Endo-L-PLS) was published in Martens *et al.* (2005) with the same focus as discussed in Example 1.4 (see also Example 8.1). It was presented as an iterative algorithm very similar to the original standard PLS method, iterating between loadings and scores for one component at a time. It was stated in the same paper that the solution can be found using the SVD of a combined matrix (matrix product) where all three matrices are involved. This matrix product is, therefore, the basis for how the Endo-L-PLS method aims to extract common variation/components among the three blocks. In this book, we will use the SVD formulation of Endo-L-PLS, but also describe the algorithm briefly.

The Endo-L-PLS method was developed primarily for interpretation purposes, but later on, as will be seen in subsequent subsections, improved prediction has become an important issue for some of the modifications. The method has been used both in sensory science and in the omics area.

The following procedure follows the description in Martens *et al.* (2005). The running index r ($r = 1, 2, \ldots, R$) refers to the number of extracted components in each step of the algorithm.

It should be mentioned that a similar type of methodology was proposed as early as in 1987 (Wold *et al.* (1987)). According to Eriksson *et al.* (2004), the method showed encouraging results, but experienced some convergence problems. That method will, therefore, not be pursued here.

Algorithm 8.7

ENDO-L-PLS ALGORITHM WITH FOCUS ON ENHANCED INTERPRETATION

The $\mathbf{X}_2$ $(I \times J)$ and $\mathbf{X}_3$ $(K \times N)$ are centred column-wise and row-wise respectively (called $\mathbf{X}_{2,0}$ and $\mathbf{X}_{3,0}$) and $\mathbf{X}_1$ $(I \times N)$ is double centred (called $\mathbf{X}_{1,0}$), i.e., both with respect to rows and columns.

The algorithm goes as follows for each component $r = 1, \ldots, R$: First the SVD of the product of all the three matrices is calculated. The first left and right singular vectors are used as weights for $\mathbf{X}_2$ and $\mathbf{X}_3$ to calculate the corresponding scores (as for PLS). Then the scores are used for calculating loadings using standard LS regression (as for PLS) before residuals for $\mathbf{X}_2$ and $\mathbf{X}_3$ are calculated by subtraction and used in a new iteration. In more detail the four steps can be formulated as

$$\mathbf{W}_{2,r}\mathbf{S}\mathbf{W}^t_{3,r} = SVD(\mathbf{X}^t_{2,r-1}\mathbf{X}_{1,0}\mathbf{X}^t_{3,r-1}) \tag{8.10}$$

Then use the first left and right eigenvectors in the following step.

$$\mathbf{t}_{2,r} = \mathbf{X}_{2,r-1}\mathbf{w}_{2,r} \text{ and } \mathbf{t}_{3,r} = \mathbf{X}^t_{3,r-1}\mathbf{w}_{3,r}, \tag{8.11}$$

$$\mathbf{p}_{2,r} = \mathbf{X}^t_{2,r-1}\mathbf{t}_{2,r}(\mathbf{t}^t_{2,r}\mathbf{t}_{2,r})^{-1} \text{ and } \mathbf{p}_{3,r} = \mathbf{X}_{3,r-1}\mathbf{t}_{3,r}(\mathbf{t}^t_{3,r}\mathbf{t}_{3,r})^{-1} \tag{8.12}$$

$$\mathbf{X}_{2,r} = \mathbf{X}_{2,r-1} - \mathbf{t}_{2,r}\mathbf{p}^t_{2,r} \text{ and } \mathbf{X}^t_{3,r} = \mathbf{X}^t_{3,r-1} - \mathbf{t}_{3,r}\mathbf{p}^t_{3,r} \tag{8.13}$$

We refer to Figure 8.10 for a visual illustration of the procedure for calculating scores based on weights in Equation 8.11.

In the original presentation in Martens *et al.* (2005), the $\mathbf{X}_{1,0}$ is left unaltered after each iteration. For an alternative, see Sæbø *et al.* (2010).

The results from the procedure are the scores and loadings for $\mathbf{X}_2$ and $\mathbf{X}_3$, i.e., $\mathbf{T}_2$, $\mathbf{T}_3$, and $\mathbf{P}_2$, $\mathbf{P}_3$. The scores obtained by this algorithm are orthogonal to each other for the different components. In Martens *et al.* (2005) an alternative to this method without orthogonal scores was discussed briefly. The weights in this variant are the eigenvectors of the same SVD without the deflation procedure. The two were compared empirically and the results were almost indistinguishable. The alternative version will not be pursued here.

Outliers for this method can be spotted using for instance the residuals in Equation 8.13.

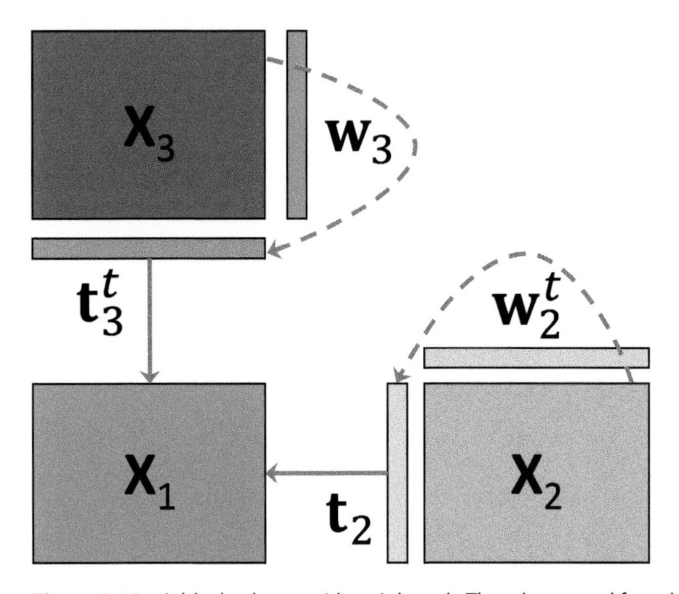

Figure 8.10 L-block scheme with weights **w**'s. The **w**'s are used for calculating scores for deflation.

In the original description of the method an alternative iterative (NIPALS) procedure was proposed for calculating the full Endo-L-PLS solution without using the SVD. The algorithm is quite involved and here we only give a brief description. We refer to Martens *et al.* (2005) for further details. Each step of the algorithm is based on projections and starts with arbitrary weights for $\mathbf{X}_2$ which are used to create $\mathbf{X}_2$ scores. The $\mathbf{X}_1$ is then projected onto the $\mathbf{X}_2$ scores to obtain $\mathbf{X}_1$ scores in the direction of $\mathbf{X}_3$. These are use for creating weights for $\mathbf{X}_3$ which are again used for calculating the $\mathbf{X}_3$ scores. The algorithm then goes back the opposite way through $\mathbf{X}_1$, creating $\mathbf{X}_1$ scores in the direction of $\mathbf{X}_2$ which again are use for calculating new $\mathbf{X}_2$ weights. The process continues until convergence. The main results from the algorithm, for interpretation and deflation, are the $\mathbf{X}_2$ scores, $\mathbf{X}_3$ scores and the $\mathbf{X}_1$ scores (two sets, one set in the direction of $\mathbf{X}_2$ and one set in the direction of $\mathbf{X}_3$). The latter two are not used in the Endo-L-PLS, only in the deflation step for Exo-L-PLS discussed in the next section.

As can be seen, the structure of the Endo-L-PLS method in Algorithm 8.7 can be interpreted in light of the standard PLS algorithm in Chapter 2. First, weight vectors $\mathbf{w}$ for the data blocks are estimated, then the scores $\mathbf{t}$ are obtained by standard multiplication of data and weights, before the scores are used together with the original data blocks to create loadings $\mathbf{p}$ (see Equation 8.12). In the end, the matrices are deflated (see Equation 8.13) before a new sequence is started.

Different sets of columns of the matrices of scores, $\mathbf{T}_2$ and $\mathbf{T}_3$, and matrices of loadings, $\mathbf{P}_2$ and $\mathbf{P}_3$, can be plotted against each other in scatter plots (as for standard PLS solutions), but in Martens *et al.* (2005) a slightly different procedure based on correlating the original data with the scores of the model is proposed. This is equivalent to what is done when making correlation loading plots (Martens and Martens (2001)). This is a way of highlighting the positioning of variables with smaller variance than others and, therefore, have less influence on the solution (see also Elaboration 5.8). For the $\mathbf{X}_1$ block two different types of correlations exist, i.e., correlation between columns of $\mathbf{X}_1$ and scores $\mathbf{t}_2$ for $\mathbf{X}_2$ and between rows of $\mathbf{X}_1$ and scores $\mathbf{t}_3$ for $\mathbf{X}_3$. This leads to two different correlation loading plots for $\mathbf{X}_1$ (see also Example 8.3).

The method indisputably contributes to the analysis of complex relations, but more research is needed in order to understand fully how to interpret the eigenvectors of a product of three matrices with possibly very different singular value structures. For more general and theoretical treatments of the SVD for products of three matrices we refer to Zha (1991), De Moor and Zha (1991), and De Moor (1992).

Since double centring is used for $\mathbf{X}_1$, the Endo-L-PLS focuses essentially on interactions between product and consumer as discussed in Example 8.2 and the preceding text. This means that no information about, for instance, average liking of the products is available from the plot. This can, for instance, be analysed by standard ANOVA procedures as done in Section 8.2.2. (See Endrizzi *et al.* (2014) and Example 8.2 in Section 8.2.3 for more information on interpreting double centred data.)

Example 8.3: Apple data analysed by Endo-L-PLS

This example is taken from Martens *et al.* (2005) and is based on a consumer study of apples. Six apples were analysed by standard sensory analysis by a trained panel ($\mathbf{X}_2$-block). Then 125 consumers gave their liking score for the same apples ($\mathbf{X}_1$-block). The consumer information data (consumer attributes or characteristics, $\mathbf{X}_3$-block) contains information about gender, age, plus a number of variables related to what type of fruit the consumer would prefer first and last (most and least preferred) among a number of options (red apple, green apple, pear, banana, or orange). The structure of the data set is then exactly the same as the one in Examples 1.4 and 8.1.

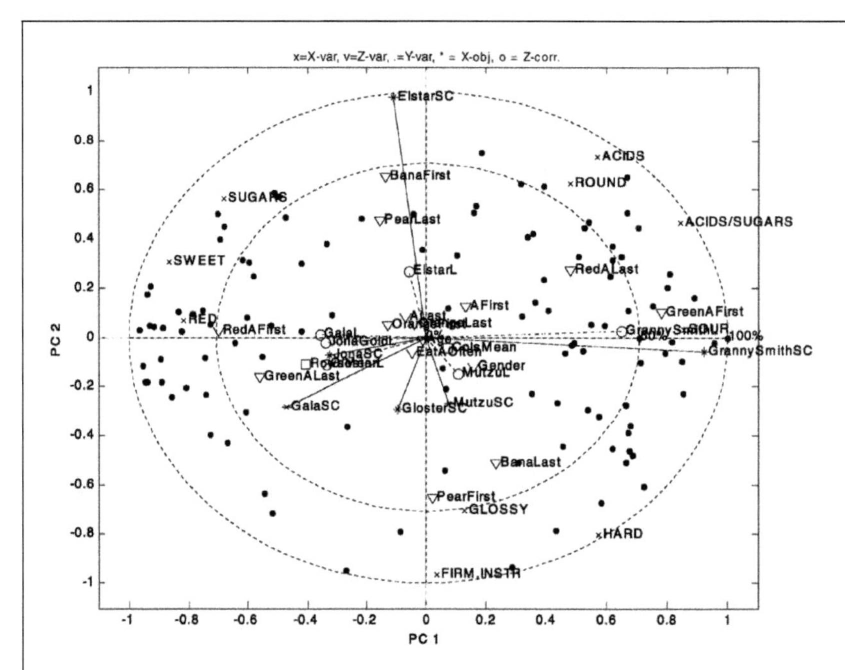

Figure 8.11 Endo-L-PLS results for fruit liking study. Source: (Martens *et al.*, 2005). Reproduced with permission from Elsevier.

The data were analysed by standard Endo-L-PLS methodology as presented above. The variables in $\mathbf{X}_2$ and $\mathbf{X}_3$ data were scaled by their standard deviations. Figure 8.11 presents all the correlation loadings plotted together. The solid lines represent the different product-s/apples (product scores) and the points represent the consumers. The sensory attributes are represented by their names. We can for instance see to the right in the plot that the ratio acid-s/sugar is in the same direction as GreenAFirst and also in the direction of Granny Smith which makes sense. The consumers lying in the same direction are those consumers who like prod-ucts in this direction more than the average consumer. On the opposite side we find Sweet and RedAFirst which is also quite natural.

Other examples can be found in Lengard and Kermit (2006), Giacalone *et al.* (2013), and in Plaehn and Lundahl (2006).

8.3.2 Exo- Versus Endo-L-PLS

In the original algorithm the deflation is done with a basis in scores for $\mathbf{X}_2$ and $\mathbf{X}_3$ (see Equation 8.13). Therefore, it represents a directional (prediction) order of the blocks $\mathbf{X}_2$ and $\mathbf{X}_3$ inwards towards $\mathbf{X}_1$. It was later called the Endo-L-PLS. In Sæbø *et al.* (2010), an alternative approach called Exo-L-PLS was described (originally proposed in Martens *et al.* (2005)). This method represents outward predictive relations from the corner block $\mathbf{X}_1$ to both $\mathbf{X}_2$ and $\mathbf{X}_3$ in order to enhance the importance of the $\mathbf{X}_1$ block in the model. This is achieved by deflation based on the two sets of $\mathbf{X}_1$ scores described immediately after the Endo-L-PLS algorithm. Apart from that the two methods are identical, i.e., the algorithm for finding a new component is the same.

The main idea behind the new variant is that in some cases we are more interested in describing variability in $\mathbf{X}_1$ and understanding how well the $\mathbf{X}_1$ predicts $\mathbf{X}_2$ and $\mathbf{X}_3$ than the opposite. This viewpoint can be based on information about causal relations or simply of more pragmatic interest. The topology for the Exo-L-PLS is illustrated in Figure 8.3d.

Algorithm 8.8

EXO-L-PLS

The Exo-L-PLS method is for each new component based on the same algorithm as the Endo-L-PLS method (Sæbø *et al.* (2010)), the only difference lies in the deflation which is now done with a basis in scores for $\mathbf{X}_1$, one set of scores in the $\mathbf{X}_2$-direction ($\mathbf{T}_{1,2}$) and one in the $\mathbf{X}_3$-direction ($\mathbf{T}_{1,3}$). The two sets of $\mathbf{X}_1$ scores are obtained by projecting $\mathbf{X}_1$ onto the scores for $\mathbf{X}_3$ and $\mathbf{X}_2$ respectively (see e.g., Equation 8.11 and Figure 8.10). These scores play a role in the algorithmic formulation of L-PLS described briefly right after Algorithm 8.7, but are not needed in the SVD formulation for Endo-L-PLS.

The Exo-L-PLS method creates non-orthogonal scores for each of the blocks. Therefore, deflation (see Equation 8.15) has to be done with respect to all previous components for both $\mathbf{X}_1$, $\mathbf{X}_2$, and $\mathbf{X}_3$, instead of only one component at a time as in Equation 8.13. Therefore, matrix notation is used for the $\mathbf{X}_1$ scores in the description of the deflation. Further details can be found in Sæbø *et al.* (2010).

For the $\mathbf{X}_2$-block, the deflation is done using the equation

$$\mathbf{X}_{2,r} = \mathbf{X}_{2,0} - \mathbf{T}_{1,2}\mathbf{P}_{1,2}^t \tag{8.14}$$

where $\mathbf{T}_{1,2}$ is the matrix of the first $r-1$ $\mathbf{X}_1$ scores in the direction of $\mathbf{X}_2$. For each component, the new column in $\mathbf{T}_{1,2}$ is obtained by projecting the deflated $\mathbf{X}_1^t$ onto the $\mathbf{t}_{3,r}$ score (see Figure 8.10). The $\mathbf{P}_{1,2}$ is the corresponding loading matrix obtained by regressing $\mathbf{X}_{2,0}$ onto scores $\mathbf{T}_{1,2}$. For the $\mathbf{X}_3$-block the same type of deflation is conducted.

The deflation for the $\mathbf{X}_1$ block is obtained by

$$\mathbf{X}_{1,r} = \mathbf{X}_{1,0} - \mathbf{T}_{1,2}\mathbf{D}_r\mathbf{T}_{1,3}^t \tag{8.15}$$

where the $\mathbf{D}$ is equal to

$$\mathbf{D}_r = (\mathbf{T}_{1,2}^t\mathbf{T}_{1,2})^{-1}\mathbf{T}_{1,2}^t\mathbf{X}_{1,0}\mathbf{T}_{1,3}(\mathbf{T}_{1,3}^t\mathbf{T}_{1,3})^{-1} \tag{8.16}$$

in other words, the deflation is obtained by subtracting the projection of $\mathbf{X}_{1,0}$ onto both the two scores matrices. All these deflations create a prediction direction outwards (with the basis in scores for $\mathbf{X}_1$), from $\mathbf{X}_1$ to $\mathbf{X}_2$ and $\mathbf{X}_3$. Note the structural similarity between the deflation operator and Equation 8.3. The scores $\mathbf{T}$ used here are, however, not the same as the $\mathbf{T}$ and $\mathbf{U}$ in the other situation.

A prediction equation for $\mathbf{X}_2$ can be obtained by, for instance, regressing the response matrix $\mathbf{X}_2$ onto the $\mathbf{X}_1$ scores $\mathbf{T}_{1,2}$ (see Sæbø *et al.* (2010) for more details on this issue).

The method was in Sæbø *et al.* (2010) applied on a liking of beer data set of the same structure as presented in Example 1.4. It was found that both the Endo- and Exo-L-PLS gave quite similar results in terms of interpretation based on correlation loadings. It was also demonstrated empirically that the Exo-L-PLS describes more of the $\mathbf{X}_1$ than the alternative. Apart from this, little has been done in the literature to compare the two variants.

8.4 Modifications of the Original L-PLS Idea

In this section we will briefly discuss two modifications of the original L-PLS method/idea. Both of them may have an important role in practice, but only a limited number of studies have yet been published.

8.4.1 Weighting Information from $\mathbf{X}_3$ and $\mathbf{X}_1$ in L-PLS Using a Parameter α

Sæbø *et al.* (2008b) proposed a modification of L-PLS for enhancing the influence of $\mathbf{X}_3$. Again, the main difference from the original L-PLS is the deflation. The topology is illustrated in Figure 8.3(c). The main idea is to create new weights for deflation which depend on a parameter α (between 0 and 1) and which can be used to enhance the importance of $\mathbf{X}_3$ in the estimation. The algorithm is quite involved and the reader is referred to the paper for details.

The applicability of this method was discussed in the context of incorporating additional information about the regressors in regression analysis for improved prediction ability. The horizontal link represents the explanatory ($\mathbf{X}_1$) and dependent variables ($\mathbf{X}_2$) and the vertical dimension represents the additional information about the explanatory variables ($\mathbf{X}_3$). Note that this type of study can also be conducted using the other L-PLS methods above.

Typical examples of additional information about $\mathbf{X}_1$ listed in the paper are variable grouping, historical data, and 'variable similarity' matrices. In the former category, each variable can, for instance, be represented by its membership values (between 0 and 1) in a number of groups. The second class refers to situations where extra samples are available from previous studies. This could in spectroscopy mean for instance so-called unlabelled samples without measurements of the response to be predicted. In such cases the extra samples may stabilise the space of regressor variables in such a way that improved predictions are obtained (see also Section 8.5.2). For the last category Sæbø *et al.* (2008b) proposed the following, directly quoted from the paper: for gene expression data the similarity could be the inverted 'distances' between regulatory genes in biological pathways. We refer to Example 8.4 for a case study based on breast cancer data.

Example 8.4: Genomics and breast cancer classification

This example is from the field of breast cancer genomics (Sæbø *et al.*, 2008b). Breast cancer status is classified from gene-expressions in blood. Micro-array expression data are available from 130 samples of which 66 comprise breast cancer cases and 64 are healthy controls. For the genes, a similarity matrix $\mathbf{V}(486 \times 486)$ is constructed encoding the closeness of these genes in terms of their function (e.g., if they belong to the same biochemical pathway). Subsequently, also the gene-expression data was reduced to this number of genes. The matrix $\mathbf{V}$ can be decomposed as $\mathbf{ZZ}^t$ with $\mathbf{Z}^t\mathbf{Z}$ being diagonal (e.g., by using the eigendecomposition of $\mathbf{V}$). By setting $\mathbf{X}_3$ equal to $\mathbf{Z}$ and letting $\mathbf{X}_1$ be the matrix of gene-expression, we are in the realm of Figure 8.3c. The $\mathbf{x}_2$ is here an indicator variable indicating breast cancer or not.

In the α-weight scheme, the value of α determines how much of $\mathbf{X}_3$ is used in the prediction of $\mathbf{x}_2$ from $\mathbf{X}_1$: $\alpha = 1$ means that only $\mathbf{X}_3$ is used and $\alpha = 0$ means that only $\mathbf{X}_1$ is used.

It is instructive to see how the classification error rates depend on the number of components and the value of α which can be studied with cross-validation. Figure 8.12 reveals that

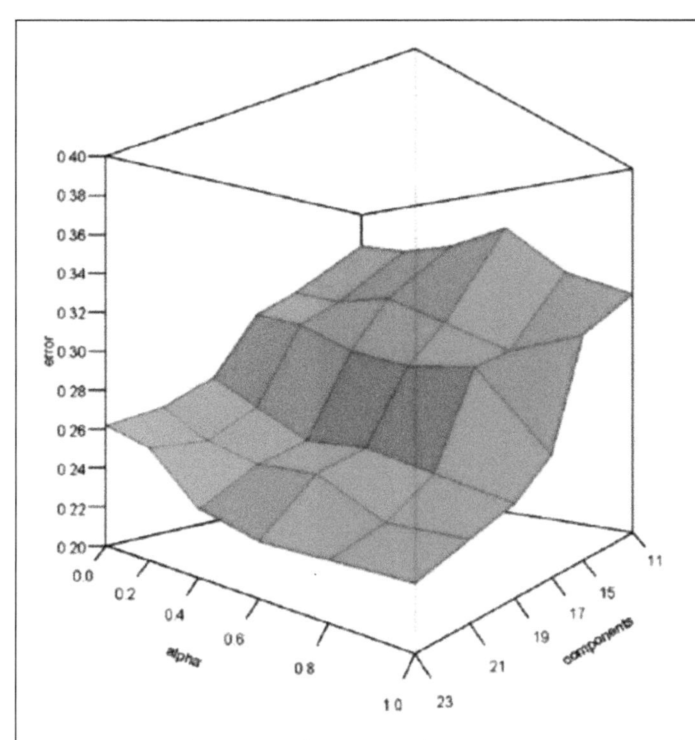

Figure 8.12 Classification CV-error as a function of the α value and the number of L-PLS components. Source: (Sæbø *et al.*, 2008b). Reproduced with permission from Elsevier.

for all components selected, the error rates decrease with increasing α from zero. This means that incorporating prior knowledge of $\mathbf{X}_3$ is favourable. It also shows that we need many components to reach a reasonably low error rate which shows the complexity of this type of data. A more detailed analysis of the regression coefficient estimates from the PLS model including their jackknifed error estimates shows that the higher the α value the lower the number of significant regression coefficients, hence, the model becomes more parsimonious which is favourable for interpretation.

8.4.2 Three-blocks Bifocal PLS

A strongly related idea to the one discussed in Sæbø *et al.* (2008b) was published in Eriksson *et al.* (2004). The method was called three-blocks bifocal PLS (3BIF-PLS) and is based on a similar data skeleton, but in this case the focus is on additional information about the responses. This means that there are two sources of information that may have an impact on the responses in $\mathbf{X}_1$, one from the vertical direction ($\mathbf{X}_3$) and one from the horizontal ($\mathbf{X}_2$). The main conceptual difference from the method in (Sæbø *et al.*, 2008b) is that now the $\mathbf{X}_1$, i.e., the block in the 'corner' of the topology, is the response matrix. The topology for this method is the same as for the Endo-L-PLS, see Figure 8.3a.

In the two examples considered in Eriksson *et al.* (2004) the $\mathbf{X}_2$ data consist of amino acid properties or DNA fingerprints, the $\mathbf{X}_1$-data are haloalkanes or concentration readings and $\mathbf{X}_3$ represents various chemical (or physico-chemical) descriptors.

The method is algorithmically oriented and has a resemblance with the original L-PLS method above. The deflation strategy is as above based on the standard PLS principles (Chapter 2). A common feature between 3BIF-PLS and the method by (Sæbø *et al.*, 2008b) is that the influence of $\mathbf{X}_3$ on the solution is controlled by a parameter. Results are presented in standard scatter plots of scores and loadings.

The important message from this is that L-shape data analysis can be useful both for utilising additional information about the independent variables and the dependent variables in regression analysis.

8.5 Alternative L-shape Data Analysis Methods

In this section we present three alternative L-shape methods which do not fit into the framework and headlines of the previous sections.

8.5.1 Principal Component Analysis with External Information

Takane and Shibayama (1991) published a framework for L-shape data which separates the joint effect of $\mathbf{X}_2$ and $\mathbf{X}_3$ on $\mathbf{X}_1$ from their individual effects (topology in Figure 8.3(a)). The model used for this is

$$\mathbf{X}_1 = \mathbf{X}_2\mathbf{H}\mathbf{X}_3 + \mathbf{X}_2\mathbf{G} + \mathbf{D}\mathbf{X}_3 + \mathbf{E}. \tag{8.17}$$

The matrices $\mathbf{H}$, $\mathbf{G}$, and $\mathbf{D}$ contain parameters to be estimated.

The first term represents the 'interaction' between $\mathbf{X}_2$ and $\mathbf{X}_3$ and the other two terms the individual effects of the two blocks $\mathbf{X}_2$ and $\mathbf{X}_3$. The model can also be interpreted as a way of splitting the decomposition of $\mathbf{X}_1$ from the two input blocks into a common (interaction) part and two distinct parts. In this sense, there is a conceptual analogy with for instance PO-PLS in Chapter 7. The splitting in Equation 8.17 also has some conceptual similarities with ANOVA modelling where the main effects of two blocks are separated from the interactions. Since decomposition of $\mathbf{X}_1$ (as a function of $\mathbf{X}_2$ and $\mathbf{X}_3$) is the focus here, the topology assumed is the one in Figure 8.3(a). Martens *et al.* (2005) discussed a similar possible splitting of contributions in connection with L-PLS, but did not elaborate it further.

There is, however, a redundancy in this model since $\mathbf{X}_2$ and $\mathbf{X}_3$ are represented separately and in a joint interaction term. In the paper by Takane and Shibayama (1991), two possible solutions were suggested, a simultaneous approach based on imposing restrictions on the relations between data and parameters and a sequential approach based on first fitting the first term in the decomposition (the interactions) and then using the residuals as a basis for estimating the other two parts. The second approach is used in the original paper, but it is stated that they give the same result. This means that the 'interaction' which contains information about the joint effect of $\mathbf{X}_2$ and $\mathbf{X}_3$ on $\mathbf{X}_1$ is estimated first before the individual contributions of the two blocks are estimated.

The least squares (LS) solution for the $\mathbf{H}$ in $\mathbf{X}_2\mathbf{H}\mathbf{X}_3$ can be written as

$$\mathbf{H} = (\mathbf{X}_2^t\mathbf{X}_2)^{-}\mathbf{X}_2^t\mathbf{X}_1\mathbf{X}_3^t(\mathbf{X}_3\mathbf{X}_3^t)^{-} \tag{8.18}$$

where the $\mathbf{X}^{-}$ represents a generalised inverse of $\mathbf{X}$, which does not need to be unique. Premultiplying $\mathbf{H}$ with $\mathbf{X}_2$ and postmultiplying $\mathbf{H}$ with $\mathbf{X}_3$ gives $\mathbf{X}_2\mathbf{H}\mathbf{X}_3 = \mathbf{P}_2\mathbf{X}_1\mathbf{P}_3$ where the $\mathbf{P}$ are projection matrices for the two spaces spanned by $\mathbf{X}_2$ and $\mathbf{X}_3^t$, respectively. Then the residuals $\mathbf{E}_1$ can be obtained as

$$\mathbf{E}_1 = \mathbf{X}_1 - \mathbf{P}_2\mathbf{X}_1\mathbf{P}_3. \tag{8.19}$$

In other words, the projection of $\mathbf{X}_1$ onto both $\mathbf{X}_2$ and $\mathbf{X}_3^t$ along the sample and variable mode is subtracted from $\mathbf{X}_1$. It is known from matrix algebra that projections are unique even though the generalised inverse is not.

The residuals $\mathbf{E}_1$ are then fitted to the two other terms in Equation 8.17 separately to obtain the LS solution for $\mathbf{G}$ and $\mathbf{D}$. They can be written as

$$\mathbf{G} = (\mathbf{X}_2^t\mathbf{X}_2)^{-}\mathbf{X}_2^t\mathbf{X}_1(\mathbf{I} - \mathbf{P}_3) \tag{8.20}$$

$$\mathbf{D} = (\mathbf{I} - \mathbf{P}_2)\mathbf{X}_1\mathbf{X}_3^t(\mathbf{X}_3\mathbf{X}_3^t)^{-}. \tag{8.21}$$

The full decomposition can then be written as

$$\mathbf{P}_2\mathbf{X}_1\mathbf{P}_3 + \mathbf{P}_2\mathbf{X}_1(\mathbf{I} - \mathbf{P}_3) + (\mathbf{I} - \mathbf{P}_2)\mathbf{X}_1\mathbf{P}_3 \tag{8.22}$$

Again the projection matrices are unique even though the generalised inverses in Equations 8.20 and 8.21 may not be so. For interpretation purposes we can use PCA to decompose the individual contributions in Equation 8.22 as was also suggested in Takane and Shibayama (1991). The method is in the original paper tested on data from consumer science.

It can be seen that the formula for the interaction effect in Equation 8.18 shares some resemblance with Equation 8.3 for the two step sequential approach in Section 8.2. A major difference is that in Section 8.2 the projections are based on PLS scores while here they are based on the original data. Comparisons between the two approaches have not been made yet. There is also a strong resemblance between the interaction part in Equation 8.18 and the deflation procedure for Exo-L-PLS (see, e.g., Equation 8.16). In other words, Equation 8.3 for the two step PLS approach, the deflation for L-PLS and Equation 8.18 share the same structure although the symbols have a different meaning in the three.

8.5.2 A Simple PCA Based Procedure for Using Unlabelled Data in Calibration

The method discussed here focuses on improved prediction ability in the $\mathbf{X}_1$-$\mathbf{X}_2$ relation using extra information in $\mathbf{X}_3$. The procedure is sequential (Figure 8.1(b)) and the topology is the one in Figure 8.3(c).

A typical example is calibration in spectroscopy (Martens and Næs (1989)), where it is easy to collect many spectra but more expensive or complicated to establish chemical response measurements. The data set with both $\mathbf{X}_2$ and $\mathbf{X}_1$ measurements (chemistry and spectroscopy) will here be referred to as labelled data (Thomas (2019)), as opposed to unlabelled data, here represented in the $\mathbf{X}_3$-block consisting of spectral measurements on additional samples. Utilising the full data set is sometimes referred to as semi-supervised learning (Thomas (2019)). We refer to Figure 8.13(a) for the data setup.

A simple two-step procedure for this, based on PCA and standard LS regression, was suggested in Isaksson and Næs (1990). The first step was based on running a PCA on all the spectra, i.e., both on $\mathbf{X}_1$ and $\mathbf{X}_3$ (see also description of the SCA in Chapter 5).

$$\begin{bmatrix}\mathbf{X}_3\\ \mathbf{X}_1\end{bmatrix} = \begin{bmatrix}\mathbf{T}_3\\ \mathbf{T}_1\end{bmatrix}\mathbf{P}^t + \mathbf{E} \tag{8.23}$$

In this case, the joint matrix $\begin{bmatrix}\mathbf{X}_3\\ \mathbf{X}_1\end{bmatrix}$ is centred without any prior centring of $\mathbf{X}_1$ and $\mathbf{X}_3$. This model creates a common set of loadings $\mathbf{P}$ and scores $\mathbf{T}_1$ and $\mathbf{T}_3$ for both $\mathbf{X}_1$ and $\mathbf{X}_3$. The

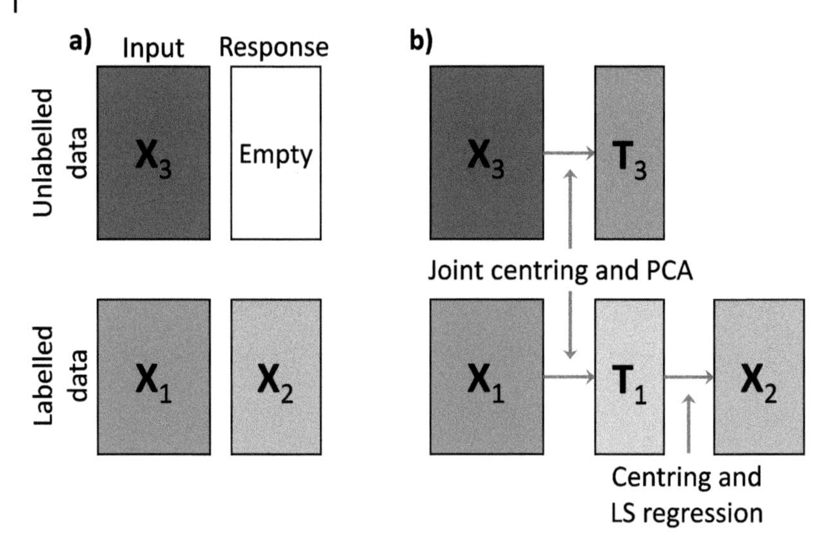

Figure 8.13 (a) Data structure for labelled and unlabelled data. (b) Flow chart for how to utilise unlabelled data

scores for the samples for which there are chemistry measurements are then used as predictor variables in a standard least squares (LS) prediction of the responses $\mathbf{X}_2$ (the $\mathbf{X}_2$ is used instead of $\mathbf{Y}$ as response here in order to stick to the same notation as in the rest of the chapter) using the (PCR like) model

$$\mathbf{X}_2 = \mathbf{T}_1\mathbf{B} + \mathbf{F} \tag{8.24}$$

Both input and output are centred based on the data in the labelled blocks (see Figure 8.13(b) for a schematic illustration of the procedure). The idea is that using the full set of spectra for calculating the scores to be used, contributes to creating a more stable and representative space that can improve predictions in the second phase. Isaksson and Næs (1990) demonstrated that in a case with only 10 labelled samples, the inclusion of additional NIR spectra (unlabelled) made the prediction results comparable to a calibration based on a much larger set of labelled samples (see also Thomas (1995)).

An overview of alternative methods for semi-supervised calibration with a more theory driven focus (and no special focus on component based methods) can be found in Thomas (2019). In that paper both so-called inverse and classical calibration (Martens and Næs (1989)) was considered. A major conclusion in the paper is that *the advantage is most likely to be realised when the quantity of labelled data is small, the quantity of unlabelled data is large, and when the values of the characteristics to be predicted are distant from the centroid of associated values of the labelled data.* The first of these conclusions is very natural since with little data available for calibration, the predictor is uncertain and the benefit of using more unlabelled data may be larger, given of course that the data are relevant for the purpose. The second part of the conclusion refers to representativity of the unlabelled data.

8.5.3 Multivariate Curve Resolution for Incomplete Data

A new multivariate curve resolution (MCR) methodology for L-shape data was proposed in Alier and Tauler (2013) and De Luca *et al.* (2014) (see also Piqueras *et al.* (2018)). As in standard MCR, the focus is on determining both chemical concentrations and spectra using information about, for instance, non-negativity of concentrations. In the situations described, one set of samples is analysed by two different instruments ($\mathbf{X}_1$ and $\mathbf{X}_2$), while another set of samples is analysed by only one of the instruments $\mathbf{X}_3$ (the same instrument as used to obtain $\mathbf{X}_1$). The $\mathbf{X}_1$ and $\mathbf{X}_2$ in Figure 8.1(a) constitute the data set with measurements from both instruments. The vertical link (along the variable mode) between $\mathbf{X}_1$ and $\mathbf{X}_3$ represents the measurements for all samples made with one of the instruments. The two parts are linked as described by Equations 8.26 and 8.27.

In principle the two data matrices ($\mathbf{X}_1$, $\mathbf{X}_2$) and ($\mathbf{X}_1$, $\mathbf{X}_3$) could be analysed separately using the standard MCR model

$$\mathbf{X} = \mathbf{C}\mathbf{S}^t + \mathbf{E} \tag{8.25}$$

where $\mathbf{X}$ is the data matrix, $\mathbf{C}$ are the concentrations and $\mathbf{S}$ are the base spectra. In both cases, the focus would be on estimating both spectra and concentrations using various constraints. This could, however, mean loss of information since there is overlap in the data sets. Therefore, a combined approach was suggested.

The two MCR models relevant here are

$$[\mathbf{X}_1, \mathbf{X}_2] = \mathbf{C}_{12}[\mathbf{S}_1^t, \mathbf{S}_2^t] + \mathbf{E}_{12} \tag{8.26}$$

for the horizontal link and

$$\begin{bmatrix}\mathbf{X}_3\\ \mathbf{X}_1\end{bmatrix} = \begin{bmatrix}\mathbf{C}_{13}\\ \mathbf{C}_{12}\end{bmatrix}\mathbf{S}_1^t + \mathbf{E}_{13} \tag{8.27}$$

for the vertical link. Here $\mathbf{C}_{12}$ represents the concentrations of the constituents for the data set based on the two instruments involved. The $\mathbf{C}_{13}$ represents the concentrations for the data block based on one of the instruments only. The $\mathbf{S}_1$ and $\mathbf{S}_2$ are the spectra for the two instruments in $\mathbf{X}_1$ and $\mathbf{X}_2$.

In the combined approach the criterion to be optimised is essentially the sum-of-squares of the residuals $\mathbf{E}_{12}$ and $\mathbf{E}_{13}$. We refer to the original papers for details about the optimisation procedure.

This method is primarily applicable in situations with a clear Beer's law linearity structure, which is different from the other methods in this chapter. The application of the method reported in De Luca *et al.* (2014) is based on combining hyphenated liquid chromatographic data and UV spectrophotometric data for drug photostability studies. The application in Alier and Tauler (2013) is from atmospheric monitoring of O_2 and NO concentrations combining UV photometry and a method using automatically operated chemiluminescence analysers.

8.5.4 An Alternative Approach in Consumer Science Based on Correlations Between $\mathbf{X}_3$ and $\mathbf{X}_1$

Thybo *et al.* (2004) proposed an alternative two-step sequential way of combining the three blocks. The context is again the sensory science example in the introduction of this chapter and in Example 1.4 and focus is interpretation. As opposed to the sequential approaches discussed in Section 8.2, this method starts with considering the vertical (Figure 8.1(b)) link and combining the information from $\mathbf{X}_3$ (consumer attributes) and $\mathbf{X}_1$ (consumer liking) before taking $\mathbf{X}_2$ (descriptive sensory data) into account. The topology is given in Figure 8.3(b).

The first step is to calculate the correlation matrix $\mathbf{G}$ of the matrices $\mathbf{X}_3$ and $\mathbf{X}_1$. In other words, the correlation between each row in $\mathbf{X}_3$ (K) and each row in $\mathbf{X}_1$ (I) is calculated and put into $\mathbf{G}$, using the consumers as units. The $\mathbf{G}$-matrix will have dimension ($I \times K$), i.e., one row for each sample and one column for each consumer attribute. This means that the element i, k of $\mathbf{G}$ can be written as

$$\mathbf{G}_{(i,k)} = \mathbf{x}_{1i}^t \mathbf{x}_{3k} / (\text{std}(\mathbf{x}_{1i}) \text{std}(\mathbf{x}_{3k})) \tag{8.28}$$

where std stands for standard deviation. Here $\mathbf{x}_{1i}^t$ is the (centred) ith row of $\mathbf{X}_1$ and the column vector $\mathbf{x}_{3k}$ corresponds to the (centred) kth row of $\mathbf{X}_3$. The matrix $\mathbf{G}$ then contains information about how well the liking of the different samples is correlated with the different consumer attributes. Then the correlation matrix is concatenated with the liking matrix $\mathbf{X}_1$, i.e.,

$$\mathbf{X}_{1,\text{New}} = [\mathbf{X}_1, \mathbf{G}] \tag{8.29}$$

before PLS regression is used for predicting $\mathbf{X}_{1,\text{New}}$ from the sensory attributes in $\mathbf{X}_2$ using the standard model (Equation 8.1). The $\mathbf{X}_{1,\text{New}}$ has I rows and $N + K$ columns and $\mathbf{X}_2$ has I rows and J columns. This model provides scores and loadings that can be plotted in scatter plots the standard way.

The underlying philosophy behind this method is to combine information about liking and the relation between liking and external consumer attributes with the sensory data in one single PLS regression. The method has, however, a heuristic character and the properties have not been much explored in the literature. Martens *et al.* (2005) and Thybo *et al.* (2004) state that the results are only slightly different from L-PLS (See Section 8.3) due to different mean centring (see also Lengard and Kermit (2006)).

8.6 Domino PLS and More Complex Data Structures

An extension of the above L-shaped framework was proposed by Martens *et al.* (2005). The extension was called domino PLS because of the resemblance of the data structure with how the pieces are put together in the game Domino. For instance, in addition to the L-shape structure above, an extra block is allowed above the $\mathbf{X}_2$-block in Figure 8.1(a), giving a shape similar to a U (Figure 8.1(d)) (see also Chapter 3 for a description of the topology). The idea behind the domino PLS method is more or less the same as for L-PLS, namely to provide scores and loadings for all the blocks that are related through the linkage structure imposed. The procedure is again algorithmically oriented and based on successive use of LS regression in the different blocks, until convergence.

Note that although a domino diagram may visually resemble a path diagram, the two structures are completely different. The matrices in the domino diagram fit together according to the rows and columns dimensions in the matrices involved. In the path diagram, however, the matrices involved share the sample mode and the focus is on finding relations between the variables in the different blocks.

A skeleton similar to the one proposed in Martens *et al.* (2005) was considered in Löfstedt *et al.* (2012). A method called bi-modal on-PLS was suggested. This is an extension of the previously published OnPLS (Löfstedt and Trygg (2011)).

As it stands now, the domino PLS represents a possibly interesting idea, but it is still unclear what types of applications that could benefit from such an approach.

8.7 Conclusions and Recommendations

The tree structure in Figure 8.14 describes the main distinctions and methods in this chapter. Two of the choices plus a general aspect related to common components are discussed below, the rest of the diagram is quite obvious given data type and situation.

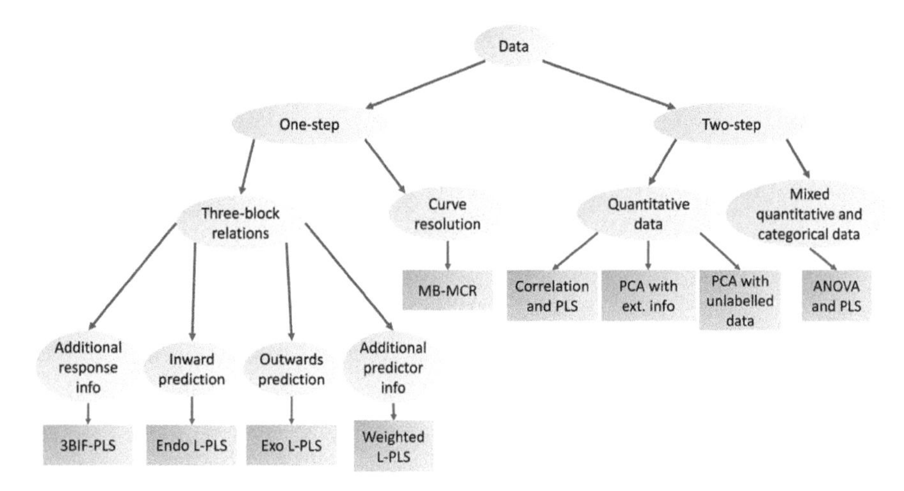

Figure 8.14 Tree for selecting methods with complex data structures.

One-step versus two-step approach

The most important distinction made in this chapter is between methods which treat all three blocks at the same time (L-PLS and variants) and methods which analyse the horizontal axis (sample mode) in the L-shape first before the vertical axis (variable mode) is analysed based on the results from the first analysis. There is no clear indication in the literature on which of them is to be preferred. The two step methods have the advantage that all steps are well understood and established and that interpretation can be done in sequence based on well-known methods. The other approach has the advantage that the whole solution is found using one method only, which can make it simpler to use. The properties of the eigenvectors of products of three matrices with different singular value structures, as used for L-PLS, are, however, not well-known/established in this area. Other than these aspects there is no scientific reason to prefer one of the approaches over the other.

Different Variants of the L-PLS Approach

The different variants of the L-PLS are all based on the SVD of a product of all matrices. They differ primarily in the way the deflation between each component is done. The Endo-L-PLS uses $\mathbf{X}_2$ and $\mathbf{X}_3$ as input in joint prediction of variability in $\mathbf{X}_1$, while Exo-L-PLS does the opposite. The Endo-L-PLS therefore deflates $\mathbf{X}_2$ and $\mathbf{X}_3$ and tries to predict the $\mathbf{X}_1$ the best possible way, while Exo-L-PLS explores how well $\mathbf{X}_1$ can simultaneously predict the $\mathbf{X}_2$ and $\mathbf{X}_3$. An empirical study indicates that the difference between the methods in practice can be quite small.

Common versus distinct components

For most methods there is an implicit search for common information in all blocks and for separating this from the distinct.

For the two-step procedures analysing the sample mode first, the results (loadings) from the first analysis are considered the carrier of common information to be matched with the information in the variable mode. The final results and corresponding plots can be considered a representation of the common information of all three blocks.

For the L-PLS methods, the common information is sought through a singular value decomposition of the products of the three matrices in the L-shape. It was shown that there are similarities in structure of the obtained results from all these methods, but it is still unclear how the mentioned methods relate to each other.

8.8 Open Issues

An important point that needs more attention is the difference between the L-PLS (Section 8.3), and the two-step PLS based method in Section 8.2.1. Some of the formulae that are involved are relatively similar, but the symbols have different meanings. Therefore, there is an open question what the actual relations between the two are. Which of the methods is more suitable for establishing the common link and which one is easier to interpret?

The importance of the model which separates the distinct effects of $\mathbf{X}_2$ and $\mathbf{X}_3$ on $\mathbf{X}_1$ from the 'interaction' between them (see Section 8.5.1, Takane and Shibayama (1991)) is another issue that needs attention. Is there any clear advantage of handling the effects of $\mathbf{X}_2$ and $\mathbf{X}_3$ on $\mathbf{X}_1$ separately? Another issue that is worth examining is whether there is any parallel to standard modelling of main effects and interactions in ANOVA.

It was stated at the beginning of the chapter that all the methods considered involve some sort of intuitive idea about common and distinct components. A formal definition is needed also in this area. Such a definition could potentially also shed light on the differences between the different methods.

For L-PLS there are several options, the exo-variant, the endo-variant and the variant for which we can vary the importance of the additional information about regression variables. How the different versions relate to each other is not studied in the literature yet.

The semi-supervised calibration procedure for spectroscopy described here has some similarities with the DIPLS method in Chapter 10. The relations between the two should be investigated.

Many of the applications presented in this chapter are based on relatively few samples, often less than 10. This means that standard validation procedures based on for instance cross-validation ideas may be difficult to use. How to validate these methods properly needs more attention.

Part IV

Alternative Methods for Unsupervised and Supervised Topologies

9

Alternative Unsupervised Methods

9.i General Introduction

In this chapter we will briefly present some unsupervised methods which are less used than the selected ones in Chapter 5. This does not mean that they are less useful, merely that some of them are relatively new and their potential has to be established still. We will again distinguish between methods focusing on common variation and methods that separate the different types of variation (common, local, distinct). Moreover, methods are available for a shared variable mode, a shared sample mode and methods for the case that both modes are shared. We will present methods for homogeneous as well as heterogeneous data fusion. Also in this chapter, we will assume column-centred data blocks unless otherwise stated. Moreover, as in other chapters the default norm for vectors and matrices is the Frobenius norm (see Section 2.8).

9.ii Relationship to the General Framework

A brief summary of the methods in this chapter and the sections in which they appear is given in Table 9.1. Table 9.1 shows a patchwork of methods based on different principles. Most methods work in a simultaneous fashion and are model based. Some methods are based on a factor analysis model and work with penalties (e.g., BIBFA, GFA, MOFA, GAS) and some of them can handle heterogeneous data. The representation method (RM) is a different method based on three-way models of specially constructed representations of the data. It is included since it gives a very different perspective and may be useful in some applications. There are also methods based on generalisations of SVD (GSVD) and one based on copulas (XPCA). Hence, these methods show a variety of approaches of a very different nature which illustrates the richness of ideas from different fields of data science that can be used to tackle similar problems.

9.1 Shared Variable Mode

For the case of a shared variable mode, in this chapter we discuss one method that can distinguish common from distinct components: generalised SVD. This method can also be used

Multiblock Data Fusion in Statistics and Machine Learning: Applications in the Natural and Life Sciences,
First Edition. Age K. Smilde, Tormod Næs, and Kristian Hovde Liland.

Table 9.1 Overview of methods. Legend: U=unsupervised, S=supervised, C=complex, HOM=homogeneous data, HET=heterogeneous data, SEQ=sequential, SIM=simultaneous, MOD=model-based, ALG= algorithm-based, C=common, CD=common/distinct, CLD=common/local/distinct, LS=least squares, ML=maximum likelihood, ED=eigendecomposition, MC=maximising correlations/covariances. For abbreviations of the methods, see Section 1.11.

		A			B		C		D		E			F			
	Section	U	S	C	HOM	HET	SEQ	SIM	MOD	ALG	C	CD	CLD	LS	ML	ED	MC
GSVD	9.1, 9.2.2.1	■			■			■		■		■				■	
DIABLO	9.2.1.1	■			■		■			■	■						■
GCTF	9.2.1.2	■				■		■	■		■				■		
RM	9.2.1.3	■				■		■	■		■			■			
XPCA	9.2.1.4	■				■		■	■		■				■		
SLIDE	9.2.2.2	■			■			■	■				■	■			
BIBFA	9.2.2.3	■			■			■	■			■			■		
GFA	9.2.2.4	■			■			■	■				■		■		
OnPLS	9.2.2.5	■			■		■			■			■				■
GAS	9.2.2.6	■				■		■	■			■			■		
MOFA	9.2.2.7	■				■		■	■				■		■		
GPA	9.3.1	■			■			■	■		■			■			
PARAFAC	9.3.2	■			■			■	■		■			■			
Tucker	9.3.2	■			■			■	■		■			■			

in the context of shared samples. We will present both for completeness and start with the shared variable mode.

We will begin by explaining the two-block case. Note that in this case the common and distinct variation is described in the variable mode, hence, in the row-spaces of the respective matrices. An eigenvalue-based method to separate common from distinct components for shared variables is the generalised singular value decomposition (GSVD) which is a generalisation of the SVD known as the quotient SVD (QSVD) (De Moor and Zha, 1991). The mathematics of the GSVD dates back some time (Van Loan, 1976; Paige and Saunders, 1981). The original GSVD is a matrix decomposition method and does not have least squares properties. To repair its sensitivity to noise, we recommend using the adapted GSVD which comes down to first filtering the data with an SCA step (Van Deun *et al.*, 2012). For the two-block case the model is:

$$\begin{aligned} \mathbf{X}_1 &= \widehat{\mathbf{X}}_1 + \mathbf{E}_1 = \mathbf{T}_1\mathbf{D}_1\mathbf{P}^t + \mathbf{E}_1 \\ \mathbf{X}_2 &= \widehat{\mathbf{X}}_2 + \mathbf{E}_2 = \mathbf{T}_2\mathbf{D}_2\mathbf{P}^t + \mathbf{E}_2 \end{aligned} \tag{9.1}$$

with $\widehat{\mathbf{X}}_m$ as the filtered data (i.e., an SCA model is run on the concatenated matrix $[\mathbf{X} = [\mathbf{X}_1^t|\mathbf{X}_2^t]^t)$. Moreover, $\mathbf{T}_m^t\mathbf{T}_m = \mathbf{I}$ $(m = 1, 2)$ and $\mathbf{P}$ is a full-rank matrix of common loadings and not necessarily with orthonormal columns. Note that these loadings are actually consensus loadings since they are not in the row-space of the original matrices. The matrix $\mathbf{D}_m$ $(m = 1, 2)$ is diagonal and such that $\mathbf{D}_1^2 + \mathbf{D}_2^2 = \mathbf{I}$. Due to the latter constraint it is possible to split the generalised singular values (the elements of $\mathbf{D}_m$ $(m = 1, 2)$) in three groups: if $d_{1r}^2 \approx 1$ the corresponding component is distinctive for $\mathbf{X}_1$, if $d_{2r}^2 \approx 1$ the corresponding component is distinctive for $\mathbf{X}_2$ and if $d_{1r}^2 \approx d_{2r}^2$ the corresponding component is common. Obviously, there is a certain amount of arbitrariness in these choices. Once such a choice is made, the GSVD model is as follows:

$$\mathbf{X}_1 = \mathbf{T}_{11}\mathbf{D}_{11}\mathbf{P}_1^t + \mathbf{T}_{12}\mathbf{D}_{12}\mathbf{P}_2^t + \mathbf{T}_{13}\mathbf{D}_{13}\mathbf{P}_3^t + \mathbf{E}_1 \tag{9.2}$$
$$\mathbf{X}_2 = \mathbf{T}_{21}\mathbf{D}_{21}\mathbf{P}_1^t + \mathbf{T}_{22}\mathbf{D}_{22}\mathbf{P}_2^t + \mathbf{T}_{23}\mathbf{D}_{23}\mathbf{P}_3^t + \mathbf{E}_2$$

where $\mathbf{D}_{11}$ represents the distinct part of $\mathbf{X}_1$; $\mathbf{D}_{22}$ represents the distinct part of $\mathbf{X}_2$ and $\mathbf{D}_{13}, \mathbf{D}_{23}$ represent the common part of the two blocks. The matrices $\mathbf{T}_{..}$ are the scores of these different parts and $\mathbf{P}_{..}$ their loadings. Ideally, the diagonal matrices $\mathbf{D}_{12}$ and $\mathbf{D}_{21}$ contain (very) small values. Note the resemblance with DISCO (Section 5.1.2.1) where this 'spill-over' was quantified (and minimised) in the non-congruence value (see Equation 5.26).
The total number of components is selected in the SCA-filtering steps by regular methods to select the number of components in SCA. For the original GSVD, the total number of components equals the rank of the system which is a serious drawback since it always ends up with too many components, some of which are also influenced by noise. This was actually one of the reasons for using the adapted GSVD version since then the maximum number of components is limited by the first SCA step. GSVD is within- and between-block scale dependent and has been used in gene-expression analysis (Alter *et al.*, 2003). It has been extended for more than two blocks (Ponnapalli *et al.*, 2011); from the pattern of elements of $\mathbf{D}_m; m = 1, \ldots, M$ it can in principle be inferred if there are common, local, and distinct components in cases of more than two blocks of data. However, this is not trivial.

9.2 Shared Sample Mode

9.2.1 Only Common Variation

9.2.1.1 DIABLO

A method used in bioinformatics for multiblock data analysis is called **D**ata **I**ntegration **A**nalysis for **B**iomarker discovery using a **L**atent component method for **O**mics studies, DIABLO for short. This method relies strongly on the RGCCA method (see Section 5.2.1.4) and has been implemented in a sparse version (Singh *et al.*, 2019). This comes down to the following problem:

$$\max_{\mathbf{w}'s} \sum_{m,m'=1}^{M} c_{m,m'} \left[\text{corr}(\mathbf{X}_m\mathbf{w}_m, \mathbf{X}_{m'}\mathbf{w}_{m'})\right] \; s.t. \; \|\mathbf{w}_m\|_2 = 1, \|\mathbf{w}_m\|_1 < \lambda_m; \tag{9.3}$$
$$m, m' = 1, \ldots, M$$

where the values $c_{m,m'}$ indicate which blocks are connected (with values zero and one) and subsequently the components $\mathbf{t}_m = \mathbf{X}_m\mathbf{w}_m$ are calculated. The values $\lambda_m > 0$ are penalty parameters to be set by the user (and these hold both for m and m'). The requirement $\|\mathbf{w}_m\|_1 < \lambda_m$ is again a lasso-type penalty. Then the matrices are deflated according to $\mathbf{X}_{m,\text{new}} = \mathbf{X}_m - \mathbf{t}_m\mathbf{w}_m^t$. A similar type of deflation is used in RGCCA, but such a deflation is not trivial (see also Section 2.8). A simple example of a DIABLO model is given in Elaboration 9.1.

ELABORATION 9.1

Example of DIABLO

A simple example of DIABLO is given to show how the method works. Suppose that we have three data blocks sharing the sample mode ($\mathbf{X}_1(I \times J_1)$, $\mathbf{X}_2(I \times J_2)$ and $\mathbf{X}_3(I \times J_3)$). Also assume that $c_{m,m'} = 1\ (m \neq m')$. Then the DIABLO model becomes

$$\max_{\mathbf{w}'s} \left[\text{corr}(\mathbf{X}_1\mathbf{w}_1, \mathbf{X}_2\mathbf{w}_2) + \text{corr}(\mathbf{X}_1\mathbf{w}_1, \mathbf{X}_3\mathbf{w}_3) + \text{corr}(\mathbf{X}_2\mathbf{w}_2, \mathbf{X}_3\mathbf{w}_3)\right] \tag{9.4}$$

$$s.t.\ \|\mathbf{w}_1\|_2 = \|\mathbf{w}_2\|_2 = \|\mathbf{w}_3\|_2 = 1;$$

$$\|\mathbf{w}_1\|_1 < \lambda_1, \|\mathbf{w}_2\|_1 < \lambda_2, \|\mathbf{w}_3\|_1 < \lambda_3$$

and upon writing $\mathbf{t}_m = \mathbf{X}_m\mathbf{w}_m$ this becomes

$$\max_{\mathbf{w}'s} \left[\text{corr}(\mathbf{t}_1, \mathbf{t}_2) + \text{corr}(\mathbf{t}_1, \mathbf{t}_3) + \text{corr}(\mathbf{t}_2, \mathbf{t}_3)\right] \tag{9.5}$$

$$s.t.\ \|\mathbf{w}_1\|_2 = \|\mathbf{w}_2\|_2 = \|\mathbf{w}_3\|_2 = 1;$$

$$\|\mathbf{w}_1\|_1 < \lambda_1, \|\mathbf{w}_2\|_1 < \lambda_2, \|\mathbf{w}_3\|_1 < \lambda_3$$

which can be understood as a generalisation of canonical correlation to multiple blocks.

The optimisation criterion of DIABLO resembles the one of GCA closely since GCA is also maximising sums of correlations. GCA is not unique in cases of high-dimensional data (i.e., more variables than objects; see Section 2.1.10); in such cases there are many trivial solutions to GCA. Whereas RGCCA solves this problem by regularising the solution through shrunken covariance matrices (see Section 5.2.1.4), DIABLO solves this problem by the penalties on $\|\mathbf{w}_m\|$, hence, the λ_m values should be positive. How large these values should be in order to avoid trivial solutions is not known. This sparse implementation of the RGCCA method may also simplify interpretation. The DIABLO method has been implemented in the software package MixOmics (Rohart *et al.*, 2017b) where they call this type of multiblock analysis N-integration (see Chapter 1).

9.2.1.2 Generalised Coupled Tensor Factorisation

The idea of coupled matrix tensor factorisation (CMTF, see Section 5.2.2.4) can also be generalised to heterogeneous data and is then called generalised coupled tensor factorisation (GCTF) (Yılmaz *et al.*, 2011). This generalisation is based on exponential dispersion models (Jørgensen, 1992) that encompass exponential family distributions. The basic idea of these models for M blocks of data coupled in the sample mode is

$$\min_{(\boldsymbol{\theta}, \boldsymbol{\theta}_m)} \sum_{m=1}^{M} v_m d_m[\mathbf{X}_m - \widehat{\mathbf{X}}_m(\boldsymbol{\theta}, \boldsymbol{\theta}_m)] \tag{9.6}$$

where d_m is a divergence measure (see Elaboration 9.2) associated with the distribution of the data in block m; the common parameters are indicated by $\boldsymbol{\theta}$ and the block-specific parameters are indicated by $\boldsymbol{\theta}_m$. As an example, the model for $\mathbf{X}_m$ may look like $\widehat{\mathbf{X}}_m = \mathbf{T}_C\mathbf{P}^t_{Cm}$ where $\boldsymbol{\theta} = \mathbf{T}_C$ and $\boldsymbol{\theta}_m = \mathbf{P}_{Cm}$. The values v_m are weights that weigh the individual contributions to the overall loss function and $\widehat{\mathbf{X}}_m$ indicates fitted values using some model for the mth tensor (two-way or three-way array). For *a priori* set values of the weights v_m this problem can be solved by dedicated algorithms. It can also be solved in a Bayesian framework with priors on the parameters and then it is also possible to learn the weights (Şimşekli *et al.*, 2013). Note that this can be seen as generalisations of GSCA and ESCA models (see Section 5.2.1.5).

ELABORATION 9.2

Divergence measures

Divergence is a weaker notion than distance to measure differences in distributions. It is not necessarily symmetric and also the triangle inequality does not need to hold. One of the prime examples from information theory is the Kullback–Leibler divergence to measure the similarity of two distributions (Murphy, 2012). It is based on the notion of entropy of a random variable X with distribution p, denoted by $H_E(p)$:

$$H_E(p) = -\sum_{k=1}^{K} p(X = k) log_2 p(X = k) = -\sum_{k=1}^{K} p_k log_2 p_k \quad (9.7)$$

for a discrete distribution with K the number of values X can have. To define a divergence between two discrete distributions p and q, the Kullback–Leibler divergence is:

$$KL(p||q) = \sum_{k=1}^{K} p_k log_2 \frac{p_k}{q_k} = -H_E(p) + H_E(p, q) \quad (9.8)$$

and

$$H_E(p, q) = -\sum_{k=1}^{K} p_k log_2 q_k \quad (9.9)$$

is called the cross entropy. This measure is clearly not symmetric and extensions for continuous distributions are available.

9.2.1.3 Representation Matrices

The standard way of representing variables in multivariate analysis is by collecting these variables in a matrix where each variable is a column in that matrix. Yet, there are other ways of representing variables, and the standard way is just one of the ways to represent variables. Other ways of representing variables give opportunities to broaden the scope of multivariate data analysis methods and also to combine different data types. Although this method is not used much in multiblock data fusion, it has potential and gives a very different view on how to handle multivariate data. Moreover, it provides a very natural way of dealing with data of different measurement scales. Hence, it is worthwhile to explain this in some detail. The approach as described below builds on the work of Janson and Vegelius (1982) and Zegers (1986).

Representation matrices for ratio-, interval-, and ordinal-scaled variables

Suppose we have a data matrix $\mathbf{X}(I \times J)$ with columns $\mathbf{x}_j$ containing the measurements of the objects on variable j. Such measurements can be a ratio-scaled value, but can also be a binary value, a categorical value or an ordinal-scaled value. A representation operator works on this vector and produces a representation matrix[1] which serves as a building block to calculate

---

1 The original name was quantification matrix but that name has already been used differently in this book. Hence our choice to rename such matrices.

associations between variables and to analyse several variables simultaneously (Zegers, 1986; Kiers, 1989). Such a representation matrix can be a vector ($I \times 1$), a rectangular matrix ($I \times R; R < I$) or a square matrix ($I \times I$). Let $\mathbf{S}_j$ and $\mathbf{S}_k$ be the representation matrices for variables j and k, respectively, then a general equation of the association between variables j and k is

$$\tilde{q}_{jk} = \frac{2tr(\mathbf{S}_j^t\mathbf{S}_k)}{tr(\mathbf{S}_j^t\mathbf{S}_j) + tr(\mathbf{S}_k^t\mathbf{S}_k)}. \tag{9.10}$$

In most cases that follow below the representation matrices are standardised in such a way that $tr(\mathbf{S}_j^t\mathbf{S}_j) = 1$ after standardisation. In these cases, Equation 9.10 simplifies to

$$q_{jk} = tr(\mathbf{S}_j^t\mathbf{S}_k) \tag{9.11}$$

since both $tr(\mathbf{S}_j^t\mathbf{S}_j)$ and $tr(\mathbf{S}_k^t\mathbf{S}_k)$ are one because of the standardisation. As will be shown in the following, Equation 9.11 can generate the familiar associations such as the Pearson correlation or the Spearman correlation. An extensive description of all kinds of representation matrices is beyond the scope of this book; we will discuss the most relevant ones for the problem of heterogeneous data fusion.

For ratio- and interval-scaled values, two types of representation matrices can be defined: vectors and square matrices. If $\mathbf{x}_j$ represents the raw measurements of the objects on variable j then the vector quantification can be this vector itself (i.e., $\mathbf{s}_j = \mathbf{x}_j$) or a standardised version of it. When the latter is used in Equation 9.11, Pearson's R-value is obtained. In standard multivariate analysis this is by far the most used representation matrix. There is also another possibility for ratio- and interval-scaled values, namely square representation matrices. Two examples are the following. Define

$$\tilde{\mathbf{S}}_j = (\mathbf{x}_j\mathbf{1}^t - \mathbf{1}\mathbf{x}_j^t) \tag{9.12}$$

where $\mathbf{x}_j$ contain the raw measurements (object-scores) and $\mathbf{1}$ is an $I \times 1$ column of ones. This $\tilde{\mathbf{S}}_j$ generates a skew-symmetric matrix enumerating all differences between the object-scores of variable j. Hence, distances between objects are obtained per variable and these distance matrices can be subjected to an INDividual Differences SCALing (INDSCAL) model (Kiers, 1989). Upon standardising $\tilde{\mathbf{S}}_j$ as $\mathbf{S}_j = (tr\tilde{\mathbf{S}}_j^t\tilde{\mathbf{S}}_j)^{-1/2}\tilde{\mathbf{S}}_j$ and using this $\mathbf{S}_j$ (and a similarly defined $\mathbf{S}_k$) in Equation 9.11 gives again Pearson's R-value. Another example is using $\mathbf{S}_j = \mathbf{s}_j\mathbf{s}_j^t$ where $\mathbf{s}_j$ is the standardised version of $\mathbf{x}_j$. Using this $\mathbf{S}_j$ (and a similarly defined $\mathbf{S}_k$) in Equation 9.11 gives Pearson's R^2 value. Note that this $\mathbf{S}_j$ is a positive (semi-)definite matrix whereas the one of Equation 9.12 is not. Sometimes positive definiteness is a desirable property.

The above exposure differs from the regular distance-based multivariate analyses which we are not going to discuss (see Chapter 1). The regular distance-based methods calculate distances between I samples of $\mathbf{X}$ *across all variables* simultaneously resulting in an $I \times I$ distance matrix and thus variable specific information is lost. The representation matrices calculate representations *per variable*. This is crucially different.

When the data are ordinal-scaled, then the vector of readings can be encoded in terms of rank-orders $\mathbf{r}_j(I \times 1)$ [2]. For the example of strongly disagree, disagree, neutral, agree, strongly agree, such a ranking may be encoded as 1 (strongly disagree) to 5 (strongly agree). Then again – as in the ratio-scaled variables – representation can be done using the vectors $\mathbf{r}_j$ or their standardised version. In the latter case, applying Equation 9.11 to this version gives Spearman's rank-order correlation coefficient. Another representation is by using $\mathbf{r}_j$

2 This vector contains scalars not to be confused with the index for components!

in Equation 9.12 instead of $\mathbf{x}_j$ and this again generates Spearman's rank-order correlation coefficient after using Equation 9.11. Some examples of representation matrices are given in Example 9.1.

Example 9.1: Representation matrices

We will illustrate some ideas of representation matrices using a small example of a (4×2) matrix $\mathbf{X} = [\mathbf{x}_1|\mathbf{x}_2]$:

$$\mathbf{X} = \begin{pmatrix} 2 & 9 \\ 4 & 9 \\ 6 & 10 \\ 8 & 12 \end{pmatrix} \tag{9.13}$$

and the standardised version of this is

$$\mathbf{X}_s = \begin{pmatrix} -0.671 & -0.408 \\ -0.224 & -0.408 \\ 0.224 & 0 \\ 0.671 & 0.816 \end{pmatrix} \tag{9.14}$$

where indeed $\mathbf{x}^t{}_{s1}\mathbf{x}_{s1} = 1$, $\mathbf{x}^t{}_{s2}\mathbf{x}_{s2} = 1$ and $\mathbf{x}^t{}_{s1}\mathbf{x}_{s2} = 0.913$ the latter being the correlation between $\mathbf{x}_1$ and $\mathbf{x}_2$. The square representation using Equation 9.12 on $\mathbf{x}_1$ gives

$$\widetilde{\mathbf{S}}_1 = \begin{pmatrix} 0 & -2 & -4 & -6 \\ 2 & 0 & -2 & -4 \\ 4 & 2 & 0 & -2 \\ 6 & 4 & 2 & 0 \end{pmatrix} \tag{9.15}$$

which is skew-symmetric ($\widetilde{\mathbf{S}}_1^t = -\widetilde{\mathbf{S}}_1$) and contains all the differences between the elements of $\mathbf{x}_1$. The standardised version of $\widetilde{\mathbf{S}}_1$ is

$$\mathbf{S}_1 = \begin{pmatrix} 0 & -0.158 & -0.316 & -0.474 \\ 0.158 & 0 & -0.158 & -0.316 \\ 0.316 & 0.158 & 0 & -0.158 \\ 0.474 & 0.316 & 0.158 & 0 \end{pmatrix} \tag{9.16}$$

and a similar matrix can be made for $\mathbf{x}_2$. Then using Equation 9.11 on the pairs $(\mathbf{S}_1, \mathbf{S}_1)$ and $(\mathbf{S}_2, \mathbf{S}_2)$ gives a value of one; and on the pair $(\mathbf{S}_1, \mathbf{S}_2)$ gives 0.913, which is the Pearson's correlation again.
Alternative square representations of $\mathbf{x}_{s1}$ and $\mathbf{x}_{s2}$ are

$$\mathbf{S}_{A1} = \mathbf{x}_{s1}\mathbf{x}^T{}_{s1} = \begin{pmatrix} 0.45 & 0.15 & -0.15 & -0.45 \\ 0.15 & 0.05 & -0.05 & -0.15 \\ -0.15 & -0.05 & 0.05 & 0.15 \\ -0.45 & -0.15 & 0.15 & 0.45 \end{pmatrix} \tag{9.17}$$

and

$$\mathbf{S}_{A2} = \mathbf{x}_{s2}\mathbf{x}^T{}_{s2} = \begin{pmatrix} 0.167 & 0.167 & 0 & -0.333 \\ 0.167 & 0.167 & 0 & -0.333 \\ 0 & 0 & 0 & 0 \\ -0.333 & -0.333 & 0 & 0.667 \end{pmatrix} \tag{9.18}$$

where we include the 'A' in the subscript to indicate this as an alternative. Now, using Equation 9.11 on $\mathbf{S}_{A1}$ and $\mathbf{S}_{A2}$ gives 0.833 which is the squared Pearson's correlation between the original variables.

Representation matrices for nominal-scaled variables

We will discuss the representation matrices for nominal-scaled variables separately for binary data and categorical data. We first discuss representation matrices for categorical data. We have to distinguish two situations: one in which all categorical variables have the same number of categories and the situation where this is not the case. Since the latter is more general and encountered more often, we will restrict ourselves to this case. Then only square representation matrices are available. These are based on indicator matrices (Zegers, 1986; Kiers, 1989; Gifi, 1990). If variable $\mathbf{x}_j$ has four categories (A,B,C,D), then this can be encoded in the rectangular matrix $\mathbf{G}_j(I \times 4)$ where each column $\mathbf{g}_{jk}$ in $\mathbf{G}_j$ represents a category and each row an object. This matrix has only zeros or ones; where g_{ijk} is one, if and only if object i belongs to the category represented by k. The representation matrix can now be built using the products $\mathbf{G}_j\mathbf{G}_j^t(I \times I)$. There are very many versions of such square representation matrices based on indicator matrices and some of them give rise to a known correlation, e.g.,

$$\mathbf{J}\mathbf{G}_j\mathbf{D}_j^{-1}\mathbf{G}_j^t\mathbf{J} \tag{9.19}$$

where $\mathbf{J}(I \times I)$ is the centring operator $(\mathbf{I} - \mathbf{1}\mathbf{1}^t/I)$ with $\mathbf{I}$ being the unity matrix and $\mathbf{1}$ a column of ones and $\mathbf{D}_j(C_j \times C_j)$ is a diagonal matrix containing the marginal frequencies of categories $1, \ldots, C_j$ for variable j. The corresponding correlation coefficient using again Equation 9.11 is the so-called T^2 coefficient (Tschuprow, 1939)[3]. For binary data (if all variables are binary) then rectangular representation matrices are possible. This comes down to the same idea as above, namely, to consider the binary variables as representing two categories. This results then in representation matrices $\mathbf{G}_j$ of sizes $(I \times 2)$. When combining with other types of variables is the goal, then a squared type of representation is needed such as in Equation 9.19. Note that the matrix of Equation 9.19 is positive (semi-)definite.

Using representation matrices in heterogeneous data fusion

To illustrate how to use representation matrices for unsupervised multiblock data analysis, we will work with three data matrices, each on a different measurement scale sharing the same set of I samples. The first matrix $\mathbf{X}_1(I \times J_1)$ contains ratio- or interval-scaled data; the second matrix $\mathbf{X}_2(I \times J_2)$ contains binary data and the last matrix $\mathbf{X}_3(I \times J_3)$ contains categorical data. The general setup of the approach is visualised in Figure 9.1.

The representation matrices $\mathbf{S}_j$ where the variable j is from any of the three blocks can now be used in a three-way model for symmetric data (assuming a symmetric version of $\mathbf{S}_j$). The basic model for a single data block with variables $j = 1, \ldots, J$ is the INDSCAL model:

$$\min_{\mathbf{T},\mathbf{P}_j} \sum_{j=1}^{J} ||\mathbf{S}_j - \mathbf{T}\mathbf{P}_j\mathbf{T}^t||^2 \tag{9.20}$$

3 Not to be confused with Hotelling's T^2.

where $\mathbf{P}_j$ is a diagonal matrix with the jth row of the loadings $\mathbf{P}$ $(J \times R)$ on its diagonal (see Figure 9.1) and the matrix $\mathbf{T}$ $(I \times R)$ contains the object scores. If the additional constraint that $\mathbf{T}^t\mathbf{T} = \mathbf{I}$ is used, then the model is called INDORT (INDscal with ORThogonal constraints, see also Elaboration 9.3) (Kiers, 1989). The scores and loadings have the usual interpretation: scores can be used to study the (dis-)similarities between samples and the loadings show the relationships between the variables.

ELABORATION 9.3

INDSCAL and INDORT

The original INDSCAL model as shown in Equation 9.20 can be used for symmetric data. Then it is often assumed that the loadings $\mathbf{P}_j$[4] are non-negative, but this is not a necessary requirement. In the INDORT model, there is an extra requirement that the scores $\mathbf{T}$ are orthogonal: $\mathbf{T}^t\mathbf{T} = \mathbf{I}$. If it is also assumed that the matrices $\mathbf{S}_j$ are positive (semi-)definite then the loadings become non-negative automatically. This can be seen by rewriting problem 9.20 as

$$\min_{\mathbf{T},\mathbf{P}_j} \sum_{j=1}^{J} ||\mathbf{S}_j - \mathbf{T}\mathbf{P}_j\mathbf{T}^t||^2 = tr(\mathbf{S}_j^2) - tr(\mathbf{T}\mathbf{T}^t\mathbf{S}_j\mathbf{T}\mathbf{T}^t\mathbf{S}_j) - ||\mathbf{T}^t\mathbf{S}_j\mathbf{T} - \mathbf{P}_j||^2. \tag{9.21}$$

and to minimise this function over diagonal matrices $\mathbf{P}_j$ we simply take $\mathbf{P}_j$ to be the diagonal values of $\mathbf{T}^t\mathbf{S}_j\ \mathbf{T}$ which are nonnegative[5].

The INDORT method can now be generalised to analyse all three blocks simultaneously by simply stacking all similarity matrices with square $(I \times I)$ representations on top of each other (see Figure 9.1):

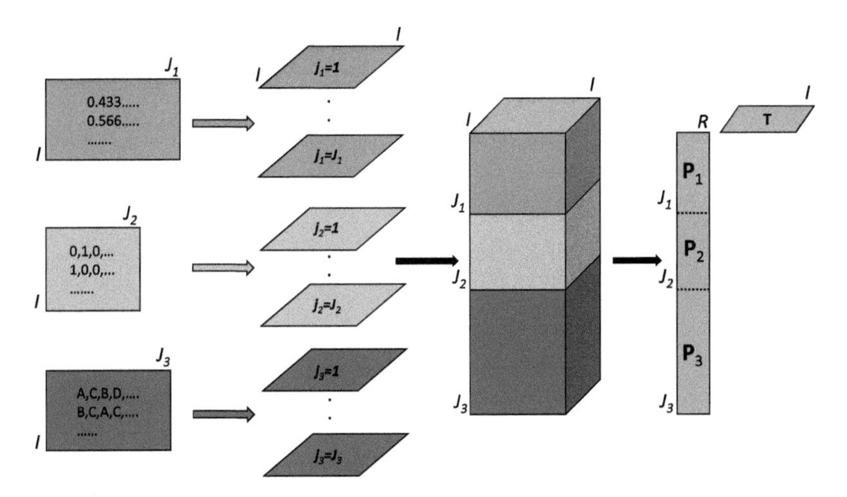

Figure 9.1 General setup for fusing heterogeneous data using representation matrices. The variables in the blocks $\mathbf{X}_1$, $\mathbf{X}_2$ and $\mathbf{X}_3$ are represented with proper $I \times I$ representation matrices which are subsequently analysed simultaneously with an IDIOMIX model generating scores and loadings. Source: Smilde *et al.* (2020). Reproduced with permission of John Wiley and Sons.

4 These are usually called weights in the INDSCAL literature but to avoid confusion we call them loadings
5 We thank Henk Kiers for pointing this out.

$$\min_{\mathbf{T},\mathbf{P}_{j_m}} \sum_{m=1}^{3} \sum_{j_m=1}^{J_m} ||\mathbf{S}_{j_m} - \mathbf{T}\mathbf{P}_{j_m}\mathbf{T}^t||^2 \tag{9.22}$$

where $\mathbf{P}_{j_m}$ is a diagonal matrix with the j_mth row of the loadings $\mathbf{P}_m$ $(J_m \times R)$ on its diagonal and the matrix $\mathbf{T}$ $(I \times R)$ containing the object scores. This model is called IDIOMIX for obvious reasons (Kiers, 1989). Note that the scores $\mathbf{T}$ are consensus scores since they are not located in the column-spaces of the individual matrices $\mathbf{X}_m$. Example 9.2 illustrates various methods in an example from genomics and is an extension of Example 5.3.

Example 9.2: Analysing heterogeneous genomics data

The example for fusing heterogeneous data is a follow-up of the earlier example of GSCA (see Example 5.3). The GSCA approach was already explained earlier. For the representation approach, we built a three-way array of size $160 \times 160 \times (410+1000)$ and performed an IDIOMIX analysis. For the binary part, this array contains the slabs $\mathbf{S}_j$ according to Equation 9.19, and for the gene-expression part the slabs $\mathbf{S}_j$ are defined by the outer products of the samples in the gene-expression data after auto-scaling the columns of that data ($\mathbf{S}_j = \mathbf{s}_j\mathbf{s}_j^t$). Optimal-scaling results are obtained by auto-scaling both raw data sets and subsequently performing

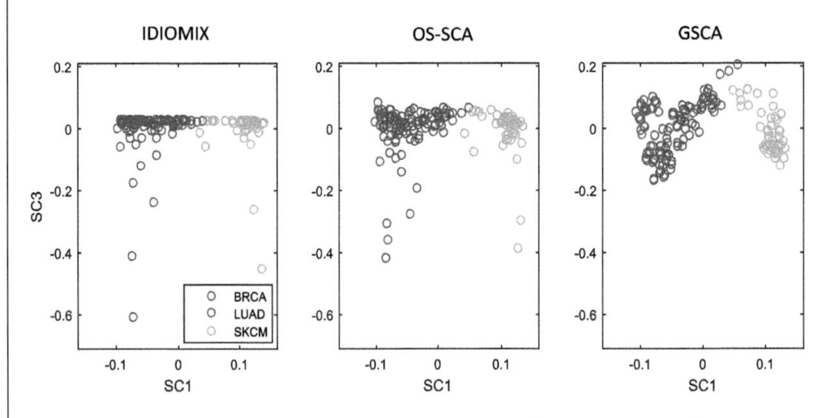

Figure 9.2 Score plots of IDIOMIX, OS-SCA and GSCA for the genomics fusion; always score 3 (SC3) on the y-axes and score 1 (SC1) on the x-axes. The third component clearly differs among the methods. Source: Smilde *et al.* (2020). Licensed under CC BY 4.0.

an (OS-)SCA on the concatenated data (see Section 5.2.1.6). The first two components were very similar in all three approaches and differences showed up in the third component (see Figure 9.2). IDIOMIX and OS-SCA are very similar and GSCA is qualitatively different from the other two. There seems to be a more compact grouping of the different types of cancer in the GSCA results. For a more detailed analysis, we refer to the original paper (Smilde *et al.*, 2020).

9.2.1.4 Extended PCA

The method extended PCA (XPCA) (Anderson-Bergman *et al.*, 2018) is based on copulas. Any J-dimensional distribution can be written as a composite of the marginal distributions of the J variables and a J-dimensional copula (Sklar, 1973). The marginal distributions of the J original variables can be very diverse; both continuous and discrete. XPCA provides a framework for handling these different types of marginal distributions and uses a Gaussian copula upon which a PCA is performed. The elegance of the method lies in its ability to deal with empirical marginal distributions based on the data. These empirical distributions can be estimated from the heterogeneous data and then can be combined using the Gaussian copula. Hence, in the multiblock situation with each block of a different measurement scale, the marginal distributions per block can be tailored according to their measurement scale and then combined using a copula. This method has not yet been applied to data from the natural or life sciences but certainly has potential to also be useful in that area.

9.2.2 Common, Local, and Distinct Variation

For a shared sample mode, several methods exist for separating common, local, and distinct components. Some of them also allow for heterogeneous data.

9.2.2.1 Generalised SVD

The GSVD method goes similarly as in the case of a shared variable mode (see Section 9.1). The main difference is that now there are consensus scores ($\mathbf{T}$) instead of consensus loadings. For convenience, the method is explained for two blocks $\mathbf{X}_1(I \times J_1)$ and $\mathbf{X}_2(I \times J_2)$ of data with a shared sample mode:

$$\begin{aligned} \mathbf{X}_1 &= \widehat{\mathbf{X}}_1 + \mathbf{E}_1 = \mathbf{T}\mathbf{D}_1\mathbf{P}_1^t + \mathbf{E}_1 \\ \mathbf{X}_2 &= \widehat{\mathbf{X}}_2 + \mathbf{E}_2 = \mathbf{T}\mathbf{D}_2\mathbf{P}_2^t + \mathbf{E}_2 \end{aligned} \tag{9.23}$$

where again $\widehat{\mathbf{X}}_m$ are the SCA-filtered data; $\mathbf{T}$ $(I \times R)$ is a full-rank matrix; $\mathbf{P}_m^t\mathbf{P}_m = \mathbf{I}$; $\mathbf{D}_m$ $(m = 1, 2)$ is diagonal and $\mathbf{D}_1^2 + \mathbf{D}_2^2 = \mathbf{I}$. Similarly as in the shared variable mode case (Section 9.1), the elements of $\mathbf{D}_m$ can be divided into three groups and this results in:

$$\begin{aligned} \mathbf{X}_1 &= \mathbf{T}_1\mathbf{D}_{11}\mathbf{P}_{11}^t + \mathbf{T}_2\mathbf{D}_{12}\mathbf{P}_{12}^t + \mathbf{T}_3\mathbf{D}_{13}\mathbf{P}_{13}^t + \mathbf{E}_1 \\ \mathbf{X}_2 &= \mathbf{T}_1\mathbf{D}_{21}\mathbf{P}_{21}^t + \mathbf{T}_2\mathbf{D}_{22}\mathbf{P}_{22}^t + \mathbf{T}_3\mathbf{D}_{23}\mathbf{P}_{23}^t + \mathbf{E}_2 \end{aligned} \tag{9.24}$$

where $\mathbf{D}_{11}$ represents the distinct part of $\mathbf{X}_1$; $\mathbf{D}_{22}$ represents the distinct part of $\mathbf{X}_2$, and $\mathbf{D}_{13}, \mathbf{D}_{23}$ represents the common (actually, consensus) part. The corresponding scores are collected in $\mathbf{T}_1$, $\mathbf{T}_2$, and $\mathbf{T}_3$, and the loadings in the matrices $\mathbf{P}_{kl}$. Ideally, the diagonal matrices $\mathbf{D}_{12}$ and $\mathbf{D}_{21}$ contain (very) small values. There is an extension of GSVD for more than two data blocks (Ponnapalli *et al.*, 2011) but, as said earlier, it is not clear how this can be used to also find local components. An example of GSVD in the `multiblock` R-package is found in Section 11.5.2.

9.2.2.2 Structural Learning and Integrative Decomposition

As already shown in Chapter 5 when presenting the methods ACMTF and PESCA, penalties can be used to invoke certain structures in the sets of parameters. Another method that uses this approach is SLIDE (Structural Learning and Integrative DEcomposition (Gaynanova and Li, 2019)) which is very similar to PESCA but only works for homogeneous data. We will explain this method using three data blocks sharing the sample mode and we will start (again) with an SCA model for three blocks of data:

$$\begin{aligned} \mathbf{X}_1 &= \mathbf{T}\mathbf{P}_1^t + \mathbf{E}_1 \\ \mathbf{X}_2 &= \mathbf{T}\mathbf{P}_2^t + \mathbf{E}_2 \\ \mathbf{X}_3 &= \mathbf{T}\mathbf{P}_3^t + \mathbf{E}_3 \end{aligned} \tag{9.25}$$

and upon using (for example) six components, we can partition the loadings matrices:

$$\mathbf{P} = \begin{bmatrix} \mathbf{P}_1 \\ \mathbf{P}_2 \\ \mathbf{P}_3 \end{bmatrix} = \left[\begin{array}{cccccc} \mathbf{p}_{11} & \mathbf{p}_{12} & \mathbf{p}_{13} & \mathbf{p}_{14} & \mathbf{p}_{15} & \mathbf{p}_{16} \\ \hdashline \mathbf{p}_{21} & \mathbf{p}_{22} & \mathbf{p}_{23} & \mathbf{p}_{24} & \mathbf{p}_{25} & \mathbf{p}_{26} \\ \hdashline \mathbf{p}_{31} & \mathbf{p}_{32} & \mathbf{p}_{33} & \mathbf{p}_{34} & \mathbf{p}_{35} & \mathbf{p}_{36} \end{array}\right]. \tag{9.26}$$

Each partitioned loading vector for block m ($\mathbf{p}_{mr}$) is now considered as a group (r is the index for component and m the index for group). Subsequently, a group penalty can be imposed on the loadings inducing sparsity. Thus the following problem is solved:

$$\min_{\mathbf{T},\mathbf{P}} \sum_m \|\mathbf{X}_m - \mathbf{T}\mathbf{P}_m^t\|^2 + GP(\mathbf{P}) \tag{9.27}$$

where the symbol $GP(\mathbf{P})$ is the group penalty to be discussed below. Hence, this is SCA with a group- or block-wise penalty. Suppose that after fitting a SLIDE model the pattern of the loadings is:

$$\mathbf{P} = \begin{bmatrix} \mathbf{P}_1 \\ \mathbf{P}_2 \\ \mathbf{P}_3 \end{bmatrix} = \left[\begin{array}{c|c|c|c|c|c} \mathbf{x} & \mathbf{x} & 0 & 0 & \mathbf{x} & \mathbf{x} \\ \hdashline \mathbf{x} & \mathbf{x} & \mathbf{x} & 0 & 0 & 0 \\ \hdashline \mathbf{x} & \mathbf{x} & \mathbf{x} & \mathbf{x} & 0 & 0 \end{array}\right] \tag{9.28}$$

where the symbol $\mathbf{x}$ means a non-zero vector. Then it can be concluded that the first two components are common; the third component is local between block 2 and 3; the fourth component is distinct for block 3; and the fifth and sixth are distinct for block 1.

It is important to realise that the method itself finds the number of components and the structure of common, local, and distinct components as a result of the penalisation. This penalisation is on a whole group of loading values within the same partial loading vector $\mathbf{p}_{mr}$. Hence, this whole partial loading vector becomes zero if it is found to be non-relevant. After having found the structure of the model (i.e., in terms of common, local, and distinct components) the model can be fitted again by assuming this structure. There are requirements for the SLIDE decomposition for obtaining uniqueness which are an extension of the ones for JIVE (see Section 5.2.2.1).

There are clearly conceptual similarities between ACMTF and SLIDE: both methods penalise a whole group of loadings either directly (SLIDE) or by the corresponding pseudo-singular values (the diagonal elements of $\mathbf{D}_1$ and $\mathbf{D}_2$ in ACMTF). There are different ways to impose the group-wise penalty (Van Deun *et al.*, 2011). In SLIDE, the sum of Euclidean norms for the partitioned loadings is used. Hence,

$$GP(\mathbf{P}) = \lambda \sum_{r=1}^{R} \sum_{m=1}^{M} \|\mathbf{p}_{mr}\| \tag{9.29}$$

which is a lasso-type penalty on the level of the Euclidean norms of the loading vectors. Hence, this lasso penalty does not work on the *individual elements* of $\mathbf{p}_{mr}$ but on its *Euclidean norm.* If this norm is forced to zero by the penalty, the whole vector $\mathbf{p}_{mr}$ becomes zero. An alternative is to use a concave penalty such as the generalised double Pareto penalty (GDP) (Song *et al.*,

2019) (see also Elaboration 5.12 for insight in different types of penalties). Applications are reported in cancer genomics (Gaynanova and Li, 2019).

9.2.2.3 Bayesian Inter-battery Factor Analysis

The Bayesian inter-battery factor analysis (BIBFA (Klami *et al.*, 2013)) model builds on the original inter-battery factor analysis model (Tucker, 1958) and the probabilistic version of that method (Browne, 1979). The model was developed for two blocks of data and can be written as:

$$\begin{aligned}
\mathbf{t} &\sim N(0, \mathbf{I}) \\
\mathbf{t}_m &\sim N(0, \mathbf{I}) \\
\mathbf{x}_m &\sim N(\mathbf{P}_{mC}\mathbf{t} + \mathbf{P}_{mD}\mathbf{t}_m, \mathbf{\Sigma}_m); \; m = 1, 2
\end{aligned} \tag{9.30}$$

where the $\mathbf{\Sigma}_m$ are diagonal matrices indicating independence of the variables between and within both blocks conditional on the structural model (as is customary in factor analysis models). This model shows that the measurement vector $\mathbf{x}_m$ has a common part ($\mathbf{P}_{mC}\mathbf{t}$) and a distinct part ($\mathbf{P}_{mD}\mathbf{t}_m$) where the matrices $\mathbf{P}$ are the loadings and the vectors $\mathbf{t}, \mathbf{t}_m$ are the scores. Note that actually the vectors $\mathbf{t}$ represent the consensus scores since these are not necessarily in the column-spaces of the individual matrices. For the two blocks, this model can be reparametrised by defining

$$\begin{aligned}
\mathbf{x} &= [\mathbf{x}_1^t | \mathbf{x}_2^t]^t \\
\mathbf{z} &= [\mathbf{t}^t | \mathbf{t}_1^t | \mathbf{t}_2^t]^t
\end{aligned} \tag{9.31}$$

and defining also

$$\mathbf{P} = \begin{bmatrix} \mathbf{P}_{1C} & \mathbf{P}_{1D} & 0 \\ \mathbf{P}_{2C} & 0 & \mathbf{P}_{2D} \end{bmatrix} \tag{9.32}$$

and

$$\mathbf{\Sigma} = \begin{bmatrix} \mathbf{\Sigma}_1 & 0 \\ 0 & \mathbf{\Sigma}_2 \end{bmatrix}. \tag{9.33}$$

Now we can write Equation 9.30 as

$$\begin{aligned}
\mathbf{z} &\sim N(0, \mathbf{I}) \\
\mathbf{x} &\sim N(\mathbf{P}\mathbf{z}, \mathbf{\Sigma})
\end{aligned} \tag{9.34}$$

which is seen to be factor analysis with restrictions. The parameters are estimated in a Bayesian framework by putting priors on the parameters and within-block variances. A group-wise prior based on automatic relevance determination (Neal, 2012) is used to arrive at the structure of Equation 9.32 including the number of components. Hence, the method resembles SLIDE apart from the estimation method. Applications are reported in genomics (Klami *et al.*, 2013) and in analytical chemistry (Acar *et al.*, 2015) (see Example 9.3). This method requires some experience in setting the priors (e.g., to arrive at a sensible number of components), some recommendations are given (Klami *et al.*, 2013). Identifiability of the model parameters should come from these imposed priors.

Example 9.3: Example of BIBFA and ACMTF

This example is a follow-up of Example 5.6 with the same experimental data. The methods BIBFA and ACMTF were tested on a real example from analytical chemistry (Acar *et al.*, 2014, 2015) to see to what extent they are able to retrieve the underlying profiles. Five chemicals were used: two peptides (Val-Tyr-Val and Trp-Gly), a single amino acid (Phe), a sugar (Malto), and an alcohol (Propanol), and 29 samples were prepared with varying concentrations according to a predetermined design. The method BIBFA cannot handle three-way data, thus the NMR-spectra were matricised keeping the sampling mode intact. From chemistry, it is known that Propanol will not give a signal in LC-MS. Hence, we expect four common components and one distinct component for the NMR block.

The selection of the number of components is a crucial issue. ACMTF returned six total components. BIBFA returned six components which were all recognised as common. Increasing the number of total components to ten in BIBFA did not help: they were all assigned to be in common. Figure 9.3 shows the recovery of the sample concentrations for BIBFA and ACMTF. The recovery of BIBFA is reasonable, but worse than ACMTF. This also illustrates one of the difficulties of applying multiblock data analysis methods. There is always an element of trial and error and especially finding a proper number of components is never trivial. For a more detailed comparison, see the original publication (Acar *et al.*, 2015).

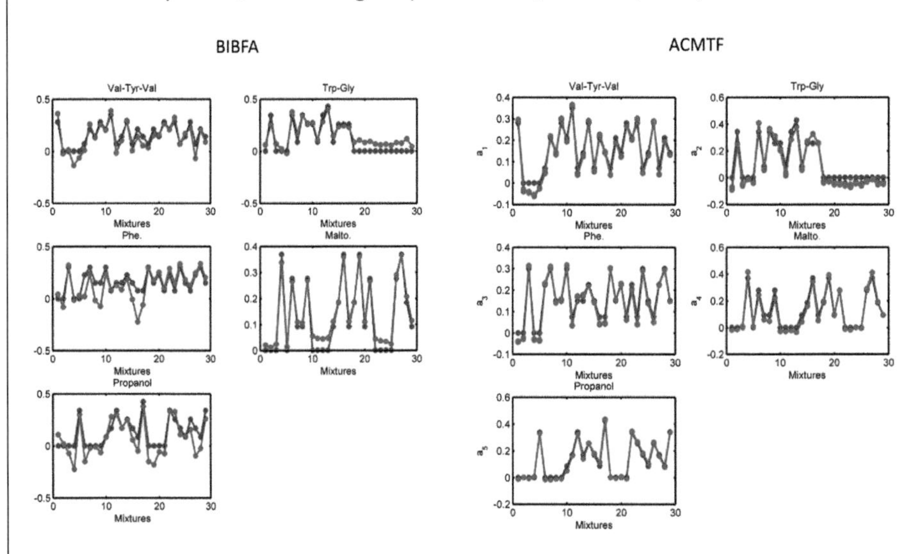

Figure 9.3 True design used in mixture preparation (blue) versus the columns of associated factor matrix corresponding to the mixture mode extracted by the BIBFA model (red) and the ACMTF model (red). Source: Acar *et al.* (2015). Reproduced with permission from IEEE.

9.2.2.4 Group Factor Analysis

Group factor analysis (GFA) is a machine learning method which can be seen as an extension of BIBFA (Virtanen *et al.*, 2012). It starts with the basic factor analysis model:

$$\mathbf{X} = \mathbf{T}\mathbf{P}^t + \mathbf{E} \tag{9.35}$$

and upon assuming that $\mathbf{X}$ is the concatenation of $[\mathbf{X}_1|\ldots|\mathbf{X}_M]$ the GFA model can be written as

$$[\mathbf{X}_1|\ldots|\mathbf{X}_M] = \mathbf{T}[\mathbf{P}_1|\ldots|\mathbf{P}_M]^t + [\mathbf{E}_1|\ldots|\mathbf{E}_M] \tag{9.36}$$

which has the same structure as the SLIDE model. For simplicity we use the symbols $\mathbf{T}$ and $\mathbf{P}$ here, but note that these have a different status than in SLIDE. In this case, a fully stochastic model is assumed where the latent variables, as collected in $\mathbf{T}$, are assumed Gaussian distributed. This is a very strong assumption for data in the natural and life sciences since there is usually structure among the samples (e.g., an experimental design). The residuals per block are assumed to be independent normally distributed with equal variances within the blocks but this variance may differ between the blocks. Similar to BIBFA, Bayesian maximum-likelihood methods are used to estimate the parameters given certain priors, and the common, local, and distinct components are found by using a group-wise ARD prior which should also generate identifiable models. Applications are reported for genomics and fMRI (neurobiology) (Virtanen *et al.*, 2012).

9.2.2.5 OnPLS

A sequential method for finding common and distinct components is called O2PLS (Trygg and Wold, 2003) and its extension OnPLS (Löfstedt and Trygg, 2011). We will first discuss O2PLS.

O2PLS has its root in OSC filters and is thus part of the family of methods such as O-PLS (Trygg and Wold, 2002). For two blocks $\mathbf{X}_1$ and $\mathbf{X}_2$ sharing the sample mode the total model can be written as:

$$\begin{aligned}
\mathbf{X}_1 &= \mathbf{T}_{1C}\mathbf{P}_{1C}^t + \mathbf{T}_{1D}\mathbf{P}_{1D}^t + \mathbf{E}_1 = \mathbf{X}_{1C} + \mathbf{X}_{1D} + \mathbf{E}_1 \\
\mathbf{X}_2 &= \mathbf{T}_{2C}\mathbf{P}_{2C}^t + \mathbf{T}_{2D}\mathbf{P}_{2D}^t + \mathbf{E}_2 = \mathbf{X}_{2C} + \mathbf{X}_{2D} + \mathbf{E}_1 \\
\widehat{\mathbf{T}}_{1C} &= \mathbf{T}_{2C}\mathbf{B}_1 + \mathbf{E}_{12} \\
\widehat{\mathbf{T}}_{2C} &= \mathbf{T}_{1C}\mathbf{B}_2 + \mathbf{E}_{12} \\
\widehat{\mathbf{X}}_1 &= \mathbf{T}_{2C}\mathbf{B}_1\mathbf{P}_{1C}^t \\
\widehat{\mathbf{X}}_2 &= \mathbf{T}_{1C}\mathbf{B}_2\mathbf{P}_{2C}^t
\end{aligned} \tag{9.37}$$

where in the original publication the parts $\mathbf{T}_{1D}\mathbf{P}_{1D}^t$ and $\mathbf{T}_{2D}\mathbf{P}_{2D}^t$ were referred to as structural noise components, but these can also be interpreted as distinct components. The subscripts refer to the data block and whether the part is common (C) or distinct (D). The part of $\mathbf{X}_1$ which can be predicted by $\mathbf{X}_2$ is given by the common part: $\widehat{\mathbf{X}}_1 = \widehat{\mathbf{T}}_{1C}\mathbf{P}_{1C}^t = \mathbf{T}_{2C}\mathbf{B}_1\mathbf{P}_{1C}^t$ and this explains the last two lines of Equation 9.37.

There is similarity between O2PLS and canonical correlation which is evident from the relationships in lines three and four of Equation 9.37. The scores, weights, and loadings of this model are obtained by an iterative algorithm much in the spirit of PLS. Applications are reported in integrating omics data (gene-expression and metabolomics) (El Bouhaddani *et al.*, 2016; Bylesjö *et al.*, 2007) and in spectroscopy-based metabolomics data (Anđelković *et al.*, 2017).

The method OnPLS is a generalisation of O2PLS for more than two blocks of data (Löfstedt and Trygg, 2011). For three data blocks, the end result of the method is a model of the form:

$$\begin{aligned} \mathbf{X}_1 &= \mathbf{T}_{1C}\mathbf{P}^t_{1C} + \mathbf{T}_{12L}\mathbf{P}^t_{12L} + \mathbf{T}_{13L}\mathbf{P}^t_{13L} + \mathbf{T}_{1D}\mathbf{P}^t_{1D} + \mathbf{E}_1 \\ \mathbf{X}_2 &= \mathbf{T}_{2C}\mathbf{P}^t_{2C} + \mathbf{T}_{21L}\mathbf{P}^t_{21L} + \mathbf{T}_{23L}\mathbf{P}^t_{23L} + \mathbf{T}_{2D}\mathbf{P}^t_{2D} + \mathbf{E}_2 \\ \mathbf{X}_3 &= \mathbf{T}_{3C}\mathbf{P}^t_{3C} + \mathbf{T}_{31L}\mathbf{P}^t_{31L} + \mathbf{T}_{32L}\mathbf{P}^t_{32L} + \mathbf{T}_{3D}\mathbf{P}^t_{3D} + \mathbf{E}_3 \end{aligned} \tag{9.38}$$

where the subscripts on the scores and loadings indicate whether the components are common (C), local (L) or distinct (D) (the subscript ijL means that this is a local component for block i and j). The algorithm for arriving at this solution is complicated and can be found in Löfstedt and Trygg (2011).

9.2.2.6 Generalised Association Study

The method published as generalised association study (GAS) is an extension of JIVE combined with exponential family distributions and thus capable of handling heterogeneous data (Li and Gaynanova, 2018). It was already introduced in Chapter 4 since it has a matrix correlation associated with it. The structural model is:

$$\begin{aligned} \boldsymbol{\Theta}_1 &= \mathbf{1}\boldsymbol{\mu}^t_1 + \mathbf{T}_C\mathbf{P}^t_{1C} + \mathbf{T}_{1D}\mathbf{P}^t_{1D} \\ \boldsymbol{\Theta}_2 &= \mathbf{1}\boldsymbol{\mu}^t_2 + \mathbf{T}_C\mathbf{P}^t_{2C} + \mathbf{T}_{2D}\mathbf{P}^t_{2D} \end{aligned} \tag{9.39}$$

where the $\boldsymbol{\Theta}_1$ and $\boldsymbol{\Theta}_2$ are the underlying parameters of the data sets $\mathbf{X}_1$ and $\mathbf{X}_2$ (see GSCA, ESCA and PESCA). This is the same structural model as JIVE, and also in this case the dimensionalities of the matrices $\mathbf{T}_C$, $\mathbf{T}_{1D}$, and $\mathbf{T}_{2D}$ have to be assumed ***a priori***. Identifiability restrictions are required for the scores and loadings matrices much in the same way as for JIVE. The full model including the stochastic part now becomes:

$$\begin{aligned} \mathbf{X}_1 &= \boldsymbol{\Theta}_1 + \mathbf{E}_1 \\ \mathbf{X}_2 &= \boldsymbol{\Theta}_2 + \mathbf{E}_2 \end{aligned} \tag{9.40}$$

and the error terms $\mathbf{E}_1$ and $\mathbf{E}_2$ can be assumed to stem from different distributions. Hence, GAS can handle heterogeneous data. Estimation is done using a block-descent algorithm alternately estimating the common and distinct parts within a generalised linear model (GLM) context under the constraints of identifiability and the proper distributional assumptions for the error terms. Like JIVE, the GAS model is not able to find local components if more than two blocks are available.

9.2.2.7 Multi-Omics Factor Analysis

Multi-omics factor analysis (MOFA) is an extension of the BIBFA and GFA models (see Sections 9.2.2.3 and 9.2.2.4) and can handle heterogeneous data (Argelaguet *et al.*, 2018). The generic equation of the MOFA model is:

$$\mathbf{X}_m = \mathbf{T}\mathbf{P}^t_m + \mathbf{E}_m;\ m = 1, \ldots, M \tag{9.41}$$

where $\mathbf{T}(I \times R)$ carries all latent variables (common, local, and distinct). This model is estimated within a Bayesian context with priors on the loadings $\mathbf{P}_m$, on the $\mathbf{T}$ and on the distributions of the noise terms in $\mathbf{E}_m$. Different levels of sparsity are imposed on the weights: group sparsity and feature sparsity. This is performed by using the ARD prior and a spike-and-slab prior for the feature-wise sparsity (Argelaguet *et al.*, 2018). Because estimation is done in a full probabilistic setting, different distributions can be used thereby accommodating different types of data. This is advanced Bayesian modelling and requires good knowledge of such methods. For three blocks of data, the final model may then look like:

$$\begin{aligned}\mathbf{X}_1 &= \mathbf{T}\mathbf{P}_1^t + \mathbf{E}_1\\ \mathbf{X}_2 &= \mathbf{T}\mathbf{P}_2^t + \mathbf{E}_2\\ \mathbf{X}_3 &= \mathbf{T}\mathbf{P}_3^t + \mathbf{E}_3\end{aligned} \tag{9.42}$$

with

$$\mathbf{P}_1 = \begin{bmatrix} 0 & x & x & 0 & 0\\ 0 & x & x & 0 & 0\\ 0 & 0 & x & 0 & 0\\ 0 & 0 & 0 & 0 & 0\\ 0 & x & x & 0 & 0 \end{bmatrix}$$

$$\mathbf{P}_2 = \begin{bmatrix} x & 0 & x & x & 0\\ x & 0 & x & 0 & 0\\ 0 & 0 & x & 0 & 0\\ 0 & 0 & 0 & x & 0\\ 0 & 0 & 0 & x & 0\\ x & 0 & 0 & 0 & 0 \end{bmatrix}$$

$$\mathbf{P}_3 = \begin{bmatrix} 0 & 0 & x & x & x\\ 0 & 0 & x & x & x\\ 0 & x & x & 0 & 0\\ 0 & x & 0 & 0 & 0\\ 0 & x & 0 & 0 & x\\ 0 & x & 0 & 0 & x\\ 0 & 0 & 0 & x & 0\\ 0 & 0 & x & x & 0 \end{bmatrix} \tag{9.43}$$

where an ‘x’ means a non-zero scalar. This shows that component one is distinct for block two; component two is local for blocks one and three; component three is common; component four is local for blocks two and three and component five is distinct for block three. Moreover, all components are sparse. It is important to realise that the structure of **P** is learned from the data (similar to SLIDE and PESCA) and not imposed *a priori* (such as in DISCO models). Hence, the **P** is a result of the estimation procedure. An extensive comparison between MOFA and PESCA was made and some results are reported in Example 9.4.

Example 9.4: PESCA versus MOFA

The example is from the MOFA paper (Argelaguet *et al.*, 2018) and reworked in the PESCA paper (Song *et al.*, 2019). The data are from chronic lymphocytic leukaemia (CLL) (Dietrich *et al.*, 2018) and consist of drug response measurements, mutation status, transcriptomics profiling, and DNA methylation data of around 200 samples with missing values (both PESCA and MOFA can handle missing values). Of those data, the mutation status is binary, the other blocks are quantitative (methylation data can be expressed as binary data but also as quantitative data).

The MOFA model was used as in the original paper (Argelaguet *et al.*, 2018) with a threshold of 2% for explained variance: a component is only kept if it explains at least 2% in one of the data blocks. The PESCA model was used as explained in Section 5.2.2.5 using cross-validation to select the penalty parameters λ_{bin} and λ_{quan}. A grid search was performed

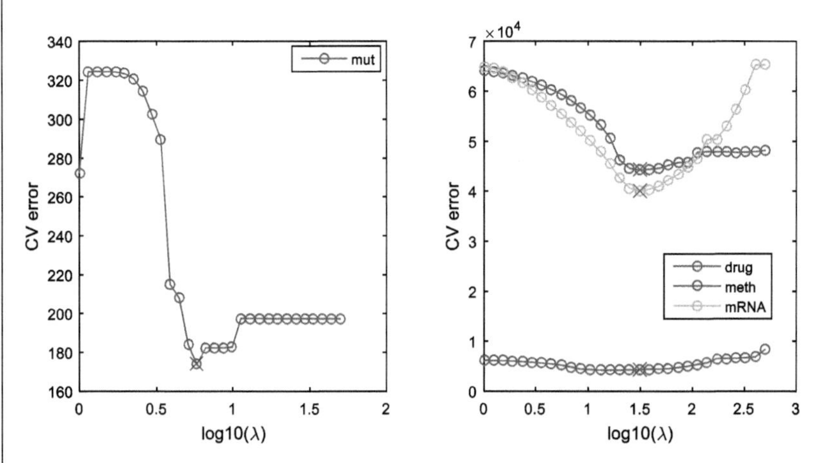

Figure 9.4 Cross-validation results for the penalty parameter λ_{bin} of the mutation block (left) and for the drug response, transcriptome, and methylation blocks (λ_{quan}, right) in the PESCA model. More explanation, see text. Adapted from Song *et al.* (2019).

on these penalty parameters which resulted in the CV-values as shown in Figure 9.4. From these plots, optimal penalty parameters were chosen and used in the PESCA model. A value of $\gamma = 1$ was chosen for the GDP penalty. The selected number of components for the PESCA model was 41; and for the MOFA model only 10. When the same threshold of 2% explained variation was used for the PESCA model, then 17 components remained. Hence, the PESCA model was a bit more complex than the MOFA model. The explained variations are shown in Figure 9.5.

There is some overlap between the two models. Both models have one strong consensus component and local components where two (PESCA) and three (MOFA) blocks participate. The consensus component is probably a predictive relationship between transcriptome and drug response; whereby the methylation and mutation contribute through the transcriptome (see also Example 4.2). Moreover, the drug response and transcriptome have many distinct components in PESCA. The main difference is in the binary part where PESCA finds one component (for the mutation) and MOFA finds two components. The PESCA model is more complex but if the threshold for the MOFA model is set to zero, then a 50 component model is found for MOFA.

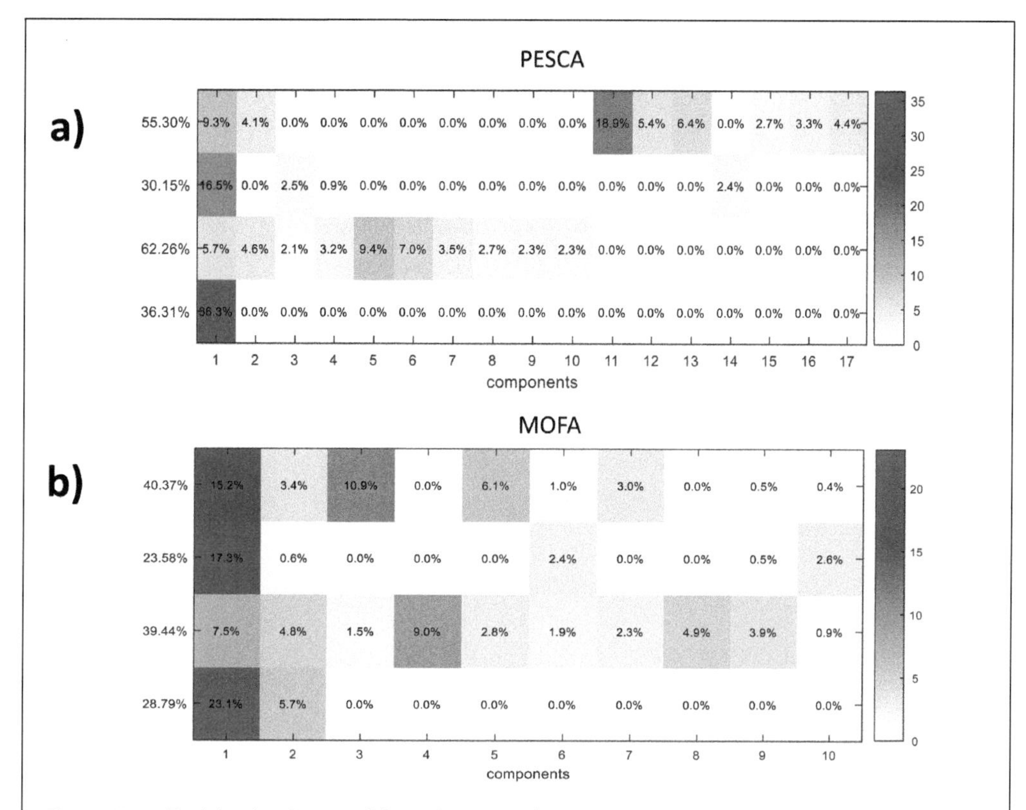

Figure 9.5 Explained variances of the PESCA (a) and MOFA (b) model on the CCL data. From top to bottom: drug response, methylation, transcriptome, and mutation data. The values are percentages of explained variation. More explanation, see text. Adapted from Song *et al.* (2019).

There are many components with a low amount of explained variation. It is questionable whether these components are stable or meaningful. In general it would be advisable to have less complex models which can be enforced by building in a more strict threshold for the minimum of explained variation.

9.3 Two Shared Modes and Only Common Variation

When multiple blocks of data are available with two shared modes, then special methods exist to analyse such data. This typically means that a set of the same samples and the same variables are measured, and then measurements are repeated over different occasions. The method of choice depends on the nature of the data and the research questions asked.

In a sensory example, it may be that different trained assessors assess the same products (mode 1) using the same sensory characteristics (mode 2). The question is then whether the assessors can be 'matched' in such a way that consensus between them is reached. This problem can be solved by generalised procrustes analysis (GPA, Gower and Dijksterhuis (2004)).

Another example is from chemistry where a sample is measured using excitation (mode 1) emission (mode 2) fluorescence spectroscopy. If multiple samples are measured using the same instrumental setup, then the samples can be analysed simultaneously to estimate the underlying concentrations of the fluorescing chemical components. This is a three-way analysis problem which can be solved with three-way methods. We will discuss both types of methods briefly in the following.

9.3.1 Generalised Procrustes Analysis

Suppose that there are M blocks of data (e.g., M assessors in a sensory experiment, assessing I samples with J attributes). The assessors may typically use the scales differently and also the attribute terms slightly differently. There is a need to find a panel consensus to be used for interpretation and further analysis. The generalised procrustes analysis (GPA) pertains to matching the configurations of the samples across the M blocks as good as possible regarding translation, dilation, and rotation. The first part – translation – can be ensured by column-centring each block of data. The dilation and rotation part can be accounted for by solving:

$$\min_{\mathbf{Q}'s,\lambda's,\mathbf{V}} \sum_{m=1}^{M} \|\lambda_m \mathbf{X}_m \mathbf{Q}_m - \mathbf{V}\|^2 \tag{9.44}$$

where the $\mathbf{Q}$s are rotation matrices and λs > 0 are dilation (isotropic scaling) constants. There exists efficient algorithms to solve this problem (Gower and Dijksterhuis, 2004). The matrix $\mathbf{V}$ represents the consensus between the assessors and can be studied with a subsequent PCA. GPA has also found its way to completely different areas of science, such as genomics (Xiong *et al.*, 2008). An example of GPA in the `multiblock` R-package is found in Section 11.6.3.

9.3.2 Three-way Methods

Three-way methods can be considered as special cases of multiblock methods with a specific topology (see Chapter 3). The three-way problem is visualised in Figure 9.6 and is concerned with finding a low-dimensional representation of the three-way array; a generalisation of PCA to three-way arrays. There are many methods to analyse such data; these are called three-way data analysis methods or, more generally, multiway data analysis methods. The two most

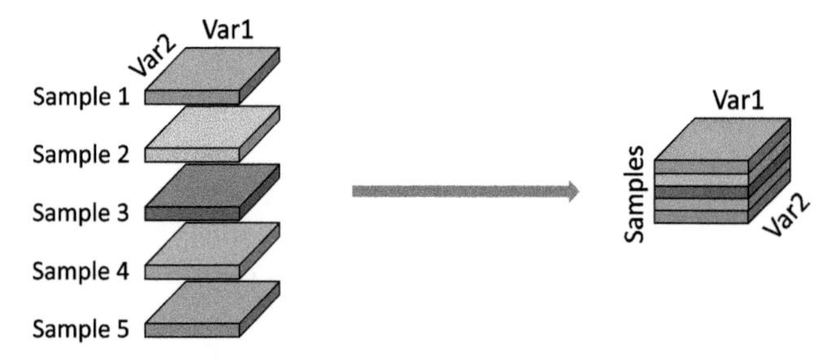

Figure 9.6 From multiblock data to three-way data.

used methods in this field are PARAFAC (Harshman, 1970) (also called CANDECOMP (Carroll and Chang, 1970)) and Tucker3 (Tucker, 1964) models. After concatenating the different blocks $\mathbf{X}_m$ $(I \times J)$ as $\mathbf{X} = [\mathbf{X}_1|, \ldots, |\mathbf{X}_M]$, the PARAFAC model is:

$$\mathbf{X} = \mathbf{A}(\mathbf{C} \odot \mathbf{B})^t + \mathbf{E} \tag{9.45}$$

where the matrices **A**, **B**, and **C** are the loading matrices of sizes $(I \times R)$, $(J \times R)$, and $(M \times R)$, respectively and $\odot$ is the Khatri-Rao product (see Section 1.10). An alternative way of writing the PARAFAC model is

$$\mathbf{X}_m = \mathbf{A}\mathbf{D}_m\mathbf{B}^t + \mathbf{E} \; ; \; m = 1, \ldots, M \tag{9.46}$$

where $\mathbf{D}_m$ is a diagonal matrix with the mth row of **C** on its diagonal. This way of writing shows the similarity of PARAFAC models and SCA models. For PARAFAC models, there are also methods that can handle non-quantitative data (Hong *et al.*, 2020).

The Tucker3 model is a bit more complicated and uses the core-array $\underline{\mathbf{G}}$ $(R_1 \times R_2 \times R_3)$ and its matricised version $\mathbf{G}$ $(R_1 \times R_2R_3)$:

$$\mathbf{X} = \mathbf{A}\mathbf{G}(\mathbf{C} \otimes \mathbf{B})^t + \mathbf{E} \tag{9.47}$$

where the loading matrices $\mathbf{A}(I \times R_1)$, $\mathbf{B}(I \times R_2)$, and $\mathbf{C}(I \times R_3)$ are different from the PARAFAC ones and can also have different numbers of components. The symbol $\otimes$ is the Kronecker product (see Section 1.10).

The Tucker3 model reduces all the modes of the data. It is also possible to reduce only two modes (Tucker2 models) or only one mode, the Tucker1 model. The latter then becomes:

$$\mathbf{X} = \mathbf{A}\mathbf{P}^t + \mathbf{E} \tag{9.48}$$

where only the first mode is reduced (for simplicity we use the symbol **E** throughout for all these models, but the actual residuals differ). This model is equivalent to the SCA model (see Equation 5.34). There are efficient algorithms to estimate the model parameters and monographs exist (Smilde *et al.*, 2004; Kroonenberg, 2008). Hence, we will not further discuss these methods.

9.4 Conclusions and Recommendations

This chapter contained a number of different ways on how to deal with unsupervised multiblock methods. As already mentioned in the introduction and shown in Table 9.1, there are methods based on singular value decompositions (GSVD), on copulas (XPCA), on sequential algorithms (DIABLO, OnPLS), and on model based methods. The methods SLIDE, BIBFA, GFA, GAS, and MOFA are all based on a similar structural model:

$$\begin{aligned}
\mathbf{X}_1 &= \mathbf{T}_{1C}\mathbf{P}_{1C}^t + [\mathbf{X}_{1Local}] + \mathbf{T}_{1D}\mathbf{P}_{1D}^t + \mathbf{E}_1 \\
\mathbf{X}_2 &= \mathbf{T}_{2C}\mathbf{P}_{2C}^t + [\mathbf{X}_{2Local}] + \mathbf{T}_{2D}\mathbf{P}_{2D}^t + \mathbf{E}_2 \\
\mathbf{X}_3 &= \mathbf{T}_{3C}\mathbf{P}_{3C}^t + [\mathbf{X}_{3Local}] + \mathbf{T}_{3D}\mathbf{P}_{3D}^t + \mathbf{E}_3
\end{aligned} \tag{9.49}$$

and the differences between the methods are the assumptions made regarding the distributions of **T** and **E**; the type of estimation and way of implementing penalties. Some methods allow for only separating distinct and common components (BIBFA and GAS) and some methods allow also for finding local components (indicated with the terms $[\mathbf{X}_{mLocal}]$, see also Table 9.1). Note that BIBFA can only handle two blocks of data. An overview of the methods is given in Figure 9.7.

To give recommendations regarding which method to use for a specific application, the following aspects should be considered:

Shared mode: which modes are shared in the multiblock data set?
Number of blocks: how many blocks are in the multiblock data set?
Measurement scales: are the data homogeneous or heterogeneous?
Common/distinct: is a separation between common, local and distinct components required?

The first distinction comes with the aspect of which modes are shared in the multiblock data. The options are: shared variables (GSVD), shared samples (most other methods) and both shared variables and samples (GPA, PARAFAC, and Tucker). Some of the methods only work for two blocks (BIBFA and GAS). There is also a clear distinction between methods that can/-cannot deal with heterogeneous data (see Table 9.1). The final point of decision is whether the analysis requires separating common from local and distinct variation.

As an example, if there are more than two blocks of homogeneous data for which common, local, and distinct components are sought then the alternatives are SLIDE, GFA, and OnPLS. Of these, SLIDE and GFA are very similar, but OnPLS is very different. SLIDE and GFA require tuning parameters to be set by the user and OnPLS requires selecting the number of common, local and distinct components. The main difference is that SLIDE and GFA are simultaneous methods whereas OnPLS is a sequential method. The choice is then a matter of taste.

9.4.1 Open Issues

Apart from the open issues already mentioned in Chapter 5 (see Section 5.5) there are a few more in this chapter.

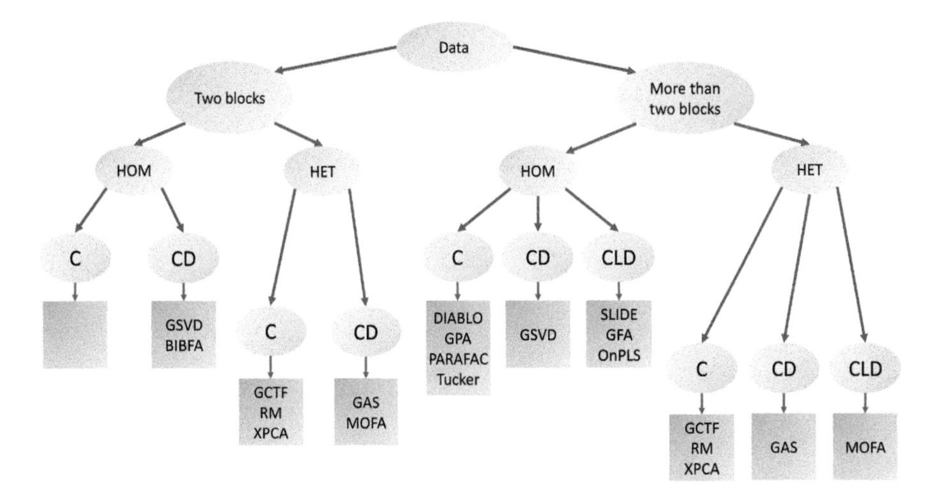

Figure 9.7 Decision tree for selecting an unsupervised method. For abbreviations, see the legend of Table 9.1. The furthest left leaf is empty but also CD methods can be used in that case. For more explanation, see text.

Priors and penalties

Some of the methods in this chapter (e.g., MOFA) make extensive use of Bayesian estimation using priors and hyper priors and advanced types of penalties. This requires quite extensive experience from the user in selecting these priors and how to judge and validate the results. Moreover, there are many types of penalties which may be used, and it is not yet clear which type of penalty to use in which application.

Properties of the estimated parameters

Due to the complexity of some of the methods in this chapter it is not always clear what the properties of the estimated scores and loadings are. These can be properties in terms of stability under resampling conditions or in terms of subspaces in which they are located. Also, for some methods it is not clear whether the solution in terms of scores and loadings is fully identified.

10

Alternative Supervised Methods

10.i General Introduction

We refer to Chapter 7 for notation and a more comprehensive introduction to supervised methodology. The main difference between this chapter and Chapter 7 is that in Chapter 7 only PLS-based methods, which are closely linked through the diagram in Figure 7.1, are considered. In this chapter on the other hand, we discuss a number of alternative optimisation criteria and also methods for situations where the samples are split into subgroups. Some of the methods described are well known, while others are new and less used. A subsection on various extensions of the SO-PLS method described in Chapter 7 is also included.

10.ii Relations to the General Framework

The relation to the general framework in Chapter 1 is briefly described in Table 10.1.

Table 10.1 Overview of methods. Legend: U=unsupervised, S=supervised, C=complex, HOM=homogeneous data, HET=heterogeneous data, SEQ=sequential, SIM=simultaneous, MOD=model-based, ALG= algorithm-based, C=common, CD=common/distinct, CLD=common/local/distinct, LS=least squares, ML=maximum likelihood, ED=eigendecomposition, MC=maximising correlations/covariances. The abbreviations for the methods follow the same order as the sections. For abbreviations (or descriptions) of the methods, see Section 1.11.

		A			B		C		D		E			F			
	Section	U	S	C	HOM	HET	SEQ	SIM	MOD	ALG	C	CD	CLD	LS	ML	ED	MC
Sparse MBCovR	10.2.1																
MWMBCovR	10.2.2																
MB-RDA	10.3.1																
MBVarPart	10.4.1																
NI-SL	10.4.2																
ComDim	10.4.3																
Multigroup PLS	10.6.1																
ClustMBR	10.6.2																
DI-PLS	10.6.3																

Multiblock Data Fusion in Statistics and Machine Learning: Applications in the Natural and Life Sciences,
First Edition. Age K. Smilde, Tormod Næs, and Kristian Hovde Liland.

The topology given attention here is the one in Figure 3.10, i.e., the supervised topology with shared sample mode. For a few of the methods the data set is split along the sample mode to allow for heterogeneity among the samples. But also in these cases, the main focus is the relation between the **X** (or **X**'s) and the **Y**. All methods except two are sequential. The only model assumption applied is an underlying assumed linear relationship between input and output. Finding common predictive information in the data blocks is important for all methods (see also Chapter 7). Maximum likelihood is not involved in any of the methods listed, only least squares and maximising covariance and variance. Maximum likelihood is, however, used in one of the extensions of SO-PLS described in Section 10.5.

10.1 Model and Focus

As in Chapter 7, only linear relations between input data and responses will be considered. The number of samples in all data blocks is equal to I, the number of columns in the output block $\mathbf{Y}$ is equal to J and the number of columns in the input blocks are $J_1, J_2, \ldots, J_M$. For two of the methods to be discussed, the **X**-blocks are allowed to have a three-way structure (see Figure 3.8). It will be assumed throughout this chapter that all blocks are centred and, therefore, the intercept is omitted from the equations. Like for standard PLS regression, no special modification is attempted for handling heteroscedastic and correlated errors. The only exception is the extension of SO-PLS presented in Section 10.5.3. All variables are assumed to be quantitative (i.e., homogeneous fusion).

As in Chapter 7, the focus will be on both prediction ability and interpretation. Since the methods are based on different criteria they will also provide different interpretation tools allowing for inspection of the data from different perspectives.

10.2 Extension of PCovR

Principal covariates regression is an alternative to PLS and already described in Section 2.1.7. Based on PCovR several extensions of multiblock methods are developed: sparse multiblock principal covariates regression and multiblock multiway covariates regression. These are presented in this section.

10.2.1 Sparse Multiblock Principal Covariates Regression, Sparse PCovR

The PCovR framework allows for extensions of sparse modelling in combination with finding common and distinctive predictive variation for a set of response variables **Y**. As such, it is an alternative to PO-PLS (see Section 7.4). We will explain the method using three predictor blocks ($\mathbf{X}_1(I \times J_1)$, $\mathbf{X}_2(I \times J_2)$ and $\mathbf{X}_3(I \times J_3)$) and one univariate response $\mathbf{y}(I \times 1)$. All blocks are assumed to be properly preprocessed (centred and possibly within- and between-block scaled). Upon defining $\mathbf{X} = [\mathbf{X}_1|\mathbf{X}_2|\mathbf{X}_3]$, $\mathbf{W} = [\mathbf{W}_1^t|\mathbf{W}_2^t|\mathbf{W}_3^t]^t$ and $\mathbf{P} = [\mathbf{P}_1^t|\mathbf{P}_2^t|\mathbf{P}_3^t]^t$, the multiblock principal covariates regression model is found by solving

$$\min_{\mathbf{W}} \left[\alpha\|\mathbf{X} - \mathbf{XWP}^t\|^2 + (1-\alpha)\|\mathbf{y} - \mathbf{XWq}\|^2\right] \tag{10.1}$$

where again $0 \leqslant \alpha \leqslant 1$ is a predefined value. The $\mathbf{q}$ and $\mathbf{P}$ are found by LS of the two terms separately. It is convenient to normalise the solution, e.g., by giving all columns of $\mathbf{T} = \mathbf{XW}$ length one.

Much in the spirit of sparse SCA (see Section 2.1.11), SLIDE (see Section 9.2.2.2), PESCA (see Section 5.2.2.5), and MOFA (see Section 9.2.2.7), we can now make a sparse version by imposing penalties on the weights $\mathbf{W}$ (Van Deun *et al.*, 2019). In contrast to SLIDE, PESCA, and MOFA, the common, local, and distinct structure in the blocks is not found by imposing group-wise penalties but by selecting *a priori* the number of distinct, local, and common components per block and imposing that by invoking a zero-block constraint on the $\mathbf{W}$. For example, in the case of three X-blocks, if the first column in $\mathbf{W}_2$ and $\mathbf{W}_3$ are forced to zero, then the first column of $\mathbf{W}_1$ will define a distinct component for block $\mathbf{X}_1$. Hence, a structure on $\mathbf{W}$ is imposed much like in DISCO by setting parts of the $\mathbf{W}$ to zero (see Section 5.1.2.1). Such a predefined pattern of zeros then defines sets of common, local, and distinct components. Apart from that, an elastic net penalty is imposed on the $\mathbf{W}$ resulting in the following problem

$$\min_{\mathbf{W}} \left[\alpha\|\mathbf{X} - \mathbf{XWP}^t\|^2 + (1-\alpha)\|\mathbf{y} - \mathbf{XWq}\|^2 + \lambda_L\|\mathbf{W}\|_1 + \lambda_R\|\mathbf{W}\|^2\right] \tag{10.2}$$

and solving this optimisation problem will give sparse weights, $\mathbf{W}$. The scores for the individual blocks are $\mathbf{T}_m = \mathbf{X}_m\mathbf{W}_m$ and the loadings for the individual blocks are the matrices, $\mathbf{P}_m$. The super-scores are simply $\mathbf{T} = \mathbf{XW}$, and these are used to predict $\mathbf{y}$.

Model selection is an issue with this modelling strategy. The values to be provided are the number of components, the parameter α, the number of common, local, and distinct components and the ridge and lasso regularisation parameters λ_R and λ_L. A full cross-validation strategy to select these meta-parameters is very time consuming, hence, some short-cuts need to be taken (Van Deun *et al.*, 2019).

10.2.2 Multiway Multiblock Covariates Regression

The criterion used here is essentially the same as in Equation 10.1, i.e., based on a weighted sum of measures of the two aspects, controlled by a parameter which can be determined for instance by cross-validation (Section 2.7.5). The main advantage here is that the method can handle both two-way as well as three-way input blocks.

The starting point is multiway covariates regression (Smilde and Kiers, 1999), here called MCovR. Having available a three-way array $\underline{\mathbf{X}}(I \times J_X \times K)$ (and its matricised version $\mathbf{X}(I \times J_X K)$ as predictors and a matrix $\mathbf{Y}(I \times J_Y)$ as responses, the multiway covariates regression model is:

$$\begin{aligned} \mathbf{T} &= \mathbf{XW} \\ \mathbf{X} &= \mathbf{TP}^t + \mathbf{E} \\ \mathbf{Y} &= \mathbf{TQ}^t + \mathbf{F} \end{aligned} \tag{10.3}$$

where the $\mathbf{W}(J_X K \times R)$ defines the scores, $\mathbf{T}(I \times R)$. The loadings, $\mathbf{P}(J_X K \times R)$, can be structured according to a Tucker1, Tucker2, Tucker3 or PARAFAC (CP) model (Smilde and Kiers, 1999):

$$\begin{aligned} \mathbf{P}_{T1} &= \mathbf{P} \\ \mathbf{P}_{T2} &= \mathbf{G}_{T2}(\mathbf{I} \otimes \mathbf{U}^t) \\ \mathbf{P}_{T3} &= \mathbf{G}_{T3}(\mathbf{V}^t \otimes \mathbf{U}^t) \\ \mathbf{P}_{CP} &= \mathbf{G}_{CP}(\mathbf{V}^t \otimes \mathbf{U}^t) \end{aligned} \tag{10.4}$$

with $\mathbf{G}_{T2}$, $\mathbf{G}_{T3}$, and $\mathbf{G}_{CP}$ matricised core-arrays of the proper size and structure. For the PARAFAC model, the core-array $\underline{\mathbf{G}}_{CP}$ is a three-way array with values on the superdiagonal and all off-diagonal elements are zero. The $\mathbf{U}$, $\mathbf{V}$ are loading matrices of the proper modes. Note that the loading matrices $\mathbf{P}_{T1}$, $\mathbf{P}_{T2}$, $\mathbf{P}_{T3}$, and $\mathbf{P}_{CP}$ all have the same sizes $J_X K \times R$. Both these loadings and weights, $\mathbf{W}$, are now estimated by solving

$$\min_{\mathbf{W}} \left[\alpha\|\mathbf{X} - \mathbf{XWP}^t\|^2 + (1-\alpha)\|\mathbf{Y} - \mathbf{XWQ}^t\|^2\right] \tag{10.5}$$

where α is between 0 and 1 and can be chosen *a priori* or using cross-validation. The solution can be obtained by an alternating least squares algorithm (Smilde and Kiers, 1999). The weights $\mathbf{W}$ are defined using the matricised version of $\mathbf{X}$ but it is also possible to use trilinear weights (Stahle, 1989) (i.e., to impose a PARAFAC structure on the weights). As can be seen, this criterion is essentially the same as the one in the previous subsection. Again, it is convenient to normalise the solution, e.g., by giving all columns of $\mathbf{T}$ length one.

Now suppose that there are two three-way blocks $\underline{\mathbf{X}}_1(I \times J_1 \times K_1)$ and $\underline{\mathbf{X}}_2(I \times J_2 \times K_2)$ (and their matricised versions $\mathbf{X}_1(I \times J_1K_1)$ and $\mathbf{X}_2(I \times J_2K_2)$) and a matrix $\mathbf{Y}(I \times J)$, and both three-way blocks are predictors of $\mathbf{Y}$. Then a multiway multiblock covariates regression (MWMBCovR) model would look like:

$$\begin{aligned}
\mathbf{T}_1 &= \mathbf{X}_1\mathbf{W}_1 \\
\mathbf{T}_2 &= \mathbf{X}_2\mathbf{W}_2 \\
\mathbf{X}_1 &= \mathbf{T}_1\mathbf{P}_1^t + \mathbf{E}_1 \\
\mathbf{X}_2 &= \mathbf{T}_2\mathbf{P}_2^t + \mathbf{E}_2 \\
\mathbf{T} &= [\mathbf{T}_1|\mathbf{T}_2] \\
\mathbf{T} &= \mathbf{T}\mathbf{W}_T\mathbf{P}_T^t + \mathbf{F}_T \\
\mathbf{Y} &= \mathbf{T}\mathbf{W}_T\mathbf{Q}_T^t + \mathbf{F}_Y
\end{aligned} \tag{10.6}$$

and again the weights are found by solving

$$\min_{\mathbf{W}} \left[\alpha_{T1}\|\mathbf{X}_1 - \mathbf{X}_1\mathbf{W}_1\mathbf{P}_1^t\|^2 + \alpha_{T2}\|\mathbf{X}_2 - \mathbf{X}_2\mathbf{W}_2\mathbf{P}_2^t\|^2 + \alpha_T\|\mathbf{T} - \mathbf{T}\mathbf{W}_T\mathbf{P}_T^t\|^2 + \alpha_Y\|\mathbf{Y} - \mathbf{T}\mathbf{W}_T\mathbf{Q}_T^t\|^2\right]. \tag{10.7}$$

The α values are chosen as a convex combination. The $\mathbf{Q}$ and the $\mathbf{P}$ are found by LS of the four terms separately. Normalisation of $\mathbf{T}$ can be done as above (i.e., columns have length 1).

This model looks complicated but the concepts are fairly simple. Components $\mathbf{T}_1$ and $\mathbf{T}_2$ are taken from the exogenous (input) blocks $\mathbf{X}_1$ and $\mathbf{X}_2$. Next, the components are combined in a matrix $[\mathbf{T}_1|\mathbf{T}_2]$ from which super-components $\mathbf{T}$ are taken. This is much in line with the original idea of MB-PLS (see Section 7.2). These super-components are then used to model $\mathbf{Y}$ in a principal covariates regression framework (last two lines of Equation 10.6). Note that also in this case there are several options for the structure of the loadings $\mathbf{P}$. An example of a multiway multiblock covariates regression is given in Example 10.1 which also illustrates a strategy on how to choose the α values.

The method can be extended to more than two input blocks and also be used to combine three-way and two-way input blocks, i.e., one or more three-way blocks can be replaced by a two-way block.

Example 10.1: Multiway multiblock covariates regression model of batch process data

The example concerns data from a real polymerisation batch process (Nomikos and MacGregor, 1995; Kosanovich *et al.*, 1996). The autoclave in this chemical batch process converts an aqueous effluent from an upstream evaporator into a polymer product. A batch process is run according to a recipe and in this case the recipe specifies reactor and heat source pressure trajectories through five different stages in order to arrive at the end-product.

Data are available from the autoclave process, $\underline{\mathbf{X}}_1$, which consists of eight process variables (such as temperatures and pressures) measured on-line. For the 42 batches these measurements had to be aligned and resulted in 116 time-point measurements for every batch (separated in 9, 43, 22, 20, and 22 points per stage). Hence, the size of $\underline{\mathbf{X}}_1$ is $(42 \times 8 \times 116)$. The autoclave history is a two-way array consisting of four variables describing the history of the autoclave and is collected in $\mathbf{X}_2(42 \times 4)$. Finally, the quality of the end-product of the batch process is characterised by molecular weight and titrated ends resulting in $\mathbf{Y}(42 \times 2)$. For more details we refer to Nomikos and MacGregor (1995); Kosanovich *et al.* (1996); Smilde and Kiers (1999).

It is important to preprocess the data, and this was done by autoscaling $\mathbf{X}_2$ and $\mathbf{Y}$ and subsequently block-scaling those matrices to a sum-of-squares one. The multiway array $\underline{\mathbf{X}}_1$ was centred across the batch mode and scaled to a sum-of-squares one within the process variable mode which is one of the recommended ways of preprocessing three-way data (Bro and Smilde, 2003).

It is a good idea to start analysing the relationships between only two blocks. For that purpose we analysed the relationship between $\underline{\mathbf{X}}_1$ and $\mathbf{Y}$ using two methods: PLS and multiway covariates regression. The first method simply builds a PLS model between the matricised three-way array ($\mathbf{X}_1$) and $\mathbf{Y}$ (to distinguish it from PLS for the other block, we here call it unfold-PLS or U-PLS). The relationship between $\mathbf{X}_2$ and $\mathbf{Y}$ was analysed by PLS and Principal Covariates Regression (see Chapter 2). The results are shown in Table 10.2.

The α values were determined using cross-validation, and the Q^2 statistic in Table 10.2 was calculated with a leave-three-batches-out cross-validation strategy. The PCovR regression model for $\mathbf{Y}$ and $\mathbf{X}_2$ was calculated with a two component model, and a low α value was found to be optimal. The Q^2 statistic is rather low both for the PLS and PCovR models indicating that the history block is not very predictive for the end-product quality.

Table 10.2 Results of the single-block regression models. PCovR is Principal Covariates Regression, U-PLS is unfold-PLS, MCovR is multiway covariates regression. The 3,2,3 components for MCovR refer to the components for the three modes of Tucker3. For more explanation, see text.

Method	Block	Components	α	R^2_X	R^2_Y	Q^2_Y
PLS	$\mathbf{X}_2$	2		0.61	0.34	0.22
PCovR	$\mathbf{X}_2$	2	0.10	0.65	0.35	0.22
U-PLS	$\underline{\mathbf{X}}_1$	2		0.28	0.46	0.16
MCovR	$\underline{\mathbf{X}}_1$	3,2,3	0.99	0.41	0.36	0.23

For the three-way array $\underline{\mathbf{X}}_1$ the U-PLS model used two components, and the multiway covariates regression used a Tucker3 model for the three-way block with three, two, and three components in the batch-, process variable-, and time-mode, respectively. A very high α value was needed to get good predictions of $\mathbf{Y}$, which is often the case to generate stable components from a three-way array for predictions (Smilde and Kiers, 1999). The multiway covariates regression model has a better predictive ability than the U-PLS model and the latter also seems to overfit (compare the R^2_Y with the Q^2_Y values). Overall, the low Q^2 values are not unusual for this type of (noisy) process data.

To start modelling the batch process with a multiway multiblock covariates regression model (MWMBCovR), a choice has to be made regarding the α values. It seems a good idea to stay close to the optimal α values as found in the analysis regarding two data blocks (see above). After some fine-tuning using cross-validation this resulted in the values α_{T2} = 0.000 32, α_{T1} = 0.96, α_T = 0.032 and α_Y = 0.0064. This shows that the main emphasis is (again) on the three-way block $\underline{\mathbf{X}}_1$ and much less to fitting $\mathbf{Y}$. The final multiway multiblock covariates regression model is compared with a multiway MB-PLS model, which is simply found by matricising the three-way block and using a MB-PLS model on $\mathbf{X}_1$, $\mathbf{X}_2$, and $\mathbf{Y}$ (see Section 7.2). The results are reported in Table 10.3.

Again, for the three-way block a Tucker3 model with 3, 2, and 3 components in the batch-, process variables-, and time-mode was selected. The multiway multiblock covariates regression model gave a slightly better predictability than the PLS counterpart and seems to overfit less.

Table 10.3 Results of the multiway multiblock models. MB-PLS is multiblock PLS, MWMBCovR is multiway multiblock covariates regression. For more explanation, see text.

Method	Comp.	$R^2_{X_1}$	$R^2_{X_2}$	R^2_Y	Q^2_Y
MB-PLS	2	0.27	0.33	0.47	0.18
MWMBCovR	2, (3,2,3), 2	0.41	0.37	0.36	0.23

The multiway multiblock covariates regression model has modelled a simultaneous relationship between $\underline{\mathbf{X}}_1$ and $\mathbf{Y}$ on the one hand and of $\mathbf{X}_2$ and $\mathbf{Y}$ on the other hand. An alternative modelling strategy for this problem would be a path model by also assuming a predictive relationship from $\mathbf{X}_2$ to $\underline{\mathbf{X}}_1$ using methods in the same style as described in Section 10.5.4. A disadvantage of the multiway multiblock model is the need to choose the α parameters; that can be become quite involved but single analyses between only two blocks can help setting the limits for such values.

10.3 Multiblock Redundancy Analysis

Redundancy analysis was already briefly described in Chapter 2, and here we present a multiblock versions of RDA. We will discuss a standard multiblock version and a sparse version.

10.3.1 Standard Multiblock Redundancy Analysis

To describe multiblock redundancy analysis we will focus on the one formulated in Bougeard *et al.* (2011a). For each component extracted, RDA with one input block can be formulated as maximising $cov^2(\mathbf{t}, \mathbf{u})$ with $\mathbf{u}$ and $\mathbf{t}$ being linear combinations of $\mathbf{X}$ and $\mathbf{Y}$, i.e., $\mathbf{t} = \mathbf{Xw}$ and $\mathbf{u} = \mathbf{Yc}$ where $||\mathbf{t}|| = ||\mathbf{c}|| = 1$ (see Section 2.1.8).

The multiblock version of the RDA (MB-RDA) is then simply defined as maximising the sum of the same type of criterion over the different blocks, i.e.,

$$\sum_m cov^2(\mathbf{t}_m, \mathbf{u}) \tag{10.8}$$

where $\mathbf{t}_m = \mathbf{X}_m\mathbf{w}_m$, $\|\mathbf{t}_m\| = 1$ and $\mathbf{u}$ is defined as above ($\mathbf{u} = \mathbf{Yc}$ with $\|\mathbf{c}\| = 1$). It was shown in the same paper that this sum can be written as

$$\sum_m \mathbf{c}^t\mathbf{Y}^t\mathbf{X}_m(\mathbf{X}_m^t\mathbf{X}_m)^{-1}\mathbf{X}_m^t\mathbf{Yc} \tag{10.9}$$

assuming full column rank of $\mathbf{X}_m$. This means that for each component in the RDA, the solution for $\mathbf{c}$ can be obtained as the normalised eigenvector associated with the largest eigenvalue of the matrix

$$\sum_m \mathbf{Y}^t\mathbf{X}_m(\mathbf{X}_m^t\mathbf{X}_m)^{-1}\mathbf{X}_m^t\mathbf{Y} \tag{10.10}$$

which is a direct generalisation of RDA (see Chapter 2). It was shown in Bougeard *et al.* (2011a) that the corresponding scores $\mathbf{t}_m$ can be written as

$$\mathbf{t}_m = \mathbf{P}_{\mathbf{X}_m}\mathbf{u}/\|\mathbf{P}_{\mathbf{X}_m}\mathbf{u}\| \tag{10.11}$$

where the $\mathbf{P}_{\mathbf{X}_m}$ is the projection operator onto $\mathbf{X}_m$ and $\mathbf{u} = \mathbf{Yc}$. If the inverse does not exist in Equation 10.9, it must be replaced by a generalised inverse.

A super-score is calculated as a weighted sum of the individual $\mathbf{t}_m$ with weights based on the covariances between $\mathbf{t}_m$ and $\mathbf{u}$. This can be written as

$$\mathbf{t} = \sum_m \mathbf{P}_{\mathbf{X}_m}\mathbf{u}/\sqrt{\|\sum_l \mathbf{P}_{\mathbf{X}_l}\mathbf{u}\|^2} \tag{10.12}$$

Deflation is done for the $\mathbf{X}_m$-blocks only using the super-score. A regression equation can be obtained by regressing $\mathbf{Y}$ onto the super-scores $\mathbf{t}$.

It is shown empirically in an example in Bougeard *et al.* (2011b) that redundancy analysis can be better than MB-PLS in fitting the data, but also less precise from a prediction point of view. This phenomenon was treated also in Chapter 2 in the discussion around Table 2.3. This shows that validation must be taken very seriously for this method.

Example 10.2: Multiblock redundancy analysis: sensory assessment of wines

We here revisit the wine study in Example 7.4, with its 21 wines, four blocks of sensory assessments (A, B, C, and D) and the two responses 'overall quality' and 'typical'. MB-RDA was not able to distinguish between the responses, so we simplified the analysis by only including 'overall quality' as a response. For easy comparison, we have plotted block-weights, $\mathbf{w}_m$, in the same format as the PCP plots in Example 7.4, and also a bar plot of $\mathbf{Y}$ scores ($\mathbf{u}_r$). The latter is chosen because all components will be identical for the response when a single variable is used. One detail which is different for the two methods is the way explained variances are calculated and presented for each block. Here they indicate the proportion of the explained variance for the current component that is attributed to the given block, e.g., 22.1% of the explained variance from the first component originates from block A.

As described before the example, deflation in MB-RDA is done per block using the super-scores. We observe that this decomposition differs in some of the blocks from what we saw when applying PCP on SO-PLS results. In that case the attributes were mostly scattered evenly away from the origin in the PCP plots, while for redundancy analysis there is more contrast between attributes close to the origin and further away, at least for block D. Regarding

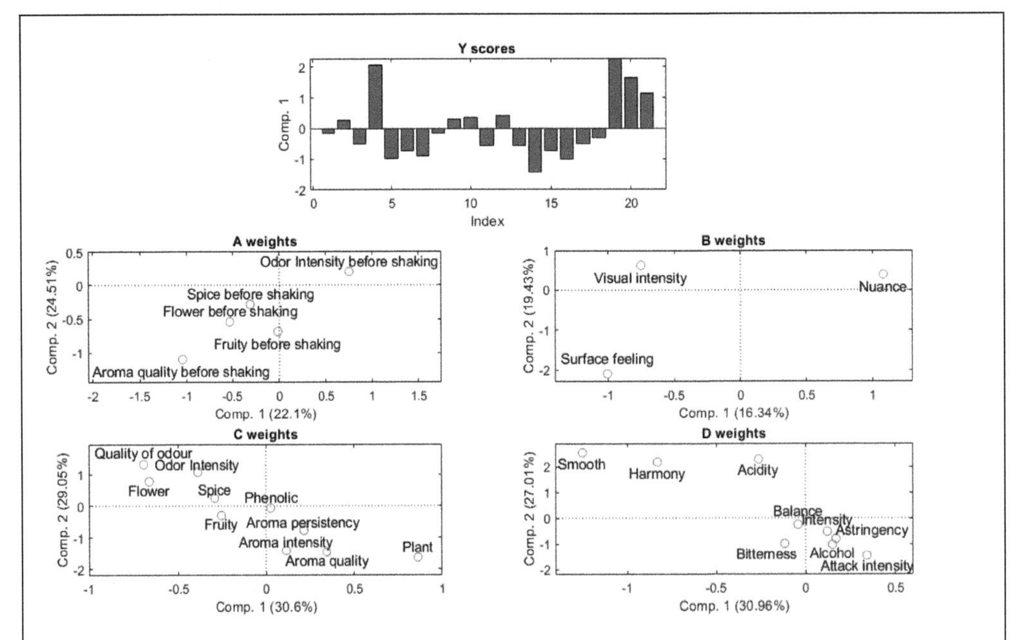

Figure 10.1 Results from multiblock redundancy analysis of the Wine data, showing **Y** scores ($\mathbf{u}_r$) and block-wise weights for each of the four input blocks (A, B, C, D).

interpretation, the latter is sometimes a nice property, but here it seems to result in attribute groups that are less intuitive, like the group of attributes 'Acidity', 'Harmony', and 'Smooth' for block D.

10.3.2 Sparse Multiblock Redundancy Analysis

As in the case of MB-PLS (see Section 7.2), there is also a sparse version of multiblock redundancy analysis. This starts with the sparse redundancy analysis model (Csala *et al.*, 2017) for relating one single **X**-block to a **Y**-block and is based on an iterative algorithm to estimate the model parameters (Fornell *et al.*, 1988), see Algorithm 2.4. The redundancy solution can be made sparse by substituting the multiple regression in step 3 of that algorithm with the following step

$$\mathbf{w}_{opt} = arg\min_{\mathbf{w}} \left[\mathbf{w}^t(\frac{\mathbf{X}^t\mathbf{X} + \lambda_2\mathbf{I}}{1 + \lambda_2})\mathbf{w} - 2\mathbf{u}^t\mathbf{X}\mathbf{w} + \lambda_1||\mathbf{w}||_1\right] \tag{10.13}$$

which is the sparse solution following the elastic net (Zou and Hastie (2005)). The λ_1 is a weight for the lasso-type constraint on the one-norm of **w** (Tibshirani, 1996) and λ_2 is a weight in the ridge-type constraint (see also sparse PCA in Section 2.1.3). The values of λ_1 and λ_2 have to be chosen *a priori* or, e.g., using cross-validation or stability selection (Meinshausen and Bühlmann, 2010).

Sparse multiblock redundancy analysis builds on the same idea. For two X-blocks predicting a single Y-block the algorithm is shown in Algorithm 10.1. This is just one version of such an algorithm. It is also possible to assume a relationship between the two X-blocks. This results in rather complicated algorithms and we refer to Csala *et al.* (2017) for details and an example from genomics.

Algorithm 10.1

SPARSE MULTIBLOCK REDUNDANCY ANALYSIS

Assume blocks $\mathbf{X}_1(I \times J_1)$, $\mathbf{X}_2(I \times J_2)$, and a block $\mathbf{Y}(I \times J)$; all centred (and possibly scaled). The iterative algorithm to calculate a sparse multiblock redundancy analysis starts with $\mathbf{c}_{start}$, $\mathbf{w}_{1,start}$, and $\mathbf{w}_{2,start}$ and then goes as follows:

Initialise vectors

1: $\mathbf{c} = \mathbf{c}_{start}$; $\|\mathbf{c}\| = 1$; $\mathbf{w}_1 = \mathbf{w}_{1,start}$; $\mathbf{w}_2 = \mathbf{w}_{2,start}$ – define starting values

2: $\mathbf{t}_1 = \mathbf{X}_1\mathbf{w}_1$; normalise $\mathbf{t}_1$ – scores for $\mathbf{X}_1$

3: $\mathbf{t}_2 = \mathbf{X}_2\mathbf{w}_2$; normalise $\mathbf{t}_2$ – scores for $\mathbf{X}_2$

4: $\mathbf{u} = \mathbf{Y}\mathbf{c}$; – scores for $\mathbf{Y}$

Loop until convergence

5: $\mathbf{u} = \mathbf{t}_1\theta_1 + \mathbf{t}_2\theta_2 + \mathbf{e}$ – solve for θ_1 and θ_2

6: $\mathbf{z}_1 = \mathbf{u}\theta_1$; $\mathbf{z}_2 = \mathbf{u}\theta_2$; $\mathbf{z}_3 = \mathbf{t}_1\theta_1 + \mathbf{t}_2\theta_2$ – auxiliary variables

7: $\mathbf{w}_1^{(new)} = arg\min_{\mathbf{w}_1}\left[\mathbf{w}_1^t\left(\frac{\mathbf{X}_1^t\mathbf{X}_1 + \lambda_2\mathbf{I}}{1+\lambda_2}\right)\mathbf{w}_1 - 2\mathbf{z}_1^t\mathbf{X}_1\mathbf{w}_1 + \lambda_1\|\mathbf{w}_1\|_1\right]$ – sparse weights for $\mathbf{X}_1$

8: $\mathbf{w}_2^{(new)} = arg\min_{\mathbf{w}_2}\left[\mathbf{w}_2^t\left(\frac{\mathbf{X}_2^t\mathbf{X}_2 + \lambda_2\mathbf{I}}{1+\lambda_2}\right)\mathbf{w}_2 - 2\mathbf{z}_2^t\mathbf{X}_2\mathbf{w}_2 + \lambda_1\|\mathbf{w}_2\|_1\right]$ – sparse weights for $\mathbf{X}_2$

9: $\mathbf{c}^{(new)} = (\mathbf{z}_3^t\mathbf{z}_3)^{-1}(\mathbf{z}_3^t\mathbf{Y})$; normalise $\mathbf{c}_1^{(new)}$ – weights for $\mathbf{Y}$

10: $\mathbf{t}_1^{(new)} = \mathbf{X}_1\mathbf{w}_1^{(new)}$; normalise $\mathbf{t}_1^{(new)}$ – scores for $\mathbf{X}_1$

11: $\mathbf{t}_2^{(new)} = \mathbf{X}_2\mathbf{w}_2^{(new)}$; normalise $\mathbf{t}_2^{(new)}$ – scores for $\mathbf{X}_2$

12: $\mathbf{u}^{(new)} = \mathbf{Y}\mathbf{c}^{(new)}$; – scores for $\mathbf{Y}$

13: $\|\mathbf{u}^{(new)} - \mathbf{u}\|^2$ – check convergence of $\mathbf{u}$

End loop

and after convergence the weights defining the latent variables for the X-blocks are sparse; subsequent components are then calculated after deflation. Step 5 is a multiple regression step; steps 7 and 8 are the sparse updates using an elastic net approach and step 9 are univariate regressions of the columns of $\mathbf{Y}$ on the auxiliary variable $\mathbf{z}_3$.

10.4 Miscellaneous Multiblock Regression Methods

Before turning to methods that handle block splitting along the sample mode, we here give a brief description of some alternative supervised multiblock methods and some extensions of SO-PLS discussed in Chapter 7. More details can be found in the original publications listed.

10.4.1 Multiblock Variance Partitioning

Skov *et al.* (2008) proposed a multiblock regression method which they called multiblock variance partitioning. The method is sequential and shares some features with PO-PLS (Section 7.4). The mathematics is different, but the conceptual idea of splitting the explained variance of $\mathbf{Y}$ into common and distinct variation is similar (in the paper, the concept unique is used instead of distinct). A special feature of this method is that it is asymmetric with respect to input blocks in the sense that one $\mathbf{X}$-block is considered the master and the other input blocks are used for calculating the unique contribution of $\mathbf{X}$. In this sense, the method also has some similarities with SO-PLS in Section 7.3.

In brief, the multiblock variance partitioning method goes as follows for three input blocks. The unique variance contribution of the master block $\mathbf{X}_1$ is obtained by first calculating predicted $\mathbf{Y}$-values from $\mathbf{X}_1$ and then orthogonalising these predicted values with respect to predicted values from the other input blocks $\mathbf{X}_2$ and $\mathbf{X}_3$, giving $\mathbf{Y}^{\text{orth}}$. Then the sum of variances of these predicted values is calculated and called $\mathbf{V}_{\text{Unique}}$.

The total variance of $\mathbf{Y}$ ($\mathbf{V}_{\text{Tot}}$) is defined as the sum of variances of the $\mathbf{Y}$-data, and the unexplained variance of $\mathbf{Y}$ ($\mathbf{V}_{\text{Unexplained}}$) is obtained from calculating the variance of the residuals from a standard PLS analysis using only $\mathbf{X}_1$ as input. The unexplained and unique variances are then subtracted from the total variance to yield the common variance as shown in Algorithm 10.2.

Algorithm 10.2

SUMMARY OF MULTIBLOCK VARIANCE PARTITIONING FOR THREE INPUT BLOCKS

1: Define master block, $\mathbf{X}_1$. The other two blocks are denoted by $\mathbf{X}_2$ and $\mathbf{X}_3$.
2: Calculate the total variance ($\mathbf{V}_{\text{Tot}}$) of all the Y variables.
3: Calculate $\mathbf{V}_{\text{Unexplained}}$ from residuals of a PLS2 model using only $\mathbf{X}_1$ as input block.
4: Calculate the predicted values of $\mathbf{Y}$ from the three input blocks ($\mathbf{X}_1$, $\mathbf{X}_2$, and $\mathbf{X}_3$) used separately by PLS, i.e., using three separate models. This gives three predicted $\mathbf{Y}$-matrices, one for each model.
5: Orthogonalise the predicted $\mathbf{Y}$-values from the model based on $\mathbf{X}_1$ with respect to the predicted $\mathbf{Y}$-values obtained from the two models based on $\mathbf{X}_2$ and $\mathbf{X}_3$. The result is denoted $\mathbf{Y}^{\text{orth}}$.
6: Calculate the unique variance $\mathbf{V}_{Unique}$ (i.e., the unique contribution of $\mathbf{X}_1$ on $\mathbf{Y}$) as $\mathbf{V}_{Unique} = tr(\mathbf{Y}^{\text{orth}\,t}\mathbf{Y}^{\text{orth}})$
7: Calculate the common variance of $\mathbf{Y}$ (between all blocks) as

$$\mathbf{V}_{\text{Common}} = \mathbf{V}_{\text{Tot}} - \mathbf{V}_{\text{Unexplained}} - \mathbf{V}_{\text{Unique}}.$$

If wanted all quantities can be divided by the total variance and multiplied by 100 to obtain values between 0 and 100. The paper presents the result in a pie chart where each sector reflects the size of variances as illustrated in Figure 10.2.

10.4.2 Network Induced Supervised Learning

Network induced supervised learning (NI-SL) is a method developed by Reis (2013) (see also Campos *et al.* (2017)). It starts by applying PLS for each $\mathbf{X}$ block separately (one Y-variable). The latent variables from each single PLS are gathered in one single super-block of all scores. A forward stepwise regression methodology is thereafter used for the training data to select

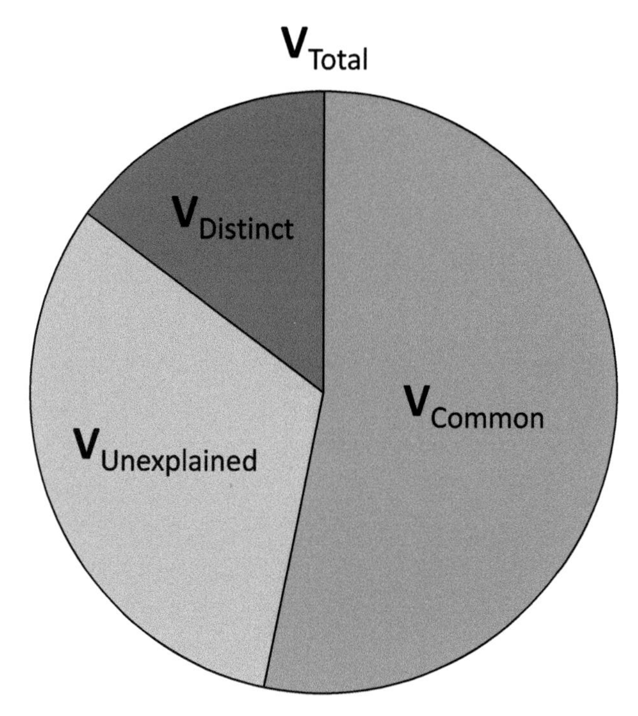

Figure 10.2 Pie chart of the sources of contribution to the total variance (arbitrary sector sizes for illustration).

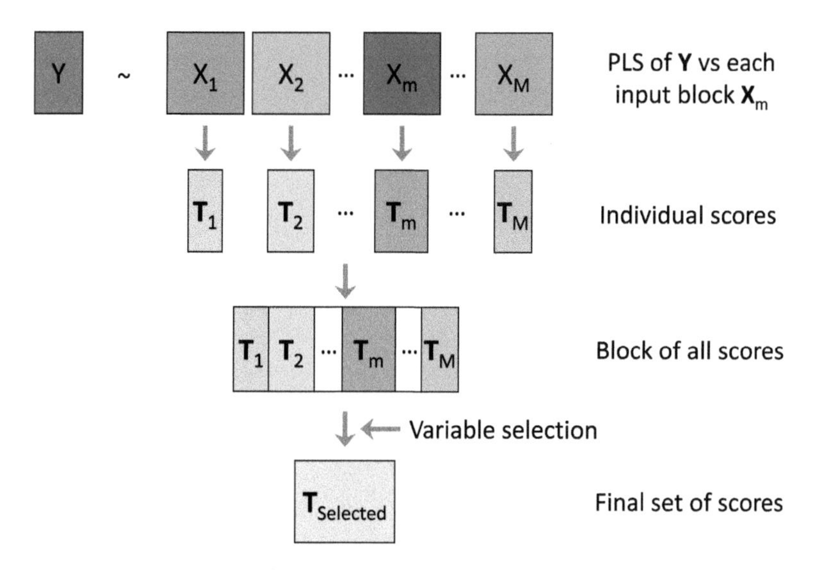

Figure 10.3 Flow chart for the NI-SL method.

the subgroups of latent variables with the best fit to **Y** (see Figure 10.3). We can then end up with only one or several components from each of the blocks in the final model.

The method is simple to understand and implement and has the advantage that it assesses the importance of the blocks. The method has some similarities with the first step of PO-PLS since it initially uses single PLS models. It is also similar to ROSA since it is based on selecting individual components from all blocks. For ROSA (Section 7.5), however, the selection is done from a new PLS regression each time, while here the PLS scores are given once they are estimated. There is a need for comparing the performance of these three strongly related methods in practice.

10.4.3 Common Dimensions for Multiblock Regression

Using column-centred data blocks, the common dimensions (ComDim) for multiblock regression (Cariou *et al.* (2019)) is for each component based on minimising the criterion

$$\sum_{m=1}^{M} ||\mathbf{R}_m - \lambda_m \mathbf{t}\mathbf{u}^t||^2 \tag{10.14}$$

where $\mathbf{R}_m = \mathbf{X}_m\mathbf{X}_m^t\mathbf{Y}\mathbf{Y}^t$ instead of $\mathbf{R}_m = \mathbf{X}_m\mathbf{X}_m^t$ as used in the unsupervised case. In other words, cross-products of the different $\mathbf{X}_m$ blocks with cross-products of the $\mathbf{Y}$ are fitted as closely as possible to an outer product of common score vectors $\mathbf{t}$ and $\mathbf{u}$. The vectors are assumed to be of length 1 and therefore a block dependent weight λ_m is also needed. The method also provides block-scores defined as $\mathbf{t}_m = \mathbf{X}_m\mathbf{X}_m^t\mathbf{t}$. Deflation of each block $\mathbf{X}_{m,\text{new}} = (\mathbf{I} - \mathbf{t}\mathbf{t}^t)\mathbf{X}_m$ and $\mathbf{Y}_{\text{new}} = (\mathbf{I} - \mathbf{t}\mathbf{t}^t)\mathbf{Y}$ is performed between each component.

Predictions can be obtained by regressing $\mathbf{Y}$ onto the scores $\mathbf{t}$. An application and comparison with MB-PLS can be found in Cariou *et al.* (2019). The results are somewhat similar, but not identical.

10.5 Modifications and Extensions of the SO-PLS Method

10.5.1 Extensions of SO-PLS to Three-Way Data

Biancolillo *et al.* (2017) proposed an extension of the SO-PLS (see Section 7.3) method to handle combinations of two-way and three-way blocks explicitly without matricising (see taxonomy in Chapter 3). This pinpoints one of the advantages of SO-PLS, namely that each new block represents a new analysis situation. When for instance a two-way block is fitted first, a subsequent three-way block, properly orthogonalised, can be fitted to the $\mathbf{Y}$ residuals using a three-way regression method, for instance N-PLS (Bro (1996)).

The orthogonalisation of a three-way array is done the same way as for two-way arrays, i.e., for each variable separately. In other words, each variable (column) in the three-way block (along the two ways not representing samples) is orthogonalised with respect to previous blocks. A three-way array provides scores that can later on be used for the next deflation. As can be seen, the approach is still based on the same iteration between orthogonalisation and regression as presented in Figure 7.11.

An illustration of regression modelling including a three-way array is shown in Figure 10.4.

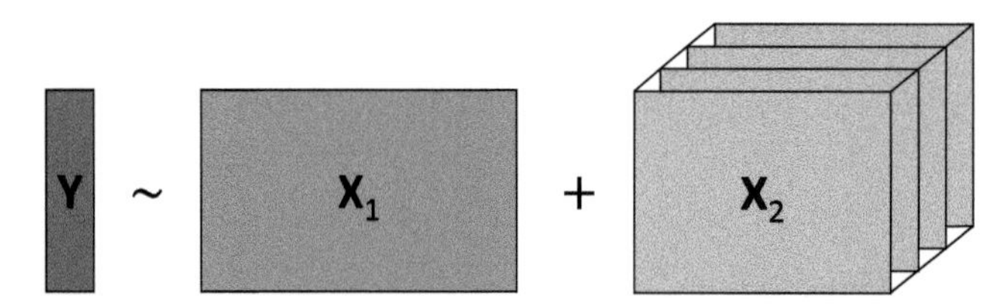

Figure 10.4 An illustration of SO-N-PLS, modelling a response using a two-way matrix, $\mathbf{X}_1$, and a three-way array, $\mathbf{X}_2$

It was shown in Biancolillo *et al.* (2017) that the explicit modelling of three-way input matrices without matricising, can have certain advantages for interpretation and possibly prediction if the underlying structure is truly three-way (Biancolillo *et al.* (2017)).

10.5.2 Variable Selection for SO-PLS

As for standard PLS regression, it is possible to select the most important variables also for SO-PLS, either for simpler interpretation or improved prediction. Variable selection in multiblock modelling can be done in different ways as shown in Biancolillo *et al.* (2016), but within the SO-PLS framework the most natural strategy is to do it sequentially. In this case variable selection is done for $\mathbf{X}_1$ first before the same is repeated for the $\mathbf{X}_2$ orthogonalised with respect to the retained variables in $\mathbf{X}_1$. We then end up with a reduced set of both $\mathbf{X}_1$ and $\mathbf{X}_2$ variables.

It was shown in Biancolillo *et al.* (2016) that this can sometimes lead to improved results, but the results in the paper were quite inconclusive with respect to which variable selection method to use. This is quite typical for variable selection methods in general since results will always depend on the data used. Variable selection is anyway an option that should definitely be tested if the situation calls for it. We refer to Biancolillo *et al.* (2020) for further results on variable selection in SO-PLS regression.

10.5.3 More Complicated Error Structure for SO-PLS

In Måge and Næs (2005) a modification of the standard SO-PLS approach was proposed based on assuming a split plot error structure for the residuals **F**. A split plot error structure is typical in industry where not all design variables can be randomised independently. An example of this is when a number of recipes have to be tested within a fixed process setting. This gives correlated errors.

In Måge and Næs (2005) a number of design variables were combined with covariates based on NIR spectra in a split plot design. The problem was then how to combine a designed experiment (represented in $\mathbf{X}_1$) with a large set of collinear covariates ($\mathbf{X}_2$) in a situation with correlated errors. The situation is similar to the one which was a starting point for the SO-PLS development, namely the LS-PLS method based on a designed experiment. The only difference is that here the errors **F** are stochastically dependent because of the split plot structure. More details are found in the original publication and in Elaboration 10.1.

ELABORATION 10.1

SO-PLS with more complicated error structure

The proposal put forward in Måge and Næs (2005) was based on first estimating the design part of the experiment ($\mathbf{X}_1$) using restricted maximum likelihood (REML) instead of LS. REML is a fitting method suitable for estimating factors effects when errors have for instance a split plot structure. The NIR spectra (or other covariates ($\mathbf{X}_2$)) are then fitted to the residuals from this REML by PLS regression, before the design factors and the scores from the PLS model are modelled together with y as response, again using a REML approach. The residuals from the design variables are then again fitted to the NIR data. The procedure is repeated until convergence.

In more detail:

1: Fit the design block $\mathbf{X}_1$ to $\mathbf{y}$ using REML.
2: Calculate the $\mathbf{y}$-residuals $\mathbf{f}$.
3: Fit the covariates $\mathbf{X}_2$ to the $\mathbf{y}$-residuals $\mathbf{f}$.
4: Fit $\mathbf{X}_1$ and the scores $\mathbf{T}_2$ to $\mathbf{y}$ using REML.

5 : Calculate the y-residuals **f** from the design block $\mathbf{X}_1$.
6 : Go to 3 and continue until convergence.

An alternative to the iterative procedure could have been to orthogonalise the spectra with respect to the design matrix ($\mathbf{X}_1$) before fitting by PLS and then combining the design and the orthogonalised scores **T** directly in a model with **y** as response in a split plot model. This option was also discussed in Måge and Næs (2005) and it was shown to improve interpretability. To our knowledge, this method has not been pursued in the literature.

10.5.4 SO-PLS Used for Path Modelling

Path modelling (see Bollen (1989a) and topology Figure 3.12 in Chapter 3) can be considered an extension of standard multiblock regression methodology where the only relation considered is the one between all the independent **X** blocks on one side and the response block **Y** on the other. In path modelling, also a number of additional relations between the input blocks are studied according to the path diagram that is set up. SO-PLS can be used to setup such a path model resulting in the SO-PLS-PM method. A major goal is to obtain insight into how the blocks are related along the paths, for instance which blocks can be predicted from the others and which blocks are important in the predictions.

The example in Figure 10.5 is taken from a tasting study of wine (see also Example 7.4). This example comes from Escofier and Pagès (1998) and is used a lot in the literature for illustration. The different blocks represent different stages in the tasting process, starting from smelling at rest (A), viewing the wine (B) before smelling after shaking (C), tasting (D), and finally assessing the overall quality (E). For each block there are a number of measured variables (in path modelling often called manifest variables). The arrows represent the paths and, therefore, how the different blocks in the sequence are theorised to be linked to each other; in this case mainly according to ordering in time. The first step of a path modelling exercise is always to establish this type of diagram.

This standard model framework for path models/diagrams, called structural equation modelling (SEM), is based on combining two model elements: a set of equations linking so-called latent variables or components in the different blocks to each other and a measurement model that defines the relation between the measured variables and the latent ones. The two parts of the joint model are frequently referred to as the structural or inner model and measurement or outer model.

In Næs *et al.* (2011b) it was proposed to use the SO-PLS as an explorative technique for studying relations between blocks in a path diagram. The SO-PLS modelling in this case is

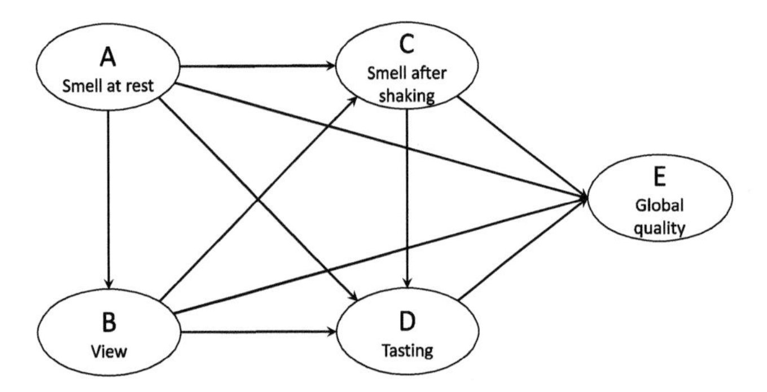

Figure 10.5 Path diagram for a wine tasting study. The blocks represent the different stages of a wine tasting experiment and the arrows indicate how the blocks are linked. Source: (Næs *et al.*, 2020). Reproduced with permission from Wiley.

not based on the standard explicit splitting into a measurement model and an inner relations model. All models are instead based directly on PLS regression analysis. The main advantage of the SO-PLS method is that it easily handles blocks with different underlying dimensionality, which is more difficult in traditional approaches. A typical example of this is when the input blocks need very different numbers of principal components to be explained properly. The method was later tested in Menichelli *et al.* (2014a) and Nguyen *et al.* (2020), and then extended and refined in Romano *et al.* (2019) and Næs *et al.* (2020). The method is only applicable for so-called DAGs (directed acyclic graph (Edwards (2012)), i.e., diagrams where it is not possible to have cyclic relationships.

The method consists of two steps. In the first step the focus is on using SO-PLS for identifying which blocks in the path diagram can be predicted from the others and which input blocks are important for the predictions. In the second step, the focus is on estimating direct effects, indirect effects, and total effects and on visualising the effects. These effects are defined in terms of explained variance of the predictions involved in the diagram. The second step is more involved and will not be considered further here.

The first step is based on standard use of the SO-PLS method for each endogenous block (i.e., a block with input paths) separately. All blocks that through a number of paths can reach the actual endogenous block are incorporated in the SO-PLS model. The focus is on the strength of the predictions for each endogenous block and which blocks contribute the most to the predictions. In this context the additional effect of adding a new block is an important tool in order to see how much each block contributes in addition to the previous (see Example 10.3). For the SO-PLS-PM (see e.g., Næs *et al.* (2020)), the order of blocks is defined by the topological order of the DAG (Edwards (2012)), as defined in Elaboration 10.2.

ELABORATION 10.2

Definition of topological order of a DAG

Let us assume that each block in the DAG is represented by a letter (see for instance Figure 10.5) and that the letters are placed beside each other on a line. If we add arrows between the letters according to the direction of the paths in the path diagram and if all arrows point from left to right, the established order of the letters corresponds to a so-called topological order of the DAG. For DAGs there is always a topological order (Edwards (2012)), but it is not necessarily unique.

In the path diagram in Figure 10.5, the topological order is equal to A, B, C, D, E. This means that for example the SO-PLS model for the smell after shaking block (C) starts with incorporating smell at rest (A) before adding the block view (B). The explained variance of A is calculated directly from the first regression, and the additional effect of B is the difference between the explained variance using both A and B as input and the model using only A. Likewise, the model for quality (E) starts with smell at rest (A), before view (B), smell after shaking (C) and taste (D). In this diagram, the topological order is unique. If the ordering is not unique, which may happen in complex diagrams, it was proposed in Næs *et al.* (2020) to select the one (for each SO-PLS model) which is most natural from a substance matter point of view.

Since the procedure is based on several PLS models, the number of plots to interpret can be quite large. It was therefore proposed in Romano *et al.* (2019) to use the PCP (principal components of prediction, see Section 7.3.3) plot for providing one single scatter plot for each model (i.e., each endogenous block) instead of one per input block.

Example 10.3: Wine example

The data for this example comes from the same tasting experiment as Example 7.4, illustrated in Figure 10.5. Table 10.4 gives the explained variance results for the four possible models. In for instance the SO-PLS model for output D, both A and B have a strong impact on prediction ability, while the contribution of C is more modest (an increase of only about 5 units). Block smell at rest (A) is the dominating in all models, but the block view (B) is also quite important for two of the models. The large effect of A is natural since it is taken in first and since the other blocks are correlated with A.

The first row in the table represents the explained variances of block A, the rest of the rows represent the accumulated explained variances obtained by combining all blocks above and including the one representing the row. For instance, the row starting with C represents the model with both C and the previous two, A and B, as input. The differences between the accumulated explained variances in a column are the additional variances. For example: the additional effect (or variability) of block C for the model for blocks D is equal to the differences between 59.65 and 54.78 which is equal to 4.87.

The PCP plots for the model for block D is given in Figure 10.6. The two components in the PCP plot explain 89.3% and 9.5% indicating that the predicted values of block tasting (D) represent a two-dimensional space (confirmed by cross-validation). The two upper plots are the loadings and scores from the PCA of the predicted values of the responses. The plot at the bottom represents the loadings for the input blocks A, B, and C.

As can be seen, the taste variables (upper plot) go in the same direction along component 1, but with variables in both directions along the second component. The second component is for the most related to the wines named T1 and T2 which stick out from the rest. For the X-blocks, we can see that rest (block A) dominates, which is in line with explained variance results. The uni-dimensionality of the loadings for shake (C) and view (B) variables reflects the uni-dimensionality of the blocks in the predictions. Most variables are to the right, with a few exceptions. We can for instance see that rest5 (from block A) and one of the shake variables lie slightly to the left of the origin. Along the second component there are rest variables (A) on both sides.

Table 10.4 SO-PLS-PM results for wine data. The four columns of numbers correspond to the explained variances for the models for the endogenous blocks B, C, D, and E (the numbers in parentheses represent the number of components used). Source: (Romano *et al.*, 2019). Reproduced with permission from Wiley.

Input	**Dependent blocks**			
	B	**C**	**D**	**E**
A	37.1% (1)	46.45% (2)	30.50 (3)	49.90% (4)
B		52.03% (1)	54.78 (1)	52.65% (1)
C			59.65 (1)	60.82% (1)
D				78.10% (1)

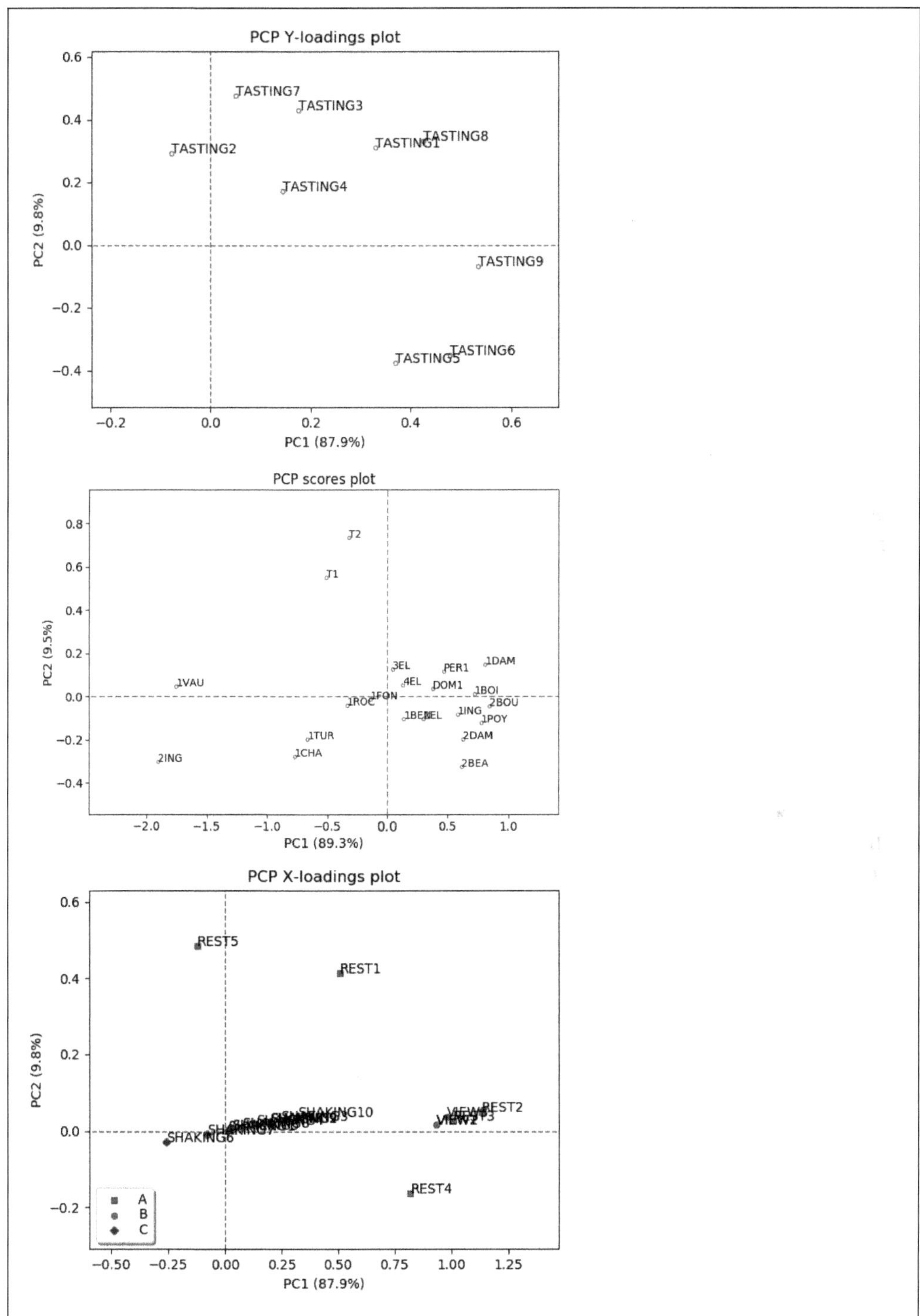

Figure 10.6 Wine data. PCP plots for prediction of block D from blocks A, B, and C. Scores and loadings from PCA on the predicted y-values on top. The loadings from projecting the orthogonalised X-blocks (except the first which is used as is) onto the scores at the bottom. Source: Romano *et al.* (2019). Reproduced with permission from Wiley & Sons.

ELABORATION 10.3

SO-PLS related methods

SO-PLS for logistic regression

A logistic variant of the SO-PLS method was published by Bazzoli and Lambert-Lacroix (2018). The method is an extension of the use of standard PLS regression in the logistic framework discussed in, for instance, Nguyen and Rocke (2002). The idea is to simply replace the standard LS method in the optimisation (IRLS, iteratively re-weighted least squares) by SO-PLS. Different solutions are considered in the paper.

Two examples are given based on real data. In the first example one of the blocks contains genetic information and the other block is based on clinical features. The Y-variable is 'the response of childhood malignant embryonal tumours of the central nervous system to therapy'. The data set in this case consists of 60 individuals; 21 of them died and 39 survived the next 24 months. The other data set is related to breast cancer and concerns prediction of status of the tumour (two groups) based on somatic and long-term clinical follow-up data. Encouraging results are obtained in both cases.

Serial PLS

Berglund and Wold (1999) proposed a method which shares some similarities with SO-PLS. It is based on some of the same underlying ideas, but it is only defined for two input blocks $\mathbf{X}_1$ and $\mathbf{X}_2$.

The method goes as follows:

1: Set $\mathbf{F}_2 = \mathbf{Y}$. First calculate the PLS model with $\mathbf{F}_2$ as dependent block and $\mathbf{X}_1$ as independent block and calculate the difference between $\mathbf{Y}$ and this model, and call it $\mathbf{F}_1$.
2: Then use PLS regression on $\mathbf{F}_1$ versus $\mathbf{X}_2$. The difference between $\mathbf{Y}$ and this model is called $\mathbf{F}_2$.
3: Go to start and continue until convergence.

In other words, the method is a repetitive use of PLS for the two blocks using the residuals from the opposite as the input. If the serial PLS (S-PLS) is stopped after the first round it is almost identical to SO-PLS except that the $\mathbf{X}_2$ is not orthogonalised. In the end the scores from the two PLS models are used as input in the full LS regression equation with $\mathbf{Y}$ as dependent block, as for SO-PLS.

The number of components has to be decided before the iteration starts since the residuals depend on the number of components used. An important difference between this method and the SO-PLS is that S-PLS does not impose any order of the blocks. In addition, S-PLS is limited to two input blocks.

10.6 Methods for Data Sets Split Along the Sample Mode, Multigroup Methods

10.6.1 Multigroup PLS Regression

The multigroup PLS regression is a supervised method for situations where both input and output blocks are naturally split along the sample mode (Eslami *et al.* (2014)). The method

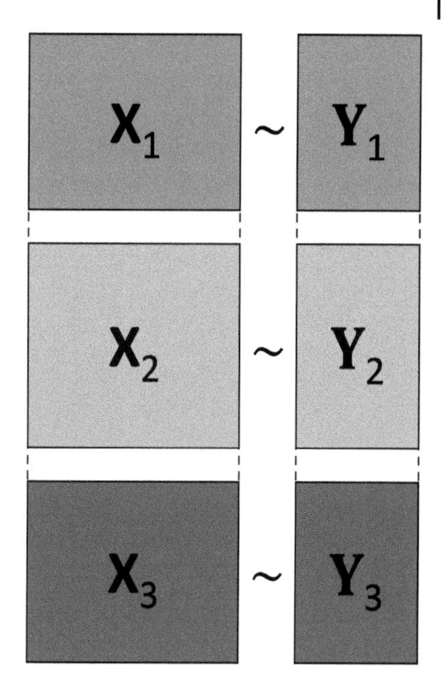

Figure 10.7 An illustration of the multigroup setup, where variables are shared among $\mathbf{X}$ blocks and related to responses, $\mathbf{Y}$, also sharing their own variables.

was later extended to classification problems also involving an L_1-penalisation (Rohart *et al.* (2017a)).

The multiblock structure for this method can be presented as in Figure 10.7. We will here call the combination of the blocks $\mathbf{X}_m$ and $\mathbf{Y}_m$ a group. This type of group structure may typically arise when the groups consist of measurements with origin in different years or in different places. The focus is to benefit from a larger data set for better predictions in each group, i.e., to utilise as much as possible of the common row structure in the groups.

For this type of data, the simplest strategy is to just use a joint PLS regression for all blocks, but this ignores the group structure. The opposite extreme is to make a new PLS model for each group separately, but in this case the number of samples for each group can easily become too small for establishing sufficient model stability. An alternative is to build a compromise PLS regression where estimation for each group benefits from the structure in the others. It is evident that such an approach is only useful when there is enough common structure in the relationship between input and output in the block groups.

A number of closely related compromise approaches were suggested in Eslami *et al.* (2014). All the approaches are based on optimising a weighted PLS-like criterion where the weights depend on the number of samples in each group. If we let $\mathbf{X}_m$ and $\mathbf{Y}_m$ denote the centred $\mathbf{X}$- and $\mathbf{Y}$-blocks for group m and I_m be the number of samples in block m, one of the criteria proposed can, for each component, be written as:

$$max \sum_{m=1}^{M} I_m^2 \text{cov}^2(\mathbf{u}_m, \mathbf{t}_m), \tag{10.15}$$

where $\mathbf{t}_m = \mathbf{X}_m\mathbf{w}$, $\mathbf{u}_m = \mathbf{Y}_m\mathbf{c}$, $||\mathbf{w}|| = ||\mathbf{c}|| = 1$. The maximisation is over the common weights $\mathbf{w}$ (X-block) and $\mathbf{c}$ (Y-block). The $\mathbf{t}_m$ and $\mathbf{u}_m$ are the scores for group m.

In other words, instead of maximising the covariance of linear functions of the whole $\mathbf{X}$ and $\mathbf{Y}$, linear functions are calculated for each block and put together in a weighted sum with the number of samples in each group as the weights. The $\mathbf{X}_m$ and $\mathbf{Y}_m$ (see Figure 10.7) are centred and possibly preprocessed group-wise prior to optimisation. A NIPALS like algorithm for finding the solution is presented in the original paper.

Deflation is done as usual for PLS-based methods, here with respect to the global component $\mathbf{t} = \mathbf{X}\mathbf{w}$. where $\mathbf{X}$ is the vertically concatenated matrix of $\mathbf{X}_m$ blocks and similarly for $\mathbf{Y}$ (see Eslami *et al.* (2014)). This means that in the next round, the original $\mathbf{X}$-s and $\mathbf{Y}$-s are replaced by:

$$\mathbf{X}_{\text{New}} = \left(\mathbf{I} - \frac{\mathbf{t}\mathbf{t}^t}{\mathbf{t}^t\mathbf{t}}\right)\mathbf{X} \quad and \quad \mathbf{Y}_{New} = \left(\mathbf{I} - \frac{\mathbf{t}\mathbf{t}^t}{\mathbf{t}^t\mathbf{t}}\right)\mathbf{Y}. \tag{10.16}$$

We also refer to Section 2.8 for a more general discussion of deflation and orthogonalisation.

In another criterion proposed in the paper by Eslami *et al.* (2014), the squared covariances are replaced by the covariances in Equation 10.15 only and the weights are I instead of I^2, thus putting less emphasis on the larger covariances and the groups with the most samples. Two alternative criteria were also considered where the $\mathbf{Y}$ weights $\mathbf{c}$ in Equation 10.15 are allowed to vary from group to group.

The methods provide scores and weights for each group as well as a prediction equation (see Eslami *et al.* (2014) for details). For a single $\mathbf{y}$-variable and corresponding $\mathbf{x}$ vector in group m, the equation can be written as:

$$\hat{y} = \bar{y}_m + \mathbf{x}_m^t\hat{\mathbf{b}} \tag{10.17}$$

where $\mathbf{x}_m$ is group centred using the x-centres from the calibration data, $\hat{\mathbf{b}}$ is the vector of regression coefficients, obtained from the parameters in the PLS model, and $\bar{y}_m$ is the mean of the $\mathbf{y}$-variable for block m in the calibration set.

The methods were in Eslami *et al.* (2014) tested and compared on data from olive oil where the blocks come from different origin of the oil. The $\mathbf{X}$-variables in this example were physico-chemical variables and the $\mathbf{Y}$-variables come from descriptive sensory analysis. The prediction results for the four strategies (replacing squared covariance by covariances and using the same or different $\mathbf{Y}$ loadings) were similar and better than standard PLS regression.

The approach was further explored and extended to classification problems in Rohart *et al.* (2017a). The criterion is essentially the same, but now focus is on classification using a dummy $\mathbf{Y}$ matrix (see Chapter 7) and an additional L_1 penalty is put on the vectors for each dimension separately. The penalty is formulated as $\lambda||\mathbf{w}||_1$ for each component. As claimed in Rohart *et al.* (2017a) this means that the method seeks variables that share common information across all the groups and which then represent a more reproducible signature of the samples. The method is applied in the context of molecular signatures and shows promising results.

10.6.2 Clustering of Observations in Multiblock Regression

Bougeard *et al.* (2018) proposed a procedure for splitting the number of samples (see taxonomy in Chapter 3) into groups with similar regression relationship. There is one output variable $\mathbf{y}$ and possibly several input blocks $\mathbf{X}_1, \ldots, \mathbf{X}_M$, i.e., the situation is a standard multiblock regression setup with one response variable. As opposed to the methods in the previous section, the groups of samples are here not given in advance. On the contrary, the purpose is to find natural groups of samples with similar relation between the input blocks and the output variable.

This method is a generalisation of ideas and approaches used in standard regression (see e.g., Næs and Isaksson (1991)) for splitting the **X** domain into sub-domains with a different relation to **y**. The extension is that now more than one input block, **X**, on the input variable side is allowed. The optimisation criterion is a least squares criterion minimising the residual between **y** and predictions of **y** based on the redundancy analysis (RDA) method. The optimisation is done over the different ways of splitting into subsets (see Chapter 5 for further discussion of clustering in multiblock analysis).

10.6.3 Domain-Invariant PLS, DI-PLS

Nikzad-Langerodi *et al.* (2018) proposed an alternative method for handling regression data that are split along the sample mode (so far only formulated for one response variable **y**). The method was called domain-invariant (DI) PLS regression. A typical example of the use of the method is in spectroscopy when an old data set, possibly collected under slightly different conditions, is to be used as a supplement for strengthening a calibration at a later stage. The old data set is in the paper called the source domain (S) and of the new data set the target (T) domain, but the procedure below is independent on this order. The question is how to combine the two sets with measurements of both **X** and **y** in a sensible way. As can be seen, the challenge here is more or less the same as for the multigroup PLS regression.

The method is based on (for each component) the following formulation for the weights for PLS

$$\mathbf{w} = \text{argmin}_{\mathbf{w}} \|\mathbf{X} - \mathbf{y}\mathbf{w}^t\|^2. \tag{10.18}$$

The modification proposed for incorporating the block source domain is by the use of a penalty. The optimisation can be formulated in the following way

$$\mathbf{w}_{DI-PLS} = \text{argmin}_{\mathbf{w}}(\|\mathbf{X} - \mathbf{y}\mathbf{w}^t\|^2 + \lambda|\frac{1}{I_S - 1}\mathbf{w}^t\mathbf{X}_S^t\mathbf{X}_S\mathbf{w} - \frac{1}{I_T - 1}\mathbf{w}^t\mathbf{X}_T^t\mathbf{X}_T\mathbf{w}|) \tag{10.19}$$

where the λ can be optimised by for instance cross-validation. The I_S and I_T are the number of rows in $\mathbf{X}_S$ and $\mathbf{X}_T$ respectively. In this case the data for both the source and target domain are centred. We refer to the original publications for a discussion of different mean centring for different usage of the method. Deflation can be done as for PLS by orthogonalisation.

As can be seen, this criterion is simply a weighted sum of the standard PLS criterion for both domains and a penalisation on the difference in variance of $\mathbf{X}_S\mathbf{w}$ and $\mathbf{X}_T\mathbf{w}$. In other words, the common variability in the variable mode is used for restricting the information used to build the PLS model. The closer the two terms in the penalty are, and the smaller the λ is, the less influence will there be from the penalty.

Another interpretation is apparent when assuming that the scores $\mathbf{X}_S\mathbf{w}$ and $\mathbf{X}_T\mathbf{w}$ are approximately normally distributed (which is not unreasonable given that they are linear combinations). These scores are both centred (because $\mathbf{X}_S$ and $\mathbf{X}_T$ are centred). Then imposing the penalty means that also the variances of these scores are forced to be similar. Hence, the first and second moments of the scores are very similar which – together with normality – means that they are equally distributed. This is known as test equating in statistics (Kolen and Brennan, 2014).

The method was tested and compared with a standard PLS regression approach on data sets from NIR spectroscopy and gave promising prediction results (Nikzad-Langerodi *et al.* (2018)). This points to the method as a good alternative for joining two data sets in regression. Comparisons between the DI-PLS and the multigroup PLS regression are needed.

10.7 Conclusions and Recommendations

The methods in this chapter are diverse in the sense that they are based on very different estimation principles/criteria, comprising PLS regression (as in Chapters 2 and 7), redundancy analysis (see Chapter 2) and principal covariates regression (MWMBCovR) where components are chosen according to both explained variance of **Y** and **X**. There is no reason for preferring one of them generally over the others, but in the following we describe some of the aspects that can help the choice. A decision tree based on type of data and handling of these is found in Figure 10.8.

Combining groups of samples in one single model

Two of the methods presented at the end of the chapter have this property which is particularly useful when the sample set is split in two or more according to possible changes in how the samples are selected. One of the methods presented is general, it is based on PLS and can handle many sets of samples. The other one handles two groups and is particularly useful when one of the sets is primarily meant for supporting (improving prediction results) based on the other.

Collinear X-data

All methods presented based on PLS handle this problem. The MWMBCovR method, where components are chosen to explain both variability in **X** and **Y**, is also able to handle this problem. The RDA method, however, is more vulnerable to collinearity, as also discussed in Chapter 2.

Combining two-way and three-way input blocks

Two of the methods presented, the MWMBCovR and one of the SO-PLS variants are able to handle this. The rest of the methods require two-way input data.

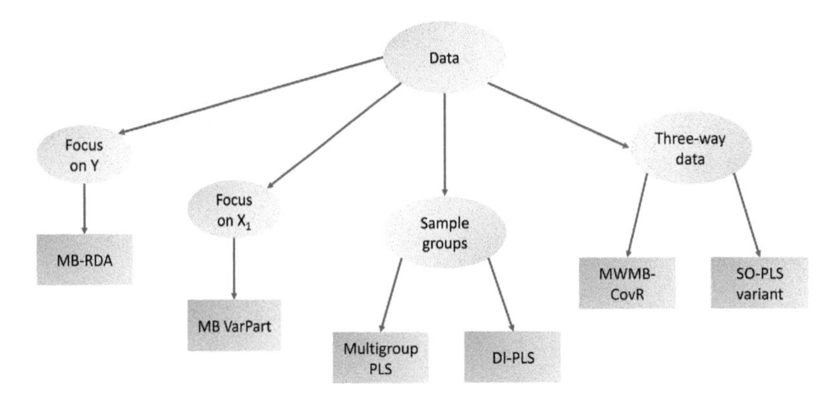

Figure 10.8 Decision tree for selecting a supervised method. For more explanation, see text.

Distinct and common variability

The only method here which focuses explicitly on the concepts of common and distinct variability is the multiblock variance partitioning. All methods have strong focus on common variability for prediction, but less focus on the distinct. The RDA based methods, however, calculate block-wise components and the SO-PLS variants calculate additional variability which are related to distinct variability. We refer to Chapter 7 for more discussion on SO-PLS and relation to these concepts.

10.8 Open Issues

The challenges here resemble the ones discussed in Chapter 7. Below we repeat some of them and indicate a few of special interest here.

Most of the methods discussed in this chapter are relatively new and tested only in a few examples. Some properties have been clarified as described in the various sections above, but much more information is needed both from a theoretical and applied perspective. One of the important points here is validation and the sensitivity to overfitting. Some of the methods involve a number of parameters to be determined, and it is not obvious how to optimise them in practice for safe and robust usage.

Another and closely related area that needs further research and investigation is comparison between methods, also with the methods in Chapter 7. Such comparisons should comprise prediction ability, usefulness of information obtained, and user-friendliness in practice. Another important point here is to investigate how the different methods handle different underlying dimensionalities of the blocks which may often happen in practice. Several of the methods are based on extracting some sort of common components, and how this functions when for instance one block has two components and another block has eight components is not easy to determine.

Outliers may occur in practice as was discussed in Chapter 2. Most methods presented in this chapter provide scatter plots and also residuals. Therefore, some of the residual and plotting procedures that were discussed in Chapter 2 also apply here. More applications and fine-tuning of these general principles in this context are, however, needed.

Missing values handling is always a difficult area. How this can and should be done, and also what are the impacts on the end result, is not studied for all the methods in this chapter.

Part V

Software

11

Algorithms and Software

11.1 Multiblock Software

A large proportion of the multiblock methods found in this book are available as software implementations both in open source software and through commercial software packages. A large, but non-exhaustive, selection is included in Section 11.10. As of 2022 the largest collection of methods is found in the programming language R, though spread over many packages with various interfaces (see Section 11.10.1). Our attempt at unifying a large amount of methods and extending with more is described below and exemplified in Section 11.2.

MATLAB (MATLAB, 2021) is also a popular programming environment for multiblock packages with many open source contributions (see Section 11.10.2) in addition to the popular commercial software in the PLS_Toolbox by Eigenvector. Further, Python (Van Rossum and Drake, 2009) is gaining popularity in multiblock analyses (11.10.3), and Excel still has a large user base among data analysts, e.g., using the XLSTAT add-on from Addinsoft.

Chapter structure

The rest of this chapter contains six sections describing the R package `multiblock` accompanying this book and a final section summarising a large proportion of the available multiblock software available through other sources. The sections pertaining to the `multiblock` package follow the organisation of methods in this book, grouping the methods into *basic*, *unsupervised*, *ASCA*, *supervised*, and *complex data structures*. Each section describes how data need to be formatted, how to perform basic modelling and how to obtain outputs and plots from more than 20 included methods. Focus is on practical use, rather than method comparison.

11.2 R package `multiblock`

For easy application of many of the methods presented in this book, we have developed an R package simply called `multiblock`, freely available from The Comprehensive R Archive Network (CRAN, R Core Team (2022)). This package includes methods that are unique to the package and imports a large range of methods from other packages. We have attempted to make the interface to the methods as unified as possible for easy testing of many methods without large changes to R scripts. Code examples with short explanations are found in Sections 11.3-11.9. A basic proficiency in R programming is a prerequisite for benefiting

Multiblock Data Fusion in Statistics and Machine Learning: Applications in the Natural and Life Sciences,
First Edition. Age K. Smilde, Tormod Næs, and Kristian Hovde Liland.

from this section. Overviews of basic package usage and included methods are split into the following subsections, largely following the chapters in this book:

- **Installing and starting the package**
  Section 11.3
- **Data handling**
  Section 11.4
- **Basic methods**
  Section 11.5, Chapter 2
- **Unsupervised methods**
  Section 11.6, Chapters 5/9
- **ASCA**
  Section 11.7, Chapter 6
- **Supervised methods**
  Section 11.8, Chapters 7/10
- **Complex data structures**
  Section 11.9, Chapters 8/10

11.3 Installing and Starting the Package

After starting the R version of choice – be it 'vanilla' R (https://cran.r-project.org/) with its vast community support, Microsoft R Open (https://mran.microsoft.com/) with its quick mathematical libraries, or R Studio with its intuitive interface and one of the mentioned R's as back-end – the `multiblock` package can be installed and started from the R terminal using the following commands.

```
# One-time installation
install.packages("multiblock")

# Starting the package
library(multiblock)
```

11.4 Data Handling

This section shows briefly how to read data from selected data formats, perform basic preprocessing, and prepare blocks of data for multiblock analysis.

11.4.1 Read From File

Data are stored in many different file formats. The following three examples cover two types of CSV-files and generic flat files.

```
# Find directory 'extdata' from the multiblock package
mbdir <- system.file('extdata/', package = "multiblock")

# Comma separated values, row names in first column
meta_data <- read.csv(paste0(mbdir, "/meta_data.csv"), row.names = 1)
# If working directory matches file location:
# meta_data <- read.csv('meta_data.csv', row.names = 1)
meta_data
```

```
        temperature  colour
John            38.0    blue
Julia           37.0   green
James           37.5    blue
Jacob           37.6     red
Jane            37.2     red
Johanna         37.9   green
```

```
# Semi-colon separated values (locales where the decimal point is
# comma), no row names
proteins <- read.csv2(paste0(mbdir, "/proteins.csv"))
proteins
```

```
        prot1       prot2      prot3
1  0.46532048  0.30183300 -1.4654414
2 -1.79802081 -0.22812232 -0.4639203
3 -1.92962434 -0.40513080  0.1767796
4  0.87437138  0.79843798  0.1234731
5 -0.62445278 -0.07975479 -1.1126332
6 -0.07493721  1.09576027  1.2656596
```

```
# Blank space separated data without labels
genes <- read.table(paste0(mbdir, "/genes.dat"))
genes
```

```
           V1         V2         V3
1  0.39033106 -0.5720390  1.9147573
2  0.55352785  0.0948703 -0.2239755
3  0.09872346 -0.1029385  0.9047138
4 -0.59213740 -0.6027739  0.6177083
5 -0.02350148  0.3572809 -0.5168416
6  0.76644845  1.2863428  1.8239298
```

11.4.2 Data Pre-processing

Before analysis, various types of preprocessing may be needed. Centring and standardising/scaling may be considered the most basic. In R, these operations are performed column-wise by default, leading to autoscaling. If these operations are performed on the rows, we perform the standard normal variate (SNV) instead.

```
# Column-centring
genes_centred <- scale(genes, scale=FALSE)
colMeans(genes_centred) # Check mean values
```

```
           V1            V2            V3
2.081668e-17 -3.006854e-17  1.850372e-17
```

```
# Autoscaling
genes_scaled <- scale(genes)
apply(genes_scaled, 2, sd) # Check standard deviations
```

```
V1 V2 V3
 1  1  1
```

```
# SNV (transpose, autoscale, re-transpose)
genes_snv <- t(scale(t(genes)))
apply(genes_snv, 1, sd) # Check standard deviations
```

```
[1] 1 1 1 1 1 1
```

11.4.3 Re-coding Categorical Data

Most analysis methods require continuous input data. The `data.frame` named `meta_data` contains a character vector (a factor in older R versions) of categories. The `multiblock` package has a function, `dummycode`, for converting categorical data to various dummy formats.

```
# Default is sum coding
dummycode(meta_data$colour)
```

```
  x1 x2
1  1  0
2  0  1
3  1  0
4 -1 -1
5 -1 -1
6  0  1
```

```
# Treatment coding
dummycode(meta_data$colour, "contr.treatment")
```

```
  xgreen xred
1      0    0
2      1    0
3      0    0
4      0    1
5      0    1
6      1    0
```

```
# Full dummy-coding (rank deficient)
dummycode(meta_data$colour, drop = FALSE)
```

```
  xblue xgreen xred
1     1      0    0
2     0      1    0
3     1      0    0
4     0      0    1
5     0      0    1
6     0      1    0
```

```
# Replace categorical with dummy-coded, use I() to index by common name
meta_data2 <- meta_data
meta_data2$colour <- I(dummycode(meta_data$colour, drop = FALSE))
meta_data2
```

```
        temperature colour.xblue colour.xgreen colour.xred
John           38.0            1             0           0
Julia          37.0            0             1           0
James          37.5            1             0           0
Jacob          37.6            0             0           1
Jane           37.2            0             0           1
Johanna        37.9            0             1           0

meta_data2$colour
  xblue xgreen xred
1     1      0    0
2     0      1    0
3     1      0    0
4     0      0    1
5     0      0    1
6     0      1    0
```

11.4.4 Data Structures for Multiblock Analysis

11.4.4.1 Create List of Blocks

A simple list of blocks can be created using the list() function. Naming of the blocks can be done directly or after creation.

```
# Direct approach
blocks1 <- list(meta = meta_data2, proteins = proteins, genes = genes)

# Two-step approach
blocks2 <- list(meta_data2, proteins, genes)
names(blocks2) <- c('meta', 'proteins', 'genes')

# Same result
identical(blocks1, blocks2)

[1] TRUE

# Access by name or number
blocks1[['meta']]
# .. or
blocks2[[1]]

        temperature colour.xblue colour.xgreen colour.xred
John           38.0            1             0           0
Julia          37.0            0             1           0
James          37.5            1             0           0
Jacob          37.6            0             0           1
Jane           37.2            0             0           1
Johanna        37.9            0             1           0
```

11.4.4.2 Create `data.frame` of Blocks

A `data.frame` is a convenient storage format for data in R and can handle many types of variables, e.g., numeric, logical, character, factor, or matrices. The latter is useful for analyses of data with shared sample mode.

```
# Construct block data.frame from list
df1 <- block.data.frame(blocks1)

# Construct block data.frame from data.frame:
# First merge blocks into data.frame
my_data <- cbind(meta_data2, proteins, genes)
# Then construct block data.frame using named
# list of indexes
df2 <- block.data.frame(my_data, block_inds =
      list(meta = 1:2,
           proteins = 3:5,
           genes = 6:8))

# Same result
identical(df1,df2)

[1] TRUE

# Access by name or number
df1[[2]]
# .. or
df2[['proteins']]
              prot1       prot2      prot3
John     0.46532048  0.30183300 -1.4654414
Julia   -1.79802081 -0.22812232 -0.4639203
James   -1.92962434 -0.40513080  0.1767796
Jacob    0.87437138  0.79843798  0.1234731
Jane    -0.62445278 -0.07975479 -1.1126332
Johanna -0.07493721  1.09576027  1.2656596
```

```
df1[c(1,3)]
# .. or
df1[-2]

$meta
        temperature colour.xblue colour.xgreen colour.xred
John           38.0            1             0           0
Julia          37.0            0             1           0
James          37.5            1             0           0
Jacob          37.6            0             0           1
Jane           37.2            0             0           1
Johanna        37.9            0             1           0

$genes
                 V1         V2         V3
John     0.39033106 -0.5720390  1.9147573
Julia    0.55352785  0.0948703 -0.2239755
James    0.09872346 -0.1029385  0.9047138
Jacob   -0.59213740 -0.6027739  0.6177083
Jane    -0.02350148  0.3572809 -0.5168416
Johanna  0.76644845  1.2863428  1.8239298

df2[c('proteins','genes')]

$proteins
              prot1       prot2      prot3
John     0.46532048  0.30183300 -1.4654414
Julia   -1.79802081 -0.22812232 -0.4639203
James   -1.92962434 -0.40513080  0.1767796
Jacob    0.87437138  0.79843798  0.1234731
Jane    -0.62445278 -0.07975479 -1.1126332
Johanna -0.07493721  1.09576027  1.2656596

$genes
                 V1         V2         V3
John     0.39033106 -0.5720390  1.9147573
Julia    0.55352785  0.0948703 -0.2239755
James    0.09872346 -0.1029385  0.9047138
Jacob   -0.59213740 -0.6027739  0.6177083
Jane    -0.02350148  0.3572809 -0.5168416
Johanna  0.76644845  1.2863428  1.8239298
```

11.5 Basic Methods

The following single- and two-block methods are available in the `multiblock` package (function names in parentheses). The corresponding theory is found in Chapters 2 and 9.

- PCA – principal component analysis (`pca`) – Section 2.1.2
- PCR – principal component regression (`pcr`) – Section 2.1.4
- PLSR – partial least squares regression (`plsr`) – Section 2.1.5
- CCA – canonical correlation analysis (`cca`) – Section 2.1.10
- IFA – interbattery factor analysis (`ifa`) – Not described in the book
- GSVD – generalised SVD (`gsvd`) – Section 9.1

The following sections will describe how to format/structure data for analysis and invoke all methods from the list above.

11.5.1 Prepare Data

We use a selection of extracts from a potato data set (Thygesen *et al.* (2001)) included in the package for the basic data analyses. The data set is stored as a named list of nine matrices with chemical, rheological, spectral (x6), and sensory measurements with measurements from 26 raw and cooked potatoes.

```
data(potato)
X <- potato$Chemical
y <- potato$Sensory[,1,drop=FALSE]
```

11.5.2 Modelling

Since the basic methods cover both single block analysis, supervised and unsupervised analysis, the interfaces for the basic methods vary a bit. Supervised methods use the formula interface and the remaining methods take input as a single matrix or list of matrices. See subsections for supervised and unsupervised analysis for details.

```
# Single block
pot.pca <- pca(X, ncomp = 2)

# Two blocks, supervised
pot.pcr <- pcr(y ~ X, ncomp = 2)
pot.pls <- plsr(y ~ X, ncomp = 2)

# Two blocks, unsupervised
pot.cca <- cca(potato[1:2])
pot.ifa <- ifa(potato[1:2])

# Variable linked decomposition (transposed matrices)
pot.gsvd <- gsvd(lapply(potato[3:4], t))
```

11.5.3 Common Output Elements Across Methods

Output from all methods include matrices called `loadings`, `scores`, `blockLoadings` and `blockScores`, or a suitable subset of these according to the method used. An `info` list describes which types of (block-) loadings/scores are in the output. There may be various extra elements in addition to the common elements, e.g., coefficients, weights etc. The `names()` and `summary()` functions below show all elements of the object and a summary based on the `info` list, respectively.

```
# GSVD returns block-scores and common loadings:
names(pot.gsvd)
summary(pot.gsvd)
```

```
[1] "loadings"     "blockScores" "GSVD"         "info"         "call"

Generalised Singular Value Decomposition
========================================

$loadings: Loadings (26x26)
$blockScores: Block-scores:
- NIRraw (1050x1050), NIRcooked (1050x1050)
```

11.5.4 Scores and Loadings

Functions for accessing scores and loadings are based on functions from the pls package, but extended with a block parameter to allow extraction of common/super-scores/loadings and their block counterparts. The default value for the parameter block is 0, corresponding to the common/global block. Block-scores/-loadings can be accessed by setting block to a number or name.

```
# Global scores plotted with object labels
scoreplot(pot.pca, labels = "names")
```

```
# Block loadings for Chemical block (with variable labels)
# in scatter format
loadingplot(pot.cca, "Chemical", labels = "names")
```

```
# Non-existing elements are swapped with existing ones with a warning.
sc <- scores(pot.cca)
```

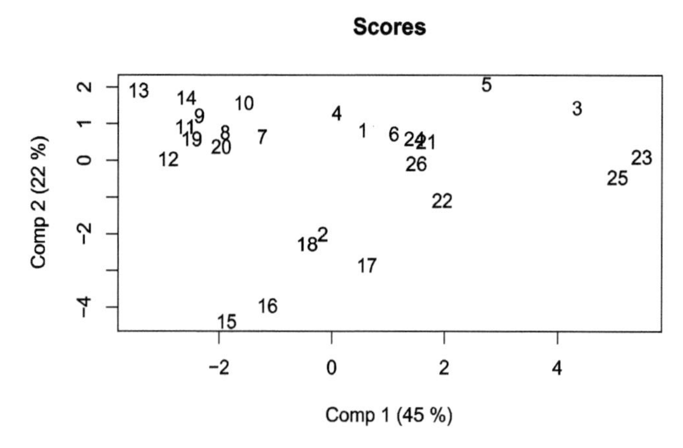

Figure 11.1 Output from use of scoreplot() on a pca object.

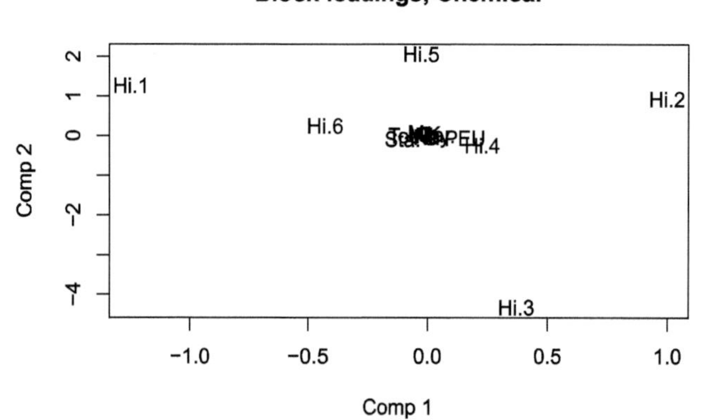

Figure 11.2 Output from use of loadingplot() on a cca object.

11.6 Unsupervised Methods

The following unsupervised methods are available in the `multiblock` package (function names in parentheses). The corresponding theory is found in Chapters 5 and 9.

- SCA – simultaneous component analysis (`sca`) – Section 2.1.11/5.1.1.1
- GCA – generalised canonical analysis (`gca`) – Section 5.2.1.3
- GPA – generalised procrustes analysis (`gpa`) – Section 9.3.1
- MFA – multiple factor analysis (`mfa`) – Section 5.2.1.2
- PCA-GCA (`pcagca`) – Section 5.2.2.3
- DISCO – distinct and common components with SCA (`disco`) – Section 5.1.2.1/5.2.2.2
- HPCA – hierarchical principal component analysis (`hpca`) – Not described in the book
- MCOA – multiple co-inertia analysis (`mcoa`) – Not described in the book
- JIVE – joint and individual variation explained (`jive`) – Section 5.2.2.1
- STATIS – Structuration des Tableaux à Trois Indices de la Statistique (`statis`) – Section 5.2.1.2
- HOGSVD – higher order generalized SVD (`hogsvd`) – Not described in the book

The following sections will describe how to format/structure data for analysis and invoke all methods from the list above.

11.6.1 Formatting Data for Unsupervised Data Analysis

Data blocks are best stored as named `lists` for use with unsupervised methods in this package. If also column names and row names are used for all blocks, these can be used for easy labelling in plots supplied by the `pls` package. See examples below for illustrations of this.

```
# Load potato data
data(potato)
class(potato)
# data.frames can contain matrices as variables,
# thus becoming sample linked lists of blocks.
str(potato[1:3])

# Explicit conversion to a list
potList <- as.list(potato[1:3])
str(potList)
```

```
[1] "data.frame"

'data.frame':        26 obs. of  3 variables:
 $ Chemical   : 'AsIs' num [1:26, 1:14] 3.21 3.26 5.18 3.75 2.92 ...
  ..- attr(*, "dimnames")=List of 2
  .. ..$ : chr [1:26] "1" "2" "3" "4" ...
  .. ..$ : chr [1:14] "PEU" "Sta." "TotN" "Phy." ...
 $ Compression: 'AsIs' num [1:26, 1:12] 2.23 1.21 1.63 2.68 2.85 ...
  ..- attr(*, "dimnames")=List of 2
  .. ..$ : chr [1:26] "1" "2" "3" "4" ...
  .. ..$ : chr [1:12] "FW20" "BW20" "ST20" "SH20" ...
 $ NIRraw     : 'AsIs' num [1:26, 1:1050] -0.831 -0.807 -0.836 ...
  ..- attr(*, "dimnames")=List of 2
  .. ..$ : chr [1:26] "1" "2" "3" "4" ...
  .. ..$ : chr [1:1050] "400" "402" "404" "406" ...

List of 3
 $ Chemical   : 'AsIs' num [1:26, 1:14] 3.21 3.26 5.18 3.75 2.92 ...
  ..- attr(*, "dimnames")=List of 2
  .. ..$ : chr [1:26] "1" "2" "3" "4" ...
  .. ..$ : chr [1:14] "PEU" "Sta." "TotN" "Phy." ...
 $ Compression: 'AsIs' num [1:26, 1:12] 2.23 1.21 1.63 2.68 2.85 ...
```

```
  ..- attr(*, "dimnames")=List of 2
  .. ..$ : chr [1:26] "1" "2" "3" "4" ...
  .. ..$ : chr [1:12] "FW20" "BW20" "ST20" "SH20" ...
 $ NIRraw      : 'AsIs' num [1:26, 1:1050] -0.831 -0.807 -0.836 ...
  ..- attr(*, "dimnames")=List of 2
  .. ..$ : chr [1:26] "1" "2" "3" "4" ...
  .. ..$ : chr [1:1050] "400" "402" "404" "406" ...
```

11.6.2 Method Interfaces

All unsupervised methods supplied by this package share a common interface which expects a list of blocks as the first input. Methods that are imported from other packages are wrapped in a function that gives the mentioned interface. Results from the imported method are stored in a separate slot in the output in case specialised `plot` or `summary` functions are available or direct inspection is needed. If default parameters are used, a single list of blocks with suitably linked matrices (shared objects or variables) will result in a basic analysis (see first code block below). In addition, all methods have parameters that control their behaviour, e.g., number of components, convergence criteria, etc.

11.6.3 Shared Sample Mode Analyses

The following block of code loads a multiblock data set, extracts three blocks, and runs through all included unsupervised methods having shared sample mode using the same interface.

```
# Attach dataset
data(potato)
potList <- as.list(potato[c(1,2,9)])

# FactoMineR <=2.3 uses recycling of length 1 array.
suppressWarnings(
# DISCOsca in package RegularizedSCA is highly verbose.
invisible({capture.output({
pot.sca <- sca(potList)
pot.gca <- gca(potList)
pot.gpa <- gpa(potList)
pot.mfa <- mfa(potList)
pot.pcagca <- pcagca(potList)
pot.disco <- disco(potList)
pot.hpca <- hpca(potList)
pot.mcoa <- mcoa(potList)
})}))
```

11.6.4 Shared Variable Mode

The following block of code loads a sensory data set (Luciano and Næs (2009), see Example 6.5 for description), extracts blocks and runs through all included unsupervised methods having shared variable mode using the same interface.

```
# Shared variable mode data
data(candies)
candyList <- lapply(1:nlevels(candies$candy),
   function(x)candies$assessment[candies$candy==x,])

# jive in package r.jive is highly verbose.
invisible({capture.output({
can.sca <- sca(candyList)
can.jive <- jive(candyList)
can.statis <- statis(candyList)
can.hogsvd <- hogsvd(candyList)
})})
```

11.6.5 Common Output Elements Across Methods

Output from all methods include slots called `loadings`, `scores`, `blockLoadings` and `blockScores`, or a suitable subset of these. An `info` slot describes which types of (block-) loadings/scores are in the output. There may be various extra elements in addition to the common elements, e.g., coefficients, weights etc.

```
# SCA used with shared variable mode data returns block-loadings and
# common scores:
names(pot.sca)
summary(pot.sca)

[1] "blockLoadings" "scores"        "samplelinked"  "info"
[3] "call"

Simultaneous Component Analysis

Score type:  Common scores
Block loadings type:  Block loadings


# MFA stores individual PCA scores and loadings as block elements:
names(pot.mfa)
summary(pot.mfa)

[1] "scores"        "loadings"      "blockScores"   "blockLoadings"
[5]  "info"         "MFA"           "call"

Multiple Factor Analysis
========================

$scores: Global scores (26x5)
$loadings: Global loadings (35x5)
$blockScores: Individual PCA scores:
- Chemical (26x5), Compression (26x5), Sensory (26x5)
$blockLoadings: Individual PCA loadings:
- Chemical (14x5), Compression (12x5), Sensory (9x5)
```

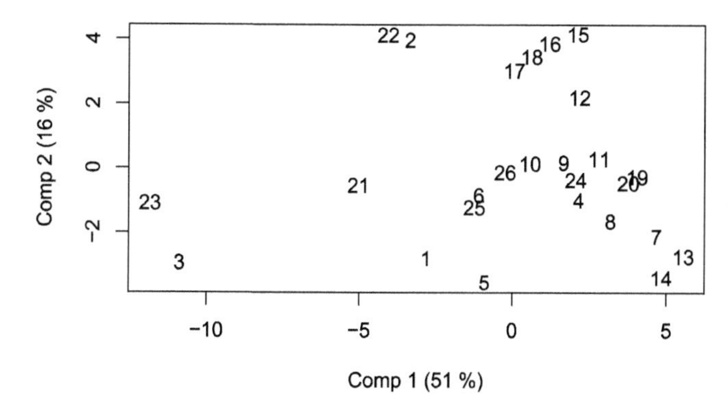

Figure 11.3 Output from use of `scoreplot(pot.sca, labels = "names")` (SCA scores in 2 dimensions).

11.6.6 Scores and Loadings

Functions for accessing scores and loadings are based on functions from the `pls` package, but extended with a `block` parameter to allow extraction of common/super-scores/loadings and their block counterparts. The default block is 0, corresponding to the common/super block. Block-scores/-loadings can be accessed by number or name.

```
# Global scores plotted with object labels
scoreplot(pot.sca, labels = "names")

# Block loadings for Sensory block with variable labels in
# scatter format
loadingplot(pot.sca, block = "Sensory", labels = "names")

# Non-existing elements are swapped with existing ones with
# a warning.
sc <- scores(pot.sca, block = 1)
```

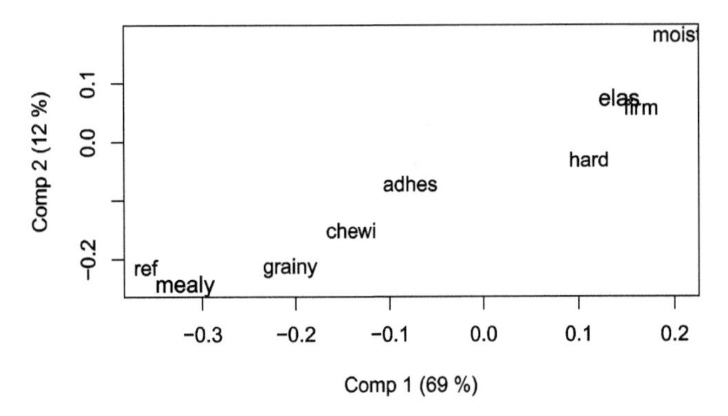

Figure 11.4 Output from use of `loadingplot(pot.sca, block = "Sensory", labels = "names")` (SCA loadings in 2 dimensions).

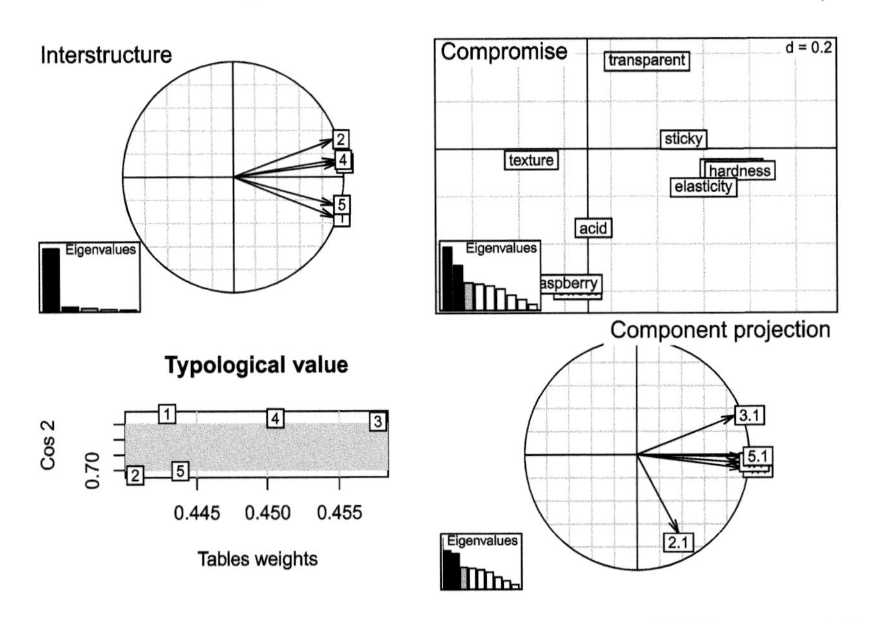

Figure 11.5 Output from use of `plot(can.statis$statis)` (STATIS summary plot).

11.6.7 Plot From Imported Package

Some methods in the package are wrappers for imported methods from other packages. Using the stored object from an imported method (see `$statis` in code below), one can exploit methods from the original package to expand on the methods available in this package. An example is the summary plot for the `statis` method in the package `ade4`.

```
# Apply a plot function from ade4 (no extra import required).
plot(can.statis$statis)
```

11.7 ANOVA Simultaneous Component Analysis

The following example uses a simulated data set for showcasing some of the possibilities of the ANOVA simultaneous component analysis (ASCA) method. The corresponding theory is found in Chapter 6.

11.7.1 Formula Interface

This ASCA implementation uses R's formula interface for model specification. This means that the first argument is a formula with response on the left and design on the right, separated by a tilde operator, e.g., `y ~ x + z` or `assessment ~ assessor + candy`. The names in the formula refer to variables in a data.frame (or list). Separation with plus (+) adds main effects to the model, while separation by stars (∗) adds main effects and interactions, e.g., `y ~ x * z`. Colons (`:`) can be used for explicit interactions, e.g., `y ~ x + z + x:z`. More complicated formulas exist, but only a simple subset is supported by `asca`.

11.7.2 Simulated Data

Two categorical factors and a covariate are simulated together with a standard normal set of 10 responses.

```
set.seed(1)
dataset <- data.frame(y = I(matrix(rnorm(24*10), ncol = 10)),
                      x = factor(c(rep(2,8), rep(1,8), rep(0,8))),
                      z = factor(rep(c(1,0), 12)), w = rnorm(24))
colnames(dataset$y) <- paste('Var', 1:10, sep = " ")
rownames(dataset) <- paste('Obj', 1:24, sep = " ")
str(dataset)
```

```
'data.frame':        24 obs. of  4 variables:
 $ y: 'AsIs' num [1:24, 1:10] -0.626 0.184 -0.836 1.595 0.33 ...
  ..- attr(*, "dimnames")=List of 2
  .. ..$ : NULL
  .. ..$ : chr [1:10] "Var 1" "Var 2" "Var 3" "Var 4" ...
 $ x: Factor w/ 3 levels "0","1","2": 3 3 3 3 3 3 3 3 2 2 ...
 $ z: Factor w/ 2 levels "0","1": 2 1 2 1 2 1 2 1 2 1 ...
 $ w: num  0.707 1.034 0.223 -0.879 1.163 ...
```

11.7.3 ASCA Modelling

A basic ASCA model having two factors is fitted and printed.

```
mod <- asca(y~x+z, data = dataset)
print(mod)
```

```
Anova Simultaneous Component Analysis fitted using 'lm' (Linear Model)
Call:
asca(formula = y ~ x + z, data = dataset)
```

11.7.4 ASCA Scores

Scores for the first factor are extracted and a score plot with confidence ellipsoids is produced.

```
sc <- scores(mod)
head(sc)
scoreplot(mod, legendpos = "topleft", ellipsoids = "confidence")
```

```
         Comp 1     Comp 2
Obj 1 0.9395791 -0.1039977
Obj 2 0.9395791 -0.1039977
Obj 3 0.9395791 -0.1039977
Obj 4 0.9395791 -0.1039977
Obj 5 0.9395791 -0.1039977
Obj 6 0.9395791 -0.1039977
```

This is repeated for the second factor.

```
sc <- scores(mod, factor = "z")
head(sc)
scoreplot(mod, factor = "z", ellipsoids = "confidence")
```

```
          Comp 1
Obj 1 -0.3831621
Obj 2  0.3831621
Obj 3 -0.3831621
Obj 4  0.3831621
Obj 5 -0.3831621
Obj 6  0.3831621
```

11.7.5 ASCA Loadings

A basic loadingplot for the first factor is generated using graphics from the `pls` package.

```
lo <- loadings(mod)
head(lo)
loadingplot(mod, scatter = TRUE, labels = 'names')
```

```
           Comp 1      Comp 2
Var 1 -0.03688007 -0.15615007
Var 2 -0.01764472 -0.05590506
Var 3  0.14250312  0.06184430
Var 4 -0.50220715 -0.35817451
Var 5 -0.54263018  0.45252899
Var 6  0.44399942 -0.01293480
```

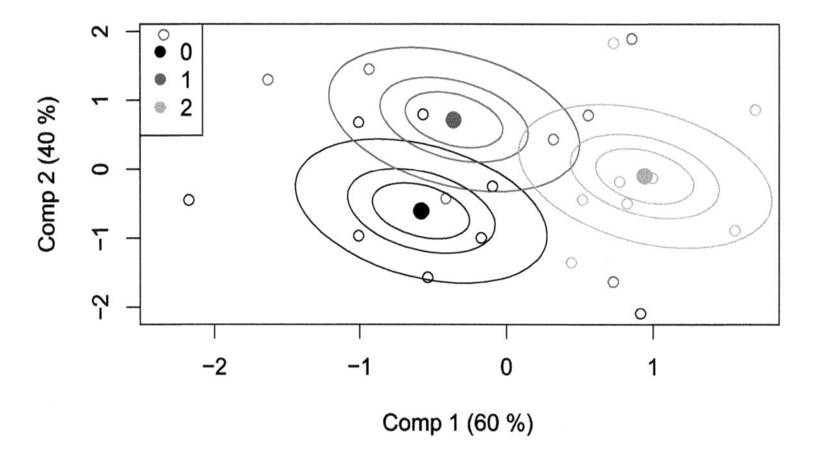

Figure 11.6 Output from use of `scoreplot()` (ASCA scores in 2 dimensions).

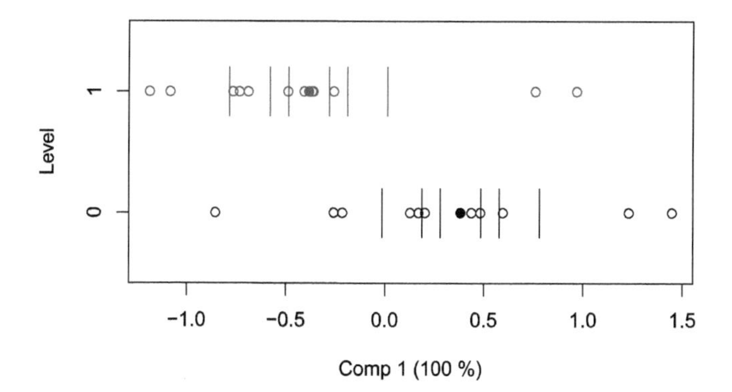

Figure 11.7 Output from use of `scoreplot()` (ASCA scores in 1 dimension).

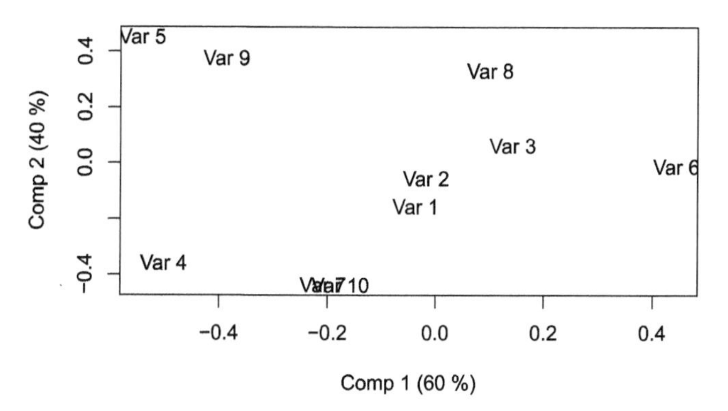

Figure 11.8 Output from use of `loadingplot()` (ASCA scores in 2 dimensions).

11.8 Supervised Methods

The following supervised methods are available in the `multiblock` package (function names in parentheses). The corresponding theory is found in Chapters 7 and 10.

- MB-PLS – multiblock partial least squares (`mbpls`) – Section 7.2
- sMB-PLS – sparse multiblock partial least squares (`smbpls`) – Section 7.2.3
- SO-PLS – sequential and orthogonalised PLS (`sopls`) – Section 7.3
- PO-PLS – parallel and orthogonalised PLS (`popls`) – Section 7.4
- ROSA – response oriented sequential alternation (`rosa`) – Section 7.5
- MB-RDA – multiblock redundancy analysis (`mbrda`) – Section 10.3

The following sections will describe how to format/structure data for analysis and invoke selected methods from the list above.

11.8.1 Formatting Data for Supervised Analyses

Data blocks are best stored as named `lists` or `data.frames` with blocks (see Section 11.4.4.2) for use with the formula interface of R. The following is an example with simulated data in one data block and one response block.

```
# Random data
n <- 30; p <- 90
```

```
X <- matrix(rnorm(n*p), nrow=n)
y <- X %*% rnorm(p) + 10

# Split X into three blocks in a named list
ABC <- list(A = X[,1:20], B = X[,21:50], C = X[,51:90], y = y)

# Model using names of blocks (see below for full SO-PLS example)
so.abc <- sopls(y ~ A + B + C, data = ABC, ncomp = c(4,3,4))
```

11.8.2 Multiblock Partial Least Squares

Multiblock PLS (MB-PLS) is presented briefly using the `potato` data. The corresponding theory is found in Section 7.2.

11.8.2.1 MB-PLS Modelling

A multi-response two-block MB-PLS model with up to 10 components in total is cross-validated with 10 random segments.

```
data(potato)
mb <- mbpls(potato[c('Chemical','Compression')], potato[['Sensory']],
            ncomp = 10, max_comps=10, validation="CV", segments=10)
print(mb)
```

```
Multiblock Partial Least Squares
Call:
mbpls(X = potato[c("Chemical", "Compression")],
      Y = potato[["Sensory"]], ncomp = 10, max_comps = 10,
      validation = "CV", segments = 10)
```

11.8.2.2 MB-PLS Summaries and Plotting

MB-PLS is implemented as a block-wise weighted concatenated ordinary PLSR. Therefore, all methods available for `plsr` are available for the super part of the MB-PLS. In addition one can extract block-`scores` and block-`loadings`.

```
data(potato)
Tb1 <- scores(mb, block=1)
scoreplot(mb, block = 1, labels = "names")

Pb2 <- loadings(mb, block=2)
loadingplot(mb, block = 1, labels = "names")
```

11.8.3 Sparse Multiblock Partial Least Squares

Sparse multiblock PLS (sMB-PLS) is presented briefly using the `potato` data. The corresponding theory is found in Section 7.2.3.

11.8.3.1 Sparse MB-PLS Modelling

A multi-response two-block sMB-PLS model with up to 10 components in total is cross-validated with 10 random segments. Here, the Soft-Threshold version is used (Truncation version also available) with parameter `shrink = 0.6` which means the weights have 60% of the largest values subtracted before setting negative values to 0.

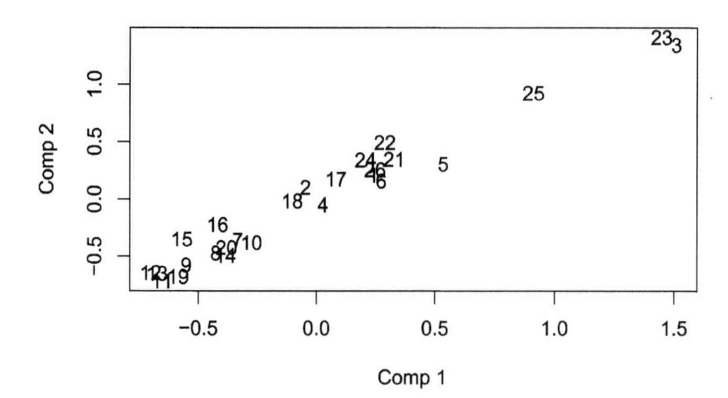

Figure 11.9 Output from use of `scoreplot()` (block-scores).

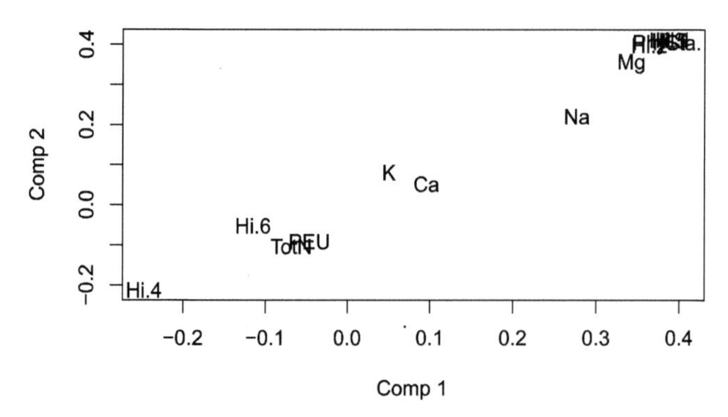

Figure 11.10 Output from use of `loadingplot()` (block-loadings).

```
data(potato)
smb <- smbpls(potato[c('Chemical','Compression')],
              potato[['Sensory']], ncomp = 10, max_comps=10,
              shrink = 0.6, validation="CV", segments=10)
print(smb)
```

```
Sparse Multiblock PLS (Soft-Threshold)

Call:
smbpls(X = potato[c("Chemical", "Compression")],
       Y = potato[["Sensory"]], ncomp = 10, shrink = 0.6,
       max_comps = 10, validation = "CV", segments = 10)
```

11.8.3.2 Sparse MB-PLS Plotting

We demonstrate the effect of shrinkage on scores and sparseness in weights by plotting results for three values of the shrinkage parameter. In the weight plots we can follow the shrinkage toward the origin of each variable, while the score plots show the effect on the sample scores.

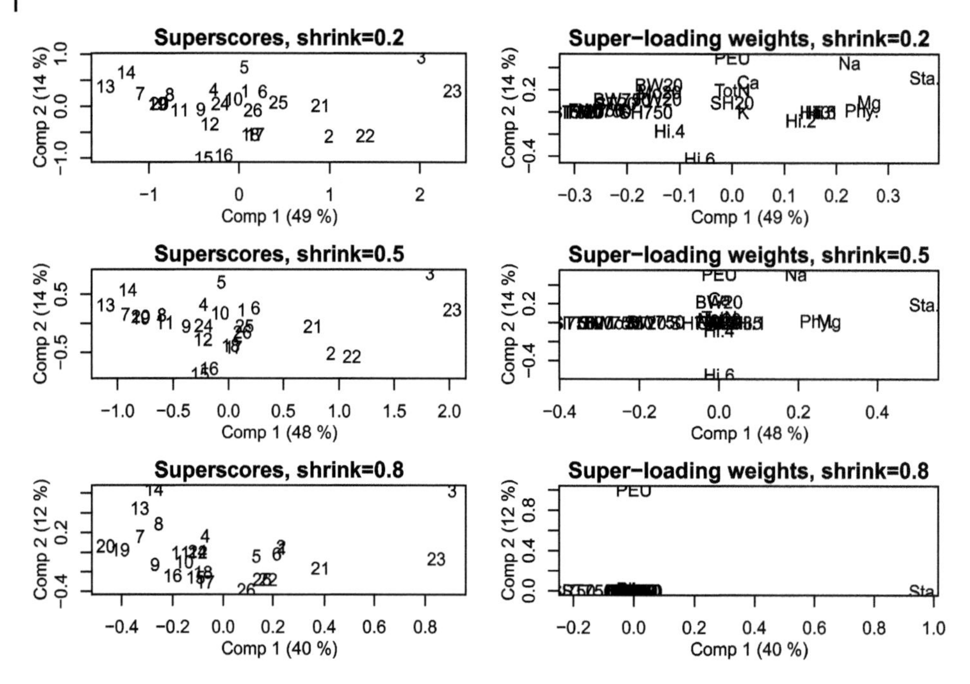

Figure 11.11 Output from use of `scoreplot()` and `loadingweightplot()` on an object from sMB-PLS.

```
old.par <- par(mfrow = c(3,2), mar = c(3.5,3.5,1.5,1), mgp = c(2,1,0))
for(shrink in c(0.2, 0.5, 0.8)){
  smb <- smbpls(potato[c('Chemical','Compression')],
           potato[['Sensory']], ncomp = 10,
           max_comps=10, shrink = shrink)
  scoreplot(smb, labels = "names", main = paste0(
                  "Super-scores, shrink=", shrink))
  loadingweightplot(smb, labels = "names", main = paste0(
                   "Super-weights, shrink=", shrink))
}
par(old.par)
```

11.8.4 Sequential and Orthogonalised Partial Least Squares

The following example uses the potato data to showcase some of the functions available for sequential and orthogonalised partial least squares (SO-PLS) analyses. The corresponding theory is found in Section 7.3.

11.8.4.1 SO-PLS Modelling

A multi-response two-block SO-PLS model with up to 10 components in total is cross-validated with 10 random segments.

```
# Load potato data and fit SO-PLS model
so.pot <- sopls(Sensory ~ Chemical + Compression, data=potato,
          ncomp=c(10,10), max_comps=10, validation="CV", segments=10)
print(so.pot)

Sequential and Orthogonalized Partial Least Squares,
   fitted with the PKPLS algorithm.
Cross-validated using 10 random segments.
Call:
```

```
sopls(formula = Sensory ~ Chemical + Compression, ncomp = c(10, 10),
   max_comps = 10, data = potato, validation = "CV", segments = 10)

summary(so.pot)

Data: X dimension: 26 0
      Y dimension: 26 9
Fit method: PKPLS
Number of components considered: 10

VALIDATION: RMSEP
Cross-validated using 10 random segments.
   0,0      0,1      0,2      0,3      0,4      0,5      0,6      0,7
1.1472  0.9974  1.1034  0.9942  1.0159  1.1523  1.0857  1.1834
   0,8      0,9     0,10      1,0      1,1      1,2      1,3      1,4
1.2494  1.3022  1.2690  0.9058  0.8860  0.9779  0.9029  0.8942
[..]

TRAINING: % variance explained
     0,0     0,1     0,2     0,3     0,4     0,5     0,6     0,7     0,8
X      0   45.87   55.51   62.90   66.46   67.73   74.95   78.26   79.85
Y1     0   42.19   54.73   65.01   66.85   73.96   74.10   74.44   77.45
Y2     0   39.11   41.97   42.23   43.80   50.95   54.87   56.78   59.50
Y3     0   42.55   57.44   59.44   61.06   66.78   68.62   69.02   71.34
Y4     0   38.64   65.31   73.51   75.63   77.23   79.25   81.19   83.71
Y5     0   16.13   18.11   26.71   26.74   29.70   42.07   44.57   46.47
Y6     0   23.21   43.78   62.23   64.02   67.18   67.72   69.83   74.79
Y7     0   35.35   41.99   57.36   58.84   67.17   70.33   73.21   77.77
Y8     0   24.48   27.71   43.10   44.27   48.41   53.39   62.37   67.60
Y9     0   21.98   22.78   48.17   55.50   59.96   68.39   72.54   76.39
[..]
```

11.8.4.2 Måge Plot

A full Måge plot (Section 7.3.5) for all combinations of components for all blocks is produced. This can be used for a global search for the best fitting cross-validated model.

Each point in Figure 11.12 is accompanied by a sequence of four numbers referring to the number of components used for each of the four blocks. Horizontal location is given by the total number of components used across all blocks, while vertical location indicates validated explained variance in percentage.

```
data(wine)
ncomp <- unlist(lapply(wine, ncol))[-5]
so.wine <- sopls('Global quality' ~ ., data=wine, ncomp=ncomp,
          max_comps=10, validation="CV", segments=10)
maage(so.wine)
```

A sequential Måge plot can be used for a sequential search for the optimal model.

```
# Sequential search for optimal number of components per block
old.par <- par(mfrow=c(2,2), mar=c(3,3,0.5,1), mgp=c(2,0.7,0))
maageSeq(so.wine)
maageSeq(so.wine, 2)
maageSeq(so.wine, c(2,1))
maageSeq(so.wine, c(2,1,1))
par(old.par)
```

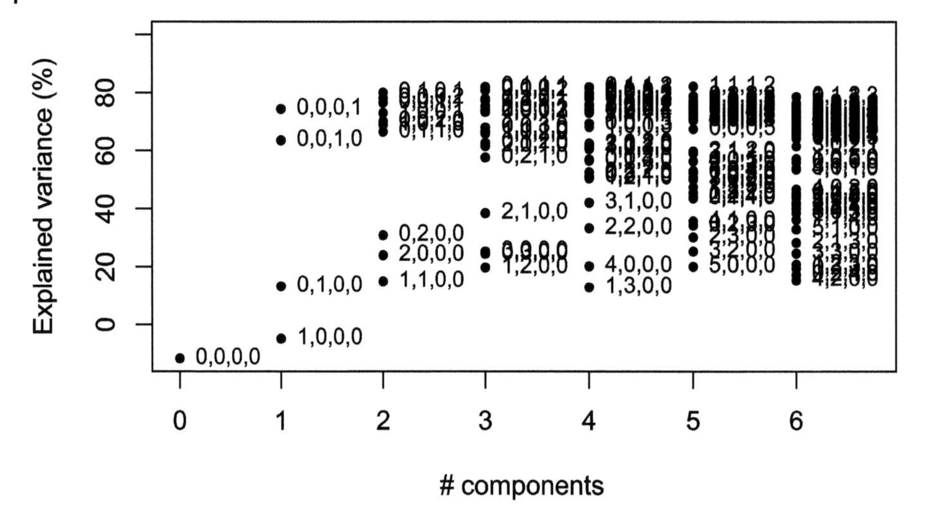

Figure 11.12 Output from use of `maage()`.

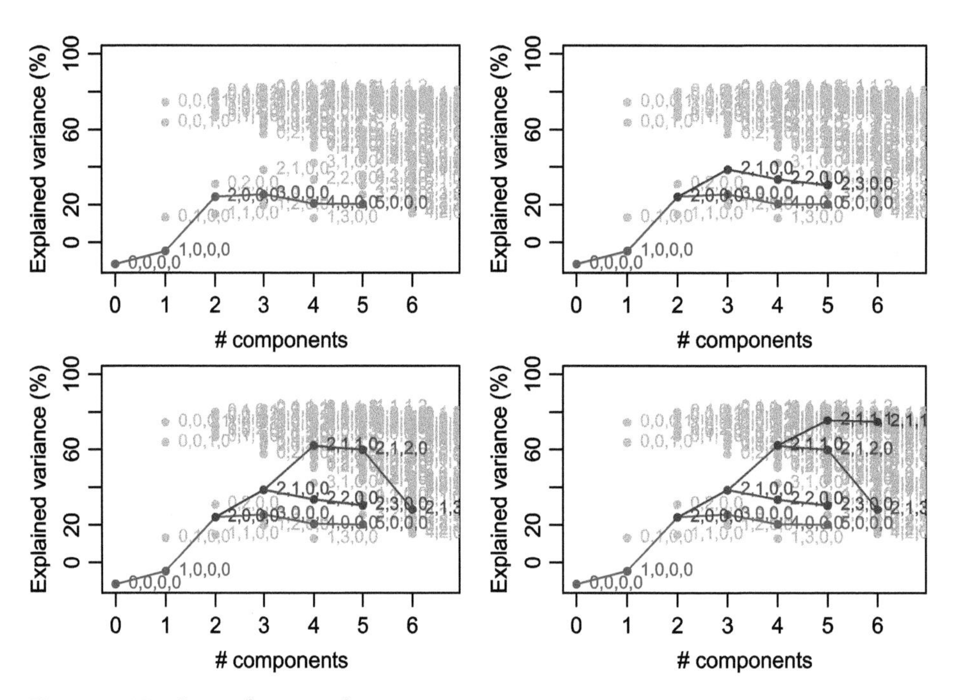

Figure 11.13 Output from use of `maageSeq()`.

11.8.4.3 SO-PLS Loadings

One set of loadings is printed and two sets are plotted to show how to select specific components from specific blocks. When extracting or plotting loadings for the second or later blocks, one must specify how many components have been used in the previous block(s) (`ncomp`) as this will affect the choice of loadings. In addition one can specify which components in the current block should be extracted (`comps`).

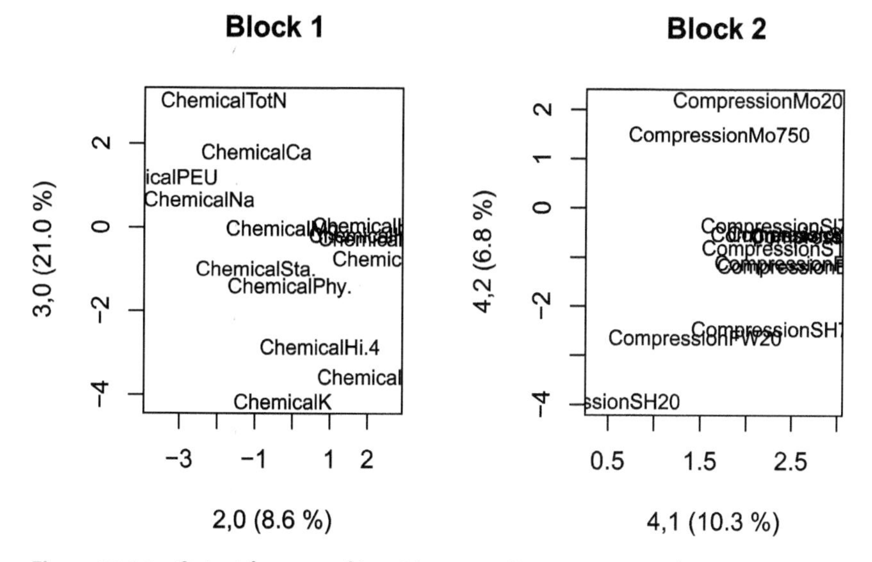

Figure 11.14 Output from use of `loadingplot()` on an `sopls` object.

```
# Display loadings up to four components for first block
loadings(so.pot, block = 1)

Loadings:
            1,0    2,0    3,0    4,0
Chemical1   0.645 -3.672  1.197  2.277
Chemical2  -4.542 -0.975 -0.984 -0.678
Chemical3   0.478 -1.848  3.046 -1.613
Chemical4  -3.970        -1.445  0.948
Chemical5  -1.365 -0.982  1.804 -1.190
Chemical6  -4.009 -0.280        -2.344
Chemical7  -3.066 -2.488  0.669  1.616
Chemical8  -0.186 -0.255 -4.161  0.149
Chemical9  -4.493  2.113         0.344
Chemical10 -4.099  2.656 -0.739  0.488
Chemical11 -4.382  2.270 -0.259  0.398
Chemical12  3.000  0.732 -2.855
Chemical13 -4.419  2.023 -0.177  0.372
Chemical14  1.580  2.266 -3.570  0.571

                  1,0    2,0    3,0    4,0
SS loadings    151.624 51.586 56.326 19.671
Proportion Var  10.830  3.685  4.023  1.405
Cumulative Var  10.830 14.515 18.538 19.943

# Plot loadings from block 1 and 2
old.par <- par(mfrow=c(1,2))
loadingplot(so.pot, comps = c(2,3), block = 1, labels = "names",
   main = "Block 1", cex = 0.8)
loadingplot(so.pot, ncomp = 4, block = 2, labels = "names",
   main = "Block 2", cex = 0.8)
par(old.par)
```

11.8.4.4 SO-PLS Scores

One set of scores is printed and two sets are plotted to show how to select specific components from specific blocks. Specification of component use in preceding blocks follows the same pattern as with loadings.

```
# Display scores up to four components for first block
scores(so.pot, block = 1)


          1,0        2,0          3,0         4,0
1  -0.078379126 -0.15756558 -0.030150796 -0.08882899
2   0.033599245  0.03345139 -0.264837392 -0.02082345
3  -0.454087240 -0.50387774 -0.257625630  0.24822188
4  -0.010144029  0.04409907  0.254739277  0.34206013
5  -0.231320589  0.02813994  0.331647163 -0.15209020
6  -0.100806072 -0.03203490  0.058511494 -0.07824489
7   0.103759602  0.05587580  0.171495995  0.13245956
8   0.142341829 -0.02952316  0.135315864  0.17766566
9   0.160175421 -0.13395363  0.067732083 -0.34567441
[..]
attr(,"class")
[1] "scores"

# Plot scores from block 1 and 2
old.par <- par(mfrow=c(1,2))
scoreplot(so.pot, comps = c(2,3), block = 1, labels = "names",
        main = "Block 1")
scoreplot(so.pot, ncomp = 4, block = 2, labels = "names",
        main = "Block 2")
par(old.par)
```

11.8.4.5 SO-PLS Prediction

A three block model is fitted using a single response, five components, and a subset of the data. The remaining data are used as test set for prediction.

```
# Modify data to contain a single response
potato1 <- potato; potato1$Sensory <- potato1$Sensory[,1]
# Model 20 first objects with SO-PLS
so.pot20 <- sopls(Sensory ~ ., data = potato1[c(1:3,9)], ncomp = 5,
               subset = 1:20)
# Predict remaining objects
testset <- potato1[-(1:20),]; # testset$Sensory <- NULL
predict(so.pot20, testset, comps=c(2,1,2))


, , 1,0,0

      Sensory
[1,] 3.922829
[2,] 3.801854
[3,] 6.409450
[4,] 2.960830
[5,] 4.433857
[6,] 3.223006

, , 2,0,0

      Sensory
[1,] 4.062985
[2,] 4.238679
[3,] 7.178207
[4,] 2.477011
[5,] 3.319856
[6,] 2.904183
[..]
```

11.8.4.6 SO-PLS Validation

Compute validation statistics; explained variance – R^2, and root mean squared error – RMSE(P/CV).

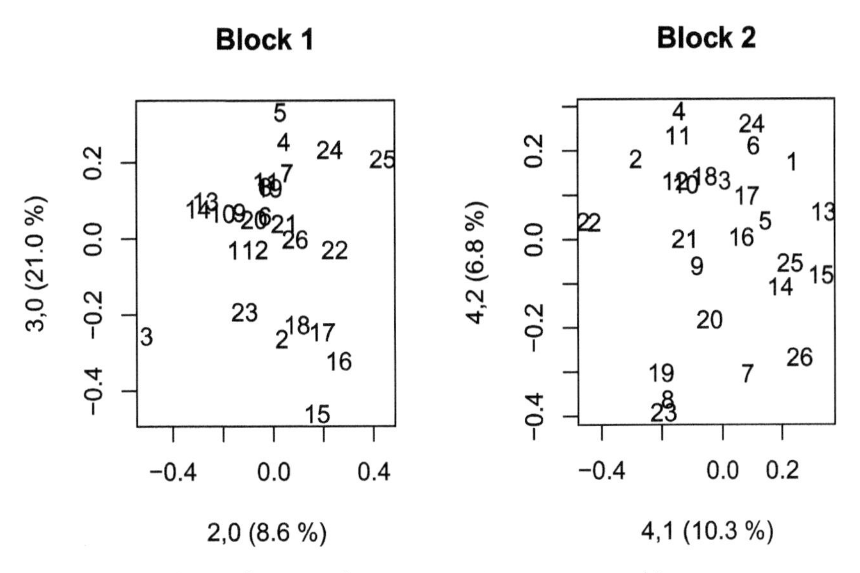

Figure 11.15 Output from use of `scoreplot()` on an `sopls` object.

```
# Cross-validation
R2(so.pot, ncomp = c(5,5))
# ... per response
R2(so.pot, ncomp = c(5,5), individual = TRUE)
# Training
R2(so.pot, 'train', ncomp = c(5,5))

# Test data
R2(so.pot20, newdata = testset, ncomp = c(2,1,2))
```

```
# Only showing first R2() output
       1,0        2,0        3,0        4,0        5,0
0.29491711 0.43937819 0.52443674 0.51731500 0.48356727
       5,1        5,2        5,3        5,4        5,5
0.51769239 0.50453909 0.37995426 0.19264375 0.08718665
```

```
# Cross-validation
RMSEP(so.pot, ncomp = c(5,5))
# ... per response
RMSEP(so.pot, ncomp = c(5,5), individual = TRUE)
# Training
RMSEP(so.pot, 'train', ncomp = c(5,5))

# Test data
RMSEP(so.pot20, newdata = testset, ncomp = c(2,1,2))
```

```
# Only showing second RMSEP() output
             1,0       2,0       3,0       4,0       5,0
ref    1.4521251 1.2875627 1.0694309 1.0651814 1.1082357
hard   0.7409166 0.8053977 0.8468041 0.9386690 0.9695054
firm   0.8153675 0.8144132 0.8027508 0.8689574 0.8348612
elas   0.5987744 0.5700821 0.5571669 0.5906552 0.5883550
adhes  0.6238669 0.6025240 0.6299503 0.5527890 0.5413816
grainy 0.9124005 0.7090731 0.5895734 0.5759413 0.6259969
mealy  1.1918516 0.9629070 0.8254784 0.7783679 0.8346700
moist  0.8218611 0.7210138 0.6804434 0.6837585 0.7346723
chewi  0.6207792 0.5133308 0.5252540 0.4739114 0.5250602
             5,1       5,2       5,3       5,4       5,5
ref    1.0241368 0.9529110 1.0187910 0.9826720 0.9433262
hard   0.9120602 0.9323634 1.1324963 1.4532307 1.6670183
```

```
firm    0.7665856 0.7521744 0.8722266 0.9742660 1.1283161
elas    0.5722469 0.5347604 0.5511978 0.5410909 0.6948752
adhes   0.5331916 0.5622102 0.6299050 0.8007815 0.8670689
grainy  0.6513774 0.6303437 0.6248834 0.6234977 0.5447751
mealy   0.8303858 0.8857717 0.9391643 0.9676215 0.8730579
moist   0.7539214 0.8511216 1.0159016 1.2562827 1.3202345
chewi   0.5372372 0.5807210 0.6379786 0.7687792 0.7621405
```

11.8.4.7 Principal Components of Predictions

A PCA is computed from the cross-validated predictions to get an overview of the SO-PLS model across all involved blocks. The blocks are projected onto the scores to form block-loadings to see how these relate to the solution. The theory is found in Section 7.3.3.

```
# PCP from so.pot object
PCP <- pcp(so.pot, c(3,2))
summary(PCP)
scoreplot(PCP)
```

```
Principal Components of Predictions
===================================

$scores: Scores (26x9)
$loadings: Loadings (9x9)
$blockLoadings: Block loadings:
- Chemical (14x9), Compression (12x9)
```

11.8.4.8 CVANOVA

A CVANOVA model compares absolute or squared cross-validated residuals from two or more prediction models using ANOVA with *Model* and *Object* as effects. Tukey's pair-wise testing is automatically computed in this implementation. The theory is found in Section 7.3.5.

```
so.pot1 <- sopls(Sensory[,1] ~ Chemical + Compression + NIRraw, data=potato,
          ncomp=c(10,10,10), max_comps=10, validation="CV", segments=10)
cva <- cvanova(so.pot1, "2,1,2")
summary(cva)
old.par <- par(mar = c(4,6,4,2))
plot(cva)
par(old.par)
```

```
Analysis of Variance Table

Response: Residual
          Df  Sum Sq Mean Sq F value     Pr(>F)
Model      2  1.0601 0.53007  5.0998  0.009647 **
Object    25 15.1295 0.60518  5.8225 6.511e-08 ***
Residuals 50  5.1969 0.10394
---
Signif. codes:  0 '***' 0.001 '**' 0.01 '*' 0.05 '.' 0.1 ' ' 1
Tukey's HSD
Alpha: 0.05

           Mean G1 G2
2,0,0 0.9696191  A
2,1,0 0.8208381  A  B
2,1,2 0.6841366     B
```

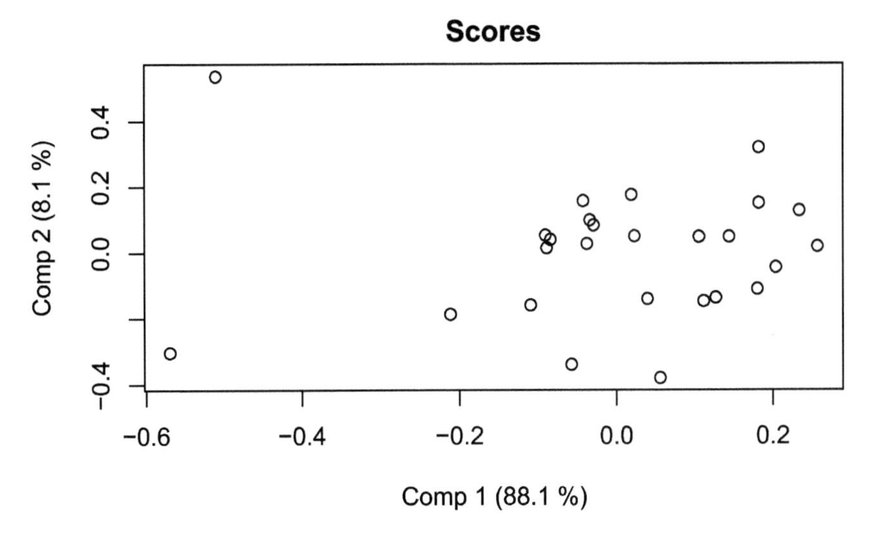

Figure 11.16 Output from use of `scoreplot()` on a `pcp` object.

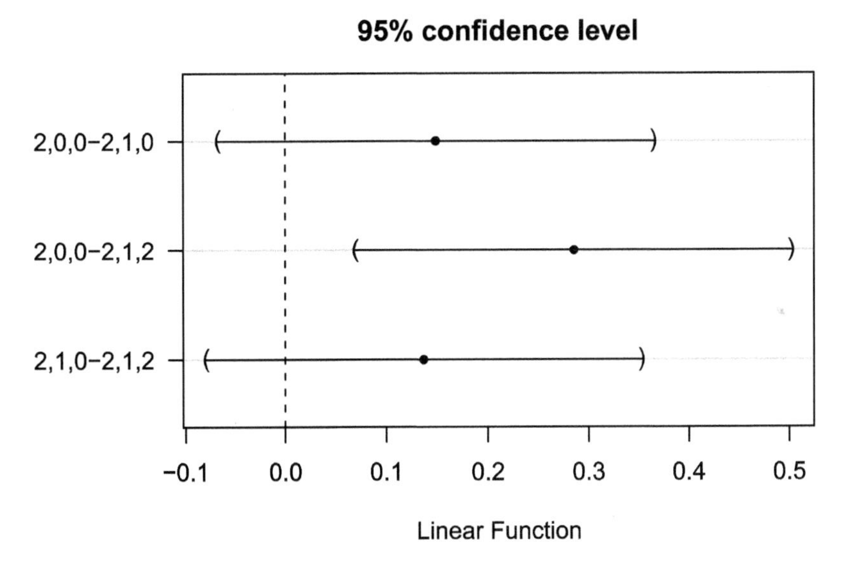

Figure 11.17 Output from use of `plot()` on a `cvanova` object.

11.8.5 Parallel and Orthogonalised Partial Least Squares

Parallel and orthogonalised partial least squares (PO-PLS) is presented briefly using the `potato` data. It is a method for separating predictive information into common, local, and distinct parts. The corresponding theory is found in Section 7.4.

11.8.5.1 PO-PLS Modelling

There are many choices with regard to numbers of components and possible local and common components. Using automatic selection, the user selects the highest number of blocks to combine into local/common components, minimum explained variance and minimum squared correlation to the response. Manual selection can be done by setting the number of initial components from the blocks and maximum number of local/common components.

```
# Automatic analysis
pot.po.auto <- popls(potato[1:3], potato[['Sensory']][,1],
                     commons = 2)

# Explained variance
pot.po.auto$explVar


$Chemical
named numeric(0)

$Compression
C(1,2), Comp 1
      68.14595

$NIRraw
named numeric(0)
```

Here the notation C(1,2) means a local component based on blocks 1 and 2.

```
# Manual choice of up to 5 components for each block and 1, 0,
# and 2 blocks, respectively from the (1,2), (1,3) and (2,3)
# combinations of blocks.
pot.po.man <- popls(potato[1:3], potato[['Sensory']][,1],
                    commons = 2, auto=FALSE,
                    manual.par = list(ncomp=c(5,5,5),
                    ncommon=c(1,0,2)))
# Explained variance
pot.po.man$explVar


$Chemical
C(1,2), Comp 1   D(1), Comp 1   D(1), Comp 2   D(1), Comp 3
      32.22944   -72871.76265   -49467.54832        0.00000

$Compression
C(1,2), Comp 1 C(2,3), Comp 1 C(2,3), Comp 2
     68.145954       5.504006       7.214911

$NIRraw
C(2,3), Comp 1 C(2,3), Comp 2   D(3), Comp 1   D(3), Comp 2
     32.606459       5.475702    -552.399938      15.775647
  D(3), Comp 3   D(3), Comp 4
      3.303994       0.000000
```

11.8.5.2 PO-PLS Scores and Loadings

Scores and loadings are stored per block. Common scores/loadings are found in each of the blocks' list of components.

```
# Score plot for local (2,3) components
scoreplot(pot.po.man, block = 3, labels="names")


# Corresponding loadings
loadingplot(pot.po.man, block = 3, labels="names", scatter = FALSE)
```

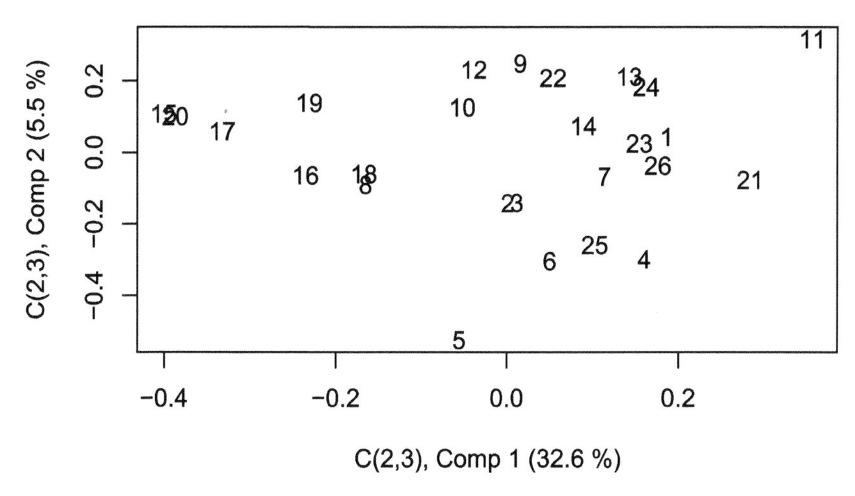

Figure 11.18 Output from use of `scoreplot()` on a `popls` object.

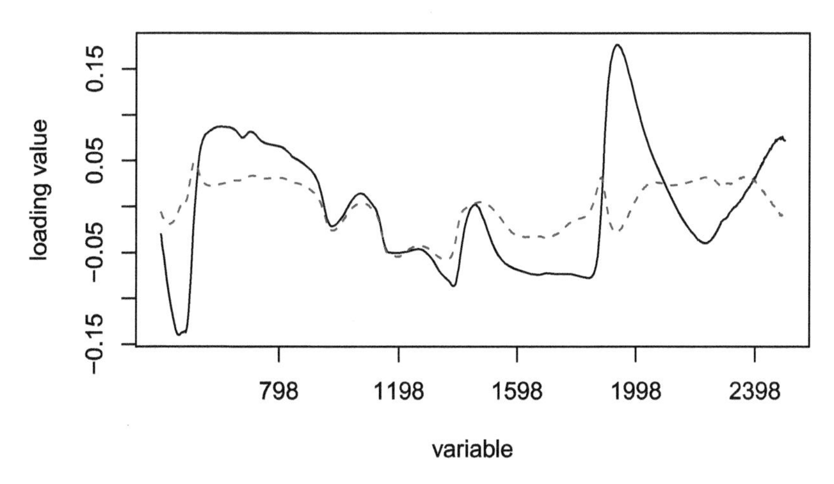

Figure 11.19 Output from use of `loadingplot()` on a `popls` object.

11.8.6 Response Optimal Sequential Alternation

The following example uses the potato data to showcase some of the functions available for response optimal sequential alternation (ROSA) analyses. The corresponding theory is found in Section 7.5.

11.8.6.1 ROSA Modelling

A multi-response two-block ROSA model with up to 10 components in total is cross-validated with 10 random segments.

```
# Model all eight potato blocks with ROSA
ros.pot <- rosa(Sensory ~ ., data = potato1, ncomp = 10,
   validation = "CV", segments = 5)

print(ros.pot)
```

```
Response Orinented Sequential Alternation, fitted with the CPPLS
    algorithm.
Cross-validated using 5 random segments.
Call:
rosa(formula = Sensory ~ ., ncomp = 10, data = potato1,
    validation = "CV", segments = 5)
```

```
summary(ros.pot)
```

```
Data:         X dimension: 26 20318
            Y dimension: 26 1
Fit method:
Number of components considered: 10

VALIDATION: RMSEP
Cross-validated using 5 random segments.
       (Intercept)  1 comps  2 comps  3 comps  4 comps
CV           1.778    1.421    1.299    1.270    1.207
adjCV        1.778    1.310    1.199    1.175    1.116
       5 comps  6 comps  7 comps  8 comps  9 comps
CV       1.198    1.193    1.191    1.219    1.196
adjCV    1.116    1.121    1.128    1.159    1.148
       10 comps
CV        1.178
adjCV     1.136

TRAINING: % variance explained
         1 comps  2 comps  3 comps  4 comps  5 comps
X          27.81    72.21    82.50    83.10    84.86
Sensory    76.53    83.36    84.22    87.91    88.06
         6 comps  7 comps  8 comps  9 comps  10 comps
X          87.07    87.22    88.36    91.15     92.81
Sensory    88.18    88.17    88.42    88.72     88.72
```

11.8.6.2 ROSA Loadings

Extract loadings (not used further) and plot two first vectors of loadings.

```
loads <- loadings(ros.pot)
loadingplot(ros.pot, comps = 1:2)
```

11.8.6.3 ROSA Scores

Extract scores (not used further) and plot two first vectors of scores.

```
sco <- scores(ros.pot)
scoreplot(ros.pot, comps = 1:2, labels = "names")
```

11.8.6.4 ROSA Prediction

A three block model is fitted using a single response, five components, and a subset of the data. The remaining data are used as test set for prediction.

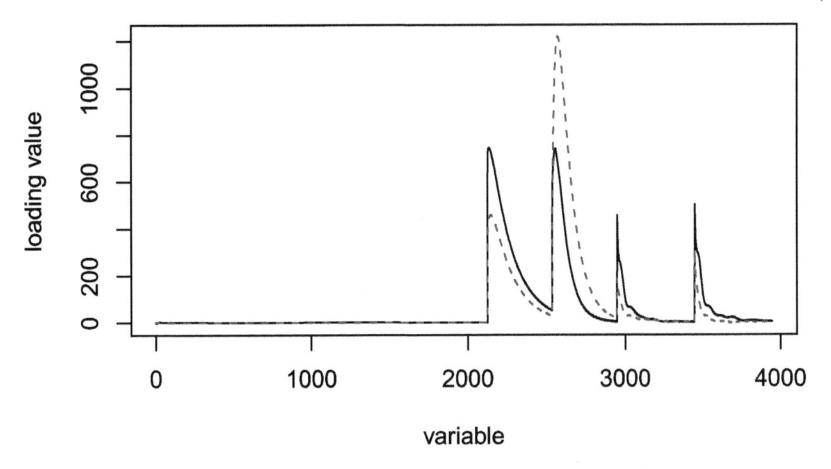

Figure 11.20 Output from use of `loadingplot()` on a `rosa` object.

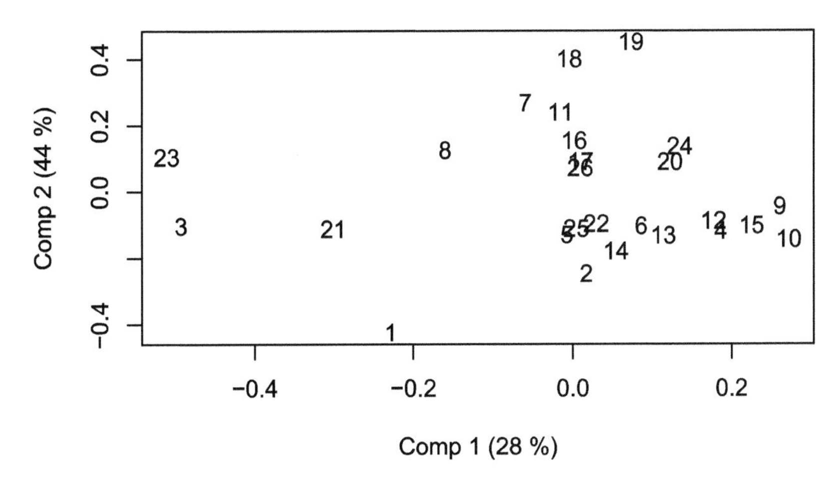

Figure 11.21 Output from use of `scoreplot()` on a `rosa` object.

```
# Model 20 first objects of three potato blocks
rosT <- rosa(Sensory ~ ., data = potatoList[c(1:3,9)], ncomp = 5,
   subset = 1:20)
testset <- potatoList[-(1:20),]; # testset$Sensory <- NULL
predict(rosT, testset, comps=2)
```

```
      Sensory
[1,] 2.2151098
[2,] 1.5123048
[3,] 2.3008727
[4,] 1.2001610
[5,] 0.8899352
[6,] 1.2181709
```

11.8.6.5 ROSA Validation

Computation of validation statistics; explained variance – R^2, and root mean squared error – RMSE(P/CV), is done exactly like with the SO-PLS methods above and give similar output (not shown here).

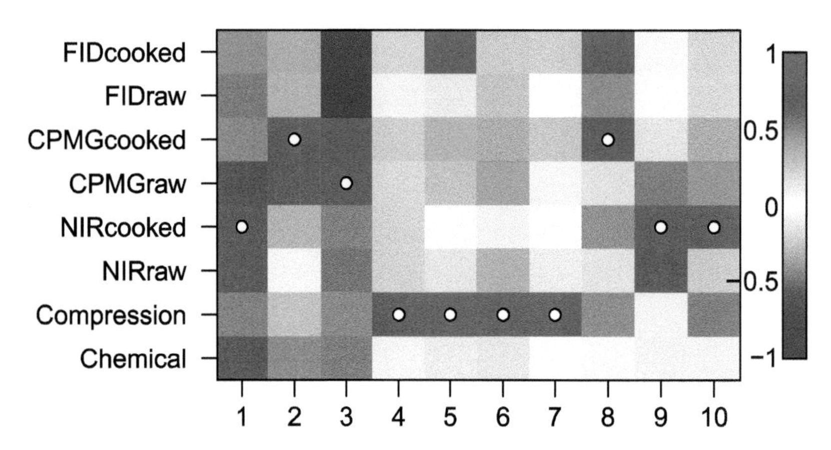

Figure 11.22 Output from use of `image()` on a `rosa` object.

11.8.6.6 ROSA Image Plots

These are plots for evaluation of the block selection process in ROSA. Correlation plots show how the different candidate scores (one candidate for each block for each component) correlate to the winning block's scores. Residual response plots show how different choices of candidate scores would affect the RMSE of the residual response. One can for instance use these plots to decide on a different block selection order than the one proposed automatically by ROSA.

```
# Correlation to winning scores
image(ros.pot)
```

```
# Residual response given candidate scores
image(ros.pot, "residual")
```

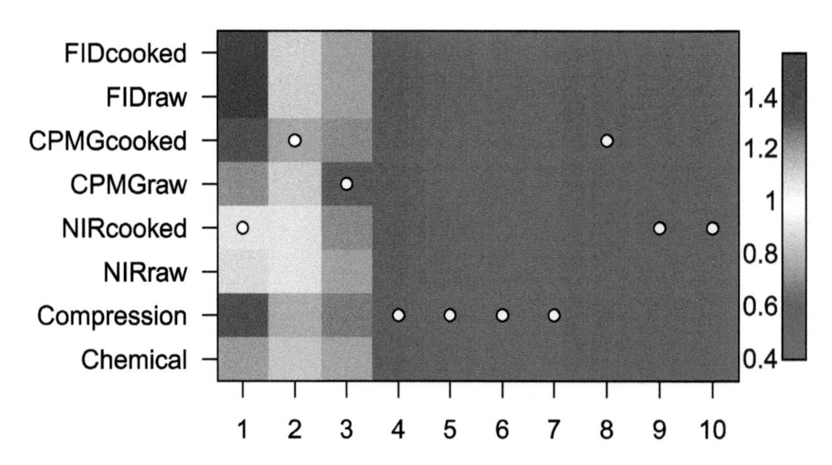

Figure 11.23 Output from use of `image()` with parameter "residual" on a `rosa` object.

11.8.7 Multiblock Redundancy Analysis

The following example uses the potato data to showcase some of the functions available for multiblock redundancy analysis (MB-RDA). The corresponding theory is found in Section 10.3.

11.8.7.1 MB-RDA Modelling

This implementation uses a wrapper for the `mbpcaiv` function in the `ade4` package to perform MB-RDA. A multi-response five component model is fitted.

```
# Convert data.frame with AsIs objects to list of matrices
potatoList <- lapply(potato, unclass)

# Perform MB-RDA with two blocks explaining sensory attributes
mbr <- mbrda(potatoList[c('Chemical','Compression')],
           potatoList[['Sensory']], ncomp = 5)
print(mbr)

Multiblock Redundancy Analysis
Call:
mbrda(X = potatoList[c("Chemical", "Compression")],
      Y = potatoList[["Sensory"]],      ncomp = 5)
```

11.8.7.2 MB-RDA Loadings and Scores

The `mbpcaiv` wrapper extracts key elements for inspection using the same format as the rest of this package. The full fitted `mbpcaiv` object is also available, e.g., through `mbr$mbpcaivObject`.

```
# Extract and view loadings
lo_mbr <- loadings(mbr)
print(head(lo_mbr))
# Plot scores
scoreplot(mbr, labels = "names")

           Ax1         Ax2        Ax3        Ax4        Ax5
V1 -0.11801083  0.45883565 -1.7606202 -0.1936109 -2.3704052
V2  1.02113965 -2.99539512  1.4968283  4.6426038 -7.6478047
V3  0.50084474 -1.40837541  1.0571404  1.8302797 -1.7127587
V4  0.24298763  0.27655908  0.3695837  0.4462580 -0.7461178
V5  0.02050271 -0.04647818  0.1311656  0.6243666 -0.9713523
V6 -0.21159305  1.41437345 -0.6562877 -1.4166021  2.0111970
```

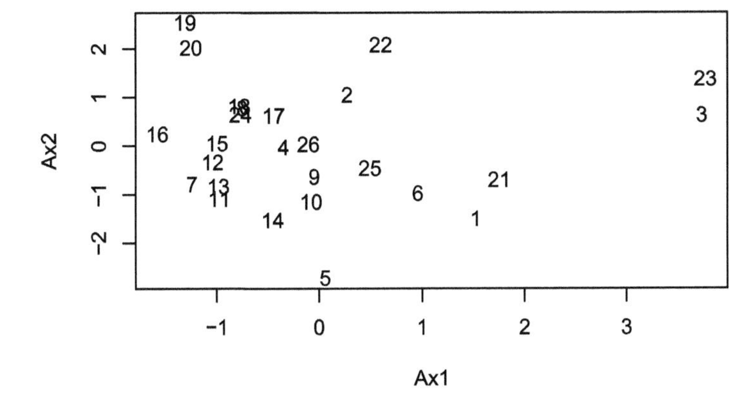

Figure 11.24 Output from use of `scoreplot()` on an `mbrda` object.

11.9 Complex Data Structures

The following methods for complex data structures are available in the `multiblock` package (function names in parentheses). The corresponding theory is found in Chapters 8 and 10.

- L-PLS – partial least squares in L configuration (`lpls`) – Section 8.3
- SO-PLS-PM – sequential and orthogonalised PLS path modelling (`sopls_pm`) – Section 10.5.4

11.9.1 L-PLS

To showcase L-PLS we will use simulated data specifically made for L-shaped data. Regression using L-PLS can be either outwards from $\mathbf{X}_1$ to $\mathbf{X}_2$ and $\mathbf{X}_3$ or inwards from $\mathbf{X}_2$ and $\mathbf{X}_3$ to $\mathbf{X}_1$. In the former case, prediction can either be of $\mathbf{X}_2$ or $\mathbf{X}_3$ given $\mathbf{X}_1$. The corresponding theory is found in Section 8.3.

11.9.1.1 Simulated L-shaped Data

We simulate two latent components in L shape with blocks having dimensions (30x20), (20x5), and (6x20) for blocks $\mathbf{X}_1$, $\mathbf{X}_2$, and $\mathbf{X}_3$, respectively.

```
set.seed(42)

# Simulate data set
sim <- lplsData(I = 30, N = 20, J = 5, K = 6, ncomp = 2)

# Split into separate blocks
X1 <- sim$X1; X2 <- sim$X2; X3 <- sim$X3
```

11.9.1.2 Exo-L-PLS

The first L-PLS will be outwards. Predictions have to be accompanied by a direction.

```
# exo-L-PLS:
lp.exo <- lpls(X1,X2,X3, ncomp = 2) # type = "exo" is default

# Predict X1
pred.exo.X2 <- predict(lp.exo, X1new = X1, exo.direction = "X2")

# Predict X3
pred.exo.X2 <- predict(lp.exo, X1new = X1, exo.direction = "X3")

# Correlation loading plot
plot(lp.exo)
```

11.9.1.3 Endo-L-PLS

The second L-PLS will be inwards.

```
# endo-L-PLS:
lp.endo <- lpls(X1,X2,X3, ncomp = 2, type = "endo")

# Predict X1 from X2 and X3 (in this case fitted values):
pred.endo.X1 <- predict(lp.endo, X2new = X2, X3new = X3)
```

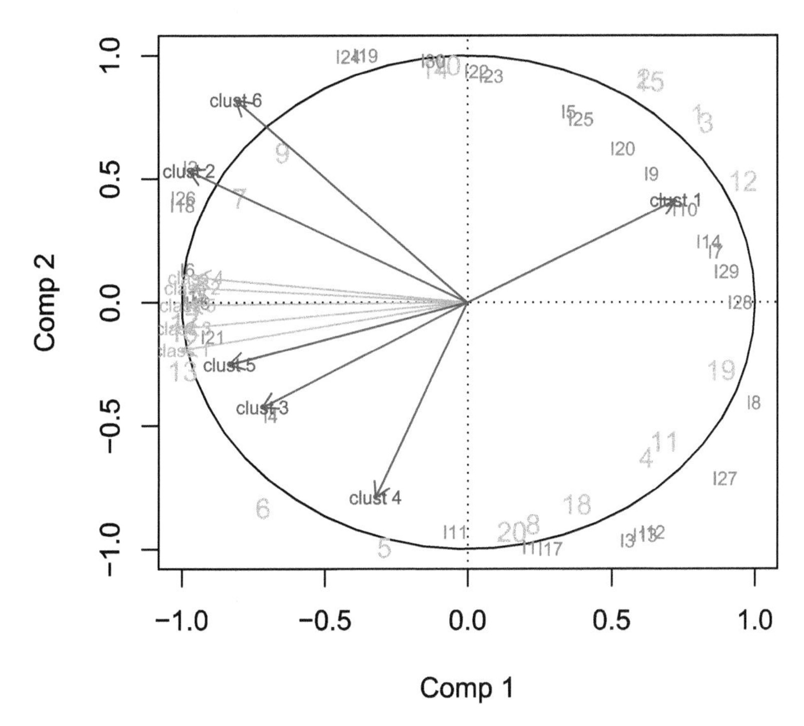

Figure 11.25 Output from use of `plot()` on an `lpls` object. Correlation loadings from blocks are coloured and overlaid each other to visualise relations across blocks.

11.9.1.4 L-PLS Cross-validation

Cross-validation comes with choices of directions when applying this to L-PLS since we have both sample and variable links. The cross-validation routines compute RMSECV values and perform cross-validated predictions.

```
# LOO cross-validation horizontally
lp.cv1 <- lplsCV(lp.exo, segments1 = as.list(1:dim(X1)[1]),
             trace = FALSE)

# LOO cross-validation vertically
lp.cv2 <- lplsCV(lp.exo, segments2 = as.list(1:dim(X1)[2]),
             trace = FALSE)

# Three-fold CV, horizontal
lp.cv3 <- lplsCV(lp.exo, segments1 = as.list(1:10, 11:20, 21:30),
             trace = FALSE)

# Three-fold CV, horizontal, inwards model
lp.cv4 <- lplsCV(lp.endo, segments1 = as.list(1:10, 11:20, 21:30),
             trace = FALSE)
```

11.9.2 SO-PLS-PM

The following example uses the `potato` data and the `wine` data to showcase some of the functions available for SO-PLS-PM analyses. The corresponding theory is found in Section 10.5.4.

11.9.2.1 Single SO-PLS-PM Model

A model with four blocks having 5 components per input block is fitted. We set `computeAdditional` to `TRUE` to turn on computation of additional explained variance per added block in the model.

```
# Load potato data
data(potato)

# Single path
pot.pm <- sopls_pm(potato[1:3], potato[['Sensory']], c(5,5,5),
                   computeAdditional=TRUE)

# Report of explained variances and optimal number of components
# (LOO cross-validation is default)
pot.pm

   direct indirect     total additional1 additional2 overall
 3.23 (2)     49.21 52.44 (3)     4.09 (3)    14.01 (2)    70.55
```

11.9.2.2 Multiple Paths in an SO-PLS-PM Model

A model containing five blocks is fitted. Explained variances for all sub-paths are estimated.

```
# Load wine data
data(wine)

# All path in the forward direction
pot.pm.multiple <- sopls_pm_multiple(wine, comps = c(4,2,9,8))

# Report of direct, indirect and total explained variance per
# sub-path.
# Bootstrapping can be enabled to assess stability.
pot.pm.multiple

$'Smell at rest->View'
    direct indirect      total
 32.68 (1)        0 32.68 (1)

$'Smell at rest->Smell after shaking'
   direct indirect     total
 6.39 (4)    33.64 40.03 (4)

$'Smell at rest->Tasting'
    direct indirect      total
  0.00 (2)    11.52 11.52 (2)

$'Smell at rest->Global quality'
    direct indirect      total
  0.00 (0)    25.25 25.25 (3)

$'View->Smell after shaking'
    direct indirect      total
 30.97 (2)        0 30.97 (2)

$'View->Tasting'
  direct indirect     total
 0.4 (1)    40.69 41.09 (2)

$'View->Global quality'
    direct indirect      total
  0.00 (0)    30.87 30.87 (2)

$'Smell after shaking->Tasting'
    direct indirect      total
 56.67 (3)        0 56.67 (3)

$'Smell after shaking->Global quality'
```

```
    direct indirect      total
  0.00 (2)     70.15 70.15 (2)

$'Tasting->Global quality'
    direct indirect      total
 78.12 (2)          0 78.12 (2)
```

11.10 Software Packages

11.10.1 R Packages

In Table 11.1 we have collected the multiblock methods we could find in the Comprehensive R Archive Network (CRAN, R Core Team (2022)) using simple search criteria. The list is probably not complete, but shows some of the variation in methods available at CRAN. Only multiblock methods are mentioned, even though the packages may contain other methods in addition. Every package on CRAN is available directly using the canonical URL: https://CRAN.R-project.org/package=packagename, where 'packagename' should be exchanged with the package of interest.

Table 11.1 R packages on CRAN having one or more multiblock methods.

Package name	Author	Multiblock methods
ade4	*Stéphanie Bougeard et al.*	MBPLS, MBPCAIV
ddsPLS	*Hadrien Lorenzo et al.*	Multi-Data-Driven Sparse Partial Least Squares
ExPosition	*Derek Beaton et al.*	GPCA, MCA, MDS
geigen	*Berend Hasselman*	GSVD
ade4	*Stéphanie Bougeard et al.*	MBPLS, MBPCAIV
ddsPLS	*Hadrien Lorenzo et al.*	Multi-Data-Driven Sparse Partial Least Squares
ExPosition	*Derek Beaton et al.*	GPCA, MCA, MDS
geigen	*Berend Hasselman*	GSVD
MatrixCorrelation	*Kristian Hovde Liland*	RV, RV_{mod}, GCD, SMI
mbclusterwise	*Stephanie Bougeard*	Clusterwise MBPLS, clusterwise MB redundancy analysis
MFAg	*Paulo Cesar Ossani*	MFA

(Continued)

Table 11.1 (Continued).

Package name Multiblock methods	Author
mixOmics DIABLO, MINT	*Kim-Anh Lê Cao*
msma Multiblock Sparse Partial Least Squares	*Atsushi Kawaguchi*
multigroup mbmgPCA	*Aida Eslami et al.*
packMBPLSDA MBPLS-DA	*Marion Brandolini-Bunlon et al.*
plspm (orphaned) PLS Path Modelling	*Gaston Sanchez et al.*
r.jive JIVE	*Michael J. O'Connell et al.*
RegularizedSCA DISCO-SCA, PCA-GCA, Regularised, sparse, and structured SCA	*Zhengguo Gu et al.*
RGCCA RGCCA (framework for many methods)	*Arthur Tenenhaus et al.*
SensoMineR/FactoMineR INDSCAL, GPA, MFA	*Francois Husson et al.*

11.10.2 MATLAB Toolboxes

Most of the multiblock code available for MATLAB (2021) comes in the form of collections of functions rather than full MATLAB toolboxes. This means that documentation is sometimes sparse and author information may be lacking, e.g., most packages marked *Nofima* below are made partly or completely by Ingrid Måge. A non-exhaustive list of open source MATLAB code is found in Table 11.2.

Table 11.2 MATLAB toolboxes and functions having one or more multiblock methods.

Package name	Company/author
CMTF Toolbox CMTF, ACMTF	*Evrim Acar*
MBToolbox CPCA, MBPCA, HPCA, SPLS, MBPLS, HPLS	*Frans van den Berg*
MCR-ALS MCR-ALS	*Romà Tauler et al.*
MultiBlock Component Analysis Clusterwise SCA-ECP, Separate PCA per data block, SCA-ECP	*Kim De Roover et al.*
Multiblock classification SO-PLS-LDA, MB-PLS-LDA	*Nofima*

(Continued)

Table 11.2 (Continued).

Package name	Company/author
Multiblock regression PO/SO-PLS	*Nofima*
Multiblock-PLS MB-PLS	*Nofima*
PCA-GCA PCA-GCA	*Nofima*
Saisir COMDIM, GCA, MFA, multiway PCA, STATIS, XCOMDIM	*Dominique Bertrand et al.*
SO-N-PLS SO-N-PLS	*Nofima*

11.10.3 Python

Very few multiblock packages for Python (Van Rossum and Drake, 2009) have been found. A non-exhaustive list is found in Table 11.3.

Table 11.3 Python packages having one or more multiblock methods.

Package name	Company/author
mbpls MB-PLS	*Andreas Baum et al.*
jive JIVE	*Iain Carmichael*
OnPLS OnPLS	*Tommy Löfstedt*
py_ddspls Multi-Data-Driven Sparse Partial Least Squares	*Hadrien Lorenzo*

11.10.4 Commercial Software

Table 11.4 Commercial software having one or more multiblock methods.

Package name	Company/author
PLS_Toolbox Data-/model-joining, multiway methods, ASCA	*Eigenvector*
SIMCA MOCA	*Sartorius Stedim Biotech*
XLSTAT GPA, MFA, STATIS	*Addinsoft*

References

Aben, N., Vis, D.J., Michaut, M., and Wessels, L.F.A. (2016) TANDEM: a two-stage approach to maximize interpretability of drug response models based on multiple molecular data types. *Bioinformatics*, **32** (17), 413–420.

Aben, N., Westerhuis, J.A., Song, Y.P., Kiers, H.A.L., Michaut, M., Smilde, A.K., and Wessels, L.F.A. (2018) iTOP: inferring the topology of omics data. *Bioinformatics*, **34** (17), 988–996.

Acar, E., Bro, R., and Smilde, A.K. (2015) Data Fusion in Metabolomics using Coupled Matrix and Tensor Factorizations. *Proceedings of the IEEE*, **103** (9), 1602–1620.

Acar, E., Papalexakis, E.E., Gurdeniz, G., Rasmussen, M.A., Lawaetz, A.J., Nilsson, M., and Bro, R. (2014) Structure-revealing data fusion. *BMC Bioinformatics*, **15**, 239.

Acar, E., Rasmussen, M., Savorani, F., Næs, T., and Bro, R. (2013) Understanding data fusion within the framework of coupled matrix and tensor factorizations. *Chemometrics and Intelligent Laboratory Systems*, **129**, 53–63.

Adams, E.W., Fagot, R.F., and Robinson, R.E. (1965) A theory of appropriate statistics. *Psychometrika*, **30** (2), 99–127.

Afseth, N.K., Segtnan, V.H., Marquardt, B.J., and Wold, J.P. (2005) Raman and near-infrared spectroscopy for quantification of fat composition in a complex food model system. *Applied Spectroscopy*, **59** (11), 1324–1332.

Alier, M. and Tauler, R. (2013) Multivariate curve resolution of incomplete data multisets. *Chemometrics and Intelligent Laboratory Systems*, **127**, 17–28.

Alinaghi, M., Bertram, H.C., Brunse, A., Smilde, A.K., and Westerhuis, J.A. (2020) Common and distinct variation in data fusion of designed experimental data. *Metabolomics*, **16** (1), 1–11.

Alinaghi, M., Jiang, P.P., Brunse, A., Sangild, P.T., and Bertram, H.C. (2019) Rapid cerebral metabolic shift during neonatal sepsis is attenuated by enteral colostrum supplementation in preterm pigs. *Metabolites*, **9** (1), 13.

Almli, V.L., Næs, T., Enderli, G., Sulmont-Rossé, C., Issanchou, S., and Hersleth, M. (2011) Consumers' acceptance of innovations in traditional cheese. a comparative study in france and norway. *Appetite*, **57** (1), 110–120.

Alter, O., Brown, P.O., and Botstein, D. (2003) Generalized singular value decomposition for comparative analysis of genome-scale expression data sets of two different organisms. *Proceedings of the National Academy of Sciences of the United States of America*, **100**, 3351–3356.

Anderson, M. and Ter Braak, C. (2003) Permutation tests for multi-factorial analysis of variance. *Journal of Statistical Computation and Simulation*, **73** (2), 85–113.

Anderson-Bergman, C., Kolda, T.G., and Kincher-Winoto, K. (2018) XPCA: Extending PCA for a Combination of Discrete and Continuous Variables. *arXiv preprint arXiv:1808.07510.*

Multiblock Data Fusion in Statistics and Machine Learning: Applications in the Natural and Life Sciences,
First Edition. Age K. Smilde, Tormod Næs, and Kristian Hovde Liland.

Anđelković, B., Vujisić, L., Vučković, I., Tešević, V., Vajs, V., and Gođevac, D. (2017) Metabolomics study of populus type propolis. *Journal of Pharmaceutical and Biomedical Analysis*, **135**, 217–226.

Argelaguet, R., Velten, B., Arnol, D., Dietrich, S., Zenz, T., Marioni, J., Buettner, F., Huber, W., and Stegle, O. (2018) Multi-omics factor analysis—a framework for unsupervised integration of multi-omics data sets. *Molecular Systems Biology*, **14** (6), e8124.

Armagan, A., Dunson, D.B., and Lee, J. (2013) Generalized double pareto shrinkage. *Statistica Sinica*, **23** (1), 119.

Asioli, D., Almli, V.L., and Næs, T. (2016) Comparison of two different strategies for investigating individual differences among consumers in choice experiments. a case study based on preferences for iced coffee in norway. *Food Quality and Preference*, **54**, 79–89.

Asioli, D., Varela, P., Hersleth, M., Almli, V.L., Olsen, N.V., and Næs, T. (2017) A discussion of recent methodologies for combining sensory and extrinsic product properties in consumer studies. *Food Quality and Preference*, **56**, 266–273.

Balcke, G.U., Bennewitz, S., Bergau, N., Athmer, B., Henning, A., Majovsky, P., Jiménez-Gómez, J.M., Hoehenwarter, W., and Tissier, A. (2017) Multi-omics of tomato glandular trichomes reveals distinct features of central carbon metabolism supporting high productivity of specialized metabolites. *The Plant Cell*, **29** (5), 960–983.

Ballabio, D., Todeschini, R., and Consonni, V. (2019) *Data fusion methodology and applications*, Elsevier, chap. Recent advances in high level fusion methods to classify multiple analytical chemical data, pp. 129–155.

Bazzoli, C. and Lambert-Lacroix, S. (2018) Classification based on extensions of LS-PLS using logistic regression: application to clinical and multiple genomic data. *BMC Bioinformatics*, **19** (1), 314.

Belsley, D., Kuh, E., and Welsch, R. (2005) *Regression diagnostics: Identifying influential data and sources of collinearity*, vol. 571, John Wiley & Sons.

Benzécri, J.P. (1980) *Pratique de l'analyse des données. vol. 1, analyse des correspondances exposé élémentaire*, Dunod.

Berget, I., Castura, J.C., Ares, G., Næs, T., and Varela, P. (2020) Exploring the common and unique variability in tds and tcata data–a comparison using canonical correlation and orthogonalization. *Food Quality and Preference*, **79**, 103 790.

Berglund, A. and Wold, S. (1999) A serial extension of multiblock PLS. *Journal of Chemometrics*, **13** (3-4), 461–471.

Bevilacqua, M., Bucci, R., Materazzi, S., and Marini, F. (2013) Application of near infrared (nir) spectroscopy coupled to chemometrics for dried egg-pasta characterization and egg content quantification. *Food Chemistry*, **140** (4), 726–734.

Bezdek, J.C. (1981) Objective function clustering, in *Pattern recognition with fuzzy objective function algorithms*, Plenum Press, New York.

Biancolillo, A., Liland, K.H., Måge, I., Næs, T., and Bro, R. (2016) Variable selection in multi-block regression. *Chemometrics and Intelligent Laboratory Systems*, **156**, 89–101.

Biancolillo, A., Måge, I., and Næs, T. (2015) Combining SO-PLS and linear discriminant analysis for multi-block classification. *Chemometrics and Intelligent Laboratory Systems*, **141**, 58–67.

Biancolillo, A., Marini, F., and Roger, J.M. (2020) SO-CovSel: A novel method for variable selection in a multiblock framework. *Journal of Chemometrics*, **34**, doi:https://doi.org/10.1002/cem.3120.

Biancolillo, A., Næs, T., Bro, R., and Måge, I. (2017) Extension of SO-PLS to multi-way arrays: SO-N-PLS. *Chemometrics and Intelligent Laboratory Systems*, **164**, 113–126.

Björck, A . and Indahl, U.G. (2017) Fast and stable partial least squares modelling: A benchmark study with theoretical comments. *Journal of Chemometrics*, **31** (8), e2898.

Bollen, K.A. (1989a) Measurement models: The relation between latent and observed variables. *Structural equations with latent variables*, pp. 179–225.

Bollen, K.A. (1989b) A new incremental fit index for general structural equation models. *Sociological Methods & Research*, **17** (3), 303–316.

Borirak, O., Rolfe, M.D., de Koning, L.J., Hoefsloot, H.C.J., Bekker, M., Dekker, H.L., Roseboom, W., Green, J., de Koster, C.G., and Hellingwerf, K.J. (2015) Time-series analysis of the transcriptome and proteome of escherichia coli upon glucose repression. *Biochimica et Biophysica Acta (BBA)-Proteins and Proteomics*, **1854** (10), 1269–1279.

Borràs, E., Ferré, J., Boqué, R., Mestres, M., Aceña, L., and Busto, O. (2015) Data fusion methodologies for food and beverage authentication and quality assessment – A review. *Analytica Chimica Acta*, **891**, 1–14.

Bougeard, S., Abdi, H., Saporta, G., and Niang, N. (2018) Clusterwise analysis for multiblock component methods. *Advances in Data Analysis and Classification*, **12**, 285–313.

Bougeard, S., Qannari, E.M., Lupo, C., and Hanafi, M. (2011a) From Multiblock Partial Least Squares to Multiblock Redundancy Analysis. A Continuum Approach. *Informatica*, **22** (1), 11–26.

Bougeard, S., Qannari, E.M., and Rose, N. (2011b) Multiblock redundancy analysis: interpretation tools and application in epidemiology. *Chemometrics and Intelligent Laboratory Systems*, **25**, 467–475.

Brink-Jensen, K., Bak, S., Jorgensen, K., and Ekstrom, C.T. (2013) Integrative analysis of metabolomics and transcriptomics data: A unified model framework to identify underlying system pathways. *Plos One*, **8** (9), e72 116.

Bro, R. (1996) Multiway calibration. multilinear pls. *Journal of chemometrics*, **10** (1), 47–61.

Bro, R. and Elden, L. (2009) PLS works. *Journal of Chemometrics*, **23** (2), 69–71.

Bro, R., Kjeldahl, K., Smilde, A.K., and Kiers, H.A.L. (2008) Cross-validation of component models: a critical look at current methods. *Analytical and Bioanalytical Chemistry*, **390** (5), 1241–1251.

Bro, R., Rinnan, A., and Faber, N.M. (2005) Standard error of prediction for multilinear pls: 2. practical implementation in fluorescence spectroscopy. *Chemometrics and Intelligent Laboratory Systems*, **75** (1), 69–76.

Bro, R. and Smilde, A.K. (2003) Centering and scaling in component analysis. *Journal of Chemometrics*, **17**, 16–33.

Bro, R. and Smilde, A.K. (2014) Principal component analysis. *Analytical Methods*, **6** (9), 2812–2831.

Browne, M. (1979) The maximum-likelihood solution in inter-battery factor analysis. *British Journal of Mathematical and Statistical Psychology*, **32** (1), 75–86.

Bruggeman, F.J. and Westerhoff, H.V. (2007) The nature of systems biology. *Trends in Microbiology*, **15** (1), 45–50.

Burnham, A., Viveros, R., and MacGregor, J.F. (1996) Frameworks for latent variable multivariate regression. *Journal of Chemometrics*, **10**, 31–45.

Bylesjö, M., Eriksson, D., Kusano, M., Moritz, T., and Trygg, J. (2007) Data integration in plant biology: the O2PLS method for combined modeling of transcript and metabolite data. *The Plant Journal*, **52** (6), 1181–1191.

Caldana, C., Degenkolbe, T., Cuadros-Inostroza, A., Klie, S., Sulpice, R., Leisse, A., Steinhauser, D., Fernie, A.R., Willmitzer, L., and Hannah, M.A. (2011) High-density kinetic analysis of the metabolomic and transcriptomic response of Arabidopsis to eight environmental conditions. *The Plant Journal*, **67** (5), 869–884.

Camacho, J. and Ferrer, A. (2012) Cross-validation in pca models with the element-wise k-fold (ekf) algorithm: theoretical aspects. *Journal of Chemometrics*, **26** (7), 361–373.

Camacho, J. and Ferrer, A. (2014) Cross-validation in pca models with the element-wise k-fold (ekf) algorithm: practical aspects. *Chemometrics and Intelligent Laboratory Systems*, **131**, 37–50.

Camacho, J., Smilde, A.K., Saccenti, E., and Westerhuis, J.A. (2020) All sparse PCA models are wrong, but some are useful. Part I: Computation of scores, residuals and explained variance. *Chemometrics and Intelligent Laboratory Systems*, **196**, 103 907.

Campos, M.P., Sousa, R., Pereira, A.C., and Reis, M.S. (2017) Advanced predictive methods for wine age prediction: Part II - A comparison study of multiblock regression approaches. *Talanta*, **171**, 132–142.

Campos, M.P., Sousa, R., and Reis, M.S. (2018) Establishing the optimal blocks' order in SO-PLS: Stepwise SO-PLS and alternative formulations. *Journal of Chemometrics*, **32**, e3032.

Cariou, V., Bouveresse, D.J.R., Qannari, E.M., and Rutledge, D. (2019) ComDim methods for the analysis of multiblock data in a data fusion perspective, in *Data Handling in Science and Technology*, vol. 31, Elsevier, pp. 179–204.

Carroll, J.D. (1968) Generalization of canonical correlation analysis to three or more sets of variables, in *Proceedings of the 76th annual convention of the American Psychological Association*, vol. 3, vol. 3, pp. 227–228.

Carroll, J.D. and Arabie, P. (1980) Multidimensional scaling. *Annual Review of Psychology*, **31** (1), 607–649.

Carroll, J.D. and Chang, J.J. (1970) Analysis of individual differences in multidimensional scaling via an N-way generalization of "Eckart-Young" decomposition. *Psychometrika*, **35** (3), 283–319.

Castura, J.C., Antúnez, L., Giménez, A., and Ares, G. (2016) Temporal check-all-that-apply (TCATA): A novel dynamic method for characterizing products. *Food Quality and Preference*, **47**, 79–90.

Cattell, R.B. (1944) "parallel proportional profiles" and other principles for determining the choice of factors by rotation. *Psychometrika*, **9** (4), 267–283.

Cederkvist, H.R., Aastveit, A.H., and Næs, T. (2005) A comparison of methods for testing differences in predictive ability. *Journal of Chemometrics*, **19**, 500–509.

Ceulemans, E. and Kiers, H.A.L. (2006) Selecting among three-mode principal component models of different types and complexities: A numerical convex hull based method. *British Journal of Mathematical and Statistical Psychology*, **59** (1), 133–150.

Christian, G.D. and O'Reilly, J.E. (1988) *Instrumental analysis*, Pearson.

Christin, C., Hoefsloot, H.C.J., Smilde, A.K., Suits, F., Bischoff, R., and Horvatovich, P.L. (2010) Time alignment algorithms based on selected mass traces for complex lc-ms data. *Journal of Proteome Research*, **9** (3), 1483–1495.

Clish, C.B., Davidov, E., Oresic, M., Plasterer, T.N., Lavine, G., Londo, T., Meys, M., Snell, P., Stochaj, W., Adourian, A., Zhang, X., Morel, N., Neumann, E., Verheij, E., Vogels, J.T.W.E., Havekes, L.M., Afeyan, N., Regnier, F., Van Der Greef, J., and Naylor, S. (2004) Integrative biological analysis of the apoe*3-leiden transgenic mouse. *Omics-A journal of integrative biology*, **8**, 3–13.

Coccia, M., Collignon, C., Herve, C., Chalon, A., Welsby, I., Detienne, S., van Helden, M.J., Dutta, S., Genito, C.J., Waters, N.C., Van Deun, K., Smilde, A.K., van den Berg, R.A., Franco, D., Bourguignon, P., Morel, S., Garcon, N., Lambrecht, B.N., Goriely, S., van der Most, R., and Didierlaurent, A.M. (2018) Cellular and molecular synergy in as01-adjuvanted vaccines results in an early ifn gamma response promoting vaccine immunogenicity. *NPJ Vaccines*, **3**, 13.

Colombo, D. and Maathuis, M.H. (2014) Order-independent constraint-based causal structure learning. *The Journal of Machine Learning Research*, **15** (1), 3741–3782.

Coombs, C.H. (1964) *A theory of data*, John Wiley & Sons, New York.

Csala, A., Voorbraak, F.P.J.M., Zwinderman, A.H., and Hof, M.H. (2017) Sparse redundancy analysis of high-dimensional genetic and genomic data. *Bioinformatics*, **33** (20), 3228–3234.

Csala, A., Zwinderman, A.H., and Hof, M.H. (2020) Multiset sparse partial least squares path modeling for high dimensional omics data analysis. *BMC Bioinformatics*, **21** (1), 1–21.

Curtis, C., Shah, S.P., Chin, S.F., Turashvili, G., Rueda, O.M., Dunning, M.J., Speed, D., Lynch, A.G., Samarajiwa, S., Yuan, Y.Y., Graf, S., Ha, G., Haffari, G., Bashashati, A., Russell, R., McKinney, S., Langerod, A., Green, A., Provenzano, E., Wishart, G., Pinder, S., Watson, P., Markowetz, F., Murphy, L., Ellis, I., Purushotham, A., Borresen-Dale, A.L., Brenton, J.D., Tavare, S., Caldas, C., and Aparicio, S. (2012) The genomic and transcriptomic architecture of 2,000 breast tumours reveals novel subgroups. *Nature*, **486** (7403), 346–352.

Dadashi, M., Abdollahi, H., and Tauler, R. (2013) Application of maximum likelihood multivariate curve resolution to noisy data sets. *Journal of Chemometrics*, **27** (1-2), 34–41.

Dahl, T. and Næs, T. (2006) A bridge between tucker-1 and carroll's generalized canonical analysis. *Computational Statistics and Data Analysis*, **50**, 3086–3098.

Dahl, T. and Næs, T. (2009) Identifying outlying assessors in sensory profiling using fuzzy clustering and multi-block methodology. *Food quality and preference*, **20** (4), 287–294.

De Jong, S. (1993) Simpls: an alternative approach to partial least squares regression. *Chemometrics and intelligent laboratory systems*, **18** (3), 251–263.

De Jong, S. and Kiers, H.A.L. (1992) Principal covariates regression. Part 1. Theory. *Chemometrics and Intelligent Laboratory Systems*, **14**, 155–164.

de Juan, A., Jaumot, J., and Tauler, R. (2014) Multivariate curve resolution (mcr). solving the mixture analysis problem. *Analytical Methods*, **6** (14), 4964–4976.

de Juan, A. and Tauler, R. (2006) Multivariate curve resolution (mcr) from 2000: progress in concepts and applications. *Critical Reviews in Analytical Chemistry*, **36** (3-4), 163–176.

de Juan, A. and Tauler, R. (2020) Multivariate curve resolution: 50 years addressing the mixture analysis problem–a review. *Analytica Chimica Acta*, **1145**, 59–78.

De Leeuw, J., Young, F.W., and Takane, Y. (1976) Additive structure in qualitative data: An alternating least squares method with optimal scaling features. *Psychometrika*, **41**, 471–503.

De Luca, M., Ragno, G., Ioele, G., and Tauler, R. (2014) Multivariate curve resolution of incomplete fused multiset data from chromatographic and spectrophotometric analyses for drug photostability studies. *Analytica Chimica Acta*, **837**, 31–37.

De Moor, B.L.R. (1992) On the structure and geometry of the product singular value decomposition. *Linear Algebra and its Applications*, **168**, 95–136.

De Moor, B.L.R. and Zha, H. (1991) A tree of generalizations of the ordinary singular value decomposition. *Linear Algebra and its Applications*, **147**, 469–500.

de Noord, O.E. and Theobald, E.H. (2005) Multilevel component analysis and multilevel PLS of chemical process data. *Journal of Chemometrics*, **19** (5-7), 301–307.

De Roover, K., Ceulemans, E., Timmerman, M.E., Nezlek, J.B., and Onghena, P. (2013a) Modeling differences in the dimensionality of multiblock data by means of clusterwise simultaneous component analysis. *Psychometrika*, **78** (4), 648–668.

De Roover, K., Ceulemans, E., Timmerman, M.E., and Onghena, P. (2013b) A clusterwise simultaneous component method for capturing within-cluster differences in component variances and correlations. *British Journal of Mathematical and Statistical Psychology*, **66** (1), 81–102.

De Roover, K., Ceulemans, E., Timmerman, M.E., Vansteelandt, K., Stouten, J., and Onghena, P. (2012) Clusterwise simultaneous component analysis for analyzing structural differences in multivariate multiblock data. *Psychological Methods*, **17** (1), 100.

Deng, L., Guo, F., Cheng, K.K., Zhu, J., Gu, H., Raftery, D., and Dong, J. (2020) Identifying significant metabolic pathways using multi-block partial least-squares analysis. *Journal of Proteome Research*, **19** (5), 1965–1974.

Dietrich, S., Oleś, M., Lu, J., Sellner, L., Anders, S., Velten, B., Wu, B., Hüllein, J., da Silva Liberio, M., Walther, T. *et al.* (2018) Drug-perturbation-based stratification of blood cancer. *The Journal of clinical investigation*, **128** (1), 427–445.

Doeswijk, T.G., Smilde, A.K., Hageman, J.A., Westerhuis, J.A., and van Eeuwijk, F.A. (2011) On the increase of predictive performance with high-level data fusion. *Analytica Chimica Acta*, **705**, 41–47.

Draper, N.R. and Smith, H. (1998) *Applied Regression Analysis*, vol. 326, John Wiley & Sons, Inc.,, New York, 3rd edn.

Driscoll, M.F. and Borror, C.M. (2000) Sums of squares and expected mean squares in SAS. *Quality and Reliability Engineering International*, **16** (5), 423–433.

Eckart, C. and Young, G. (1936) The approximation of one matrix by another of lower rank. *Psychometrika*, **1**, 211–218.

Edwards, D. (2012) *Introduction to graphical modelling*, Springer Science & Business Media.

Efron, B. and Tibshirani, R.J. (1993) *An Introduction to the Bootstrap*, Chapman & Hall, New York.

El Bouhaddani, S., Houwing-Duistermaat, J., Salo, P., Perola, M., Jongbloed, G., and Uh, H. (2016) Evaluation of O2PLS in omics data integration, in *BMC Bioinformatics*, vol. 17, vol. 17, pp. 117–132.

El Ghaziri, A. and El Qannari, M. (2015) Measures of association between two datasets; application to sensory data. *Food Quality and Preference*, **40**, 116–124.

El Ghaziri, A., El Qannari, M., Moyon, T., and Alexandre-Gouabau, M. (2015) AoV-PLS: a new method for the analysis of multivariate data depending on several factors. *Electronic Journal of Applied Statistical Analysis*, **8** (2), 214–235.

Endrizzi, I., Gasperi, F., Rodbotten, M., and Næs, T. (2013) Permutation testing for validating PCA, in *SIS 2013 Statistical Conference Advances in Latent Variables; Methods, Models and Applications.*

Endrizzi, I., Gasperi, F., Rødbotten, M., and Næs, T. (2014) Interpretation, validation and segmentation of preference mapping models. *Food Quality and Preference*, **32**, 198–209.

Endrizzi, I., Menichelli, E., Johansen, S.B., Olsen, N.V., and Næs, T. (2011) Handling of individual differences in rating-based conjoint analysis. *Food Quality and Preference*, **22**, 241–254.

Eriksson, L., Byrne, T., Johansson, E., Trygg, J., and Vikström, C. (2013) *Multi-and megavariate data analysis basic principles and applications*, Umetrics Academy.

Eriksson, L., Damborsky, J., Earll, M., Johansson, E., Trygg, J., and Wold, S. (2004) Three-block bi-focal PLS (3BIF-PLS) and its application in QSAR. *SAR and QSAR in Environmental Research*, **15** (5-6), 481–499.

Escofier, B. and Pagès, J. (1994) Multiple factor analysis. *Computational Statistics and Data Analysis*, **18**, 121–140.

Escofier, B. and Pagès, J. (1998) Analyses factorielles simples et multiples. *Dunod, Paris.*

Eslami, A., Qannari, E.M., Kohler, A., and Bougeard, S. (2014) Algorithms for multi-group PLS. *Journal of Chemometrics*, **28** (3), 192–201.

Farias, R.C., Cohen, J.E., Jutten, C., and Comon, P. (2015) Joint decompositions with flexible couplings, in *International Conference on Latent Variable Analysis and Signal Separation*, Springer, pp. 119–126.

Fazelzadeh, P., Hangelbroek, R.W.J., Tieland, M., de Groot, L.C.P.G.M., Verdijk, L.B., Van Loon, L.J.C., Smilde, A.K., Alves, R.D.A.M., Vervoort, J., Muller, M. *et al.* (2016) The muscle metabolome differs between healthy and frail older adults. *Journal of Proteome Research*, **15** (2), 499–509.

Fisher, R.A. (1921) On the "probable error" of a coefficient of correlation deduced from a small sample. *Metron*, **1**, 1–32.

Fisher, R.A. (1937) *The design of experiments*, Oliver & Boyd, Edinburgh & London., 2nd edn.

Fornell, C., Barclay, D.W., and Rhee, B.D. (1988) A model and simple iterative algorithm for redundancy analysis. *Multivariate behavioral research*, **23** (3), 349–360.

Franzosa, E. A .and Hsu, T., Sirota-Madi, A., Shafquat, A., Abu-Ali, G., Morgan, X.C., and Huttenhower, C. (2015) Sequencing and beyond: integrating molecular 'omics' for microbial community profiling. *Nature Reviews Microbiology*, **13** (6), 360.

Freund, Y. (1995) Boosting a weak learning algorithm by majority. *Information and computation*, **121** (2), 256–285.

Friendly, M., Monette, G., Fox, J. *et al.* (2013) Elliptical insights: understanding statistical methods through elliptical geometry. *Statistical Science*, **28** (1), 1–39.

Fu, W.J. (1998) Penalized regressions: the bridge versus the lasso. *Journal of Computational and Graphical Statistics*, **7** (3), 397–416.

Galindo-Prieto, B., Eriksson, L., and Trygg, J. (2014) Variable influence on projection (vip) for orthogonal projections to latent structures (opls). *Journal of Chemometrics*, **28** (8), 623–632.

Gaynanova, I. and Li, G. (2019) Structural learning and integrative decomposition of multi-view data. *Biometrics*, **75** (4), 1121–1132.

Giacalone, D., Bredie, W.L.P., and Frøst, M.B. (2013) "all-in-one test" (ai1): A rapid and easily applicable approach to consumer product testing. *Food Quality and Preference*, **27** (2), 108–119.

Gifi, A. (1990) *Nonlinear multivariate analysis*, John Wiley & Sons, Chichester.

Gomez-Cabrero, D., Tarazona, S., Ferreirós-Vidal, I., Ramirez, R.N., Company, C., Schmidt, A., Reijmers, T., von Saint Paul, V., Marabita, F., Rodríguez-Ubreva, J. *et al.* (2019) Stategra, a comprehensive multi-omics dataset of b-cell differentiation in mouse. *Scientific Data*, **6** (1), 1–15.

Gower, J.C. and Dijksterhuis, G.B. (2004) *Procrustes problems*, vol. 30, Oxford University Press on Demand.

Gustafsson, A., Herrmann, A., and Huber, F. (2007) *Conjoint measurement: methods and applications*, Springer Science & Business Media.

Hanafi, M. and Kiers, H.A.L. (2006) Analysis of k sets of data, with differential emphasis on agreement between and within sets. *Computational Statistics & Data Analysis*, **51** (3), 1491–1508.

Hand, D.J. (1996) Statistics and the theory of measurement. *Journal of the Royal Statistical Society Series A-statistics in Society*, **159**, 445–473.

Hand, D.J. (2004) *Measurement Theory and Practice: The World Through Quantification*, John Wiley & Sons.

Harrington, P.d.B., Vieira, N.E., Espinoza, J., Nien, J.K., Romero, R., and Yergey, A.L. (2005) Analysis of variance–principal component analysis: A soft tool for proteomic discovery. *Analytica chimica acta*, **544** (1-2), 118–127.

Harshman, R.A. (1970) Foundations of the parafac procedure: models and conditions for an 'explanation' multi-modal factor analysis. *UCLA Working Papers in Phonetics*, **16**, 1–84.

Harshman, R.A. (1984) How can I know it's real? A catalog of diagnostics for use with three-mode factor analysis and multidimensional scaling multidimensional scaling, in *Research Methods for Multimode Data Analysis* (eds H.G. Law, C.W. Snyder, J. Hattie, and R.P. McDonald), Preager, New York, pp. 566–591.

Harshman, R.A. and Lundy, M.E. (1984) Data preprocessing and the extended PARAFAC model, in *Research Methods for Multimode Data Analysis* (eds H.G. Law, C.W. Snyder, J. Hattie, and R.P. McDonald), Praeger, New York, pp. 216–284.

Hastie, T., Tibshirani, R., and Friedman, J. (2009) *The elements of statistical learning: data mining, inference, and prediction*, Springer Science & Business Media.

Hastie, T., Tibshirani, R., and Wainwright, M. (2015) *Statistical learning with sparsity: the lasso and generalizations*, CRC press.

Hector, A., Von Felten, S., and Schmid, B. (2010) Analysis of variance with unbalanced data: an update for ecology & evolution. *Journal of Animal Ecology*, **79** (2), 308–316.

Heijne, W.H.M., Lamers, R.J.A.N., van Bladeren, P.J., Groten, J.P., van Nesselrooij, J.H.J., and van Ommen, B. (2005) Profiles of metabolites and gene expression in rats with chemically induced hepatic necrosis. *Toxicology Pathology*, **33**, 425–433.

Helgesen, H., Solheim, R., and Næs, T. (1997) Consumer preference mapping of dry fermented lamb sausages. *Food Quality and Preference*, **8** (2), 97–109.

Hirst, D. and Næs, T. (1994) A graphical technique for assessing differences among a set of rankings. *Journal of Chemometrics*, **8** (1), 81–93.

Hoefsloot, H., Verouden, M., Westerhuis, J., and Smilde, A. (2006) Maximum likelihood scaling (mals). *Journal of Chemometrics*, **20** (3-4), 120–127.

Hong, D., Kolda, T., and Duersch, J. (2020) Generalized canonical polyadic tensor decomposition. *SIAM Review*, **62** (1), 133–163.

Höskuldsson, A. (1988) Pls regression methods. *Journal of chemometrics*, **2** (3), 211–228.

Hotelling, H. (1933) Analysis of a complex of statistical variables into principal components. *Journal of Educational Psychology*, **24**, 417–441.

Hotelling, H. (1936a) Relations between two sets of variates. *Biometrika*, **28**, 321–377.

Hotelling, H. (1936b) Simplified calculation of principal components. *Psychometrika*, **1**, 27–35.

Huang, H., Vangay, P., McKinlay, C.E., and Knights, D. (2014) Multi-omics analysis of inflammatory bowel disease. *Immunology letters*, **162** (2), 62–68.

Indahl, U.G. and Næs, T. (1998) Evaluation of alternative spectral feature extraction methods of textural images for multivariate modelling. *Journal of Chemometrics*, **12**, 261–278.

Indahl, U.G. (2014) The geometry of PLS1 explained properly: 10 key notes on mathematical properties of and some alternative algorithmic approaches to PLS1 modelling. *Journal of Chemometrics*, **28** (3), 168–180.

Indahl, U.G., Liland, K.H., and Næs, T. (2009) Canonical partial least squares-a unified pls approach to classification and regression problems. *Journal of Chemometrics*, **23**, 495–504.

Indahl, U.G., Næs, T., and Liland, K.H. (2018) A similarity index for comparing coupled matrices. *Journal of Chemometrics*, **32** (10), e3049.

Iorio, F., Knijnenburg, T.A., Vis, D.J., Bignell, G.R., Menden, M.P., Schubert, M., Aben, N., Gonçalves, E., Barthorpe, S., Lightfoot, H. *et al.* (2016) A landscape of pharmacogenomic interactions in cancer. *Cell*, **166** (3), 740–754.

Isaksson, T. and Næs, T. (1990) Selection of samples for calibration in near-infrared spectroscopy. Part II: Selection based on spectral measurements. *Applied Spectroscopy*, **44** (7), 1152–1158.

Jacobs, D.M., Gaudier, E., van Duynhoven, J., and Vaughan, E.E. (2009) Non-digestible food ingredients, colonic microbiota and the impact on gut health and immunity: a role for metabolomics. *Current drug metabolism*, **10** (1), 41–54.

Jansen, J.J., Bro, R., Hoefsloot, H.C.J., van den Berg, F.W.J., Westerhuis, J.A., and Smilde, A.K. (2008) PARAFASCA: ASCA combined with PARAFAC for the analysis of metabolic fingerprinting data. *Journal of Chemometrics*, **22** (1-2), 114–121.

Jansen, J.J., Hoefsloot, H.C.J., van der Greef, J., Timmerman, M.E., and Smilde, A.K. (2005a) Multilevel component analysis of time-resolved metabolic fingerprinting data. *Analytica Chimica Acta*, **530** (2), 173–183.

Jansen, J.J., Hoefsloot, H.C.J., van der Greef, J., Timmerman, M.E., Westerhuis, J.A., and Smilde, A.K. (2005b) ASCA: analysis of multivariate data obtained from an experimental design. *Journal of Chemometrics*, **19** (9), 469–481.

Jansen, J.J., Szymanska, E., Hoefsloot, H.C.J., and Smilde, A.K. (2012) Individual differences in metabolomics: individualised responses and between-metabolite relationships. *Metabolomics*, **8** (1), 94–104.

Janson, S. and Vegelius, J. (1982) Correlation-coefficients for more than one scale type. *Multivariate Behavioral Research*, **17** (2), 271–284.

Jaumot, J., de Juan, A., and Tauler, R. (2015) MCR-ALS GUI 2.0: new features and applications. *Chemometrics and Intelligent Laboratory Systems*, **140**, 1–12.

Jolliffe, I.T. (2010) *Principal Component Analysis*, Springer, Berlin.

Jöreskog, K.G. and Wold, H. (1982) *Systems under Indirect Observation*, North-Holland, Amsterdam.

Jørgensen, B. (1992) Exponential dispersion models and extensions: A review. *International Statistical Review/Revue Internationale de Statistique*, pp. 5–20.

Jørgensen, K., Mevik, B.H., and Næs, T. (2007) Combining designed experiments with several blocks of spectroscopic data. *Chemometrics and Intelligent Laboratory Systems*, **88**, 154–166.

Jørgensen, K. and Næs, T. (2004) A design and analysis strategy for situations with uncontrolled raw material variation. *Journal of Chemometrics*, **18** (2), 45–52.

Jørgensen, K. and Næs, T. (2008) The use of LS-PLS for improved understanding, monitoring and prediction of cheese processing. *Chemometrics and Intelligent Laboratory Systems*, **93**, 11–19.

Kaiser, H.F. (1958) The varimax criterion for analytic rotation in factor analysis. *Psychometrika*, **23** (3), 187–200.

Kanehisa, M. and Goto, S. (2000) Kegg: kyoto encyclopedia of genes and genomes. *Nucleic Acids Research*, **28** (1), 27–30.

Karaman, I., Nørskov, N.P., Yde, C.C., Hedemann, M.S., Knudsen, K.E.B., and Kohler, A. (2015) Sparse multi-block PLSR for biomarker discovery when integrating data from LC–MS and NMR metabolomics. *Metabolomics*, **11** (2), 367–379.

Kardinaal, A.F.M., van Erk, M.J., Dutman, A.E., Stroeve, J.H.M., van de Steeg, E., Bijlsma, S., Kooistra, T., van Ommen, B., and Wopereis, S. (2015) Quantifying phenotypic flexibility as the response to a high-fat challenge test in different states of metabolic health. *The FASEB Journal*, **29** (11), 4600–4613.

Kedem, B., Oliveira, V.D., and Sverchkov, M. (2017) Statistical data fusion. World Scientific, Singapore.

Kettenring, J.R. (1971) Canonical analysis of several sets of variables. *Biometrika*, **58** (3), 433–451.

Kiers, H.A.L. (1989) *Three-way methods for the analysis of qualitative and quantitative two-way data*, DSWO Press.

Kiers, H.A.L. (1991) Hierarchical relations among three-way methods. *Psychometrika*, **56**, 449–470.

Kiers, H.A.L. and Ten Berge, J.M.F. (1994) Hierarchical relations between methods for simultaneous component analysis and a technique for rotation to a simple simultaneous structure. *British Journal of Mathematical and Statistical Psychology*, **47** (Part 1), 109–126.

Kiers, H.A.L., Ten Berge, J.M.F., and Bro, R. (1999) PARAFAC2—Part I. A direct fitting algorithm for the PARAFAC2 model. *Journal of Chemometrics*, **13** (3-4), 275–294.

Klami, A., Virtanen, S., and Kaski, S. (2013) Bayesian canonical correlation analysis. *Journal of Machine Learning Research*, **14** (Apr), 965–1003.

Kleemann, R., Verschuren, L., van Erk, M., Nikolsky, Y., Cnubben, N., Verheij, E., Smilde, A.K., Hendriks, H., Zadelaar, S., Smith, G., Kaznacheev, V., Nikolskaya, T., Melnikov, A., Hurt-Camejo, E., van der Greef, J., van Ommen, B., and Kooistra, T. (2007) Atherosclerosis and liver inflammation induced by increased dietary cholesterol intake: a combined transcriptomics and metabolomics analysis. *Genome Biology*, **8**, R200.

Kolen, M.J. and Brennan, R.L. (2014) *Test equating, scaling, and linking: Methods and practices*, Springer Science & Business Media.

Kosanovich, K.A., Dahl, K.S., and Piovoso, M.J. (1996) Improved process understanding using multiway principal component analysis. *Industrial & Engineering Chemistry Research*, **35** (1), 138–146.

Kourti, T. and MacGregor, J.F. (1995) Process analysis, monitoring and diagnosis, using multivariate projection methods. *Chemometrics and intelligent laboratory systems*, **28** (1), 3–21.

Kourti, T., Nomikos, P., and MacGregor, J.F. (1995) Analysis, monitoring and fault diagnosis of batch processes using multiblock and multiway pls. *Journal of Process Control*, **5**, 277–284.

Krämer, N. and Sugiyama, M. (2011) The Degrees of Freedom of Partial Least Squares Regression. *Journal of the American Statistical Association*, **106**, 697–705, doi:10.1198/jasa.2011.tm10107.

Krantz, D.H., Luce, R.D., Suppes, P., and Tversky, A. (1971) *Foundations of Measurement (Volume I)*, Dover.

Kreutzmann, S., Svensson, V.T., Thybo, A.K., Bro, R., and Petersen, M.A. (2008) Prediction of sensory quality in raw carrots (Daucus carota L.) using multi-block LS-ParPLS. *Food Quality and Preference*, **19** (7), 609–617.

Krijnen, W.P., Dijkstra, T.K., and Stegeman, A. (2008) On the non-existence of optimal solutions and the occurrence of "degeneracy" in the Candecomp/Parafac model. *Psychometrika*, **73** (3), 431–439.

Kroonenberg, P.M. (2008) *Applied Multiway Data Analysis*, John Wiley & Sons.

Landgraf, A.J. and Lee, Y. (2015) Dimensionality reduction for binary data through the projection of natural parameters. *arXiv*, 1510.06112.

Langsrud, Ø. and Næs, T. (2003) Optimised score plot by principal components of predictions. *Chemometrics and Intelligent Laboratory Systems*, **68**, 61–74, doi:10.1016/s0169-7439(03)00088-1.

Lavit, C., Escoufier, Y., Sabatier, R., and Traissac, P. (1994) The ACT (STATIS method. *Computational Statistics and Data Analysis*, **18**, 97–119.

Law, H.G., Snyder, C.W., A., H.J., and McDonald, R.P. (1984) *Research methods for multimode data analysis*, Praeger Publishers.

Lawless, H.T. and Heymann, H. (2010) *Sensory evaluation of food: Principles and practices*, Springer Science and Business Media, New York.

Lawton, W.H. and Sylvestre, E.A. (1971) Self modeling curve resolution. *Technometrics*, **13**, 617–633.

Lê Cao, K.A., Rossouw, D., Robert-Granié, C., and Besse, P. (2008) A sparse PLS for variable selection when integrating omics data. *Statistical Applications in Genetics and Molecular Biology*, **7** (1).

Ledoit, O. and Wolf, M. (2004) A well-conditioned estimator for large-dimensional covariance matrices. *Journal of Multivariate Analysis*, **88** (2), 365–411.

Legendre, P. and Legendre, L. (2012) *Numerical ecology*, Elsevier.

Lemanska, A., Grootveld, M., Silwood, C., and Brereton, R. (2012) Chemometric variance analysis of 1 H NMR metabolomics data on the effects of oral rinse on saliva. *Metabolomics*, **8** (1), 64–80.

Lengard, V. and Kermit, M. (2006) 3-Way and 3-block PLS regressions in consumer preference analysis. *Food Quality and Preference*, **17** (3-4), 234–242.

Lepore, A., Palumbo, B., and Capezza, C. (2019) Orthogonal LS-PLS approach to ship fuel-speed curves for supporting decisions based on operational data. *Quality Engineering*, **31** (3), 386–400.

Levin, J. (1966) Simultaneous factor analysis of several gramian matrices. *Psychometrika*, **31** (3), 413–419.

Li, G. and Gaynanova, I. (2018) A general framework for association analysis of heterogeneous data. *The Annals of Applied Statistics*, **12** (3), 1700–1726.

Li, W., Zhang, S., Liu, C.C., and Zhou, X.J. (2012) Identifying multi-layer gene regulatory modules from multi-dimensional genomic data. *Bioinformatics*, **28** (19), 2458–2466.

Liland, K., Høy, M., Martens, H., and Sæbø, S. (2013) Distribution based truncation for variable selection in subspace methods for multivariate regression. *Chemometrics and Intelligent Laboratory Systems*, **122**, 103–111.

Liland, K.H., Næs, T., and Indahl, U.G. (2016) ROSA-a fast extension of partial least squares regression for multiblock data analysis. *Journal of Chemometrics*, **30**, 651–662, doi:10.1002/cem.2824.

Liland, K.H., Smilde, A.K., Marini, F., and Næs, T. (2018) Confidence ellipsoids for ASCA models based on multivariate regression theory. *Journal of Chemometrics*, **32** (e2990), 1–13.

Liland, K.H., Stefansson, P., and Indahl, U.G. (2020) Much faster cross-validation in PLSR-modelling by avoiding redundant calculations. *Journal of Chemometrics*, **34** (3), e3201.

Lips, M. A .and Van Klinken, J.B., van Harmelen, V., Dharuri, H.K., AC't Hoen, P., Laros, J.F.J., van Ommen, G.J., Janssen, I.M., Van Ramshorst, B., Van Wagensveld, B.A. *et al.* (2014) Roux-en-y gastric bypass surgery, but not calorie restriction, reduces plasma branched-chain amino acids in obese women independent of weight loss or the presence of type 2 diabetes. *Diabetes Care*, **37** (12), 3150–3156.

Little, R. and Rubin, D. (2019) *Statistical analysis with missing data*, vol. 793, John Wiley & Sons.

Lock, E.F., Hoadley, K.A., Marron, J.S., and Nobel, A.B. (2013) Joint and individual variation explained (JIVE) for integrated analysis of multiple data types. *Ann Appl Stat*, **7** (1), 523–542.

Löfstedt, T., Eriksson, L., Wormbs, G., and Trygg, J. (2012) Bi-modal OnPLS. *Journal of Chemometrics*, **26** (6), 236–245.

Löfstedt, T., Hoffman, D., and Trygg, J. (2013) Global, local and unique decompositions in onpls for multiblock data analysis. *Analytica chimica acta*, **791**, 13–24.

Löfstedt, T. and Trygg, J. (2011) OnPLS—a novel multiblock method for the modelling of predictive and orthogonal variation. *Journal of Chemometrics*, **25** (8), 441–455.

Lopes, J.A., Menezes, J.C., Westerhuis, J.A., and Smilde, A.K. (2002) Multiblock PLS analysis of an industrial pharmaceutical process. *Biotechnology and Bioengineering*, **80** (4), 419–427.

Luce, R.D. and Narens, L. (1987) Measurement scales on the continuum. *Science*, **236** (4808), 1527–1532.

Luciano, G. and Næs, T. (2009) Interpreting sensory data by combining principal component analysis and analysis of variance. *Food Quality and Preference*, **20** (3), 167–175.

Ly-Verdú, S., Gröger, T.M., Arteaga-Salas, J.M., Brandmaier, S., Kahle, M., Neschen, S., de Angelis, M.H., and Zimmermann, R. (2015) Combining metabolomic non-targeted gc× gc–tof–ms analysis and chemometric ASCA-based study of variances to assess dietary influence on type 2 diabetes development in a mouse model. *Analytical and Bioanalytical Chemistry*, **407** (1), 343–354.

MacGregor, J.F., Jaeckle, C., Kiparissides, C., and Koutoudi, M. (1994) Process monitoring and diagnosis by multiblock pls methods. *AIChE Journal*, **40** (5), 826–838.

Madssen, T.S., Giskeødegård, G.F., Smilde, A.K., and Westerhuis, J.A. (2020) Repeated measures asca+ for analysis of longitudinal intervention studies with multivariate outcome data. *medRxiv*.

Maeder, M. and Zuberbuehler, A.D. (1986) The resolution of overlapping chromatographic peaks by evolving factor analysis. *Analytica Chimica Acta*, **181**, 287–291.

Måge, I., Menichelli, E., and Næs, T. (2012) Preference mapping by PO-PLS: Separating common and unique information in several data blocks. *Food Quality and Preference*, **24** (1), 8–16.

Måge, I., Mevik, B.H., and Næs, T. (2008) Regression models with process variables and parallel blocks of raw material measurements. *Journal of Chemometrics*, **22**, 443–456.

Måge, I. and Næs, T. (2005) Split-plot regression models with both design and spectroscopic variables. *Journal of Chemometrics*, **19**, 521–531.

Måge, I., Smilde, A.K., and van der Kloet, F.M. (2019) Performance of methods that separate common and distinct variation in multiple data blocks. *Journal of Chemometrics*, **33** (1), e3085.

Manne, R. (1995) On the resolution problem in hyphenated chromatography. *Chemometrics and Intelligent Laboratory Systems*, **27** (1), 89–94.

Manne, R., Pell, R.J., and Ramos, L.S. (2009) The PLS model space: the inconsistency persists. *Journal of Chemometrics: A Journal of the Chemometrics Society*, **23** (2), 76–77.

Mardia, K.V., Kent, J.T., and Bibby, J.M. (1979) *Multivariate Analysis*, Academic Press, New York.

Martens, H., Anderssen, E., Flatberg, A., Gidskehaug, L.H., Høy, M., Westad, F., Thybo, A., and Martens, M. (2005) Regression of a data matrix on descriptors of both its rows and of its columns via latent variables: L-PLSR. *Computational Statistics & Data Analysis*, **48** (1), 103 – 123.

Martens, H. and Martens, M. (2001) *Multivariate Analysis of Quality. An Introduction*, John Wiley & Sons, Chichester.

Martens, H. and Næs, T. (1989) *Multivariate calibration*, John Wiley & Sons, Chichester.

Martin, M. and Govaerts, B. (2021) LiMM-PCA: Combining ASCA+ and linear mixed models to analyse high-dimensional designed data. *Journal of Chemometrics*, **34** (6), e3232.

MATLAB (2021) *R2021a*, The MathWorks Inc., Natick, Massachusetts.

Mayer, C.D., Lorent, J., and Horgan, G.W. (2011) Exploratory analysis of multiple omics datasets using the adjusted RV coefficient. *Statistical Applications in Genetics and Molecular Biology*, **10** (1).

McKnight, D., Huerlimann, R., Bower, D., Schwarzkopf, L., Alford, R., and Zenger, K. (2019) Methods for normalizing microbiome data: an ecological perspective. *Methods in Ecology and Evolution*, **10** (3), 389–400.

McMurdie, P. and Holmes, S. (2014) Waste not, want not: why rarefying microbiome data is inadmissible. *PLoS Computational Biology*, **10** (4), e1003 531.

Meinshausen, N. and Bühlmann, P. (2010) Stability selection. *Journal of the Royal Statistical Society: Series B (Statistical Methodology)*, **72** (4), 417–473.

Menichelli, E., Almøy, T., Tomic, O., Olsen, N.V., and Næs, T. (2014a) SO-PLS as an exploratory tool for path modelling. *Food Quality and Preference*, **36**, 122–134.

Menichelli, E., Hersleth, M., Almøy, T., and Næs, T. (2014b) Alternative methods for combining information about products, consumers and consumers' acceptance based on path modelling. *Food Quality and Preference*, **31**, 142–155.

Menichelli, E., Kraggerud, H., Olsen, N.V., and, Næs, T. (2013) Analysing relations between specific and total liking scores. *Food Quality and Preference*, **23**, 148–159.

Menichelli, E., Olsen, N.V., Meyer, C., and Næs, T. (2012) Combining extrinsic and intrinsic information in consumer acceptance studies. *Food Quality and Preference*, **23** (2), 148–159.

Michailidis, G. and de Leeuw, J. (1998) The gifi system of descriptive multivariate analysis. *Statistical Science*, **13** (4), 307–336.

Michell, J. (1986) Measurement scales and statistics - a clash of paradigms. *Psychological Bulletin*, **100** (3), 398–407.

Mitchell, H.B. (2012) Data fusion: Concepts and ideas, doi:10.1007/978-3-642-27222-6.

Moco, S., Martin, F.P.J., and Rezzi, S. (2012) Metabolomics view on gut microbiome modulation by polyphenol-rich foods. *Journal of Proteome Research*, **11** (10), 4781–4790.

Murphy, K.P. (2012) *Machine learning: a probabilistic perspective*, MIT Press.

Næs, T., and Mevik, B. (2001) Understanding the collinearity problem in regression and discriminant analysis. *Journal of Chemometrics*, **15** (4), 413–426.

Næs, T., Brockhoff, P.B., and Tomic, O. (2010) *Statistics for sensory and consumer science*, Wiley, Chichester, UK.

Næs, T. and Isaksson, T. (1991) Splitting of calibration data by cluster analysis. *Journal of Chemometrics*, **5**, 49–65.

Næs, T., Måge, I., and Segtnan, V.H. (2011a) Incorporating interactions in multi-block sequential and orthogonalised partial least squares regression. *Journal of Chemometrics*, **25**, 601–609.

Næs, T. and Mevik, B.H. (1999) The flexibility of fuzzy clustering illustrated by examples. *Journal of Chemometrics,* **13** (3-4), 435–444.

Næs, T., Romano, R., Tomic, O., Måge, I., Smilde, A.K., and Liland, K.H. (2020) Sequential and orthogonalized PLS (SO-PLS) regression for path analysis: Order of blocks and relations between effects. *Journal of Chemometrics,* p. e3243.

Næs, T., Tomic, O., Afseth, N., Segtnan, V., and Måge, I. (2013) Multi-block regression based on combinations of orthogonalisation, pls-regression and canonical correlation analysis. *Chemometrics and Intelligent Laboratory Systems,* **124**, 32–42.

Næs, T., Tomic, O., Greiff, K., and Thyholt, K. (2014) A comparison of methods for analyzing multivariate sensory data in designed experiments–a case study of salt reduction in liver paste. *Food Quality and Preference,* **33**, 64–73.

Næs, T., Tomic, O., Mevik, B.H., and Martens, H. (2011b) Path modelling by sequential PLS regression. *Journal of Chemometrics,* **25**, 28–40.

Næs, T., Varela, P., and Berget, I. (2018) *Individual differences in sensory and consumer science,* Elsevier, UK.

Nakaya, H., Wrammert, J., Lee, E., Racioppi, L., Marie-Kunze, S., Haining, W., Means, A., Kasturi, S., Khan, N., Li, G. *et al.* (2011) Systems biology of vaccination for seasonal influenza in humans. *Nature immunology,* **12** (8), 786.

Narens, L. and Luce, R.D. (1986) Measurement - the theory of numerical assignments. *Psychological Bulletin,* **99** (2), 166–180.

Neal, R.M. (2012) *Bayesian learning for neural networks,* vol. 118, Springer Science & Business Media.

Nguyen, D.V. and Rocke, D.M. (2002) Tumor classification by partial least squares using microarray gene expression data. *Bioinformatics,* **18** (1), 39–50.

Nguyen, Q.C., Liland, K.H., Tomic, O., Tarrega, A., Varela, P., and Næs, T. (2020) SO-PLS as an alternative approach for handling multi-dimensionality in modelling different aspects of consumer expectations. *Food Research International,* **133** (109189).

Niimi, J., Tomic, O., Næs, T., Bastian, S.E., Jeffery, D.W., Nicholson, E.L., Maffei, S.M., and Boss, P.K. (2020) Objective measures of grape quality: From Cabernet Sauvignon grape composition to wine sensory characteristics. *LWT,* **123**, 109 105.

Niimi, J., Tomic, O., Næs, T., Jeffery, D.W., Bastian, S.E.P., and Boss, P.K. (2018) Application of sequential and orthogonalised-partial least squares (SO-PLS) regression to predict sensory properties of Cabernet Sauvignon wines from grape chemical composition. *Food Chemistry,* **256**, 195–202.

Nikzad-Langerodi, R., Zellinger, W., Lughofer, E., and Saminger-Platz, S. (2018) Domain-invariant partial-least-squares regression. *Analytical Chemistry,* **90** (11), 6693–6701.

Nomikos, P. and MacGregor, J.F. (1995) Multivariate SPC charts for monitoring batch processes. *Technometrics,* **37** (1), 41–59.

Nørgaard, L., Saudland, A., Wagner, J., Nielsen, J.P., Munck, L., and Engelsen, S.B. (2000) Interval Partial Least-Squares Regression (iPLS): A Comparative Chemometric Study with an Example from Near-Infrared Spectroscopy. *Applied Spectroscopy,* **54**, 413–419.

Nueda, M.J., Conesa, A., Westerhuis, J.A., Hoefsloot, H.C.J., Smilde, A.K., Talón, M., and Ferrer, A. (2007) Discovering gene expression patterns in time course microarray experiments by ANOVA–SCA. *Bioinformatics,* **23** (14), 1792–1800.

Olivieri, A.C. and Tauler, R. (2021) N-BANDS: A new algorithm for estimating the extension of feasible bands in multivariate curve resolution of multicomponent systems in the presence of noise and rotational ambiguity. *Journal of Chemometrics,* **35** (3), e3317.

Pagès, J. (2005) Collection and analysis of perceived product inter-distances using multiple factor analysis: Application to the study of 10 white wines from the loire valley. *Food Quality and Preference,* **16** (7), 642–649.

Paige, C.C. and Saunders, M.A. (1981) Towards a generalized singular value decomposition. *SIAM J. Numer. Anal.*, **18** (3), 398–405.

Pearl, J. (2009) *Causality*, Cambridge University Press.

Pearson, K. (1901) On lines and planes of closest fit to points in space. *Philosophical Magazine*, **2**, 559–572.

Pell, R.J., Ramos, L.S., and Manne, R. (2007) The model space in partial least squares regression. *Journal of Chemometrics*, **21** (3-4), 165–172.

Pellis, L., van Erk, M.J., van Ommen, B., Bakker, G.C.M., Hendriks, H.F.J., Cnubben, N.H.P., Kleemann, R., van Someren, E.P., Bobeldijk, I., Rubingh, C.M. *et al.* (2012) Plasma metabolomics and proteomics profiling after a postprandial challenge reveal subtle diet effects on human metabolic status. *Metabolomics*, **8** (2), 347–359.

Piqueras, S., Bedia, C., Beleites, C., Krafft, C., Popp, J., Maeder, M., Tauler, R., and de Juan, A. (2018) Handling different spatial resolutions in image fusion by multivariate curve resolution-alternating least squares for incomplete image multisets. *Analytical Chemistry*, **90** (11), 6757–6765.

Pitman, E.J.G. (1937) Significance tests which may be applied to samples from any populations. *Journal of the Royal Statistical Society, Series B*, **4**, 119 – 130.

Plaehn, D. and Lundahl, D.S. (2006) An L-PLS preference cluster analysis on French consumer hedonics to fresh tomatoes. *Food Quality and Preference*, **17** (3-4), 243–256.

Ponnapalli, S.P., Saunders, M.A., Van Loan, C.F., and Alter, O. (2011) A higher-order generalized singular value decomposition for comparison of global mRNA expression from multiple organisms. *PLoS One*, **6** (12), e28 072.

Quenouille, M.H. (1949) Problems in plane sampling. *The Annals of Mathematical Statistics*, **20** (3), 355–375.

Quenouille, M.H. (1956) Notes on bias in estimation. *Biometrika*, **43**, 353 – 360.

R Core Team (2022) *R: A Language and Environment for Statistical Computing*, R Foundation for Statistical Computing, Vienna, Austria. URL https://www.R-project.org/.

Raîche, G., Walls, T.A., Magis, D., Riopel, M., and Blais, J.G. (2013) Non-graphical solutions for cattell's scree test. *Methodology*.

Rajko, R. (2009) Some surprising properties of multivariate curve resolution-alternating least squares (MCR-ALS) algorithms. *Journal of Chemometrics*, **23** (4), 172–178.

Rajko, R. (2010) Rejoinder to 'Comments on a recently published paper "Some surprising properties of multivariate curve resolution-alternating least squares (MCR-ALS) algorithms"'. *Journal of Chemometrics*, **24** (2), 91–93.

Ramsay, J.O., ten Berge, J.M.F., and Styan, G.P.H. (1984) Matrix correlation. *Psychometrika*, **49** (3), 403–423.

Rao, C.R. (1964) The use and interpretation of principal component analysis in applied research. *Sankhyā: The Indian Journal of Statistics, Series A*, pp. 329–358.

Reis, M.S. (2013) Network-induced supervised learning: Network-induced classification (ni-c) and network-induced regression (ni-r). *Process Systems Engineering*, **59**, 1570–1587.

Rennie, J.D.M. and Srebro, N. (2005) Loss functions for preference levels: Regression with discrete ordered labels, in *Proceedings of the IJCAI multidisciplinary workshop on advances in preference handling*, vol. 1, Kluwer Norwell, MA, vol. 1.

Richards, S.E., Dumas, M.E., Fonville, J.M., Ebbels, T.M.D., Holmes, E., and Nicholson, J.K. (2010) Intra-and inter-omic fusion of metabolic profiling data in a systems biology framework. *Chemometrics and Intelligent Laboratory Systems*, **104** (1), 121–131.

Richards, S.E. and Holmes, E. (2014) Chemometrics methods for the analysis of genomics, transcriptomics, proteomics, metabolomics, and metagenomics datasets, in *Metabolomics as a Tool in Nutrition Research*, Woodhead Publishing, pp. 37–60.

Ripley, B.D. (2007) *Pattern Recognition and Neural Networks*, Cambridge University Press.

Robert, P. and Escoufier, Y. (1976) A unifying tool for linear multivariate statistical methods: the RV-coefficient. *Journal of the Royal Statistical Society: Series C (Applied Statistics)*, **25** (3), 257–265.

Roberts, F.S. (1985) *Measurement theory, Encyclopedia of Mathematics and its applications*, vol. 7, Cambridge University Press.

Roger, J.M., Biancolillo, A., and Marini, F. (2020) Sequential preprocessing through ORThogonalization (SPORT) and its application to near infrared spectroscopy. *Chemometrics and Intelligent Laboratory Systems*, **199**, 103 975.

Rohart, F., Eslami, A., Matigian, N., Bougeard, S., and Le Cao, K. (2017a) MINT: a multivariate integrative method to identify reproducible molecular signatures across independent experiments and platforms. *BMC Bioinformatics*, **18** (1), 128.

Rohart, F., Gautier, B., Singh, A., and Lê Cao, K. (2017b) mixOmics: An R package for 'omics feature selection and multiple data integration. *PLoS Computational Biology*, **13** (11), e1005 752.

Romano, R., Tomic, O., Liland, K.H., Smilde, A.K., and Næs, T. (2019) A comparison of two PLS-based approaches to structural equation modeling. *Journal of Chemometrics*, **33** (3), e3105.

Rousseeuw, P.J., Trauwaert, E., and Kaufman, L. (1995) Fuzzy clustering with high contrast. *Journal of Computational and Applied Mathematics*, **64** (1-2), 81–90.

Rubingh, C.M., Bijlsma, S., Derks, E.P.P.A., Bobeldijk, I., Verheij, E.R., Kochhar, S., and Smilde, A.K. (2006) Assessing the performance of statistical validation tools for megavariate metabolomics data. *Metabolomics*, **2** (2), 53–61.

Rubingh, C.M., Martens, H., van der Voet, H., and Smilde, A.K. (2013) The costs of complex model optimization. *Chemometrics and Intelligent Laboratory Systems*, **125**, 139–146.

Rubingh, C.M., van Erk, M.J., Wopereis, S., van Vliet, T., Verheij, E.R., Cnubben, N.H.P., van Ommen, B., van der Greef, J., Hendriks, H.F.J., and Smilde, A.K. (2011) Discovery of subtle effects in a human intervention trial through multilevel modeling. *Chemometrics and Intelligent Laboratory Systems*, **106** (1), 108–114.

Saccenti, E., Smilde, A.K., and Camacho, J. (2018) Group-wise anova simultaneous component analysis for designed omics experiments. *Metabolomics*, **14** (6), 73.

Saccenti, E., Tenori, L., Verbruggen, P., Timmerman, M.E., Bouwman, J., van der Greef, J., Luchinat, C., and Smilde, A.K. (2014) Of monkeys and men: A metabolomic analysis of static and dynamic urinary metabolic phenotypes in two species. *PloS One*, **9** (9), e106 077.

Sæbø, S., Almøy, T., Aarøe, J., and Aastveit, A. (2008a) ST-PLS: a multi-directional nearest shrunken centroid type classifier via PLS. *Journal of Chemometrics*, **22** (1), 54–62.

Sæbø, S., Almøy, T., Flatberg, A., Aastveit, A.H., and Martens, H. (2008b) LPLS-regression: a method for prediction and classification under the influence of background information on predictor variables. *Chemometrics and Intelligent Laboratory Systems*, **91**, 121–132.

Sæbø, S., Martens, M., and Martens, H. (2010) Three-block data modeling by endo-and exo-lpls regression, in *Handbook of Partial Least Squares*, Springer, pp. 359–379.

Schäfer, J. and Strimmer, K. (2005) A shrinkage approach to large-scale covariance matrix estimation and implications for functional genomics. *Statistical Applications in Genetics and Molecular Biology*, **4** (1).

Schein, A., Saul, L., and Ungar, L. (2003) A generalized linear model for principal component analysis of binary data, in *International Workshop on Artificial Intelligence and Statistics*, PMLR, pp. 240–247.

Schott, J.R. (1997) *Matrix Analysis for Statistics*, John Wiley & Sons, New York.

Schouteden, M., Van Deun, K., Wilderjans, T.F., and Van Mechelen, I. (2014) Performing DISCO-SCA to search for distinctive and common information in linked data. *Behavior Research Methods*, **46** (2), 576–587.

Searle, S.R. (1971) *Linear models*, John Wiley & Sons, New York.

Shahzad, K. and Loor, J. (2012) Application of top-down and bottom-up systems approaches in ruminant physiology and metabolism. *Current Genomics*, **13** (5), 379–394.

Simon, N., Friedman, J., Hastie, T., and Tibshirani, R. (2013) A sparse-group lasso. *Journal of Computational and Graphical Statistics*, **22** (2), 231–245.

Şimşekli, U., Ermiş, B., Cemgil, A., and Acar, E. (2013) Optimal weight learning for coupled tensor factorization with mixed divergences, in *21st European Signal Processing Conference (EUSIPCO 2013)*, IEEE, pp. 1–5.

Singh, A., Shannon, C., Gautier, B., Rohart, F., Vacher, M., Tebbutt, S.J., and Lê Cao, K.A. (2019) Diablo: an integrative approach for identifying key molecular drivers from multi-omics assays. *Bioinformatics*, **35** (17), 3055–3062.

Sklar, A. (1973) Random variables, joint distribution functions, and copulas. *Kybernetika*, **9** (6), 449–460.

Skov, T., Ballabio, D., and Bro, R. (2008) Multiblock variance partitioning: A new approach for comparing variation in multiple data blocks. *Analytica Chimica Acta*, **615**, 18–29.

Smilde, A.K., Bro, R., and Geladi, P. (2004) *Multiway analysis. Applications in the Chemical Sciences*, John Wiley & Sons, New York.

Smilde, A.K. and Hankemeier, T. (2020) Numerical representations of metabolic systems. *Analytical Chemistry*, **92**, 13 614–13 621.

Smilde, A.K., Hoefsloot, H.C.J., Kiers, H.A.L., Bijlsma, S., and Boelens, H.F.M. (2001) Sufficient conditions for unique solutions within a certain class of curve resolution models. *Journal of Chemometrics*, **15** (4), 405–411.

Smilde, A.K., Jansen, J.J., Hoefsloot, H.C.J., Lamers, R.J.A.N., Van Der Greef, J., and Timmerman, M.E. (2005a) ANOVA-simultaneous component analysis (ASCA): a new tool for analyzing designed metabolomics data. *Bioinformatics*, **21** (13), 3043–3048.

Smilde, A.K. and Kiers, H. (1999) Multiway covariates regression models. *Journal of Chemometrics*, **13**, 31–48.

Smilde, A.K., Kiers, H.A.L., Bijlsma, S., Rubingh, C.M., and Van Erk, M.J. (2009) Matrix correlations for high-dimensional data: the modified rv-coefficient. *Bioinformatics*, **25** (3), 401–405.

Smilde, A.K., Måge, I., Næs, T., Hankemeier, T., Lips, M.A., Kiers, H.A.L., Acar, E., and Bro, R. (2017) Common and distinct components in data fusion. *Journal of Chemometrics*, **31** (7), e2900.

Smilde, A.K., Song, Y.P., Westerhuis, J.A., Kiers, H.A.L., Aben, N., and Wessels, L.F.A. (2020) Heterofusion: Fusing genomics data of different measurement scales. *Journal of Chemometrics*, p. e3200.

Smilde, A.K., Timmerman, M.E., Hendriks, M.M.W.B., Jansen, J.J., and Hoefsloot, H.C.J. (2012) Generic framework for high-dimensional fixed-effects ANOVA. *Briefings in Bioinformatics*, **13** (5), 524–535.

Smilde, A.K., Timmerman, M.E., Saccenti, E., Jansen, J.J., and Hoefsloot, H.C.J. (2015) Covariances simultaneous component analysis: A new method within the framework of modeling covariances. *Journal of Chemometrics*, **29**, 277–288.

Smilde, A.K., van der Werf, M.J., Bijlsma, S., van der Werff-van-der Vat, B.J.C., and Jellema, R.H. (2005b) Fusion of mass spectrometry-based metabolomics data. *Analytical Chemistry*, **77**, 6729–6736.

Smilde, A.K., Westerhuis, J.A., and de Jong, S. (2003) A framework for sequential multiblock component methods. *Journal of Chemometrics*, **17**, 323–337.

Song, Y., Westerhuis, J., Aben, N., Wessels, L., Groenen, P., and Smilde, A. (2018) Generalized simultaneous component analysis of binary and quantitative data. *arXiv:1807.04982*.

Song, Y., Westerhuis, J.A., Aben, N., Michaut, M., Wessels, L.F.A., and Smilde, A.K. (2017) Principal Component Analysis of binary genomics data. *Briefings in Bioinformatics*, pp. 1–13.

Song, Y., Westerhuis, J.A., Aben, N., Wessels, L.F.A., Groenen, P.J.F., and Smilde, A.K. (2021) Generalized simultaneous component analysis of binary and quantitative data. *Journal of Chemometrics*, **35** (3), e3312.

Song, Y., Westerhuis, J.A., and Smilde, A.K. (2019) Separating common (global and local) and distinct variation in multiple mixed types data sets. *Journal of Chemometrics*, p. e3197.

Stahle, L. (1989) Aspects of the analysis of three-way data. *Chemometrics and Intelligent laboratory systems*, **7** (1-2), 95–100.

Stanimirova, I., Walczak, B., Massart, D.L., Simeonov, V., Saby, C.A., and Di Crescenzo, E. (2004) STATIS, a three-way method for data analysis. application to environmental data. *Chemometrics and Intelligent Laboratory Systems*, **73** (2), 219–233.

Stasinopoulos, D.M., Rigby, R.A., Heller, G., Voudouris, V., and De Bastiani, F. (2017) *Flexible regression and smoothing*, Chapman and Hall/CRC, New York.

Stevens, S. (1946) On the theory of scales of measurement. *Science*, **103** (2684), 677–680.

Stone, M. (1974) Cross-validatory choice and assessment of statistical predictions. *Journal of the Royal Statistical Society, Series B Statistical Methodology*, **36**, 111–148.

Szymańska, E., Saccenti, E., Smilde, A.K., and Westerhuis, J.A. (2012) Double-check: validation of diagnostic statistics for pls-da models in metabolomics studies. *Metabolomics*, **8** (1), 3–16.

Takane, Y. and Shibayama, T. (1991) Principal component analysis with external information on both subjects and variables. *Psychometrika*, **56** (1), 97–120.

Takane, Y., Young, F., and De Leeuw, J. (1977) Nonmetric individual differences multidimensional scaling: an alternating least squares method with optimal scaling features. *Psychometrika*, **42** (1), 7–67.

Tarazona, S., Prado-López, S., Dopazo, J., Ferrer, A., and Conesa, A. (2012) Variable selection for multifactorial genomic data. *Chemometrics and Intelligent Laboratory Systems*, **110** (1), 113–122.

Tauler, R. (2001) Calculation of maximum and minimum band boundaries of feasible solutions for species profiles obtained by multivariate curve resolution. *Journal of Chemometrics*, **15** (8), 627–646.

Tauler, R. (2010) Comments on a recently published paper 'some surprising properties of multivariate curve resolution-alternating least squares (mcr-als) algorithms'. *Journal of Chemometrics*, **24** (2), 87–90.

Tauler, R., Smilde, A.K., and Kowalski, B.R. (1995) Selectivity, local rank, three-way data analysis and ambiguity in multivariate curve resolution. *Journal of Chemometrics*, **9** (1), 31–58.

Tayrac, M.d., Etcheverry, A., Aubry, M., Saïkali, S., Hamlat, A., Quillien, V., Treut, A.L., Galibert, M., and Mosser, J. (2009) Integrative genome-wide analysis reveals a robust genomic glioblastoma signature associated with copy number driving changes in gene expression. *Genes, Chromosomes and Cancer*, **48** (1), 55–68.

Tellegen, P.J. and Laros, J.A. (2017) *SON-R 2-8: Snijders-Oomen nonverbal intelligence test*, Hogrefe, Amsterdam.

Ten Berge, J.M.F. (1993) *Least Squares Optimization in Multivariate Analysis*, DSWO Press, Leiden.

Ten Berge, J.M.F., Kiers, H.A.L., and van der Stel, V. (1992) Simultaneous component analysis. *Statistica Applicata*, **4** (4), 377–392.

Tenenhaus, A. and Tenenhaus, M. (2014) Regularized generalized canonical correlation analysis for multiblock or multigroup data analysis. *European Journal of Operational Research*, **238** (2), 391–403.

Tenenhaus, M., Tenenhaus, A., and Groenen, P.J.F. (2017) Regularized generalized canonical correlation analysis: a framework for sequential multiblock component methods. *Psychometrika*, **82** (3), 737–777.

Tenenhaus, M., Vinzi, V., Chatelin, Y.M., and Lauro, C. (2005) PLS path modeling. *Computational Statistics & Data Analysis*, **48** (1), 159–205.

Ter Braak, C.J.F. and de Jong, S. (1998) The objective function of partial least squares regression. *Journal of Chemometrics*, **12**, 41–54.

Thiel, M., Féraud, B., and Govaerts, B. (2017) ASCA+ and APCA+: Extensions of ASCA and APCA in the analysis of unbalanced multifactorial designs. *Journal of Chemometrics*, **31** (6), e2895.

Thomas, E.V. (1995) Incorporating auxiliary predictor variation in principal component regression models. *Journal of Chemometrics*, **9** (6), 471–481.

Thomas, E.V. (2019) Semi-supervised learning in multivariate calibration. *Chemometrics and Intelligent Laboratory Systems*, **195**, 103 868.

Thybo, A.K., Kühn, B.F., and Martens, H. (2004) Explaining danish children's preferences for apples using instrumental, sensory and demographic/behavioural data. *Food Quality and Preference*, **15** (1), 53–63.

Thygesen, L.G., Thybo, A.K., and Engelsen, S.B. (2001) Prediction of sensory texture quality of boiled potatoes from low-field1H NMR of raw potatoes. The role of chemical constituents. *LWT-Food Science and Technology*, **34** (7), 469–477.

Tibshirani, R. (1996) Regression shrinkage and selection via the lasso. *Journal of the Royal Statistical Society: Series B (Methodological)*, **58** (1), 267–288.

Timmerman, M.E. (2006) Multilevel component analysis. *British Journal of Mathematical and Statistical Psychology*, **59** (Part 2), 301–320.

Timmerman, M.E., Hoefsloot, H.C.J., Smilde, A.K., and Ceulemans, E. (2015) Scaling in anova-simultaneous component analysis. *Metabolomics*, **11** (5), 1265–1276.

Timmerman, M.E. and Kiers, H.A.L. (2003) Four simultaneous component models for the analysis of multivariate time series from more than one subject to model intraindividual and interindividual differences. *Psychometrika*, **68** (1), 105–121.

Timmerman, M.E., Kiers, H.A.L., and Smilde, A.K. (2007) Estimating confidence intervals for principal component loadings: a comparison between the bootstrap and asymptotic results. *British Journal of Mathematical and Statistical Psychology*, **60** (2), 295–314.

Timmerman, M.E., Kiers, H.A.L., Smilde, A.K., Ceulemans, E., and Stouten, J. (2009) Bootstrap confidence intervals in multilevel simultaneous component analysis. *British Journal of Mathematical & Statistical Psychology*, **62**, 299–318.

Tipping, M.E. and Bishop, C.M. (1999) Probabilistic principal component analysis. *Journal of the Royal Statistical Society: Series B (Statistical Methodology)*, **61** (3), 611–622.

Tomczak, K., Czerwińska, P., and Wiznerowicz, M. (2015) The cancer genome atlas (tcga): an immeasurable source of knowledge. *Contemporary Oncology*, **19** (1A), A68.

Trygg, J. and Wold, S. (2002) Orthogonal projections to latent structures (O-PLS). *Journal of Chemometrics*, **16** (3), 119–128.

Trygg, J. and Wold, S. (2003) O2-PLS, a two-block (X–Y) latent variable regression (LVR) method with an integral OSC filter. *Journal of Chemometrics*, **17** (1), 53–64.

Tschuprow, A. (1939) *Principles of the mathematical theory of correlation*, William Hodge.

Tucker, L. (1958) An inter-battery method of factor analysis. *Psychometrika*, **23** (2), 111–136.

Tucker, L. (1964) The extension of factor analysis to three-dimensional matrices. *Contributions to mathematical psychology*, **110119**.

Tukey, J.W. (1949) Comparing individual means in the analysis of variance. *Biometrics*, **5** (2), 99–114.

Tukey, J.W. (1958) Bias and confidence in not quite large samples. *Ann. Math. Statist.*, **29** (2), 614.

Tukey, J.W. (1980) We need both exploratory and confirmatory. *The American Statistician*, **34** (1), 23–25.

Ulfenborg, B. (2019) Vertical and horizontal integration of multi-omics data with miodin. *BMC Bioinformatics*, **20** (1), 649.

Van de Geer, J.P. (1984) Linear relations among k sets of variables. *Psychometrika*, **49** (1), 79–94.

van den Berg, R.A., Rubingh, C.M., Westerhuis, J.A., van der Werf, M.J., and Smilde, A.K. (2009) Metabolomics data exploration guided by prior knowledge. *Analytica Chimica Acta*, **651**, 173–181.

Van den Wollenberg, A.L. (1977) Redundancy analysis an alternative for canonical correlation analysis. *Psychometrika*, **42** (2), 207–219.

Van der Burg, E. and Dijksterhuis, G. (1996) *Generalised canonical analysis of individual sensory profiles and instrument data*, Elsevier, pp. 221–258.

van der Kloet, F.M., Sebastián-León, P., Conesa, A., Smilde, A.K., and Westerhuis, J.A. (2016) Separating common from distinctive variation. *BMC Bioinformatics*, **17** (S5), S195.

Van Deun, K., Crompvoets, E.A.V., and Ceulemans, E. (2018) Obtaining insights from high-dimensional data: sparse principal covariates regression. *BMC Bioinformatics*, **19** (1), 104.

Van Deun, K., Smilde, A.K., Thorrez, L., Kiers, H.A.L., and Van Mechelen, I. (2013) Identifying common and distinctive processes underlying multiset data. *Chemometrics and Intelligent Laboratory Systems*, **129**, 40–51.

Van Deun, K., Smilde, A.K., van der Werf, M.J., Kiers, H.A.L., and Van Mechelen, I. (2009) A structured overview of simultaneous component based data integration. *BMC Bioinformatics*, **10**, 246.

Van Deun, K., Thorrez, L., Coccia, M., Hasdemir, D., Westerhuis, J.A., Smilde, A.K., and Van Mechelen, I. (2019) Weighted sparse principal component analysis. *Chemometrics and Intelligent Laboratory Systems*, **195**, 103 875.

Van Deun, K., Van Mechelen, I., Thorrez, L., Schouteden, M., De Moor, B., Van Der Werf, M.J., De Lathauwer, L., Smilde, A.K., and Kiers, H.A.L. (2012) DISCO-SCA and properly applied GSVD as swinging methods to find common and distinctive processes. *PLoS One*, **7** (5), e37 840.

Van Deun, K., Wilderjans, T.F., Van den Berg, R.A., Antoniadis, A., and Van Mechelen, I. (2011) A flexible framework for sparse simultaneous component based data integration. *BMC Bioinformatics*, **12** (1), 448.

Van Duynhoven, J., Vaughan, E.E., Jacobs, D.M., Kemperman, R.A., Van Velzen, E.J.J., Gross, G., Roger, L.C., Possemiers, S., Smilde, A.K., Doré, J. *et al.* (2010) Metabolic fate of polyphenols in the human superorganism. *Proceedings of the National Academy of Sciences*, p. 201000098.

Van Loan, C.F. (1976) Generalizing the singular value decomposition. *SIAM Journal on Numerical Analysis*, **13**, 76–83.

van Loon, W., Fokkema, M., Szabo, B., and de Rooij, M. (2020) Stacked penalized logistic regression for selecting views in multi-view learning. *Information Fusion*.

Van Mechelen, I. and Schepers, J. (2007) A unifying model involving a categorical and/or dimensional reduction for multimode data. *Computational Statistics & Data Analysis*, **52**, 537–549.

Van Mechelen, I. and Smilde, A.K. (2010) A generic linked-mode decomposition model for data fusion. *Chemometrics and Intelligent Laboratory Systems*, **104**, 83–94.

Van Mechelen, I. and Smilde, A.K. (2011) Comparability problems in the analysis of multiway data. *Chemometrics and Intelligent Laboratory Systems*, **106** (1), 2–11.

Van Rijsbergen, C. (1979) *Information Retrieval*, Butterworth-Heinemann, Newton MA, USA.

Van Rossum, G. and Drake, F.L. (2009) *Python 3 Reference Manual*, CreateSpace, Scotts Valley, CA.

van Velzen, E., Westerhuis, J., van Duynhoven, J., van Dorsten, F., Hoefsloot, H., Jacobs, D., Smit, S., Draijer, R., Kroner, C., and Smilde, A. (2008) Multilevel data analysis of a crossover designed human nutritional intervention study. *Journal of proteome research*, **7** (10), 4483–4491.

Varela, P. and Ares, G. (2012) Sensory profiling , the blurred line between sensory and consumer science. a review of novel methods for product characterization. *Food. Res. Int.*, **48**, 893–908.

Velleman, P. and Wilkinson, L. (1993) Nominal, ordinal, interval, and ratio typologies are misleading. *The American Statistician*, **47** (1), 65–72.

Vervloet, M., Van Deun, K., Van den Noortgate, W., and Ceulemans, E. (2013) On the selection of the weighting parameter value in principal covariates regression. *Chemometrics and Intelligent Laboratory Systems*, **123**, 36–43.

Vinzi, V.E., Guinot, C., and Squillacciotti, S. (2007) Two-step pls regression for l-structured data: an application in the cosmetic industry. *Statistical methods and applications*, **16** (2), 263–278.

Virtanen, S., Klami, A., Khan, S., and Kaski, S. (2012) Bayesian group factor analysis, in *Artificial Intelligence and Statistics*, pp. 1269–1277.

Vis, D.J., Westerhuis, J.A., Smilde, A.K., and van der Greef, J. (2007) Statistical validation of megavariate effects in asca. *BMC Bioinformatics*, **8** (1), 322.

Vitale, R., Westerhuis, J.A., Næs, T., Smilde, A.K., de Noord, O.E., and Ferrer, A. (2017) Selecting the number of factors in principal component analysis by permutation testing. numerical and practical aspects. *Journal of Chemometrics*, **31** (12), e2937.

Waaijenborg, S. and Zwinderman, A.H. (2009) Sparse canonical correlation analysis for identifying, connecting and completing gene-expression networks. *BMC Bioinformatics*, **10** (1), 315.

Wangen, L.E. and Kowalski, B.R. (1988) A multiblock partial least squares algorithm for investigating complex chemical systems. *Journal of Chemometrics*, **3**, 3–20.

Webster, J.T., Gunst, R.F., and Mason, R.L. (1974) Latent root regression analysis. *Technometrics*, **16** (4), 513–522.

Weir, T.L., Manter, D.K., Sheflin, A.M., Barnett, B.A., Heuberger, A.L., and Ryan, E.P. (2013) Stool microbiome and metabolome differences between colorectal cancer patients and healthy adults. *PloS One*, **8** (8), e70 803.

Welch, B.L. (1937) On the z-test in randomized blocks and latin squares. *Biometrika*, **29**, 21 – 52.

Wentzell, P.D., Andrews, D.T., Hamilton, D.C., Faber, K., and Kowalski, B.R. (1997) Maximum likelihood principal component analysis. *Journal of Chemometrics*, **11** (4), 339–366.

Wentzell, P.D., Karakach, T.K., Roy, S., Martinez, M.J., Allen, C.P., and Werner-Washburne, M. (2006) Multivariate curve resolution of time course microarray data. *BMC Bioinformatics*, **7** (1), 1–19.

Westad, F. and Martens, H. (2000) Variable selection in near infrared spectroscopy based on significance testing in partial least squares regression. *Journal of Near Infrared Spectroscopy*, **8** (2), 117–124.

Westerhuis, J.A., van Velzen, E., Hoefsloot, H., and Smilde, A. (2010) Multivariate paired data analysis: multilevel plsda versus oplsda. *Metabolomics*, **6** (1), 119–128.

Westerhuis, J.A., Hoefsloot, H.C.J., Smit, S., Vis, D.J., Smilde, A.K., van Velzen, E.J.J., van Duijnhoven, J.P.M., and van Dorsten, F.A. (2008) Assessment of PLSDA cross validation. *Metabolomics*, **4** (1), 81–89.

Westerhuis, J.A., Kourti, T., and MacGregor, J.F. (1998) Analysis of multiblock and hierarchical pca and pls models. *Journal of Chemometrics*, **12**, 301–321.

Westerhuis, J.A. and Smilde, A.K. (2001) Deflation in multiblock PLS. *Journal of Chemometrics*, **15** (5), 485–493.

Wilderjans, T.F., Ceulemans, E., and Meers, K. (2013) CHull: A generic convex-hull-based model selection method. *Behavior Research Methods*, **45** (1), 1–15.

Wilderjans, T.F., Ceulemans, E., and Van Mechelen, I. (2009) Simultaneous analysis of coupled data blocks differing in size: A comparison of two weighting schemes. *Computational Statistics and Data Analysis*, **53**, 1086–1098.

Wise, B.M. and Gallagher, N.B. (1996) The process chemometrics approach to process monitoring and fault detection. *Journal of Process Control*, **6** (6), 329–348.

Witten, D.M., Tibshirani, R., and Hastie, T. (2009) A penalized matrix decomposition, with applications to sparse principal components and canonical correlation analysis. *Biostatistics*, **10** (3), 515–534.

Wold, H. (1975) Path models with latent variables: The NIPALS approach, in *Quantitative sociology*, Elsevier, pp. 307–357.

Wold, S. (1978) Cross-validatory estimation of the number of components in factor and principal components models. *Technometrics*, **20**, 397–405.

Wold, S., Hellberg, S., Lundstedt, T., Sjöström, M., and Wold, H. (1987) PLS modelling with latent variables in two or more dimensions, in *PLS model building: Theory and application*. Frankfurt am main, F.R.G., 23-25 September 1987.

Wold, S., Høy, M., Martens, H., Trygg, J., Westad, F., MacGregor, J., and Wise, B. (2009) The PLS model space revisited. *Journal of Chemometrics: A Journal of the Chemometrics Society*, **23** (2), 67–68.

Wold, S., Johansson, E., Cocchi, M. *et al.* (1993) *3D QSAR in Drug Design: Theory, Methods and Applications.*, Kluwer ESCOM Science Publisher, chap. PLS: partial least squares projections to latent structures, pp. 523–550.

Wold, S., Ruhe, A., Wold, H., and Dunn, W.J. (1984) The collinearity problem in linear regression. The partial least squares (PLS) approach to generalized inverses. *SIAM Journal of Scientific and Statistical Computing*, **5**, 735–743.

Wopereis, S., Rubingh, C.M., van Erk, M.J., Verheij, E.R., van Vliet, T., Cnubben, N.H.P., Smilde, A.K., van der Greef, J., van Ommen, B., and Hendriks, H.F.J. (2009) Metabolic profiling of the response to an oral glucose tolerance test detects subtle metabolic changes. *PloS One*, **4** (2), e4525.

Wright, S. (1918) On the nature of size factors. *Genetics*, **3** (4), 367.

Wright, S. (1934) The method of path coefficients. *The Annals of Mathematical Statistics*, **5** (3), 161–215.

Xia, J., Sinelnikov, I.V., and Wishart, D.S. (2011) Metatt: a web-based metabolomics tool for analyzing time-series and two-factor datasets. *Bioinformatics*, **27** (17), 2455–2456.

Xiong, H., Zhang, D., Martyniuk, C.J., Trudeau, V.L., and Xia, X. (2008) Using generalized procrustes analysis (GPA) for normalization of cdna microarray data. *BMC Bioinformatics*, **9** (1), 25.

Yanai, H. (1974) Unification of various techniques of multivariate analysis by means of generalized coefficient of determination (GCD). *J. Behaviormetrics*, **1** (1), 45–54.

Yılmaz, K., Cemgil, A., and Simsekli, U. (2011) Generalised coupled tensor factorisation. *Advances in Neural Information Processing Systems*, **24**, 2151–2159.

Young, F.W. (1981) Quantitative-analysis of qualitative data. *Psychometrika*, **46** (4), 357–388.

Young, F.W., Takane, Y., and De Leeuw, J. (1978) The principal components of mixed measurement level multivariate data: an alternating least squares method with optimal scaling features. *Psychometrika*, **43** (2), 279–281.

Yuan, M.and Lin, Y. (2006) Model selection and estimation in regression with grouped variables. *Journal of the Royal Statistical Society: Series B (Statistical Methodology)*, **68** (1), 49–67.

Yule, G.U. (1912) On the methods of measuring association between two attributes. *Journal of the Royal Statistical Society*, **75** (6), 579–652.

Zegers, F.E. (1986) *A general family of association coefficients*, Boomker, Groningen/Haren.

Zha, H. (1991) The product-product singular value decomposition of matrix triplets. *BIT Numerical Mathematics*, **31** (4), 711–726.

Zhao, C. and Gao, F. (2012) Two-step Multiset Regression Analysis (MsRA) Algorithm. *Industry and Engineering Chemistry Research*, **51**, 1337–1354.

Zou, H. and Hastie, T. (2003) Regression shrinkage and selection via the elastic net, with applications to microarrays. *JR Stat Soc Ser B*, **67**, 301–20.

Zou, H. and Hastie, T. (2005) Regularization and variable selection via the elastic net. *Journal of the Royal Statistical Society: Series B (Statistical Methodology)*, **67** (2), 301–320.

Zou, H., Hastie, T., and Tibshirani, R. (2006) Sparse principal component analysis. *Journal of Computational and Graphical Statistics*, **15** (2), 265–286.

Zwanenburg, G., Hoefsloot, H.C., Westerhuis, J.A., Jansen, J.J., and Smilde, A.K. (2011) ANOVA–principal component analysis and ANOVA–simultaneous component analysis: a comparison. *Journal of Chemometrics*, **25** (10), 561–567.

Index

Multiblock Data Fusion in Statistics and Machine Learning: Applications in the Natural and Life Sciences, First Edition. Age K. Smilde, Tormod Næs, and Kristian Hovde Liland.